Lecture Notes on Data Engineering and Communications Technologies

180

Series Editor

Fatos Xhafa, *Technical University of Catalonia, Barcelona, Spain*

The aim of the book series is to present cutting edge engineering approaches to data technologies and communications. It will publish latest advances on the engineering task of building and deploying distributed, scalable and reliable data infrastructures and communication systems.

The series will have a prominent applied focus on data technologies and communications with aim to promote the bridging from fundamental research on data science and networking to data engineering and communications that lead to industry products, business knowledge and standardisation.

Indexed by SCOPUS, INSPEC, EI Compendex.

All books published in the series are submitted for consideration in Web of Science.

Zhengbing Hu · Qingying Zhang · Matthew He
Editors

Advances in Artificial Systems for Logistics Engineering III

Set 1

Editors
Zhengbing Hu
Faculty of Applied Mathematics
National Technical University of Ukraine
"Igor Sikorsky Kyiv Polytechnic Institute"
Kyiv, Ukraine

Qingying Zhang
College of Transportation and Logistics
Engineering
Wuhan University of Technology
Wuhan, China

Matthew He
Halmos College of Arts and Sciences
Nova Southeastern University
Fort Lauderdale, FL, USA

ISSN 2367-4512   ISSN 2367-4520 (electronic)
Lecture Notes on Data Engineering and Communications Technologies
ISBN 978-3-031-36114-2   ISBN 978-3-031-36115-9 (eBook)
https://doi.org/10.1007/978-3-031-36115-9

© The Editor(s) (if applicable) and The Author(s), under exclusive license
to Springer Nature Switzerland AG 2023

This work is subject to copyright. All rights are solely and exclusively licensed by the Publisher, whether the whole or part of the material is concerned, specifically the rights of translation, reprinting, reuse of illustrations, recitation, broadcasting, reproduction on microfilms or in any other physical way, and transmission or information storage and retrieval, electronic adaptation, computer software, or by similar or dissimilar methodology now known or hereafter developed.
The use of general descriptive names, registered names, trademarks, service marks, etc. in this publication does not imply, even in the absence of a specific statement, that such names are exempt from the relevant protective laws and regulations and therefore free for general use.
The publisher, the authors, and the editors are safe to assume that the advice and information in this book are believed to be true and accurate at the date of publication. Neither the publisher nor the authors or the editors give a warranty, expressed or implied, with respect to the material contained herein or for any errors or omissions that may have been made. The publisher remains neutral with regard to jurisdictional claims in published maps and institutional affiliations.

This Springer imprint is published by the registered company Springer Nature Switzerland AG
The registered company address is: Gewerbestrasse 11, 6330 Cham, Switzerland

Preface

The development of artificial intelligence (AI) systems and their applications in various fields is one of the modern science and technology's most pressing challenges. One of these areas is AI and logistics engineering, where their application aims to increase the effectiveness of AI generation and distribution for the world's population's life support, including tasks such as developing industry, agriculture, medicine, transportation, and so on. The rapid development of AI systems necessitates an increase in the training of an increasing number of relevant specialists. AI systems have a lot of potential for use in education technology to improve the quality of training for specialists by taking into account the personal characteristics of these specialists as well as the new computing devices that are coming out.

As a result of these factors, the 3rd International Conference on Artificial Intelligence and Logistics Engineering (ICAILE2023), held in Wuhan, China, on March 11–12, 2023, was organized jointly by Wuhan University of Technology, Nanning University, the National Technical University of Ukraine "Igor Sikorsky Kyiv Polytechnic Institute", Huazhong University of Science and Technology, the Polish Operational and Systems Society, Wuhan Technology and Business University, and the International Research Association of Modern Education and Computer Science. The ICAILE2023 brings together leading scholars from all around the world to share their findings and discuss outstanding challenges in computer science, logistics engineering, and education applications.

Out of all the submissions, the best contributions to the conference were selected by the program committee for inclusion in this book.

March 2023

Zhengbing Hu
Qingying Zhang
Matthew He

Organization

General Chairs

Q. Y. Zhang	Wuhan University of Technology, China
Ivan Dychka	National Technical University of Ukraine "Igor Sikorsky Kyiv Polytechnic Institute", Ukraine

Online conference Organizing Chairs

Z. B. Hu	National Technical University of Ukraine "Igor Sikorsky Kyiv Polytechnic Institute", Ukraine
C. L. Wang	Anhui University, China
Y. Wang	Wuhan University of Science and Technology, China

Program Chairs

Q. Y. Zhang	Wuhan University of Technology, China
Matthew He	Nova Southeastern University, USA
G. E. Zhang	Nanning University, China

Publication Chairs

Z. B. Hu	National Technical University of Ukraine "Igor Sikorsky Kyiv Polytechnic Institute", Ukraine
Q. Y. Zhang	Wuhan University of Technology, China
Matthew He	Nova Southeastern University, USA

Publicity Chairs

Y. Z. Pang	Nanning University, China
Q. L. Zhou	Shizuoka University, Japan

viii   Organization

O. K. Tyshchenko	University of Ostrava, Czech Republic
Vadym Mukhin	National Technical University of Ukraine "Igor Sikorsky Kyiv Polytechnic Institute", Ukraine

Program Committee Members

Margherita Mori	University of L'Aquila, Italy
X. H. Tao	School of Intelligent Manufacturing of Nanning University, China
Felix Yanovsky	Delft University of Technology, Netherlands
D. Y. Fang	Beijing Technology and Business University, China
Ivan Izonin	Lviv Polytechnic National University, Ukraine
L. Luo	Sichuan University, China
Essa Alghannam	Tishreen University, Syria
L. Xu	Southwest Jiaotong University, China
H. F. Yang	Waseda University, Japan
Y. C. Zhang	Hubei University, China
A. Sachenko	Kazimierz Pułaski University of Technology and Humanities in Radom, Poland
S. Gnatyuk	National Aviation University, Ukraine
X. J. Zhou	Wuhan Textile University, China
Oleksandra Yeremenko	Kharkiv National University of Radio Electronics, Ukraine
Rabah Shboul	Al-albayt University, Jordan
W. B. Hu	Wuhan University, China
Yurii Koroliuk	Chemivtsi Institute of Trade and Economics, Ukraine
O. K. Tyshchenko	University of Ostrava, Czech Republic
H. L. Zhong	South China University of Technology, China
J. L. Zhang	Huazhong University of Science and Technology, China
G. K. Tolokonnikova	FNAT VIM of RAS, Moscow, Russia

Contents

Advances of Computer Algorithms and Methods

AI Chatbots for Banks: Evolving Trends and Critical Issues 3
  Margherita Mori and Lijing Du

Traffic Flow Characteristics and Vehicle Road Coordination
Improvement in Subterranean Interweaving 14
  Enshi Wang, Bihui Huang, and Bing Liu

Influential Factors and Implementation Path of Talent Digital Evaluation
Based on ISM Model: Taking Electric Power Enterprises as an Example 25
  Wei Luo, Jiwei Tang, Saixiao Huang, and Yuan Chen

Knowledge Associated with a Question-Answer Pair 35
  Igor Chimir

Research on Low Complexity Differential Space Modulation Detection
Algorithm 45
  Shuiping Xiong and Xia Wu

Mathematical Model of the Process of Production of Mineral Fertilizers
in a Fluidized Bed Granulator 55
  Bogdan Korniyenko and Andrii Nesteruk

Simulation Study on Optimization of Passenger Flow Transfer
Organization at Nanning Jinhu Square Metro Station Based on Anylogic 65
  *Yan Chen, Chenyu Zhang, Xiaoling Xie, Zhicheng Huang,
  and Jinshan Dai*

Engine Speed Measurement and Control System Design Based
on LabVIEW 79
  Chengwei Ju, Geng E. Zhang, and Rengshang Su

Research and Design of Personalized Learning Resources Precise
Recommendation System Based on User Profile 90
  Tingting Liang, Zhaomin Liang, and Suzhen Qiu

Mixed Parametric and Auto-oscillations at Nonlinear Parametric
Excitation 101
  Alishir A. Alifov

x   Contents

Spectrum Analysis on Electricity Consumption Periods by Industry
in Fujian Province 109
  Huawei Hong, Lingling Zhu, Gang Tong, Peng Lv, Xiangpeng Zhan,
  Xiaorui Qian, and Kai Xiao

Determination of the Form of Vibrations in Vibratory Plow's Moldboard
with Piezoceramic Actuator for Maximum Vibration Effect 120
  Sergey Filimonov, Sergei Yashchenko, Constantine Bazilo,
  and Nadiia Filimonova

Evaluating Usability of E-Learning Applications in Bangladesh:
A Semiotic and Heuristic Evaluation Approach 129
  Samrat Kumar Dey, Khandaker Mohammad Mohi Uddin,
  Dola Saha, Lubana Akter, and Mshura Akhter

Perceptual Computing Based Framework for Assessing Organizational
Performance According to Industry 5.0 Paradigm 141
  Danylo Tavrov, Volodymyr Temnikov, Olena Temnikova,
  and Andrii Temnikov

Petroleum Drilling Monitoring and Optimization: Ranking the Rate
of Penetration Using Machine Learning Algorithms 152
  Ijegwa David Acheme, Wilson Nwankwo, Akinola S. Olayinka,
  Ayodeji S. Makinde, and Chukwuemeka P. Nwankwo

Application of Support Vector Machine to Lassa Fever Diagnosis 165
  Wilson Nwankwo, Wilfred Adigwe, Chinecherem Umezuruike,
  Ijegwa D. Acheme, Chukwuemeka Pascal Nwankwo,
  Emmanuel Ojei, and Duke Oghorodi

A Novel Approach to Bat Protection IoT-Based Ultrasound System
of Smart Farming 178
  Md. Hafizur Rahman, S. M. Noman, Imrus Salehin,
  and Tajim Md. Niamat Ullah Akhund

Synthesis and Modeling of Systems with Combined Fuzzy P + I -
Regulators 187
  Bohdan Durnyak, Mikola Lutskiv, Petro Shepita, Vasyl Sheketa,
  Nadiia Pasieka, and Mykola Pasieka

Protection of a Printing Company with Elements of Artificial
Intelligence and IIoT from Cyber Threats 197
  Bohdan Durnyak, Tetyana Neroda, Petro Shepita, Lyubov Tupychak,
  Nadiia Pasieka, and Yulia Romanyshyn

AMGSRAD Optimization Method in Multilayer Neural Networks 206
S. Sveleba, I. Katerynchuk, I. Kuno, O. Semotiuk, Ya. Shmyhelskyy, S. Velgosh, N. Sveleba, and A. Kopych

Regional Economic Development Indicators Analysis and Forecasting:
Panel Data Evidence from Ukraine 217
Larysa Zomchak, Mariana Vdovyn, and Olha Deresh

An Enhanced Performance of Minimum Variance Distortionless
Response Beamformer Based on Spectral Mask 229
Quan Trong The and Sergey Perelygin

Modelling Smart Grid Instability Against Cyber Attacks in SCADA
System Networks 239
John E. Efiong, Bodunde O. Akinyemi, Emmanuel A. Olajubu, Isa A. Ibrahim, Ganiyu A. Aderounmu, and Jules Degila

Program Implementation of Educational Electronic Resource
for Inclusive Education of People with Visual Impairment 251
Yurii Tulashvili, Iurii Lukianchuk, Valerii Lishchyna, and Nataliia Lishchyna

A Version of the Ternary Description Language with an Interpretation
for Comparing the Systems Described in it with Categorical Systems 261
G. K. Tolokonnikov

A Hybrid Centralized-Peer Authentication System Inspired
by Block-Chain 271
Wasim Anabtawi, Ahmad Maqboul, and M. M. Othman Othman

Heuristic Search for Nonlinear Substitutions for Cryptographic
Applications 288
Oleksandr Kuznetsov, Emanuele Frontoni, Sergey Kandiy, Oleksii Smirnov, Yuliia Ulianovska, and Olena Kobylianska

Dangerous Landslide Suspectable Region Forecasting
in Bangladesh – A Machine Learning Fusion Approach 299
Khandaker Mohammad Mohi Uddin, Rownak Borhan, Elias Ur Rahman, Fateha Sharmin, and Saikat Islam Khan

New Cost Function for S-boxes Generation by Simulated Annealing
Algorithm 310
Oleksandr Kuznetsov, Emanuele Frontoni, Sergey Kandiy, Tetiana Smirnova, Serhii Prokopov, and Alisa Bilanovych

xii   Contents

Enriched Image Embeddings as a Combined Outputs from Different
Layers of CNN for Various Image Similarity Problems More Precise
Solution   321
  Volodymyr Kubytskyi and Taras Panchenko

A Novel Approach to Network Intrusion Detection with LR Stacking
Model   334
  Mahnaz Jarin and A. S. M. Mostafizur Rahaman

Boundary Refinement via Zoom-In Algorithm for Keyshot Video
Summarization of Long Sequences   344
  Alexander Zarichkovyi and Inna V. Stetsenko

Solving Blockchain Scalability Problem Using ZK-SNARK   360
  Kateryna Kuznetsova, Anton Yezhov, Oleksandr Kuznetsov,
  and Andrii Tikhonov

Interactive Information System for Automated Identification of Operator
Personnel by Schulte Tables Based on Individual Time Series   372
  Myroslav Havryliuk, Roman Kaminskyy, Kyrylo Yemets,
  and Taras Lisovych

DIY Smart Auxiliary Power Supply for Emergency Use   382
  Nina Zdolbitska, Mykhaylo Delyavskyy, Nataliia Lishchyna,
  Valerii Lishchyna, Svitlana Lavrenchuk, and Viktoriia Sulim

A Computerised System for Monitoring Water Activity in Food
Products Using Wireless Technologies   393
  Oksana Honsor and Roksolana Oberyshyn

Reengineering of the Ukrainian Energy System: Geospatial Analysis
of Solar and Wind Potential   404
  Iryna Doronina, Maryna Nehrey, and Viktor Putrenko

Complex Approach for License Plate Recognition Effectiveness
Enhancement Based on Machine Learning Models   416
  Yakovlev Anton and Lisovychenko Oleh

The Analysis and Visualization of CEE Stock Markets Reaction
to Russia's Invasion of Ukraine by Event Study Approach   426
  Andrii Kaminskyi and Maryna Nehrey

The Same Size Distribution of Data Based on Unsupervised Clustering
Algorithms   437
  Akbar Rashidov, Akmal Akhatov, and Fayzullo Nazarov

Investigation of Microclimate Parameters in the Industrial Environments 448
Solomiya Liaskovska, Olena Gumen, Yevgen Martyn, and Vasyl Zhelykh

Detection of Defects in PCB Images by Separation and Intensity
Measurement of Chains on the Board 458
Roman Melnyk and Ruslan Tushnytskyy

Tropical Cyclone Genesis Forecasting Using LightGBM 468
*Sabbir Rahman, Nusrat Sharmin, Md. Mahbubur Rahman,
and Md. Mokhlesur Rahman*

Optimization of Identification and Recognition of Micro-objects Based
on the Use of Specific Image Characteristics 478
Isroil I. Jumanov and Rustam A. Safarov

Development and Comparative Analysis of Path Loss Models Using
Hybrid Wavelet-Genetic Algorithm Approach 488
Ikechi Risi, Clement Ogbonda, and Isabona Joseph

Mathematical Advances and Modeling in Logistics Engineering

Risk Assessment of Navigation Cost in Flood Season of the Upper
Reaches of the Yangtze River Based on Entropy Weight Extension
Decision Model 503
Jun Yuan, Peilin Zhang, Lulu Wang, and Yao Zhang

Influencing Factors and System Dynamics Analysis of Urban Public
Bicycle Projects in China Based on Urban Size and Demographic
Characteristics 514
Xiujuan Wang, Yong Du, Xiaoyang Qi, and Chuntao Bai

Cross-border Logistics Model Design e-Tower Based on Blockchain
Consensus Algorithm 524
Shujun Li, Anqi He, Bin Li, Fang Ye, and Bin Cui

Research on the Service Quality of JD Daojia's Logistics Distribution
Based on Kano Model 536
Yajie Xu and Xinshun Tong

CiteSpace Based Analysis of the Development Status, Research
Hotspots and Trends of Rural E-Commerce in China 547
Yunyue Wu

The Optimization and Selection of Deppon Logistics Transportation
Scheme Based on AHP 558
  Long Zhang and Zhengxie Li

Pharmacological and Non-pharmacological Intervention in Epidemic
Prevention and Control: A Medical Perspective 573
  Yanbing Xiong, Lijing Du, Jing Wang, Ying Wang, Qi Cai,
  and Kevin Xiong

Passenger Flow Forecast of the Section of Shanghai-Kunming
High-Speed Railway from Nanchang West Station to Changsha South
Station 583
  Cheng Zhang, Puzhe Wei, and Xin Qi

The Effect of Labor Rights on Mental Health of Front-Line Logistics
Workers: The Moderating Effect of Social Support 598
  Yi Chen and Ying Gao

An Investigation into Improving the Distribution Routes of Cold Chain
Logistics for Fresh Produce 608
  Mei E. Xie, Hui Ye, Lichen Qiao, and Yao Zhang

Development of Vulnerability Assessment Framework of Port Logistics
System Based on DEMATEL 618
  Yuntong Qian and Haiyan Wang

Application Prospect of LNG Storage Tanks in the Yangtze River Coast
Based on Economic Model 628
  Jia Tian, Hongyu Wu, Xi Chen, Xunran Yu, and Li Xv

Multi-depot Open Electric Truck Routing Problem with Dynamic
Discharging 640
  Xue Yang and Ning Chen

Analysis on the Selection of Logistics Distribution Mode of JD Mall
in the Sinking Market 651
  Weihui Du, Xiaoyu Zhang, Saipeng Xing, and Can Fang

Research on Cruise Emergency Organization Based on Improved
AHP-PCE Method 664
  Long Zhang and Zhengxie Li

Fresh Agricultural Products Supplier Evaluation and Selection
for Community Group Purchases Based on AHP and Entropy Weight
VIKOR Model 681
    Gong Feng, Jingjing Cao, Qian Liu, and Radouani Yassine

Logistics of Fresh Cold Chain Analysis of Joint Distribution Paths
in Wuhan 696
    Weihui Du, Donglin Rong, Saipeng Xing, and Jiawei Sun

Optimization of Logistics Distribution Route Based on Ant Colony
Algorithm – Taking Nantian Logistics as an Example 708
    Zhong Zheng, Shan Liu, and Xiaoying Zhou

Optimization Research of Port Yard Overturning Operation Based
on Simulation Technology 719
    Qian Lin, Yang Yan, Ximei Luo, Lingxue Yang, Qingfeng Chen,
    Wenhui Li, and Jiawei Sun

A Review of Epidemic Prediction and Control from a POM Perspective 734
    Jing Wang, Yanbing Xiong, Qi Cai, Ying Wang, Lijing Du,
    and Kevin Xiong

Application of SVM and BP Neural Network Classification in Capability
Evaluation of Cross-border Supply Chain Cooperative Suppliers 745
    Lei Zhang and Jintian Tian

Research on Port Logistics Demand Forecast Based on GRA-WOA-BP
Neural Network 754
    Zhikang Pan and Ning Chen

Evaluation and Optimization of the Port *A* Logistics Park Construction
Based on Fuzzy Comprehensive Method 764
    Xin Li, Xiaofen Zhou, Meng Wang, Rongrong Pang, Hong Jiang,
    and Yan Li

Advances in Technological and Educational Approaches

OBE Oriented Teaching Reform and Practice of Logistics Information
System Under the Background of Emerging Engineering Education 777
    Yanhui Liu, Jinxiang Lian, Xiaoguang Zhou, and Liang Fang

Teaching Practice of "Three Integration" Based on Chaoxing Learning
Software – Taking the Course of "Complex Variable Function
and Integral Transformation" as an Example 787
    Huiting Lu and Xiaozhe Yang

Transformation and Innovation of E-Commerce Talent Training in the Era of Artificial Intelligence ... 801
Lifang Su and Ke Liu

Talent Training Mode Based on the Combination of Industry-Learning-Research Under the Background of Credit System Reform ... 811
Shanyong Qin and Minwei Liu

Analysis of the Innovation Mechanism and Implementation Effect of College Students' Career Guidance Courses Based on Market Demand ... 822
Jingjing Ge

Comparative Study on the Development of Chinese and Foreign Textbooks in Nanomaterials and Technology ... 833
Yao Ding, Jin Wen, Qilai Zhou, Li Liu, Guanchao Yin, and Liqiang Mai

Practical Research on Improving Teachers' Teaching Ability by "Train, Practice and Reflect" Mode ... 844
Jing Zuo, Yujie Huang, and Yanxin Ye

College Foreign Language Teacher Learning in the Context of Artificial Intelligence ... 854
Jie Ma, Pan Dong, and Haifeng Yang

The Innovation Integration Reform of the Course "Single Chip Microcomputer Principle and Application" ... 865
Chengquan Liang

Discussion and Practice on the Training Mode of Innovative Talents in Economics and Management in Women's Colleges ... 876
Zaitao Wang, Ting Zhao, Xiujuan Wang, and Chuntao Bai

Cultivation and Implementation Path of Core Quality of Art and Design Talents Under the Background of Artificial Intelligence ... 886
Bin Feng and Weinan Pang

Reform and Innovation of International Logistics Curriculum from the Perspective of Integration of Industry and Education ... 899
Xin Li, Meng Wang, Xiaofen Zhou, Jinshan Dai, Hong Jiang, Yani Li, Sida Xie, Sijie Dong, and Mengqiu Wang

An Analysis of Talent Training in Women's Colleges Based on the Characteristics of Contemporary Female College Students ... 911
Ting Zhao, Zaitao Wang, Xiujuan Wang, and Chuntao Bai

Solving Logistical Problems by Economics Students as an Important
Component of the Educational Process 921
   Nataliya Mutovkina

Exploration and Practice of Ideological and Political Construction
in the Course of "Container Multimodal Transport Theory and Practice"
for Application-Oriented Undergraduate Majors—Taking Nanning
University as an Example 931
   Shixiong Zhu, Liwei Li, and Zhong Zheng

A Study on Learning Intention of Digital Marketing Micro Specialty
Learners Under the Background of New Liberal Arts—Based
on Structural Equation Model 944
   Yixuan Huang, Mingfei Liu, Jiawei You,
   and Aiman Magde Abdalla Ahmed

Comparisons of Western and Chinese Textbooks for Advanced
Electronic Packaging Materials 954
   Li Liu, Guanchao Yin, Jin Wen, Qilai Zhou, Yao Ding, and Liqiang Mai

Innovation and Entrepreneurship Teaching Design
in Application-Oriented Undergraduate Professional Courses – Taking
the Transportation Enterprise Management Course as an Example 963
   Yan Chen, Liping Chen, Hongbao Chen, Jinming Chen,
   and Haifeng Yang

The Construction of University Teachers' Performance Management
System Under the Background of Big Data Technology 974
   Fengcai Qin and Chun Jiang

Curriculum Evaluation Based on HEW Method Under the Guidance
of OBE Concept 983
   Chen Chen and Simeng Fan

The Relevance of a Systematic Approach to the Use of Information
Technologies in the Educational Process 995
   Nataliya Mutovkina and Olga Smirnova

Construction and Practice of "CAD/CAM Foundation" Course Based
on Learning Outcome 1006
   Ming Chang, Wei Feng, Zhenhua Yao, and Qilai Zhou

Research and Practice of Ideological and Political Education
in the Context of Moral Education and Cultivating People 1016
   Geng E. Zhang and Liuqing Lu

xviii   Contents

A Quantitative Study on the Categorized Management of Teachers'
Staffing in Colleges and Universities 1028
   Zhiyu Cui

Course Outcomes and Program Outcomes Evaluation
with the Recommendation System for the Students 1039
   Khandaker Mohammad Mohi Uddin, Elias Ur Rahman,
   Prantho kumar Das, Md. Mamun Ar Rashid, and Samrat Kumar Dey

Methodology of Teaching Educational Disciplines to Second (Master's)
Level Graduates of the "Computer Science" Educational Program 1054
   Ihor Kozubtsov, Lesia Kozubtsova, Olha Myronenko, and Olha Nezhyva

Professional Training of Lecturers of Higher Educational Institutions
Based on the Cyberontological Approach and Gamification 1068
   Oleksii Silko, Lesia Kozubtsova, Ihor Kozubtsov,
   and Oleksii Beskrovnyi

Exploring the Perceptions of Technical Teachers Towards Introducing
Blockchain Technology in Teaching and Learning 1080
   P. Raghu Vamsi

Author Index 1091

Advances of Computer Algorithms and Methods

AI Chatbots for Banks: Evolving Trends and Critical Issues

Margherita Mori[1]([✉]) and Lijing Du[2,3]

[1] University of L'Aquila (Retd), L'Aquila, Italy
morimargherita@gmail.com
[2] School of Management, Wuhan University of Technology, Wuhan 430072, China
[3] Research Institute of Digital Governance and Management Decision Innovation,
Wuhan University of Technology, Wuhan 430072, China

Abstract. The paper aims at providing a conceptual framework for analyzing evolutionary trends in the financial arena that have to do with banking on chatbots designed to take advantage of most recent developments in the domain of Artificial Intelligence (AI). Attention is focused on the positive effects that are associated with its applications and, at the same time, on critical issues that include potential risks and limitations, such as those faced by physically impaired bank customers. The starting point is an overview of how chatbots have changed the business areas of banking and financial services, as well as what can be expected in terms of future strategic shifts and behavioral changes of both banks and their customers. The next step is to conduct a more in-depth investigation of the role that AI-powered chatbots have begun to play - and are likely to continue to play - in the areas under scrutiny, particularly in the banking industry. Conclusions are based on success stories that should be replicated not only to improve customer experience and service support, but also to help modernize the banking industry, which is the most traditional subset of the financial sphere of the economy.

Keywords: AI applications · Banking and financial services · Chatbots

1 Introduction

The recent hype around ChatGPT and its competitors leads to shed unprecedented light upon cutting-edge technologies that keep accelerating the pace of significant developments in the financial arena and that include artificial intelligence (AI). Actually, technological advances have fueled financial innovation worldwide in the last decades: it has involved new types of financial firms (such as fintechs, that combine finance and technology), new financial products (such as digital payments and cryptoassets), and new ways of doing financial business (such as those reshaping banking into neobanking); as such, what seemed incredible a few months ago can be perceived as foundational, even in the historically change-resistant banking industry.

Within this framework, AI applications have rapidly gained momentum. Among others, financial institutions have begun to use them to improve their customer experiences

© The Author(s), under exclusive license to Springer Nature Switzerland AG 2023
Z. Hu et al. (Eds.): ICAILE 2023, LNDECT 180, pp. 3–13, 2023.
https://doi.org/10.1007/978-3-031-36115-9_1

by enabling frictionless, 24/7 interactions while saving money; certain AI applications have been widely disseminated in the financial industry, particularly by banks, which are thought to offer the greatest cost savings opportunities through front- and middle-office applications. Chatbots, for example, enable two-way communication by allowing an online conversation via text or text-to-speech as a cost-effective alternative to direct contact with live human agents.

These considerations pave the way for emphasizing the role of AI-powered chatbots in the financial arena, particularly in the banking sector, which is gradually becoming more self-service oriented in order to meet the needs of digitally savvy customers. Given the wider adoption of these first-response tools that greet, engage, and serve customers in a friendly and familiar way, it sounds appealing to overview the implied advantages and risks: therefore, this paper is aimed at providing a conceptual framework for analyzing the challenges that banks – as well as other financial institutions – have to take by embracing AI chatbots and by ultimately making conversational banking an automated process; conclusions draw upon success stories that are worth replicating not only to benefit those directly involved, but also to contribute to evolving trends in the financial sphere of the economy. The roadmap of this research is shown in Fig. 1.

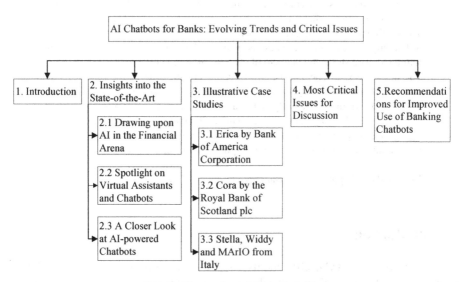

Fig. 1. The roadmap of this research

2 Insights into the State-of-the-Art

Due to the underlying cross-cultural issues, it seems appropriate to explore the most recent literature on several interrelated topics: on one hand, AI applications need to be investigated in order to devise those that are most useful to banks and other financial institutions; on the other hand, it is a matter of scrutinizing the distinguishing features of virtual assistants and chatbots. Finally, attention should be paid to how they can benefit from AI to better satisfy the needs on both the demand and the supply side.

2.1 Drawing upon AI in the Financial Arena

According to a shared view, AI is one of the technologies that can fundamentally change industries and banking makes no exception. Areas of major interest tend to be identified with investing, financial crime reporting, and complying with environmental, social and governance (shortly ESG) criteria: for example, AI can help banks to know their customers better and to provide them with suggestions that better align with their needs and preferences; algorithms can continuously analyze the portfolios of wealth management clients for risks while searching for the "next best offer" and can recommend a switch when its expected benefits exceed its costs [1].

As far as AI applications designed to help combating financial crime, they analyze transactions and record suspicious cases. For instance, various criteria are applied to each movement of capital, such as the amount, the currency, and the country to which the money either is going to or coming from: if a criterion does not match the typical patterns, the algorithm reports the anomaly to the account manager for further inquiry; as the feedback increases, the AI-powered model learns how to classify transactions correctly and ultimately reports only those that imply a real threat of a crime.

Furthermore, the ability to quickly process large amounts of data makes AI models attractive to provide banks with the support they need to comply with the Taxonomy Regulation and delegated acts that have been adopted by the European Commission since 2020: these rules have requested European Union (EU) banks to sort out which transactions can be deemed "environmentally sustainable", in an attempt at reducing greenwashing risks; in a few words, "the implementation of the Taxonomy signals an enhancement of mandatory sustainability reporting in the EU by driving capital towards activities that are 'irrefutably' green" [2]. Therefore, banks have to cope with a lot of new data from their corporate customers to properly classify their transactions and adhere to the evolving rules that fall within the scope of the climate and energy targets set by the EU for 2030 and that reflect its aim to be climate-neutral by 2050.

2.2 Spotlight on Virtual Assistants and Chatbots

The increasing reliance on virtual offices – and to home banking – has initially shaped the recourse to virtual assistants, who are self-employed workers specialized in providing administrative services from a remote location: these independent contractors usually work from a home office and can access the necessary documents remotely; tasks that are typically performed by virtual assistants include scheduling appointments, making phone calls, and managing email accounts. However, the specific duties to be performed depend on the terms of each contract, that in some cases may involve clerical and bookkeeping tasks, while in others may be centered upon posting regular updates to social media or writing articles for a blog [3].

Intelligent virtual assistants (IVAs) look like the most conspicuous way in which AI has modernized so far – and can still upgrade – the financial system by adding "intelligent" features to the human version of these assistants and by ultimately contributing to the development of the so-called "intelligent banks" [4]: to make short a long story, IVAs are engineered entities residing in software that interfaces with humans in a human way; the embedded technology incorporates elements of interactive voice response and other modern AI applications to deliver full-fledged "virtual identities" that are enabled to converse with users [5]. As such, IVAs can perform tasks or provide services for an individual that are based on commands or questions and that may detect both intent and sentiment.

IVAs tend to be compared to chatbots (as shown in Table 1), which deploy AI and natural language processing (NLP) to simulate human-like conversations with users via chat: the key task to be performed by chatbots is to understand user questions and automate responses to them with instant messages; in other words, chatbots are designed to make it easier for users to find the information they are searching for by responding to their requests and questions – through text input, audio input, or both – without the need for human intervention [6]. By way of comparison, due to their advanced natural language understanding and artificial emotional intelligence, IVAs can automate both complicated and repetitive tasks whereas chatbots mainly rely on rule-based conversational AI and tend to be associated with easier deployment [7].

Table 1. Comparison between virtual assistants and chatbots

Virtual Assistants	Chatbots
Follow commands and complete daily tasks for users	Interact with people and answer their questions
Perform daily tasks, such as setting alarms, making phone calls, etc	Help people interact with websites and solve problems
Wide range of functions and terminology	Only understand some specific terms
More interactivity	Low interactivity

2.3 A Closer Look at AI-Powered Chatbots

Actually, some AI-enhanced chatbots use complex algorithms that allow to provide more detailed responses: for instance, the deep learning capabilities of AI chatbots enable interactions to become more accurate over time, thus allowing to build a web of appropriate responses via the interactions between the chatbots themselves and humans; not occasionally conversational chatbots are programmed and trained in NLP, which enables them to facilitate a 2-way communication with the customers and their banks thanks to machine learning and AI. Accordingly, Conversational Banking is considered a smart way to boost customer loyalty by offering quick responses to clients' queries and AI chatbots are thought of making up one of the most promising strategic areas for banks "to win the satisfaction vote of their loyal customers" [8].

By common opinion, AI chatbots bring about several benefits not only to the banks that rely on them but also to the banking sector as a whole: for instance, these strategic tools allow for improved service to customers by answering their questions round the clock, contribute to boost employee productivity by solving minor issues posed by customers, and can assist them with paying down debt, checking account status, and getting credit score insights; AI chatbots do not only serve customers but can also support employees, for instance to schedule meetings and send messages. Personalized marketing is another area of interest, as these chatbots can be used to suggest customer-tailored investment options and offer promotions based upon available customer's data, such as the user's location, preferences and interests [9].

However, as previously stated, "the success of banking chatbots will be effective when customers are satisfied with the chatbots and engage in using them," and efforts have been made to investigate both the antecedents and consequences of customer brand engagement in using banking chatbots: the antecedents include interactivity, time convenience, compatibility, complexity, observability, and trialability; and the consequences deal with satisfaction with the brand experience. "Trialability, compatibility, and interactivity positively influence customer brand engagement through a chatbot, thereby influencing satisfaction with the brand experience and customer brand usage intention," according to clear empirical evidence [10].

3 Illustrative Case Studies

In sight of fully exploiting the potential of banking on AI chatbots, the financial arena – and particularly the banking industry – can be conceived as a huge laboratory with both physical and virtual features that provide useful insights. Interesting case studies emerge while surfing the web, since most banks have embraced the trend towards digitalization by setting up online facilities, even when traditional branches continue to be operating: therefore, these chatbots are getting part of the overall strategies developed within the banking industry and offer a promising avenue to both engage their actual customers and attract new ones; a look at the websites of major banks sounds like a reinforcement, as AI-powered chatbots are often deployed for connecting and communicating with customers, besides internally.

3.1 Erica by Bank of America Corporation (BofA)

Not surprisingly, BofA is credited towards becoming the industry leader in the market segment populated with AI chatbots, thanks to Erica that was introduced in 2018 and improved in early 2023 to accentuate its human touch. Erica is marketed as personalized, proactive, and predictive, which has helped it become one of the most frequently used virtual banking assistants, assisting 32 million customers [11]: this is described as "a powerful virtual assistant that helps you stay on top of your finances" due to its ability to consider a variety of data within BofA, such as cash flows, balances, individual transaction history, and upcoming bills; for example, you may "be alerted when merchant refunds post to your account", "receive bill reminders when payments are scheduled to be made", and "review weekly updates on monthly spending".

Moreover, Erica is available 24/7 to help you make the most of your money: this virtual financial assistant provides insights and guidance to help put cash to work, assists customers with exploring investment strategies, and refers them to a specialist to discuss different opportunities; support is also provided to Merrill Investments and hence BofA customers who hold Merrill Investments accounts can resort to Erica to access quotes, track performances, place orders, as well as to get connected with Merrill advisors from the Mobile Banking app. However, Erica has been made exclusively available in the Mobile Banking, which may not allow to satisfy all needs, including those stemming from potential (rather than actual) customers.

It is sufficient to "enter your U.S. mobile number and we'll text you a link to download the Bank of America Mobile Banking app so you can get started," though plans are in the works to provide services available through Erica in Online Banking in the future. As part of an ongoing improvement process, this AI bot can now use more methods to quickly answer customer questions, including a live chat with a BofA specialist: looking ahead, it appears relevant that Erica learns from customer conversations and has a dedicated team that is constantly working to expand the capabilities offered; to emphasize this point, "Erica is only available in English, but it is expected to learn Spanish" [12].

3.2 Cora by the Royal Bank of Scotland plc (RBS)

Heading to Europe, it was July 20, 2018 when RBS, part of the NatWest Group plc, posted on Facebook an introduction to Cora: it is a digital chatbot that has been continuously trained to provide instant answers to a quite large set of banking queries and that is "here to help you 24/7", no matter whether customers are using the app, are logged into digital banking or are getting access to the institutional website [13]. As shown on it, both actual and potential customers are invited to "meet Cora, your digital assistant".

Several tasks can be performed thanks to Cora, that are listed together their technicalities: for instance, you can "change your details", since "Cora can update your address and help you get your phone number, email address or name updated too"; you can also retrieve your "balance and transactions", provided that "in your app or Digital Banking, Cora can securely see your spending to search your transactions". This tool can even help if "you want a debit or credit card refund", as the section centered upon "retail disputes" showcases.

The "how to chat with Cora" section suggests to "keep your question simple and straight to the point" and provides a few examples: "I want to set up a standing order", "How do I make a payment" and "How do I download a statement"; during the interaction, "Cora might ask you a few questions to make sure you get the right answer, and may bring in a human colleague to help if necessary". Last, but not least, this chatbot is part of the marketing activities carried out by RBS, as Cora has been nominated for Innovation of the Year and potential users are encouraged to "vote for Cora at the British Bank Awards 2023" so as to get a chance to win £1000 in this Awards' prize draw (though "T&C's apply", besides being this initiative reserved to "UK Residents" and "over 18s only").

3.3 Stella, Widdy and MArIO from Italy

Personification stands out as the underlying asset of several Italian banks that have adopted AI chatbots, such as Stella by Banca Sella Group, Widdy by Banca Widiba SpA and MArIO by UBI Banca (Unione di Banche Italiane SpA). Stella dates back to 2009 as a chatbot that was adopted to support customers engaged in home banking, and evolved in 2017 as an updated version that took advantage of a thorough analysis of the interactions recorded up to that point: this empirical approach has allowed to identify the areas that seem most convenient to focus on and has enabled to improve the quality of the responses that customers are provided with; in the pandemic era, Stella has helped to devise new areas that it could deal with, thus paving the way to the development of new skills, such as those regarding loans, in sight of better and sooner satisfying customers' needs by both human and virtual service assistance [14].

Widdy is a conversational digital employee who is capable of sophisticated understanding of complex issues and who not only assists customers but also learns from these interactions: Widiba chose Teneo to build its intelligent online assistant, which has been in use since January 2014 to provide high-quality service and create a positive customer experience, and that can be easily updated with new knowledge and capabilities. Widdy can assist users with every step on the Widiba website, such as walking them through all stages of opening a new account step by step, and if users suspend their activity and return to it later, Widdy can recognize what steps have already been completed, allowing the account to be finalized without wasting the individual customer's time. Widiba recently expanded and upgraded its use of the Teneo platform, and can now provide "predictive help" to customers who email the contact center via the website, using natural language understanding to automatically answer their questions even before they hit the send button [15].

As far as MArIO, it has been defined as the smartest banking bot in Italy [16]: MArIO stands for Migration Artificial Intelligence Operator and was adopted in 2017 in order to support the process developed by UBI Banca – a leading Italian retail bank, acquired by Intesa Sanpaolo SpA in 2020 – to integrate 3 other banks (Nuova Banca dell'Etruria e del Lazio SpA, Nuova Cassa di Risparmio di Chieti SpA and Nuova Banca delle Marche SpA); MArIO was the first AI chatbot deployed for internal customer support in an Italian bank and allowed for great user engagement (with about 4 thousand bank employees involved). Actually, lots of improvements have been achieved once in production, with the number of different answers grown from 89 to 227 and the number of topics recognized increased from 121 to 609, which led to consider extending MArIO bot for day-to-day customer support.

4 Most Critical Issues for Discussion

Unquestionable empirical evidence surfaces, which leads to identify several reasons why AI-powered chatbots should be more and more widely adopted by banks but also to grasp critical issues to be addressed without delay, in sight of bridging the gap between supply and demand factors, as well as between theory and practice. Some of these issues stem from questions that sound all but surprising and that sometimes are mentioned in the FAQs, just as in the case of Erica: they include questions and answers such as is the chatbot secure and private? Yes, your interactions with it are protected by the same industry-leading privacy and security features as the mobile app and Online Banking. Can someone else but the customers activate the chatbot while using their phones? No, customers need to be authenticated through the app in order to be able to use the chatbot itself.

Other issues can be referred to in terms of digital accessibility: on one hand, the chatbots under investigation tend not to be available to potential (versus actual) customers of the banks involved, which may confine them into a "missing out" situation, provided that struggling to retain their customer base should be combined with trying to enlarge it by attracting new customers; on the other hand, chatbots that are only available on digital banking apps exclude both actual customers who prefer to otherwise avail themselves of their digital banking facilities and potential customers who want to learn more of the targeted banks by accessing their institutional websites. Success stories include the Spanish-speaking virtual assistant that was launched in 2018 by Banpro (Grupo Promerica) – Nicaragua's largest bank, headquartered in Managua – in cooperation with Canadian banking technology firm Finn.ai and that has been made available via Facebook Messenger, a channel reportedly being used daily at that time by 50% of Nicaraguans [17].

Critical issues may also surface with regard to diversity, equity and inclusion (or DE&I) as closely linked values that imply support to different groups of individuals, including people of different races, ethnicities, religions, abilities, genders, and sexual orientations. Actually, valuable opportunities may be missed by underscoring the needs that pertain to people with disabilities and that could be eventually satisfied by specific AI chatbots: with disability being part of the diversity conversation activity, it sounds like a must – rather than an option – to adapt these strategic tools to make them fully operational even in the case of special circumstances and situations, as it is the case when users are physically (for instance, visually) impaired or when they are affected by invisible illnesses, as it may happen in the case of neurodiversity; understanding neurodivergent people (just like those who are affected by a quite low stress tolerance) means embracing and being open to other ways of learning and collaborating, which leads to propose AI chatbots trained to support people with cognitive differences, that reflect natural variations in how the brain is wired.

Above all, the major critical issues are illustrated in Fig. 2.

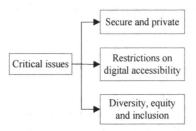

Fig. 2. The critical issues

5 Recommendations for Improved Use of Banking Chatbots

Focusing on recommendations, relevant issues can be referred to linguistic minorities, such as Spanish speaking bank customers in several parts of the United States, who cannot interact with AI chatbots in English. No wonder that some of these tools can be trained to work in several languages, as shown by Nordea, the leading Nordic bank that covers Sweden, Denmark, Norway and Finland: this bank has been using several virtual agents in production on the Boost.ai platform, operating in each market's local language (plus English in some countries); to this end, a comprehensive conversational AI strategy has been developed to scale customer service across the four markets served – for a total of over 9 million private customers and more than 500,000 active corporate customers – with a varied channel mix of customer service touchpoints (via in-person, phone, email and both live and automated chat) [18].

Improvements can also be made by solving some critical but frequent problems, since being smart – as AI-powered chatbots promise to be – should not allow for answers like "I did not understand" and banks cannot afford reputational risks, encompassing the risk of losing their credibility by relying upon poor performing chatbots: same questions may be written in many different ways, misspellings may occur not occasionally, items may be referred to with different names; moreover, questions may be very – even too – complex, or may be out of scope. Solutions include the ability to use partial understanding and to tell the user how to rephrase, as well as to recognize and be explicit on out of scope.

It must also be acknowledged that the implementation of chatbot technology is evolving rapidly in the banking system, just like across all other industries, while customer acceptance is lagging behind: from a theoretical point of view, it has been highlighted "the importance of perceived compatibility and perceived usefulness in the adoption of banking chatbot technology", with "awareness of the service" postively impacting "perceived ease of use" and "perceived privacy risk" and indirectly affecting "usage intention of banking chatbots through perceived usefulness"; moreover, "perceived ease of use influences perceived usefulness, and perceived compatibility has an effect on both perceived ease of use and perceived usefulness" [19]. Therefore, it makes sense to arrange promotional campaigns designed to encourage customers to rely on AI chatbots, as shown by Zenith Bank Plc – a large financial service provider in Nigeria and Anglophone West Africa – that has launched a marketing initiative between October 3 and December 30, 2022 to reward users of its Whatsapp Banking through ZiVA (Zenith Intelligent Virtual Assistant) "your personal banking assistant on WhatsApp" [20].

6 Conclusion

To summarize, the successful implementation of AI-powered chatbots in the banking industry should be part of comprehensive strategies: these tools can assist in sensing, comprehending, acting, and learning, thus envisioning a system that can perceive the world around them, analyze, and understand the information they receive, take actions based on that understanding, and improve their own performance by learning from what happened; recommended strategies have the potential to support and drive growth, in sight of a more sustainable and inclusive financial industry. Several trends in digital engagement have accelerated during the corona pandemic, and the bleak picture to be confronted should push all financial institutions to take advantage of AI technology by fully utilizing its potential, particularly in providing distinct customer experiences.

Needless to say, AI is unlikely to replace skilled human workers, and banks, by the way, require a dedicated team to manage and improve their chatbots: because computational algorithms lack the ability to reason cognitively, it appears that humans will always play a role in the emotional aspects of bank-customer relationships; at the same time, human agents are required to monitor the training of AI models to ensure that banks do not run the risk of providing unsuitable solutions to their customers or, worse, that the models do not create unintentional bias against certain types of consumers. As a result, as technology advances, AI can be expected to augment – but not replace – the human elements of the banking relationship.

All in all, customer demand for both digital banking and human interaction requires banks to find the right balance between them, not only to preserve their market shares but also to try to increase them. To this end, making AI chatbots available to some extent to potential customers (besides the full operability reserved to actual ones) may prove a win-win strategy: success stories sound like a reinforcement to acknowledge the key role that these tools can play, though critical issues should not be underestimated; likewise, recommendations stemming from use cases developed should be carefully checked and eventually complied with, in order to fully exploit the potential of AI applications in the market segment under scrutiny, thereby contributing to shape sustainable pathways to prosperity.

References

1. Bremke, K., Cavus, M., Mindt, M., et al.: How AI is changing banking. Frankfurt am Main: Deutsche Bank AG (2023). https://www.db.com/what-next/digital-disruption/better-than-humans/how-artificial-intelligence-is-changing-banking/index. Accessed 13 Feb 2023
2. Pettingale, H., de Maupeou, S., Reilly, P.: EU Taxonomy and the Future of Reporting. Harvard Law School Forum on Corporate Governance. Harvard College, Cambridge, MA (April 4, 2022). https://corpgov.law.harvard.edu/2022/04/04/eu-taxonomy-and-the-future-of-reporting/. Accessed 13 Feb 2023
3. Will, K.: What is a Virtual Assistant, and What Does One Do?. Investopedia, New York, NY. https://www.investopedia.com/terms/v/virtual-assistant.asp. Accessed 15 Feb 2023
4. Jubray, E., Graham, T., Ryan, E.: The Intelligent Bank - Redefining Banking with Artificial Intelligence. Accenture, Dublin (2018). https://www.accenture.com/t00010101t000000z__w__/gb-en/_acnmedia/pdf-71/accenture-banking-aw-jf-web.pdf. Accessed 19 Feb 2023
5. Intelligent Virtual Assistant. Techopedia. Edmonton, AB: Janalta Interactive (June 28, 2022). https://www.techopedia.com/definition/31383/intelligent-virtual-assistant. Accessed 15 Feb 2023
6. What is a chatbot? Armonk, NY: IBM. https://www.ibm.com/topics/chatbots. Accessed 15 Feb 2023
7. Dilmegani, C.: Chatbot vs Intelligent Virtual Assistant; Use Cases Comparison. AIMultiple, Tallin (January 20, 2023). https://research.aimultiple.com/chatbot-vs-intelligent-virtual-assistant/. Accessed 15 Feb 2023
8. Sayiwal, S.: Chatbots in Banking Industry: A Case Study. Journal of Emerging Technologies and Innovative Research (JETIR) 7(6), 1498–1502 (2020). https://www.jetir.org/papers/JETIR2006221.pdf. Accessed 19 Feb 2023
9. Shashank, B.R., Rashmi, R.: A Review of Chatbots in the Banking Sector. Int. J. Eng. Res. Technol. (IJERT), 10(06), 428–430 (2021). https://www.ijert.org/research/a-review-of-chatbots-in-the-banking-sector-IJERTV10IS060203.pdf. Accessed 15 Feb 2023
10. Hari, H., Iyer, R., Sampat, B.: Customer brand engagement through chatbots on bank websites – examining the antecedents and consequences. Int. J. Human-Comp. Interac. 38(13), 1212–1227 (2022). https://doi.org/10.1080/10447318.2021.1988487

11. McNamee, J.: Bank of America adds a human touch to its virtual assistant, Erica. Insider Intelligence. NY, NY: Insider Inc (December 14, 2022). Erica becomes a little more human - Insider Intelligence Trends, Forecasts & Statistics. Accessed 16 Feb 2023
12. Personalized. Proactive. Predictive. Charlotte, NC: Bank of America Corporation (2023). https://promotions.bankofamerica.com/digitalbanking/mobilebanking/erica. Accessed 16 Feb 2023
13. Meet Cora – Here to help you 24/7. Edinburgh: The Royal Bank of Scotland plc (2023). https://www.rbs.co.uk/support-centre/cora.html. Accessed 15 Feb 2023
14. Gaudiuso, R.: Intelligenza artificiale e assistenti virtuali: come cambia la relazione con i clienti. Sella Insights. Biella: Banca Sella Holding SpA (February 21, 2022) (in Italian). https://sellainsights.it/-/intelligenza-artificiale-e-assistenti-virtuali-come-cambia-la-relazione-con-i-clienti. Accessed 16 Feb 2023
15. Walker, T.: Widdy (Widiba). Amsterdam: Chatbors.org (2023). https://www.chatbots.org/chatbot/widdy/. Accessed 16 Feb 2023
16. Vitale, A., Boido, M.: The Story Behind the Smartest Banking Chatbot in Italy!, Chatbot Summit, Milan (January 31, 2018). https://www.slideshare.net/pmontrasio/il-pi-intelligente-chatbot-bancario-in-italia. Accessed 16 Feb 2023
17. Banpro and Finn.ai roll out virtual banking assistant in Central America. Retailer Banker International (March 7, 2018). https://www.retailbankerinternational.com/news/banpro-finn-ai-roll-ai-powered-virtual-banking-assistant-central-america/. Accessed 18 Feb 2023
18. Case Study Nordea. Stavanger: Boost.ai (2023). https://www.boost.ai/case-studies/nordea-employs-comprehensive-conversational-ai-strategy-to-scale-customer-service. Accessed 18 Feb 2023
19. Alt, M., Vizeli, I., Săplăcan, Z.: Banking with a chatbot – a study on technology acceptanc. Studia Universitatis Babeş-Bolyai Oeconomica **66**(1), 13–35 (2021). https://doi.org/10.2478/subboec-2021-0002
20. Terms and Conditions. Lagos: Zenith Bank Plc (2022). https://www.zenithbank.com/images/TERMS_AND_CONDITIONS.pdf. Accessed 18 Feb 2023

Traffic Flow Characteristics and Vehicle Road Coordination Improvement in Subterranean Interweaving

Enshi Wang[1(✉)], Bihui Huang[1], and Bing Liu[2]

[1] CCCC Second Expressway Consultant Co. Ltd., Wuhan CCCC Traffic Engineering CO. Ltd., Wuhan 430050, China
cueeha@163.com

[2] School of Transportation and Logistics Engineering, Wuhan University of Technology, Wuhan 430063, China

Abstract. Subterranean interchange accidents are frequent, which frequently occur in weaving segments. Among them, rear-end collisions are their main types of accidents, and rear-end collisions are one of the typical phenomena of traffic instability. In view of this, this project adopts the method of empirical analysis to obtain the characteristics of traffic flow in the subterranean interchange weaving segment. Through the Lyapunov index to determine its stability, it is found that the stability of the weaving segment is significantly reduced, resulting in multiple rear-end collision accidents. The main causes of instability are: The limited space for the layout of the traffic signs and the lack of GPS information lead to inadequate access to the driver's route information, which makes it easier to make mistakes in decision-making, emergency braking, and frequent sudden lane changes, and imposes random strong interference on the traffic flow. Based on this, the project improves traffic flow stability by reducing random strong interference: Pushing position information to the navigation system through the Beacon base station to achieve self-position information compensation, inducing the drivers to change lanes reasonably through the driving instructions, to reduce random strong interference, and at the same time rerouting and pushing routes for real-time relocation of vehicles.

Keywords: Traffic safety · Subterranean interchange · Weaving segment · Traffic flow stability · Information compensation · Vehicle road coordination

1 Introduction

As the urban underground road gradually forms a network, a large number of subterranean interchanges will be formed, and the subterranean interchange will be the key node of the road network, which will affect the operation efficiency of the overall road network, and its traffic safety issues need more attention. According to statistics, the annual average traffic accident of Xiamen Wanshishan-Zhonggushan Mountain subterranean interchange is as high as 42, and the proportion of rear-end accidents is 94%

© The Author(s), under exclusive license to Springer Nature Switzerland AG 2023
Z. Hu et al. (Eds.): ICAILE 2023, LNDECT 180, pp. 14–24, 2023.
https://doi.org/10.1007/978-3-031-36115-9_2

[1]. Weaving segment is the most common accidents area, accounting for 42.3% of the accidents. At present, the safety of the weaving segments on the ground has been attached much importance. Many scholars have proposed improvement programs from the aspects of ramp control, driver characteristics, traffic flow characteristics, etc. For example, Aries van Beinum mainly revealed that lane changes caused by merging and diverging vehicles create most turbulence [2]; Mohamed Abdel-Aty's research showed the reduction of upstream speed is the key to improve the safety of weaving segment and lower upstream variable speed limit would better improve the safety of the whole weaving segment [3]; David Sulejic proposed a optimization technique for the lane- changing concentration [4]; Ling Wang studied crash likelihood using real-time crash data with the objection, identified hazardous conditions, but the above research mainly focuses on the ground [5]. Different from the ground, the underground environment has obvious disadvantages in terms of light, line shape and information conditions. It is more likely to interfere with traffic flow, and the system stability is poor, which in turn increases the accident frequency and intensity. Therefore, the underground environment needs further exploration.

Traffic flow stability is an important indicator to measure the anti-jamming capability of the system [6, 7]. According to relevant research, improving the stability of the traffic flow system can significantly improve the traffic safety level [8, 9]. On this basis, research is carried out to obtain the driving behavior characteristic data of subterranean interchange weaving segment through real vehicle experiment and surveillance video. The stability is judged by Lyapunov exponent, and the traffic loss is caused by the characteristics of underground traffic environment [10]. The Beacon base station disposed on the road side is used to construct a navigation system to provide position information for the drivers, improve the stability of the traffic flow from the perspective of vehicle road coordination, and improve the traffic safety level of the subterranean interchange.

This article is divided into five parts. In the second part, through the real vehicle experiment and monitoring video, acquire the vehicle characteristics. The third part analysis collected data and draw some conclusions. The fourth part proposes improvement measures, that is, using iBeacon technology to build a navigation system to improve traffic flow stability. Finally, the fourth part summarizes and discusses the results of this paper.

2 Experiment and Result

In the research, we got the trajectory, speed and space headway data through two ways: vehicle experiment and Monitor Video Recognition.

2.1 Vehicle Experiment

Carry out vehicle experiments on the road at the of Wanshishan and Zhonggushan underground intersection in Xiamen from August 4, 2017 to August 12, 2017. This intersection is the first subterranean interchange project in China. The subterranean interchange tunnels consist of 7 tunnels on the left and right of Wanshishan Tunnel, A and B tunnels of Zhonggushan Tunnel and three ramps A, B and C connecting the upper and lower tunnels. The tunnels have six bifurcations, one parallel section, four through sections, 15 kinds of tunnel clearance sections, the maximum excavation depth is 26.4 m and the excavation height is 15.4 m. It is designed according to the urban expressway standard 2 lanes, the design speed is 60 km/h, the single tunnel net width is 11.5 m, the net height is 5.0 m, the left tunnel is 2850.5 m, the right tunnel is 2805 m. The space inside the ramp of the interchange is designed 8.5 m according to the width of the single lane, and the design speed is 30 km/h. The A ramp is annular and plane curve radius is 100 m. Figure 1 is the subterranean intersection.

1) A ramp: Zhonggushan Tunnel from south to east to Wanshishan Tunnel right turn ramp.
2) B ramp: Wanshishan Tunnel from east to north to Zhonggushan Tunnel right turn ramp.
3) C ramp: Zhonggushan Tunnel from north to east to Wanshishan Tunnel right turn annular ramp.

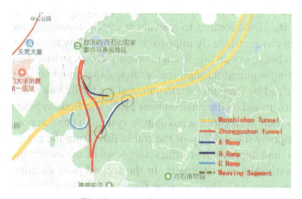

Fig. 1. Experimental location

The experiment selected 5 men and 2 women drivers as the experimenters, driving for more than 3 years as same as less than 3 years. Table 1 shows 5 periods to collect traffic flow statistics in different drafficc characteristics and visual environments. Table 2 is the equipment and collected data.

The choice of experimental path is to avoid drivers from taking repeated paths and eliminating adaptability. Figure 2 shows the experiment path. Firstly, starting from the A tunnel, drove along the A-c-C-d path, drove through the diversion into the blue ramp and out of the d-tunnel; then, drove along the d-D-b-B-d-D-c path, halfway through the Yellow ramp to the main line after the convergence of c-tunnel. Finally, it traveled along the c-C-d-D-b-B-c-c-a path, crossed the shunt lane into the red ramp, then ran out of the c-tunnel, after drove through the C portal, it returned to a portal along the C-a path to complete the operation of a loop. The time required to complete an annular is nearly 1 h, a driver need to complete six times a day of the above six circuits.

Table 1. Five periods to carry out the experiment

Number	Time	Traffic Flow Character
1	7:00 ~ 9:00	Morning Rush Hour
2	9:00 ~ 11:00	Morning Flat Peak
3	15:00 ~ 17:00	Afternoon Flat Peak
4	17:00 ~ 19:00	Evening Rush Hour
5	21:00 ~ 22:30	Free Flow

Table 2. Specific collected data

Number	Component	Collected Data
1	Mobileye ME630 System	Lane Departure, Headway, Front Vehicle Speed
2	Vehicle OBD Data Parsing	Steering Wheel Angle and Speed, Speed, Throttle, Brake Pedal
3	RT2500 Inertial Navigation System	Vehicle Coordinates, Trajectory
4	Illuminance Instrument	Visual Environment
5	Tachograph	Traffic Video Recording

Fig. 2. Experiment path

2.2 Monitor Video Recognition

Through MATLAB image recognition programming, analyze the monitor video of the test section. And through the background-based target detection method, target recognition and tracking of vehicles were made to finally get the vehicle trajectory data.

2.3 Trajectory Data Analysis

Figure 3 is the trajectory characteristics of the vehicle at the diversion point combined with real vehicle experiments and monitoring video. The curve represents the vehicle trajectory, and the bar graph represents the vehicle occupancy. Among them, up to 26.7% of the vehicles suddenly change lanes 50 m before the split nose, and bring great obstruction to the traffic flow.

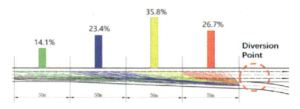

Fig. 3. Vehicle trajectory

3 Further Analysis

3.1 Space Headway Analysis

According to the analysis of the experimental data, the relationship between the actual space headway(S) and stopping sight distance (ST) is shown in Fig. 4. Among them, up to 86% of the space headway is smaller than the safe distance. Overall, in the subterranean interweaving section, the space headway is close, and the risk of following the vehicle is large.

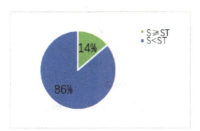

Fig. 4. Space headway and stopping sight distance

The maximum Lyapunov exponent in chaos theory is used to judge the stability of traffic flow [11]. If the index is positive a, the system is unstable, and when it is negative, the system is stable. Taking the time-varying data of the vehicle as the input, using the wolf reconstruction method and being calculated by MATLAB, the result is shown in Fig. 6. It can be seen that the Lyapunov exponent of space headway is greater than 0, so the system is unstable.

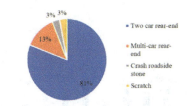

Fig. 5. Traffic accident type distribution

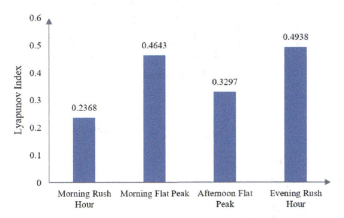

Fig. 6. Lyapunov Index of space headway

According to analysis the space headway and the traffic accident (Fig. 5), we found that the stability of the interweaving section system was poor, and the space headway was significantly lower than the safety distance, resulting in frequent rear-end collisions. Further analysis of the reasons for the small space headway, by consulting the relevant literature, we found the following explanation: Saffarian pointed out that when the visibility is extremely low, the driver will tend to reduce the headway to ensure that the front car is within the visible distance [12]. Leeuwen studied the effects of vertical information loss on driving behavior through perceptual occlusion paradigms and simulation experiments [13]. It was found that when the driver only presented the near-end visual region (similar to the underground intercommunication feature), the driver's gaze point was even more Focusing on the front car, it is closer to the car. It is also found through driving simulation studies that when the visibility is very low, the driver usually chooses a smaller headway.

3.2 Speed Analysis

The experimental data was collated and analyzed, and the overspeed ratio was obtained as shown in Fig. 7. According to the analysis we found that the overspeed ratio was higher than 20% in each period, so the speed risk was high.

Calculating the Lyapunov exponents of the space headway, the value of each time period was greater than 0, so the system was in an unstable state (see Fig. 8).

Fig. 7. Overspeed ratio

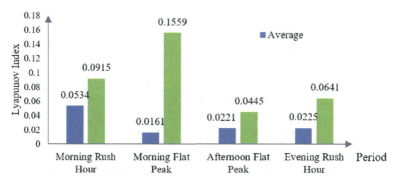

Fig. 8. Lyapunov Index of speed

It was found that overspeed vehicles accounts for higher than 20% in each period. Hogendoorn's and Champion's research showed that the low contrast of visual stimuli leads to Drivers' underestimating the speed of the car, which results in overspeed [14, 15]. Therefore, the low illuminance of the subterranean environment and the high concentration of pollutants can be the reason of overspeed phenomenon.

3.3 Result Analysis

Through the trajectory analysis, it was found that the emergency braking and sudden lane change occurred frequently in the subterranean interchange. According to the car-following theory, the emergency braking and sudden lane change of the preceding vehicle can generate strong stimulation to drivers of following vehicles, which leads to drivers' drastic driving behaviors [16]. Those behaviors destroy the stability of the system. When the system is unstable, the emergency braking or sudden lane change behavior of the preceding vehicle can cause the disturbance to propagate in the reverse direction of traffic, increasing the probability of the rear-end collision.

Further analysis of the subterranean environment, the reasons for this phenomenon are as follows: (1) The horizon is limited in subterranean environment, resulting in inadequate signs and information. As a result, it is difficult to effectively guide the driving route and drivers unfamiliar with the road cannot change lanes in advance; (2) GPS in the subterranean space doesn't work, so vehicles that rely on navigation cannot obtain effective navigation information, and it is difficult to determine an accurate driving route; (3) Low subterranean space illumination and high concentration of pollutants decrease the contrast of visual stimuli in the space, so drivers cannot identify traffic signs in time and eventually miss the best lane change position.

4 Solution

4.1 Some Solutions and Their Disadvantages

In traditional navigation, GPS positioning, WIFI positioning, base station positioning, etc. are usually used. In the subterranean tunnel, GPS positioning is invalid because of the lack of GPS signals. The WIFI positioning works well for indoor positioning, but it is expensive and cannot be used independently and permanently. The difficulty of subterranean environment positioning has led people to resort to other ways to obtain route information, mainly through two. One is the inferred navigation, that is, the inertial navigation. The principle is to collect the acceleration of the moving object and automatically perform the integral operation to obtain the instantaneous velocity and instantaneous position data of the moving object for navigation. However, since the sampling frequency is generally high (tens or even hundreds of times in one second), the error is easily to accumulate, so inertial navigation is only suitable for use in a short time; the other is to obtain information through the traffic signage. However, due to the limited subterranean space and signage layout, it is unable to provide sufficient traffic information.

4.2 Introducing IBeacon Technology

In order to solve the above problems, we designed roadside facilities to meet the driver's information needs, that is, road coordination. The iBeacon technology was chosen, which is a low-energy Bluetooth technology that works like the previous Bluetooth technology [17]. The Beacon base station continuously transmits signals in an area of tens of meters, and the device receiving the signal can respond accordingly. UUID, minor, and major are some of the parameters in iBeacon signal packets. They are used to identify Beacon devices and pass information [18]. The iBeacon positioning was used for indoor navigation and indoor preferential information push because of the small consumption of Beacon equipment, but it has not been involved in subterranean navigation. In this research, this technology was used to improve the navigational dilemma in places where GPS doesn't work like subterranean tunnels.

4.3 Key Problems

The specific method is as follows: the subterranean navigation module based on the iBeacon technology sends the location information by the Beacon base station, and the drivers' device pushes the driving route in real time after receiving the location information to realize the compensation of the route information in the subterranean environment (as shows in Fig. 9). To achieve this goal, a set of methods was designed to convert the received beacon information into navigation instructions to be presented to the driver. In order to remind the lane change and turn in advance, the Beacon base station was arranged at a certain distance before the lane change line and the intersection. After the device obtains the position information, voice prompt is performed to realize the effect of promptly changing the lane change.

Fig. 9. How does the Beacon work

Another key issue is the placement to place beacon base stations, which affects the accuracy and effectiveness of navigation. Flat layout: The base station was divided into two types by their function: positioning base station and broadcasting base station. The positioning base station was disposed every 20 m for path navigation, and the distance between broadcasting base station and intersection, S was calculated from the distance formula $S = V_{85\%} * T$. T is the time a navigation activity cost. For example, the characteristic driver's perception reaction time and braking time total 0.7s. In addition, the voice prompt time is 1.5 s (considering the broadcast speed and content), so T is 2.2 s. Combined with operating speed as 50 km/h(13.9/s), S is 31.97 m. Therefore, it is laid at a distance of 30 m from the intersection to provide the driver with voice prompt (see Fig. 10). Section layout: Considering the signal coverage of the Beacon base station, for the two-lane section, the Beacon base station was placed at the top of the tunnel; for the three-lane section, Beacon base stations was placed on both sides of the tunnel to ensure the effect (see Fig. 11).

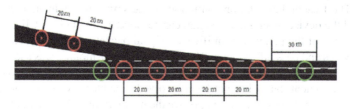

Fig. 10. Flat layout

A test has been carried out to evaluate the performance of the Beacon system. Beacon interval of broadcasting was set as 100 ms and Beacon base station covered an area of 20 m [19]. The experimenter drove toward Beacon base station from 30 m away and recorded the period from the time that the car 50 m was away from Beacon to the time

Fig. 11. Section layout

that the signal was received. Finally, the average interval was 0.2 s, negligible compared with the time one navigation activity cost.

5 Conclusion

In the above study, through the analysis of the data, it was concluded that up to 26.7% of the vehicles suddenly changing lane 50 m before the split nose. By analyzing the environment of the weaving segments, it was proposed that the lack of route information results in the driver unable to reasonably change the lane in advance, which leads to accidents. In the actual environment, the driver's own conditions, such as psychological load, driving experience, road line type and traffic marking settings, may affect the occurrence of traffic accidents, this will be the goal of future research.

In terms of solutions, the above application for iBeacon technology is not yet mature. In order to achieve good results, there must be a unified communication protocol standard to represent various traffic information, and the accuracy and post-care of Beacon base stations require further research. In addition, there are many short-range wireless communication technologies, such as UWB technology, Z-Wave technology, and wireless Mesh network. It is also the direction of the next step to test which one is more suitable for the compensation of underground intercommunication information.

References

1. Li, X., Yan, X., et al.: A rear-end collision risk assessment model based on drivers' collision avoidance process under influences of cell phone use and gender—A driving simulator based study[J]. Accid. Anal. Prev. **97**, 1–18 (2016)
2. van Beinum, A., Farah, H., et al.: Driving behaviour at motorway ramps and weaving segments based on empirical trajectory data. Transportation Research Part C: Emerging Technologies **92**, 426–441 (2018)
3. Abdel-Aty, M., Wang, L.: Implementation of variable speed limits to improve safety of congested expressway weaving segments in microsimulation. Transportation Research Procedia **27**, 577–584 (2017)
4. Sulejic, D., Jiang, R., et al.: Optimization of lane-changing distribution for a motorway weaving segment. Transportation Research Procedia **21**, 227–239 (2017)
5. Wang, L., Abdel-Aty, M., et al.: Real-time crash prediction for expressway weaving segments. Transportation Research Part C: Emerging Technologies **61**, 1–10 (2015)

6. Derbel, O., Peter, T., et al.: Modified intelligent driver model for driver safety and traffic stability improvement. IFAC Proceedings Volumes **46**(21), 744–749 (2013)
7. Peng, G., Liu, C., et al.: Influence of the traffic interruption probability on traffic stability in lattice model for two-lane freeway. Physica A **436**, 952–959 (2015)
8. Mhaskar, P., El-Farra, N.H., et al.: Stabilization of nonlinear systems with state and control constraints using Lyapunov-based predictive control. Syst. Control Lett. **55**(8), 650–659 (2006)
9. Haddad, J., Geroliminis, N.: On the stability of traffic perimeter control in two-region urban cities. Transportation Research Part B: Methodological **46**(9), 1159–1176 (2012)
10. Lee, S., Heydecker, B.G., et al.: Stability analysis on a dynamical model of route choice in a connected vehicle environment. Transportation Research Part C: Emerging Technologies **94**, 67–82 (2018)
11. Johansson, M., Rantzer, A.: Computation of piecewise quadratic lyapunov functions for hybrid systems. European Control Conference (ECC) (1997). https://doi.org/10.23919/ecc.1997.7082399
12. Saffarian, M., Happee, R., et al.: Why do drivers maintain short headways in fog? A driving-simulator study evaluating feeling of risk and lateral control during automated and manual car following. Ergonomics **55**(9), 971–985 (2012)
13. van Leeuwen, P.M., Happee, R., et al.: Vertical field of view restriction in driver training: A simulator-based evaluation. Transport. Res. F: Traffic Psychol. Behav. **24**, 169–182 (2014)
14. Hogendoorn, H., Alais, D., et al.: Velocity perception in a moving observer. Vision. Res. **138**, 12–17 (2017)
15. Champion, R.A., Warren, P.A.: Contrast effects on speed perception for linear and radial motion. Vision. Res. **140**, 66–72 (2017)
16. Wagner, P.: Analyzing fluctuations in car-following. Transportation Research Part B: Methodological **46**(10), 1384–1392 (2012)
17. Varsamou, M., Antonakopoulos, T.: A bluetooth smart analyzer in iBeacon networks. In: 2014 IEEE Fourth International Conference on Consumer Electronics Berlin (ICCE-Berlin), pp. 288–292. IEEE (2014)
18. Lin, X.-Y.: A mobile indoor positioning system based on iBeacon technology. Annual International Conference of the IEEE Engineering in Medicine and Biology Society. IEEE Engineering in Medicine and Biology Society. Annual Conference, 4970–4973 (2015)
19. de Henares, A.: Towards accurate indoor localization using iBeacons, fingerprinting and particle filtering. International Conference on Indoor Positioning and Indoor Navigation (IPIN) **2016**, 4–7 (2016)

Influential Factors and Implementation Path of Talent Digital Evaluation Based on ISM Model: Taking Electric Power Enterprises as an Example

Wei Luo, Jiwei Tang[✉], Saixiao Huang, and Yuan Chen

CHN Energy Dadu River Big Data Services Co., Ltd., Chengdu 610000, China
547272841@qq.com

Abstract. The digital talent evaluation index system of electric power enterprises is usually affected by multiple factors. Clarifying the relationship between various factors is of great significance for the digital transformation and upgrading of enterprises and talent storage and allocation. Therefore, based on the ISM model, this paper constructs the adjacency matrix with the power enterprise as the representative framework, calculates the accessibility matrix with the help of Matlab programming, and finally analyzes the implementation path of the impact factor. In the study, the factors that affect the digitalization of enterprise talents are summarized into 20 sub - factors, which are divided into six progressive levels: top - level factor, core factor, critical factor, hub factor, key factor, and bottom - level factor according to the hierarchical factors. This research based on hierarchical structure can provide guidance and reference for the future digital evaluation of enterprise talents.

Keywords: Power transmission enterprises · Digitization · Talent characteristics · ISM model · Accessibility matrix

1 Introduction

The digital economy has become an essencial driving force for the transformation and upgrading of industrial structures and the promotion of high-quality economic development [1–3]. According to the White Paper on China's Digital Economy Development (2021), the scale of China's digital economy has reached 39.2 trillion yuan, with an annual growth rate of 9.7% [4]. As the mainstay of the national economy, the digital construction of state-owned enterprises is crucial to the innovative development of the national economy [5]. In the report of the 20th National Congress of the Communist Party of China, the General Secretary pointed out that we should speed up the construction of enterprise digitalization, promote the deep integration of digital economy and the real economy, and build a digital industrial cluster with international competitiveness [6]. The digital transformation and upgrading of enterprises is an inevitable requirement for the implementation of the decision and deployment of the CPC Central Committee, and also a new concept, strategy and measure for talent construction in the new era [7, 8].

© The Author(s), under exclusive license to Springer Nature Switzerland AG 2023
Z. Hu et al. (Eds.): ICAILE 2023, LNDECT 180, pp. 25–34, 2023.
https://doi.org/10.1007/978-3-031-36115-9_3

To deepen the talent supply - side structural reform, the construction of talent digitalization is an essential basis for improving the innovation ability of enterprises [9, 10]. In the context of the digital economy, the information base of the digital economy has impacted the development of the entire industry, resulting in "disruptive" innovation and the subversion of the organizational structure and business model [11–13]. On the one hand, information exchange, data flow and knowledge diffusion pose new challenges to the construction of enterprise human resources; On the other hand, in the digital wave, how to better select, cultivate, evaluate and motivate employees through human resource management practice has become a problem that enterprise managers need to consider [14]. Although the digital talent construction of enterprises has been put forward as an important strategy at the national level, from the content of the study, most of the research only stays at the level of policy formulation, or a more macro level of industry development, while ignoring the theoretical level of digital talent realization mechanism and logical analysis [15]. In particular, there is a lack of talent in digital model research focusing on specific industries and specific industrial types [16]. In addition, as a unique resource of the enterprise, human resources itself is complex and have an important impact on the business activities of the enterprise [17]. Therefore, the research on the digital evaluation index system of talents needs to consider not only the general ability items of skills, but also the professional ability items of talents, and build a systematic and comprehensive analysis framework [18].

Based on the above considerations, this paper takes the electric power industry as the representative to explore the digital talent impact factor set in the electric power industry, and uses the Interpretative Structural Model theory to explore the impact factors of the digital talent evaluation of power transmission enterprises and research on the implementation path, providing countermeasures and suggestions for building a reasonable digital talent evaluation standard in the electric power industry.

2 Construction of Digital Influence Factor Set for Talents in Power Transmission Industry

According to the theory of system dynamics and game theory, the transformation of economic and social systems is the result of the interaction of complex elements, and the transformation and upgrading of systems are usually closely related to policy environment, market environment, technological development and other factors. Abroad, Afha and others analyzed the development process of European power system based on socio - technical theory [19]. Wang Jifeng and others pointed out that the application of information technology has always played an essential role in promoting the development of China's power system, and the digitalization of power system runs through the whole process of production, operation and sales of power enterprises [20]. In fact, the process of digital transformation of electric power enterprises is essentially the transformation and upgrading of labor force based on the digitalization of talents, which is affected by the knowledge, ability and attitude of employees [21]. Factor analysis is one of the important methods [22, 23]. As a consequence, to determine the critical factors of talent digitalization in power enterprises, this part uses the literature research method and expert consultation method to sort out the factors that affect talent digitalization. The specific analysis contents are shown in Table 1.

Influential Factors and Implementation Path of Talent Digital Evaluation 27

Table 1. Influential factor set of talent digitalization

Level I indicators	Level II indicators	Level III indicators	Level I indicators	Level II indicators	Level III indicators
Knowledge and skills	Compliance knowledge	Legal knowledge S_1	Ability	Innovation	Awards S_{15}
		Rules and regulations S_2			Patents/Property S_{16}
		Network risk awareness S_3		Executive force	Performance appraisal S_{17}
	Managerial skills	Major S_4		Loyalty	Personal honor S_{18}
		Degree S_5	Attitude	Responsibility	Annual performance S_{19}
		Training S_6		Execution effect	Performance result S_{17}
	Informatization skills	ERP application S_7		Service awareness	Ratio of service objects S_{20}
		Computer operation level S_8		Compressibility	Psychological resilience S_{21}
Ability	Communication ability	Cross department work S_9	Professionalism	Aggressiveness	Part-time work S_{22}
	Writing ability	Press Publicity S_{10}		Professional judgment	Risk control ability S_{23}
		Official writing S_{11}		Refine on	Labor awards S_{24}
		Thesis and books S_{12}		Technical Writing	Writing technical manuals S_{25}
	Learning ability	Professional Certificate S_{13}			
		Highest education S_{14}			

In order to improve the reliability and validity of digital talent evaluation elements, 30 experts were invited to screen 25 factors in combination with Delphi technology and SPOT discussion method, remove the factors that do not meet the requirements and repeated factors, and adjust the factors with low correlation.

Specifically, the, characteristics S_{10} and S_{11} in Table 1 are merged into F_{10}, i.e. Official document writing;

Delete factor S_{14}, S_{17}, S_{18}, S_{21}, S_{22} and S_{25};
New factor F_{14} - improvement of on-the-job education is added.

Finally, a talent digitalization evaluation index system database consisting of 20 factors ($F_1 - F_{20}$) is obtained, as shown in Table 2 for details.

Among them, knowledge consists of three secondary hands, namely, "compliance knowledge, management skills and informatization skills", including eight tertiary evaluation indicators; Competence consists of four secondary indicators, which are "communication ability, writing ability, innovation ability and learning ability", including seven secondary indicators in total; Attitude is composed of three secondary indicators: "conscientious performance, executive ability and service awareness"; The specialty consists of two secondary indicators, see Table 2.

Table 2. Digital talent evaluation index system base

Level I indicators	Level II indicators	Level III indicators	Level I indicators	Level II indicators	Level III indicators
Knowledge and skills	Compliance knowledge	Legal knowledge F_1	Ability	Innovation	Patents/Property F_{12}
		Rules and regulations F_2			Awards F_{13}
		Network risk awareness F_3		Learning	on-the-job education F_{14}
	Managerial skills	Major F_4			Professional Certificate F_{15}
		Degree F_5	Attitude	Responsibility	Annual assessment ranking F_{16}
		Training F_6		Execution effect	Performance Appraisal Results F_{17}
	Informatization skills	ERP application F_7		Service awareness	Ratio of service objects F_{18}
		Computer operation level F_8	Professionalism	Professional judgment	Risk control ability F_{19}
Ability	Communication ability	Cross department work F_9		Refine on	Labor awards F_{20}
	Writing ability	Official writing F_{10}			
		Thesis and books F_{11}			

3 Analysis of Impact Factors of Power Transmission Industry

Interpretative Structural Model (ISM) is a theoretical system used to explain the relationship between systems and among various factors within a plan, and is widely used in system dynamics. Specifically, the operation of ISM is divided into three steps: first, invite an expert group to analyze the correlation between factors and build an Adjacency

Matrix; second, phase change is carried out based on Matlab to create the Accessibility Matrix; finally, an interpretation model is formed through hierarchical extraction.

3.1 Adjacency Matrix

Adjacency matrix is used to test whether there is correlation between factors. In general, if adjacency matrix A is expressed as a square array of $i \times j$:

$$A = \begin{bmatrix} a_{11} \cdots a_{1j} \\ \cdots \cdots \\ a_{i1} \cdots a_{ij} \end{bmatrix} \tag{1}$$

Then, the value range of a_{ij} in (1) is:

$$a_{ij} = \begin{cases} 0 & S_i \text{ and } S_j \text{ has no direct relationship} \\ 1, & S_i \text{ and } S_j \text{ has direct relationshop} \end{cases} \tag{2}$$

For this reason, in this paper, the initial relational matrix can be obtained by consulting the experts in Table 2 again through a 0–1 matrix questionnaire.

Table 3. Accessibility Matrix satisfying conditions (1) - (3)

	F_1	F_2	F_3	F_4	F_5	F_6	F_7	F_8	F_9	F_{10}	F_{11}	F_{12}	F_{13}	F_{14}	F_{15}	F_{16}	F_{17}	F_{18}	F_{19}	F_{20}
F_1	0	0	1	0	1	0	0	0	0	0	1	0	0	0	1	0	0	0	0	0
F_2	1	0	0	0	0	0	1	0	0	0	0	0	0	1	0	0	1	0	1	0
F_3	0	0	0	0	0	0	0	0	1	0	0	0	0	0	0	0	0	0	0	0
F_4	0	0	0	0	0	1	0	0	0	0	0	1	0	0	0	1	0	0	0	0
F_5	1	0	0	0	0	0	0	0	0	0	1	0	0	0	0	0	0	0	0	0
F_6	0	1	0	0	0	0	0	1	0	0	0	0	0	0	1	0	0	0	0	1
F_7	0	0	0	1	0	0	0	0	0	0	0	0	1	0	0	0	0	0	0	0
F_8	0	0	0	0	1	0	0	1	0	0	0	0	0	0	0	0	0	0	0	0
F_9	0	0	1	0	0	0	0	0	0	0	0	0	1	0	0	0	1	0	0	0
F_{10}	0	0	0	0	0	0	1	0	0	0	0	0	0	0	1	0	0	0	1	0
F_{11}	0	0	1	0	0	0	0	0	0	0	0	0	0	0	0	0	1	1	0	0
F_{12}	0	0	0	0	0	0	0	0	0	1	0	0	0	0	0	1	1	1	0	0
F_{13}	0	0	0	0	0	0	0	1	0	0	0	0	0	0	0	1	1	1	0	1
F_{14}	0	0	0	0	0	1	0	0	0	0	0	0	0	0	0	1	1	0	0	0
F_{15}	0	1	0	0	0	0	0	0	0	0	1	0	1	0	0	0	0	1	0	0
F_{16}	0	0	0	0	0	0	1	0	0	0	0	0	0	0	0	0	1	0	0	0
F_{17}	0	0	0	1	0	0	0	0	0	0	0	1	0	0	0	0	0	0	0	0
F_{18}	0	0	0	0	1	0	0	0	1	0	0	0	0	1	0	0	0	0	1	0
F_{19}	0	0	1	0	0	0	0	0	0	0	0	0	0	0	0	0	0	1	0	0
F_{20}	1	0	0	0	0	0	0	0	0	0	1	0	0	0	0	1	0	0	0	0

3.2 Accessibility Matrix

The reachable matrix is a skeleton matrix (1) measured through the Boolean operation. The reachable matrix reflects whether there is an influence path and the degree of one element S_i in the adjacency matrix for another element S_j. According to the definition of accessibility matrix, if there is:

$$(A + I)^{k-1} \neq (A + I)^k = (A + I)^{k+1} = M \tag{3}$$

Then, $M = (A + I)^{k+1}$ is the reachable matrix of adjacency matrix A.

3.3 Result Derivation

In this study, the $eye(\cdot)$ function and $isequal$ $(\cdot)$ function are used in MATLAB for programming calculation, and the 20 factors in total, from F_1 to F_{20} in Table 2 are considered, the accessibility matrix satisfying conditions (1) - (3) can be obtained as shown in Table 3.

4 Construction of Interpretation Model for Power Transmission Industry

In the reachable matrix (Table 3), if the reachable set is assumed to be $C(F_i)$, it represents the factor set from row i to column j; The antecedent set is $P(F_j)$, representing the factor set from column j to row i; The symbol of intersection of reachable set $C(F_i)$ and antecedent set $P(F_j)$ is: $C(F_i) \bigcap P(F_j)$.

If, the accessibility matrix is decomposed into hierarchical relations, and if the factors of each row and lattice meet the conditions (4) and (5):

$$C(F_i) = C(F_i) \bigcap P(F_j) \tag{4}$$

$$P(F_j) = C(F_i) \bigcap P(F_j) \tag{5}$$

Then, when condition (4) is met, factor F_i can be considered as the upper - level factor; while if condition (5) is met, then F_i is a low - level factor.

Considering the upper level factors, when they (the upper - level factors) appear, the corresponding rows and columns should be crossed out in the accessibility matrix, and then new upper - level factors should be found from the remaining rows and columns. This cycle continues until all the upper elements have been cleared.

In this study, through the above methods, it can be got the factor set that affects the digital evaluation of talents in the power transport industry, as shown in Table 4.

Specifically, there are 6 levels, including $L_1 = \{F_3, F_7, F_8, F_{17}\}$; $L_2 = \{F_1, F_9, F_{19}\}$; $L_3 = \{F_2, F_{10}, F_{16}, F_{18}\}$; $L_4 = \{F_6, F_{13}\}$; $L_5 = \{F_{11}, F_{12}\}$; $L_6 = \{F_4, F_5, F_{14}, F_{15}, F_{20}\}$. Combining Table 2 and Table 4, an explanatory structure model diagram is drawn, shown in Fig. 1.

Table 4. Influential Factors of Accessibility Matrix for Digital Evaluation of Talents in Electric Power Transportation Industry

Hierarchy	Factor set
L1	F_3, F_7, F_8, F_{17}
L2	F_1, F_9, F_{19}
L3	$F_2, F_{10}, F_{16}, F_{18}$
L4	F_6, F_{13}
L5	F_{11}, F_{12}
L6	$F_4, F_5, F_{14}, F_{15}, F_{20}$

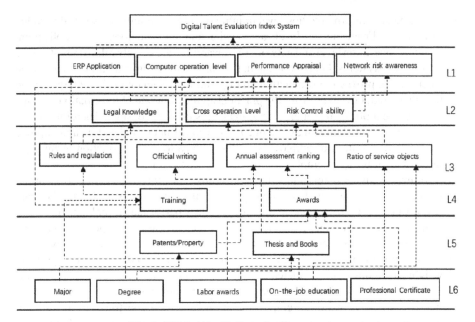

Fig. 1. ISM - based influencing factor system

5 Research Results and Enlightenment

5.1 Research Results

Through Matlab and Boolean operation, this paper divides the factor set that affects the digital evaluation of talents in the electric power transportation industry into six progressive levels, specifically:

(1) Top - level factors: ERP application, computer operation level, performance appraisal results, and network risk awareness;
(2) Core factors: legal knowledge, number of cross departmental work, risk control ability;

(3) Important factors: rules and regulations, official document writing, annual performance ranking, service object ratio;
(4) Key factors: training and awards;
(5) Principal factors: patents/knowledge, papers/works;
(6) Underlying factors: major, academic degree, labor awards, in-service education promotion, professional qualification certificate.

5.2 Research Enlightenment

The index system that affects the digital evaluation of talents in the electric power transportation industry includes six progressive levels. According to the interaction between factors, this paper proposes that the digital level of mastery in enterprises can be improved from the following aspects:

(1) Strengthen the informatization level training of enterprise staff and strengthen value recognition. From the top level, among the four main factors, ERP application, computer operation level and network risk awareness are closely related to enterprise information systems. On the one hand, the application of information system in the enterprise improves the production and operation efficiency of the enterprise; On the other hand, how to improve the practical application rate of information systems has become a problem that enterprise managers need to consider. Therefore, it is a feasible method to build an excellent corporate atmosphere and culture and improve the value recognition of employees on the application of information technology through the training of employees' informatization level.
(2) Optimize the assessment indicators and improve the staff's sense of participation. It can be seen from Fig. 1 that among the influencing factors at the middle level (L2-L5 in Fig. 1), performance appraisal, annual performance ranking, risk management and control, and the ratio of service objects play an important role. This shows that the construction of digitalization of enterprise talents requires extensive participation of employees. In the performance evaluation indicators, the appraising weight of digitalization - related standards should be increased, such as "technical competition awards, ERP security sharing, frequency of use of OA system", to improve the participation of employees in digitalization construction. In addition, in the assessment indicators, it is also necessary to improve the weight of indicators related to information security and operational security to ensure the steady progress of enterprise digital construction.
(3) Strengthen infrastructure construction and improve the internal digital talent training system of the enterprise. From the perspective of the underlying factors, the non-external factors such as the significant and academic degree of the staff have had a fundamental and fundamental impact on the digital talent system. Therefore, in the enterprise, it is necessary and capable to distinguish and compare the differences between the two types of employees by identifying "excellent employees" and "ordinary employees" under the digital talents, forming a mutual matching support mechanism for "excellent ordinary" employees, improving the operating ability of ordinary employees in computer operation, network security, enterprise ERP application, and thus indirectly improve the level of digital talents in the enterprise.

5.3 Research Conclusion

The digital talent construction of enterprises is a very complex project, which is affected by the interaction of multiple system elements. With the help of the ISM model, taking electric power transportation enterprises as an example, this paper preliminarily explores the influencing factors of the evaluation of enterprise talent digitalization and deduces a six - level progressive influencing factor system, which can guide the development of enterprise talent digitalization evaluation scale in the future. However, it should be noted that the results of this study are obtained by electric power transport enterprises, and whether the study is applicable to other enterprises needs further analysis.

Acknowledgment. This project is supported by Application of Talent Management Wisdom Based on ERP (E654900065).

References

1. Aliyev, A.G.: Technologies ensuring the sustainability of information security of the formation of the digital economy and their perspective development directions. Int. J. Info. Eng. Elect. Bus. (IJIEEB) **14**(5), 1–14 (2022)
2. Aliyev, A.G., Shahverdiyeva, R.O.: Scientific and methodological bases of complex assessment of threats and damage to information systems of the digital Economy. Int. J. Info. Eng. Elect. Bus. (IJIEEB) **14**(2), 23–38 (2022)
3. Ravenelle, A.J.: The digital economy. Contemporary Sociology: A Journal of Reviews **50**(5), 416–417 (2021)
4. Wei, Z.: Analysis on the key factors and guarantees of enterprise digital transformation-taking the science and technology innovation board listed companies as an example. People's Forum Academic Frontier (18), 70–78 (2022). (in Chinese)
5. Qiwei, Z., Xin, L., Donghong, L.: Research framework on multiple functions and openness of enterprise digital transformation. Journal of Xi'an Jiaotong University (Social Science Edition) **42**(03), 10–19 (2022). (in Chinese)
6. Jinping, X.: Hold high the great banner of socialism with Chinese characteristics and work together to build a socialist modern country in an all-round way[EB/OL] [2022-10-16]. https://www.12371.cn/2022/10/25/ARTI1666705047474465.shtml
7. Jing, X., Ping, Z.: Can digitalization realize the "quality improvement and increment" of enterprise green innovation-Based on the resource perspective. Scientific research, 1–19. (in Chinese)
8. Xia, F., Yun, Z., Ping, Z.: Research on the difficulties and countermeasures of financial talents' data literacy training in the digital economy era. China University Teaching **09**, 23–27 (2022). (in Chinese)
9. Yanping, L., Le, L., Xiang, H.: Digital human resource management: integration framework and research outlook. Sci. Technol. Progr. Countermeas. **38**(23), 151–160 (2021) (in Chinese)
10. Pei, Z., Yong, S.: Spatial pattern and influencing factors of coordinated development of information infrastructure and integrated infrastructure. Economic Issues Exploration **10**, 94–104 (2022). (in Chinese)
11. Liyan, C.: Research on innovation choices and destructive innovation paths of latecomers. Harbin University of Technology, Harbin (2013). (in Chinese)
12. Jianhua, H., Yongduan, G., Lijun, C.: Digital platform enterprise network contract psychological contract: content, measurement and service performance impact verification. Business Economics and Management **03**, 5–15 (2022). (in Chinese)

13. Yumei, L., Luyan, M., Yi, Z.: Research on the impact of innovative use behavior of enterprise information system - based on social impact theory and individual innovation traits. Science and Technology Management Research **39**(10), 177–184 (2019). (in Chinese)
14. Roberts, C., Geels, F.W.: Conditions and intervention strategies for the deliberate acceleration of sociotechnical transitions: lessons from a comparative multi-level analysis of two historical case studies in dutch and danish heating. Technology Analysis & Strategic Management **31**(9), 1081–1103 (2019)
15. Järvi, K., Khoreva, V.: The role of talent management in strategic renewal. Employee Relations **42**(1), 75-89 (2020)
16. Laia, Y., Ishizakab, A.: The application of multi-criteria decision analysis methods into talent identification process: a social psychological perspective. J. Bus. Res. **109**, 637–647 (2020)
17. Yijie, W., Wenjun, M., Jianli, F.: Impact factors and realization path of carbon neutralization target under new development pattern. Resource Development and Market **38**(08), 915–920+985 (2022) (in Chinese)
18. Ping, L., Yi, Z.: Wang yijie research on the stability of low carbon technological innovation alliance under punishment mechanism - based on the perspective of stochastic evolutionary game. Resource Development and Market **38**(1), 16–22 (2022). (in Chinese)
19. Afha, B., Scc, D., Edce, F., Bp, G., et al.: From global to national scenarios: bridging different models to explore power generation decarbonizatio based on insights from socio-technical transition case studies. Technol. Forecast. Soc. Chang. **151**, 119882 (2020)
20. Jifeng, W., Peng, L., Jinzhao, L., Yufei, S.: The course and development trend of power system digitalization. China Southern Power Grid Technology **15**(11), 1–8 (2021). (in Chinese)
21. Liu, P., Zhang, Y., Xiong, Z., Wang, Y., Qing, L.: Judging the emotional states of customer service staff in the workplace: A multimodal dataset analysis. Frontiers in Psychology **13**, 1001885. https://doi.org/10.3389/fpsyg.2022.1001885
22. Ahuja, S., Kaur, P., Panda, S.N.: Identification of influencing factors for enhancing online learning usage model: evidence from an Indian University. Int. J. Edu. Manage. Eng. (IJEME), **9**(2), 15–24 (2019)
23. Matthew, O., Buckley, K., Garvey, M.: A framework for multi-tenant database adoption based on the influencing factors. Int. J. Info. Technol. Comp. Sci. (IJITCS) **8**(3), 1–9 (2016)

Knowledge Associated with a Question-Answer Pair

Igor Chimir[(✉)]

Odessa State Environmental University, Lvovskaya Street 15, Odessa 65016, Ukraine
`chimirigor@gmail.com`

Abstract. The article is devoted to the modeling and presentation of declarative knowledge, which exchanged by participants of a question-answer dialogue. The knowledge of the reactive (responding) agent of the dialogue is considered from the point of view of the classical epistemological understanding of knowledge. It is shown in what cases the epistemological formula "the subject R knows that the proposition p'' is not complete and must be supplemented with a question for which the proposition p is the true answer. Further, the article examines the logical connection between the question and the relevant answer in the context of the interrogative formula of the question, considered in erotetic logic. Declarative knowledge expressed by the structural component of the interrogative formula, which is called the subject of the question, is proposed to be modeled using linguistic-independent semantic entities in the categories of the Ternary Description Language. The use of the Ternary Description Language makes it possible to construct models of the subject of the question that do not depend on a specific natural language. The final part of the article describes a set of eight patterns that can be used to represent declarative knowledge associated with a question-answer pair.

Keywords: Dialogue · Declarative knowledge · Ternary Description Language

1 Introduction

The words "conversation" and "dialogue" are often understood as synonyms. However, in what follows we will prefer the word "dialogue", and the participants in the dialogue will be called dialogue agents. We are primarily interested in the dialogue that takes place between an artificial dialogue agent (for example, a chatbot) and a human (a chatbot user).

In the process of the dialogue, the dialogue agents form a dialogue transaction, which is an elementary complete cycle of knowledge exchange between the agents. Although there can be many participants in a dialogue, only two dialogue agents form a dialogue transaction.

A dialogue agent, in the process of dialogue interaction with its partner, can perform one of two alternative roles: (1) the role of an active dialogue agent; (2) the role of a reactive dialogue agent. An active agent is an inquiring agent. The part of the dialogue

© The Author(s), under exclusive license to Springer Nature Switzerland AG 2023
Z. Hu et al. (Eds.): ICAILE 2023, LNDECT 180, pp. 35–44, 2023.
https://doi.org/10.1007/978-3-031-36115-9_4

transaction that the active agent forms is not necessarily a single verbal question, but always has the status of an interrogative. In goal-oriented dialogues, the main motivation for a dialogue agent to play the role of an active agent is the lack of knowledge necessary to continue the dialogue. An agent plays the role of an active dialogue agent in case when he needs additional knowledge that he expects to receive from his dialogue partner. A reactive agent is a responding agent, and the part of a dialogue transaction that it forms has a response status with respect to the active agent. A dialogue agent who honestly plays the role of a reactive agent provides the partner with the knowledge that, from his point of view, is relevant to the active agent question.

A chatbot can be viewed as an artificial agent participating in an unstructured verbal question-answering dialogue. Often a chatbot is a reactive agent and must be able to generate answers to human questions. In a dialogue transaction, the chatbot and its partner exchange knowledge. Thus, a chatbot is a knowledge-based system, and when designing it, it is important to rely on a model for representing knowledge in a question-answer transaction. Knowledge bases, which are used by modern chatbots, are connected in one way or another with natural language [1, 2]. Representing knowledge using natural language sentences seems to be the norm since the dialogue between the chatbot and its user is verbal. However, this should not be imperative. We will consider a possible way of representing declarative knowledge associated with a question-answer transaction in semantic categories that are invariant to a particular natural language. The article develops the direction of research published in [3]. In the initial part of the article, attention is focused on understanding the knowledge of the reactive agent of dialogue from the point of view of the classic epistemological formula "true and justified belief". The subsequent part of the article analyzes the logical connection between the structure of the question and the structure of the relevant answer from the point of view of the interrogative formula of Belnap and Steel [4]. The structural component of the interrogative formula, called the subject of the question, is considered an indefinite relevant answer. The subject of the question, as a rule, is modeled by a set of propositions and is represented by sentences of a natural language. The article discusses the possibility of modeling the subject of the question using linguistically independent entities in the categories of the Language of Ternary Description [5–8]. The final part of the article describes the declarative knowledge models and patterns associated with a question-answer transaction.

2 Knowledge of the Reactive Agent

What is the declarative knowledge that a reactive agent operates and how can it be understood and represented? Epistemologists propose to understand this knowledge as a proposition and call it knowledge-*that*. The following verbal formula is known: "knowledge-*that* is a Justified and True Belief (JTB)" [9]. According to this verbal formula, the necessary and sufficient conditions for the reactive agent R to know that the proposition p takes place can be formulated as follows:

1) proposition p is true,
2) reactive agent R believes that p, and

3) reactive agent R is justified in believing that p.

Epistemology is a philosophical science, and therefore, when interpreting the formula JTB and conditions of the reactive agent R knowledge, that p, pays much attention to philosophical issues, for example, how the phrase "justified belief" should be understood. The only thing that is not questioned is the statement that the proposition must be true. False propositions are not considered knowledge. Thus, the truth of the proposition is understood in the absolute, and not in the relative (in relation to the reactive agent) sense. One can agree with this, however, one can find examples when a proposition that is true for one person is not the same for another person. For example, the proposition represented by the sentence "Tobacco smoking is a virtue" may be true for one reactive agent, but false for another.

The knowledge of the agent R that p may be represented by a diagrammatic formula in the form of a UML class diagram using the relation of association between the classes R and p. The UML formula of knowledge-*that* is shown in Fig. 1.

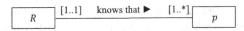

Fig. 1. Diagrammatic formula of knowledge-that of a reactive agent

The multiplicity [1..1] of the pole of the association adjacent to the class R means that only one object of the class R participates in the association, and the multiplicity [1..*] of the pole adjacent to the class p means that one or more objects of the class p participate in the association. In other words, one reactive agent can know one or more propositions.

The memory of a reactive agent cannot store all possible knowledge-*that*. This knowledge is formed by the reactive agent in the process of dialogue as a response to the question of the active agent. From the point of view of the way of forming knowledge-*that* by the reactive agent, all questions of the active agent can be divided into two classes: (1) questions of the "search instruction" type and (2) questions of the "task" type.

A question of the "search instruction" type assumes that at the moment the question is received, the requested knowledge is already in the memory of the reactive agent, and the structural elements of the question position the memory to the required area. To form an answer to a question of the "search instruction" type, the use of the attention resource is not required. An example is the following question: "What is your name?" When the active agent, in order to gain access to the knowledge of the reactive agent, uses a question of the "search instruction" type and the requested knowledge is already in the memory of the reactive agent, the diagrammatic formula of knowledge-that shown in Fig. 1 is adequate. Let us show that this formula needs to be refined in the case when a question of the "task" type is used to gain access to the knowledge of a reactive agent.

To obtain the knowledge that is requested using a question of the "task" type, the reactive agent must solve the task associated with this question. The answer is a variant of the solution of the task, obtained by the reactive agent. An example of a question "task" type is: "If Socrates was born in 469 BC, how old would Socrates be today?" It is clear that, most likely, the knowledge requested by this question is not stored in the memory of the reactive agent in a ready-made form, and the use of a mental resource is required for their formation.

From the point of view of the declarative-procedural dichotomy of knowledge, knowledge-*that* is declarative. When forming declarative knowledge, which is the answer to a question of the "task" type, the reactive agent must use procedural knowledge. The procedural knowledge that a reactive agent uses depends on how a question of the "task" type is formulated. Let us illustrate the last statement by the example of convergent questions of the "task" type. Convergent questions are different questions that assume the same true answer. In other words, the same true proposition of the reactive agent can be the answer to different questions of the active agent. Schaffer analyzes various convergent questions, including the following two [10].

(1) Is there a goldfinch in the garden, or a raven?
(2) Is there a goldfinch in the garden, or a canary?

If it is true that there is a goldfinch in the garden, then the reactive agent should give the same answer to both questions: "There is a goldfinch in the garden." However, in order to get this answer in the case of question (1), the reactive agent needs relatively simple procedural knowledge (knowledge that makes it possible to distinguish a goldfinch from a raven), and in order to get the same answer in the case of question (2), the reactive agent needs much more complex procedural knowledge (knowledge that makes it possible to distinguish a goldfinch from a canary).

Thus, the diagrammatic formula for the knowledge-*that* of a reactive agent shown in Fig. 1, in the general case, is incomplete and must be supplemented by a question, the answer to which is knowledge-*that* of the reactive agent. Figure 2 shows a refined formula for the knowledge-*that*.

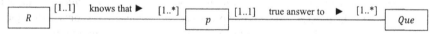

Fig. 2. A refined formula for the knowledge-that of a reactive agent

In Fig. 2, *Que* denotes a class of questions. The class diagram shown in Fig. 2 differs from the class diagram shown in Fig. 1 in that it contains one more relation of the association type between the class of propositions *p* and the class of questions *Que*. The multiplicity [1..*] of the pole of the association adjacent to the class *Que* means that one or more objects of the class *Que* participate in the association. In other words, the same proposition can be the answer to several different questions.

3 Interrogative Formula of the "Search Instruction" Type Question

The logical connection between the question of the "search instruction" type and the answer to this question can be revealed by explicating the question formula proposed by Belnap and Steel [4]. The question formula, which Belnap and Steel consider, and which they call the interrogative formula, defines a question as a composition of two logical components: the request of the question and the subject of the question. In Fig. 3, the interrogative formula is presented in the form of a UML class diagram.

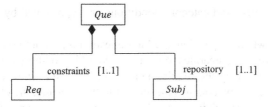

Fig. 3. The interrogative formula in the form of a UML class diagram

The class diagram in Fig. 3 represents the class of questions *Que* as a composite of two classes: the class of question's requirements *Req* and the class of question's subjects *Subj*. The multiplicity [1..1] of poles adjacent to the *Req* and *Subj* classes means that exactly one object of each of these classes participates in the composition. The names of the poles reflect the roles of objects of these classes in the composition. The pole adjacent to the *Subj* class is named "repository", since the subject of the question points to that part of the reactive agent's memory where the knowledge that includes the answer to the question is located. The pole adjacent to the *Req* class is named "constraints" since the requirement of the question sets constraints that the reactive agent must take into account when forming an answer.

Belnap and Steel's theory assumes that the knowledge determined by the subject of the question is propositions that are represented by sentences in natural language, and the requirement of the question may, for example, require that these sentences be considered as an alternative.

Questions are classified with the cardinality of the subject of the question. The subject of the question can be: (1) a finite and small number of propositions; (2) an infinite or very large number of propositions. In the first case, the relevant questions are called whether-questions, and in the second case – which-questions.

An example of a whether-question is the question of tobacco smoking, which was, allegedly, formulated by the English king James the first [4]:

«Tobacco smoking: a vice, a virtue, a vagary, an extravagance, a cure for all ills?»

The subject of this question consists of exactly five elements: "tobacco smoking is a vice"; "tobacco smoking is a virtue"; "tobacco smoking is a vagary"; "tobacco smoking is an extravagance"; "tobacco smoking is a cure for all ills".

An example of a which-question is the following question:

«Which positive integer is the smallest prime greater than 45?»

The subject of this question consists of an infinite number of elements and can be represented by two clauses with one variable.

x is the smallest prime greater than 45.
x is a positive integer.

It is clear that whether-question is a special case of which-question, and whether-subject is a special case of which-subject. Belnap and Steel propose a notation for whether-subject, according to which whether-subject is represented by a clause with variables (such a clause is called the subject matrix) and a set of constraints that define the values of variables in the matrix (such constraints are called category conditions).

40   I. Chimir

Both the subject matrix and category conditions are represented by means of natural language.

The answer to whether-question or which-question is part of the subject of the question that the reactive agent separates from the whole subject of the question in accordance with the constraints set by the requirement of the question. The subject of whether-question or which-question contains the answer to the question, therefore the subject of a question of the type "search instruction" can be understood as an indefinite answer, or an answer with some (often significant) degree of indefiniteness.

4 Knowledge Structure in Question-Answer Transactions

The concept of "proposition" is not identical to the concept of "sentence" in natural language. A proposition can be represented by a natural language sentence. Representation of propositions in the form of sentences in natural language is convenient when writing articles on the interpretation and modeling of knowledge-*that*, but, from the author's point of view, it limits the developers of artificial reactive agents using such kinds of models.

If the knowledge-*that* used by an artificial reactive agent is tied to a specific natural language, then the agent can maintain a dialogue only in a specific language and is monolingual. In the case when the representation of the knowledge-that of a reactive agent is carried out in categories that are not related to a specific natural language, then there is a potential opportunity to design an artificial reactive agent that can support a dialogue in several languages and be multilingual. It is very important to be multilingual, especially for an artificial dialogue agent "dwelling" on the Internet.

4.1 The Language of Ternary Description

It is advisable to represent declarative knowledge in a question-answer transaction by means, of the ontology which is based on non-linguistic entities. One of these means is the Language of Ternary Description (LTD), proposed by Uyemov [5–8].

The initial entity in LTD is an object, in the most general sense of the word. An object, depending on its place in the structure of knowledge, can exist in one of three forms: object-thing, object-property, and object-relation.

The categories "thing" and "property" have the traditional meaning in the LTD, which is accepted in classical logic, and the meaning of the category "relation" differs from the traditional one. It is generally accepted to use the concept of "relation" to denote the mutual influence of several things, or the relationship between things. In the context of the LTD, a relation is understood as what constitutes a thing, or the relationship that takes place in a thing. In other words, the relation in LTD is, in some sense, another name for the internal logical organization of a thing.

The binary association of an object-thing with an object-property generates two prototypes for representing entities in LTD.

1) The name of the first prototype is "a thing, which possesses a property", and the formal notation is: (∗)∗.

2) The name of the second prototype is: "a property, which attributed to a thing", and the formal notation is: $(*))*$.

The binary association of an object-thing with an object-relation generates two more prototypes for representing entities in LTD.

3) The name of the third prototype is: "a thing, in which a relation takes place", and the formal notation is: $*(*)$.
4) The name of the fourth prototype is "a relation, which takes place in a thing", and the formal notation is: $*((*)$.

The object-thing symbol is written inside the brackets, the object-property symbol is written to the right of the brackets, and the object-relation symbol is written to the left of the brackets.

The association of an object-thing with an object-property or with an object-relation has a direction. If the symbol of an object-thing is enclosed in ordinary (single) parentheses, then this means that the association is directed from the object-thing to the object-property or object-relation. In words, this is expressed as: "a thing, which possesses a property ", or "a thing in which a relation takes place." An asymmetric (double) parenthesis means that the association is directed from an object-property or an object-relation to an object-thing. Verbal formulation: "a property, which attributed to a thing ", or "a relation, which takes place in a thing."

An object, depending on the degree of indefiniteness of knowledge about it, exists in one of three forms:

1) a definite object (the asterisk in the prototype substitutes by symbol t),
2) an indefinite object (the asterisk in the prototype substitutes by symbol a),
3) an arbitrary object (the asterisk in the prototype substitutes by symbol A).

The categories "thing, property and relation", as well as "definiteness, indefiniteness, and arbitrariness" are independent and form nine classes of objects: (1) a definite object-thing, (2) an indefinite object-thing, (3) an arbitrary object-thing, (4) a definite object-property, (5) an indefinite object-property, (6) an arbitrary object-property, (7) a definite object-relation, (8) an indefinite object-relation, (9) an arbitrary object-relation.

Substituting into prototypes for representing entities in LTD instead of the asterisk one of the symbols t, a or A, we get a set of possible models of knowledge-*that* represent in the ontological basis of LTD. Since there are four prototypes, in each of which we can substitute two symbols of the object, and the total number of object symbols is three, there are 24 possible models of knowledge-that.

4.2 Representation of Declarative Knowledge in Question and Answer

The question-answer transaction, in the context of the LTD-representation of declarative knowledge, will be considered as a development of the idea of the interrogative formula, shown in Fig. 3.

The subject of the question is the key element of the question-answer transaction. The reactive agent, when constructing an answer, is essentially engaged in transforming the indefinite knowledge pointed to by the subject of the question into definite knowledge-*that* of the answer.

When modeling the knowledge pointed to by the subject of the question, we will restrict ourselves to only the following four alternative models.

$$K_{subj} = (t)a \qquad (1)$$

$$K_{subj} = (a)t \qquad (2)$$

$$K_{subj} = (t))a \qquad (3)$$

$$K_{subj} = (a))t \qquad (4)$$

There are several reasons for choosing models (1)–(4) to represent the knowledge of the subject of the question. First, the subjects of most of the examples of questions described in works devoted to the logic of questions and answers can be represented by one of the models (1)–(4) [4, 10, 11]. Secondly, these are exactly the models that correspond to the idea that a reactive agent, when constructing an answer, transforms the indefinite knowledge of the subject of the question into definite knowledge-that of the answer. One of the objects in models (1)–(4) is indefinite, and the active agent expects to receive more specific knowledge about it from the reactive agent.

The model $K_{subj} = (t)a$ represents knowledge about a specific definite thing, which possesses indefinite property. A question with such a subject is formed by an active agent in the case when he wants to know which properties a given thing possesses.

The model $K_{subj} = (a)t$ represents knowledge about an indefinite thing, which possesses specific definite property. A question with such a subject is formed by an active agent in the case when he wants to know which things possess a given property.

The model $K_{subj} = (t))a$ represents knowledge about an indefinite property, which is attributed to a specific definite thing. A question with such a subject is formed by an active agent in the case when he wants to know which specific property is attributed to a given thing.

The model $K_{subj} = (a))t$ represents knowledge about a definite property, which is attributed to some indefinite thing. A question with such a subject is formed by an active agent in the case when he wants to know to which things a given property is attributed.

The disadvantage of models (1)–(4), from the point of view of software engineering, is their poor adaptability for mapping into relevant data structures. These models could be useful for the development of software systems if we could find a way to transform them into types or data structures of modern programming systems. This is, first of all, about the datalogical interpretation of an indefinite object.

One of the possible datalogical interpretations of indefiniteness is multiplicity. An indefinite object can be understood as a set of definite objects, and the cardinality of this set as a measure of indefiniteness. Then a decrease in the degree of indefiniteness of an object is equivalent to a decrease in the cardinality of the corresponding set. An indefinite object turns into a definite one when the cardinality of a set becomes equal to one, or when the set is represented by one specific definite object.

Taking into account such interpretation of indefiniteness, we can replace indefinite objects with lists of specific objects and describe models (1)–(4) with the following patterns.

$$K_{subj} = (thing) \rightarrow list\ of\ properties \qquad (5)$$

$$K_{subj} = (list\ of\ things) \rightarrow propery \qquad (6)$$

$$K_{subj} = (thing)) \leftarrow list\ of\ properties \qquad (7)$$

$$K_{subj} = (list\ of\ things)) \leftarrow property \qquad (8)$$

Patterns (5)–(8) are datalogic analogs of models (1)–(4), which describe declarative knowledge that can be transferred to a reactive agent by means of the subject of the question.

Since the subject of the question is, in fact, an indefinite answer, and the reactive agent, when constructing the answer, reduces the degree of indefiniteness of the subject of the question to a level acceptable for the answer, then the knowledge-that patterns of the answer should be similar to patterns (5)–(8). The patterns of knowledge-that of the answer differ from the patterns of knowledge of the subject of the question by the cardinality of the set of objects-things or objects-properties. We may describe the knowledge-that of the answer with the following patterns.

$$K_{ans} = thing\ POSSESSES\ PROPERTIES\ list\ of\ properties \qquad (9)$$

$$K_{ans} = list\ of\ things\ POSSESSES\ PROPERTY\ property \qquad (10)$$

$$K_{ans} = list\ of\ properties\ ATTRIBUTED\ TO\ thing \qquad (11)$$

$$K_{ans} = property\ ATTRIBUTED\ TO\ list\ of\ things \qquad (12)$$

5 Conclusion

Chatbots are capable of maintaining long-term dialogue with a human. The dialogue protocol between a chatbot and a human can be represented as a sequence of transactions. The human partner forms the interrogative part of the transaction in order to receive from the chatbot a portion of the declarative knowledge necessary to continue the dialogue. The portion of declarative knowledge that is requested from the chatbot belongs to the class of knowledge-*that* that is a proposition. When forming an answer in the form of a proposition, the chatbot uses procedural knowledge, which depends on how the question is formulated. Thus, a chatbot is a knowledge-based system, and during its development, it is necessary to rely on one or another model of knowledge representation associated with a dialogue transaction. The logical connection between question and answer in a single transaction is modeled by Belnap and Steel's interrogative formula. The key component of the interrogative formula is the subject of the question, which for the question of "search instruction" type can be considered as knowledge-*that* of the answer, but with a significant degree of indefiniteness.

It is advisable to build models of knowledge on an ontological basis, which is not associated with linguistic categories but represents the semantic essence of knowledge. In this case, there is a potential opportunity to develop multilingual chatbots, which is especially important for chatbots operating on the Internet. The article illustrates the applicability of Uyemov's Ternary Description Language for representing portions of declarative knowledge associated with a question-answer transaction. The patterns for representing the knowledge of the subject of the question and the corresponding answer, represented by formulas (5)–(12), explain the internal logical essence of the question-answer transaction in the context of knowledge representation and are not associated with a specific natural language.

Further development of this work is supposed to be carried out in the direction of synthesizing a model of a dialogue knowledge base, which includes not only declarative knowledge associated with dialogue transactions but also procedural knowledge necessary for conducting a question-answer dialogue focused on solving problems.

References

1. Serban, I.V., Lowe, R., Henderson, P., Charlin, L., Pineau, J.: A survey of available corpora for building data-driven dialogue systems. Dialogue Discourse **9**(1), 1–49 (2018)
2. Rashkin, H., Smith, E.M., Li, M., Boureau, Y.-L.: Towards empathetic open-domain conversation models: a new benchmark and dataset. ACL (2019)
3. Chimir, I., Ghazi, A., Abu-Dawwas, W.: Modeling human dialogue transactions in the context of declarative knowledge. Int. Arab J. Inf. Technol. **10**(3), 305–315 (2013)
4. Belnap, N.B., Steel, T.: The Logic of Questions and Answers. Yale University Press, New Haven and London (1976)
5. Uyemov, A.: The ternary description language as a formalism for the parametric general systems theory. Part I. Int. J. General Syst. **28**(4–5), 351–366 (1999)
6. Uyemov, A.I.: The ternary description language as a formalism for the parametric general systems theory. Part II. Int. J. General Syst. **31**(2), 131–151 (2010)
7. Uyemov, A.I.: The ternary description language as a formalism for the parametric general systems theory. Part III. Int. J. General Syst. **32**(6), 583–623 (2003)
8. Leonenko, L.: The language of ternary description and its founder. Modern Logic **8**(3,4), 31-52 (2000-2001)
9. Armstrong, D.M.: Belief, Truth, and Knowledge. Cambridge University Press, Cambridge (1973)
10. Schaffer, J.: Knowing the answer. Philos. Phenomenol. Res. **75**(2), 383–403 (2007)
11. Wisniewski, A.: The Posing of Questions. Logical Foundations of Erotetic Inferences. Kluwer Academic Publishers, Dordrecht (1995)

Research on Low Complexity Differential Space Modulation Detection Algorithm

Shuiping Xiong[1(✉)] and Xia Wu[2]

[1] School of Artificial Intelligence and Manufacturing, Hechi University, Yizhou 546300, Guangxi, China
601421650@qq.com
[2] Lenovo Information Products (Shenzhen) Co. Ltd., Shenzhen, China

Abstract. SM (Spatial Modulation) technology is a key research direction of MIMO (Multiple Input Multiple Output) technology in recent years. Because SM uses a single RF module, only one transmitting antenna is selected for activation at any time slot, Thus, SM avoids the inter-channel interference (ICI), but SM technology requires channel state information (CSI) for signal detection at the receiver, so channel estimation must be performed. The channel estimation process is very complex and needs a lot of resources, and the error of channel estimation will cause serious loss of detection performance. In this case, differential spatial modulation (DSM) technology is proposed. As a new multi-antenna transmission technology, Differential Space Modulation (DSM) can improve the transmission rate of wireless communication to a certain extent, but DSM system has high complexity of signal detection algorithm. In order to reduce the complexity of DSM signal detection, this paper proposes a better low-complexity detection algorithm, called time-division combination (TDC) detection algorithm. Theoretical analysis and simulation results based on MATLAB show that the improved time-sharing merging algorithm not only has the best detection performance, but also greatly reduces the computational complexity, thus improving the theoretical and practical significance.

Keywords: Differential space modulation · TDC · Detection algorithm · ML

1 Introduction

SISO (Single In Single Out) technology has no inter channel interference (ICI), but its spectral efficiency is low. The traditional MIMO (Multiple Input Multiple Output) technology solves the problem of low spectral efficiency. But on the other hand, the inter channel interference is severe, the decoding complexity for the receiver is high, while some technologies are not suitable for asymmetric antenna systems. A new MIMO technology, according to Spatial Modulation (SM) technology, has emerged and gradually became a research hotspot in recent years [1]. However, SM technology has some disadvantages. First, the traditional SM system hides part of the bit information in the serial number of the active antenna, so it requires the transmit antenna N_t to be the power

---

© The Author(s), under exclusive license to Springer Nature Switzerland AG 2023
Z. Hu et al. (Eds.): ICAILE 2023, LNDECT 180, pp. 45–54, 2023.
https://doi.org/10.1007/978-3-031-36115-9_5

of 2. Second, in order to avoid inter channel interference, the modulation channels from the transmitter to the different receivers of SM systems need to be uncorrelated. Finally, since the receiver of SM system assumes that the channel state information is known, that is, channel estimation must be performed before detection, and the complexity of channel estimation process is very high, and there will be corresponding errors in the estimation, which will have a serious impact on the performance of SM system. In order to solve the problems in the SM system, the DSM technology was proposed by Bian Y et al. in 2015. Each time slot of the DSM also only activates one antenna, which eliminates the need for inter antenna synchronization and avoids inter channel interference [2]. DSM technology, on the basis of SM, additionally introduces a time domain differential algorithm to transmit data in the form of space-time blocks. Each space-time block contains information about multiple symbols. The information carried by the serial number of the DSM antenna maps the bit information to the activation sequence of the transmission antenna. DSM makes difference in time domain, perfectly avoiding the problem of channel estimation. DSM technology mainly includes two algorithms: maximum likelihood detection and time-sharing combining (TDC) detection [3].

2 Maximum Likelihood Detection Algorithm

The maximum likelihood detection algorithm ML (Maximum Likehood) is a classical detection algorithm with good bit error rate performance in the current DSM system.

2.1 ML Detection Algorithm Process

The maximum likelihood detection algorithm ML detection algorithm of the DSM system. The ML detection process is as follows [4]:

(1) Step 1: Search all possible N_t antenna activation sequence/symbol sequence pairs (u_{order}, s_{N_t}) within a time slot to get a set of all candidate solution information block matrices $\Re_M$. According to formula (1), the euclidean distance between the received signal matrix at the receiver and the post-processing matrix of all candidate solution information block matrices can be calculated in turn;

$$d^X_{DSM\_ML} = \|Y_t - Y_{t-1}X\|^2_F, (X \in \Re_M) \tag{1}$$

(2) The second step is to compare all Euclidean distances $d^{\mathbf{X}}_{DSM\_ML}$; $d^{\mathbf{X}}_{DSM\_ML}$ to get the minimum;

(3) The third step is to obtain the optimal information block matrix corresponding to the minimum Euclidean distance, which will be used as the recovery signal X, and finally obtain the corresponding antenna activation sequence and symbol sequence.

The above process is simplified as [5]:

$$\hat{X}_t = \arg \min_{\forall \mathbf{X} \in \Re_M} \|Y_t - Y_{t-1}X\|^2 \tag{2}$$

For DSM system, since only one transmit antenna is active in each time slot, there is only one element not 0 in each column vector of the information block matrix

X. Then the ML detection algorithm can be further deduced from Eq. (2) as follows [6].

$$\left[\hat{u}^t_{order\_ML}, \hat{s}^t_{N_t\_ML}\right] = \underset{(u_{order}, s_{N_t}) \in \Re_M}{\arg\min} \sum_{1 \le n \le N_t} \left\| \mathbf{y}_t(n) - \mathbf{y}_{t-1}(u_{order}(n)) \cdot s_{N_t}(n) \right\|^2$$

(3)

N_t Indicates the number of transmitting antennas, $\hat{u}^t_{order\_ML}$ the estimated sequence of antenna activation, $\hat{s}^t_{N_t\_ML}$ the estimated sequence of symbols, the nth time slot, $y_t(n)$ the nth column vector, that is, the reception vector of the nth time slot, $u_{order}(n)$ the sequence number of the nth time slot activation antenna, $s_{N_t}(n)$ the constellation point symbol transmitted by the nth time slot activation antenna, and. Represents the column of $u_{order}(n)$.

Equation (3) can be further expressed [7]:

$$\left[\hat{u}^t_{order\_ML}, \hat{s}^t_{N_t\_ML}\right] = \underset{(u_{order}, s_{N_t}) \in \Re_M}{\arg\min} \sum_{1 \le n \le N_t} \sum_{1 \le w \le N_r} \left\| y_t(w, n) - y_{t-1}(w, u_{order}(n)) \cdot s_{N_t}(n) \right\|^2$$

(4)

Although the DSM system reduces the process of channel estimation and the total complexity of the receiver compared with the SM system, it can be seen from Formula (2) that every detection requires searching all possible antenna activation sequence/symbol sequence pairs, which shows that the complexity of the ML detection algorithm based on the DSM system is still high [8].

2.2 Complexity Analysis of ML Detection Algorithm

According to Eqs. (2) and (4), where $y_{t-1}(w, u_{order}(n))$, $s_{N_t}(n)$ and are complex numbers, four real number multiplication operations are required for each $y_{t-1}(w, u_{order}(n)) \cdot s_{N_t}(n)$ calculation, and $\left\| y_t(w, n) - y_{t-1}(w, u_{order}(n)) \cdot s_{N_t}(n) \right\|^2$. It represents the sum of the squares of the real part and the imaginary part of the element $\left\| y_t(n, w) - y_{t-1}(u_{order}(n), w) \cdot s_{N_t}(n) \right\|^2$. The square of the element in requires 2 real number multiplication [9]. Then calculate once $\left\| y_t(w, n) - y_{t-1}(w, u_{order}(n)) \cdot s_{N_t}(n) \right\|^2$, 6real number multiplications are required. For a candidate solution information block matrix X, $N_r N_t$ times of $\left\| y_t(n, w) - y_{t-1}(u_{order}(n), w) \cdot s_{N_t}(n) \right\|^2$ calculation is required, so $6N_r N_t$ times of $\left\| Y_t - Y_{t-1} X \right\|^2$ real number multiplication is required. In one ML detection of the DSM system, for the N_r root transmit antenna, there are $2^{\lfloor \log_2(N_t!) \rfloor}$ different antenna activation sequence. Assuming that the modulation mode of each time slot is the same, there are $2^{N_t b}$ different symbol sequences, that is, there are $\Re_M$ different symbols X in the set. Therefore, the computational complexity of the ML detection algorithm based on the DSM system can be expressed as [10]:

$$C_{DSM\_ML} = 6N_r N_t \cdot 2^{\lfloor \log_2(N_t!) \rfloor + N_t b}$$

(5)

The relative complexity reduction of ML detection algorithm is expressed by the following formula:

$$Cre(\%) = 100 \times \left(C_{ML} - C_{propose}\right) / C_{ML}$$

(6)

48 S. Xiong and X. Wu

In formula (6), Cpose is the computational complexity of other detection algorithms [11].

3 TDC Detection Algorithm

3.1 TDC Detection Algorithm Process

Equation (5) describes that the complexity C_{DSM_ML} of ML detection algorithm in DSM system in one detection is $6N_r N_t \cdot 2^{\lfloor \log_2 (N_t!) \rfloor + N_t b}$.

Table 1. Detection process of TDC detector

---

input: $Y(t), Y(t-1), N_r, N_t$

output: Demodulated optimal information block $\hat{X}$

1. For: $i = 1 : N_t$

(1) For: $x_i \in \Theta_i$

(i) Calculate the distance of all possible transmission vectors in i slot: $D(x_i) = \left\| y_t(i) - Y_{t-1} x_i \right\|^2$

(ii) All distances $D(x_i)$ form a set: $D(x_i) \in o_i$

(2) For: $u = 1 : N_t$

(i) Get the minimum distance corresponding to the active antenna when the serial number u is: $D_u(i) = \min\limits_{\forall D(x_i(u)) \in o_i, 1 \leq u \leq N_t} (D(x_i(u)))$

(ii) All minimum distance update sets: $D_u(i) \in o_i$

2. For: $u_{order} = \begin{bmatrix} u_1 & u_2 & \cdots & u_{N_t} \end{bmatrix} \in \mho$

(1) Get the current u_{order} corresponding minimum distance: $D_{u_{order}} = \sum\limits_{1 \leq i \leq N_t} D_{u_i}(i)$

3. Get the best information block: $\hat{X} = \arg\min\limits_{\forall u_{order} \in \mho} (D_{u_{order}})$

In order to make DSM system more advantageous in practical application, its detection complexity needs to be greatly reduced. A TDC detection algorithm is proposed to directly detect the candidate solution information block X. Instead, the candidate solution vectors of each slot of the candidate solution Nr information block are calculated separately, and then the optimal information block is obtained by comprehensively detecting each slot. The detection process is shown in Table 1 [12].

3.2 Complexity Analysis of TDC Detection Algorithm

TDC detection algorithm breaks through the idea that traditional detection algorithms in DSM system directly detect candidate solution information blocks, and calculates the candidate solution vectors x_i of each N_t slot of the candidate solution information block. Firstly, the optimal symbol vector is obtained by traversing all modulation symbol vectors of different active antennas; Then, all the antenna activation sequences are traversed to obtain the optimal information block. The distance corresponding to the possible candidate solution vector x_i of each time slot calculated by TDC detection algorithm can be expressed as

$$D(x_i) = \|y_t(i) - Y_{t-1}x_i\|^2, x_i \in \Theta_i \tag{7}$$

Equation (7): The set $(u, s) \in \Theta_i$ of candidate solution vectors x_i is a complex matrix of size $N_r \times N_t$, and the vector is a complex vector of size $N_t \times 1$, and there is only one non-zero complex element, so $Y_{t-1}x_i$ can be simplified as formula (8) [13].

$$Y_{t-1}x_i = s_m y_{t-1}(u_i), x_i \in \Theta_i \tag{8}$$

Therefore, the number of $4N_r$ times of real number multiplication to be calculated once is time $Y_{t-1}x_i$. Since a complex $N_r \times 1$ vector of size is obtained by $Y_{t-1}x_i$ and is also a complex vector of size, the number of times of real number multiplication to be calculated once is times. Therefore, the total number of times of real number multiplication to be performed for a candidate solution vector is times. In addition, there is a possible candidate solution vector in a time slot, so in one detection of TDC detection algorithm, the distance corresponding to the possible candidate solution vector in any time slot is calculated, and the computational complexity required can be expressed as:

$$C_{TDC\_i} = 6N_rN_t \cdot 2^b \tag{9}$$

One detection of TDC detection algorithm includes calculating the distance corresponding to the possible candidate solution vectors x_i of N_t time slots. When all time slots of the DSM system adopt the same modulation method, the computational complexity at this time can be expressed as:

$$C_{TDC} = 6N_rN_t^2 \cdot 2^b \tag{10}$$

When all N_t time slots adopt the same modulation method, the computational complexity of the TDC algorithm is shown in Formula (10), that is, the TDC detection algorithm proposed in this paper theoretically reduces the computational complexity of the DSM system so that the DSM system can be used in more practical scenarios. The relative complexity reduction and relative complexity of TDC detection algorithm proposed by the system are as follows:

$$Cre_{TDC}(\%) = 100 \times \left(1 - \frac{N_t}{2^{\lfloor \log_2(N_t!) \rfloor + (N_t-1)b}}\right) \tag{11}$$

$$Cre_{TDC} = \frac{N_t}{2^{\lfloor \log_2(N_t!) \rfloor + (N_t-1)b}} \tag{12}$$

It can be seen from the above formula that in the DSM system, the relative reduction N_t of the complexity of TDC detection algorithm compared with the ML detection algorithm shows a strong upward trend with the increase of the modulation order b. In the case of large N_t and b, the reduction of complexity is very significant [14].

4 Simulation Analysis

The bit error rate performance and computational complexity of ML and TDC detection algorithms are simulated and verified through MATLAB 2019.

4.1 Simulation Analysis of Bit Error Rate Performance

The SNR value of the simulation is 0–30 dB. At the same SNR, the TDC detection algorithm is simulated. The length of the data block included in each simulation is 10. The performance curves of the ML and TDC algorithms are given. Figure 1 shows the bit error rate performance curve of TDC algorithm under different N_t modulation combinations when the system spectral efficiency is the same. The modulation methods of, N_t and N_r are indicated in the figure. Under different conditions, the BER performance curve of TDC detection algorithm and ML algorithm always coincide [15]. When N_t and the modulation mode are the same as the modulation mode, the bit error rate performance of the system becomes better with the increase of N_r. As shown in the figure, when the signal to noise ratio is low, the bit error rate performance curves of $N_r.N_t = 3$, 8816 modulation mode and $N_t = 5$, 8888Q modulation mode show that the former has better bit error rate performance than the latter [16]. With the increase of the signal to noise ratio, the bit error rate performance of the two will gradually approach or even coincide, and then with the further increase of the signal to noise ratio, the latter has better bit error rate performance, and the performance difference will gradually increase, And with the increase of the number of receiving antennas, the signal-to-noise ratio corresponding to the intersection of the two curves gradually decreases. The small figure at the bottom left can clearly see the current $N_r = 2$ trend. The two curves cross and coincide before the signal noise ratio is 8. When the spectral efficiency is the same, the BER performance curve of the system is mainly affected in the case of low SNR. In the case of high SNR, the BER performance curve is mainly affected by the modulation method N_t, and the smaller the number of receiving antennas, the greater the impact of the number of transmitting antennas [17].

It can be seen from the above simulation that in any case, the TDC detection algorithm proposed by the system always maintains the same performance as the ML detection algorithm, which is consistent with the theoretical analysis. In general, under the N_r same conditions, the bit error rate performance of DSM systems decreases with the increase of spectral efficiency; In the case of the same spectral efficiency, the bit error rate performance of the system is determined by N_r and modulation methods; The bit error rate performance of the DSM system will improve with the N_r increase of under the same conditions as the modulation mode [18].

4.2 Complexity Simulation Analysis

Figure 2 shows the complexity curve of TDC algorithm. Figures 2 and 3 respectively show the relative complexity reduction curve and its relative complexity curve of TDC algorithm when N_t and modulation modes change. It can be clearly seen from the two figures that, under the same modulation mode, the relative complexity of TDC algorithm decreases with the increase of N_t, and decreases faster and faster. Similarly, at the same

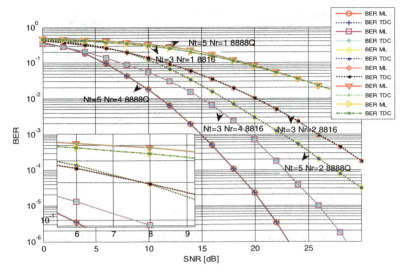

Fig. 1. BER performance of TDC in same spectral efficiency

N_t time, when the modulation order continues to rise, the relative complexity of the TDC algorithm also continues to decrease, and the greater the number, the faster the reduction. Figure. When $N_t = 4$, the modulation mode of and N_t is QPSK, the relative complexity of TDC algorithm has been reduced to 10^{-2}. When $N_t = 8$, the modulation mode of and N_t is 16PSK, the relative complexity of TDC algorithm has been reduced to 10^{-12}. It can be seen that when the modulation order and N_t are large and the modulation order is high,.the complexity of TDC algorithm is significantly reduced and the advantages are obvious, which is consistent with the theoretical analysis [19].

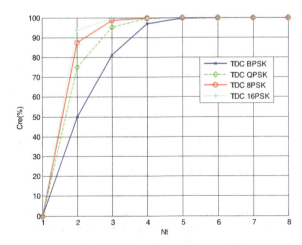

Fig. 2. Relevant complexity reduction of TDC in different N_t and modulation

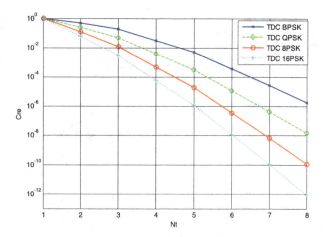

Fig. 3. Relevant complexity of TDC in different N_t and modulation

Figure 4 shows the relative complexity curve of the improved detection algorithm TDC algorithm with the change of N_t and modulation methods when the number of receiving antennas N_r is different As shown in the figure, with the change of N_t and modulation modes, the curves are completely coincident at different times, which shows that the relative complexity of TDC detection algorithm is independent of the size of N_r, which verifies its theoretical derivation [20].

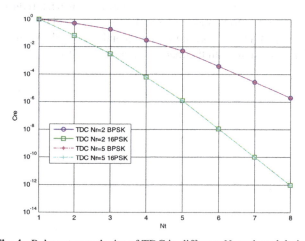

Fig. 4. Relevant complexity of TDC in different N_r and modulation

Through the experimental simulation of the complexity, the proposed TDC detection algorithm significantly reduces the detection complexity of the DSM system, and with the increase of the modulation order and N_t, the relative complexity decreases exponentially [21], which is consistent with the theoretical analysis. TDC algorithm enables DSM system to be applied in higher and larger modulation order scenarios and N_t and broadens the application scenarios of DSM system.

5 Conclusion

This paper analyzes the idea and detection process of signal detection algorithm in DSM system, and proposes a low complexity detection algorithm. TDC algorithm with optimal performance, aiming at the reason of high complexity of ML detection algorithm in DSM system. The complexity of TDC algorithm is theoretically analyzed, and the formula of the relative reduction of TDC detection algorithm complexity is derived. Then through the simulation and comparison of TDC algorithm and ML algorithm, it is found that in any case, TDC algorithm can guarantee the same bit error rate performance as ML algorithm and reduce the computational complexity. And the relative complexity of the algorithm decreases exponentially with the increase of N_t and modulation order, so TDC algorithm has more advantages when N_t and modulation order are large.

Acknowledgment. This project is supported by the School-Level Scientific Research Project "Research on Signal Detection Algorithm of Differential Space Modulation System (2019XJZD004).

References

1. Mugen, P., et al.: Introduction to Wireless Communication, pp. 1–27. Beijing University of Posts and Telecommunications Press, Beijing (2011). (in Chinese)
2. You, X., Cao, S., Li, J.: Development status and prospect of the third generation mobile communication system. Acta Electronica Sinicas **27**(Z1), 3–8 (1999)
3. Xie, X.: Long term evolution plan and comparison of the third generation mobile communication. Telecommun. Sci. **22**(2), 1–4 (2006). (in Chinese)
4. Hu, D., et al.: Application status and prospect of 4G technology. Pract. Electron. (1), 30–30 (2016). (in Chinese)
5. Zhu, H.: Development prospect of the fourth generation mobile communication. China Sci. Technol. Inform. **17**, 111–112 (2010). (in Chinese)
6. Yao, Y.: The fourth generation mobile communication system and its key technologies. Xi'an Univ. Posts Telecommun. J. Sci. **12**(5), 25–29 (2007). (in Chinese)
7. Hu, H., Dong, S., Jiang, Y., et al.: Analysis of the fourth generation mobile communication technology. Comput. Eng. Des. **32**(5), 1563–1567 (2011). (in Chinese)
8. You, X., Pan, Z., Gao, X., et al.: 5G mobile communication development trend and several issues key technology. Sci. China: Inform. Sci. (5): 551–563 (2014). (in Chinese)
9. Bian, Y., Cheng, X., Wen, M., et al.: Differential spatial modulation. IEEE Trans. Veh. Technol. **64**(7), 3262–3268 (2015)
10. Afif, O.: Mobile and wireless communications system for 2020 and beyond (5G) (2014)
11. Osseiran, A., Braun, V., Hidekazu, T., et al.: The foundation of the mobile and wireless communications system for 2020 and beyond: challenges, enablers and technology solutions. IEEE Veh. Technol. Conf. **14**(2382), 1–5 (2014)
12. Wang, J.: Research and Simulation of Multi-user and Multi carrier Spatial Modulation Technology, pp. 25–40. University of Electronic Science and Technology, Chengdu (2013). (in Chinese)

13. Zhi, L., Cheng, X., et al.: A low-complexity optimal sphere decoder for differential spatial modulation. In: IEEE Global Communications Conference, pp. 1–6 (2015)
14. Jun, L., Wen, M., et al.: Differential spatial modulation with gray coded antenna activation order. IEEE Commun. Lett. **20**(6), 1100–1103 (2016)
15. Wang, Y.: Research on Signal Detection Algorithm of Spatial Modulation System Based on Spherical Decoding. Chongqing University, Chongqing (2016). (in Chinese)
16. Men, H., Jin, M.: A low-complexity ml detection algorithm for spatial modulation systems With PSK constellation. IEEE Commun. Lett. **18**(8), 1375–1378 (2014)
17. Mengna, L.: Enhanced Differential Space Modulation. South China University of Technology, Guangzhou (2016). (in Chinese)
18. Kumar, P., Singh, A.K.: Mapreduce algorithm for single source shortest path problem. Int. J. Comput. Netw. Inform. Secur. **12**(3), 11–21 (2020)
19. Das, N.R., Rai, S.C., Nayak, A.: Intelligent scheduling of demand side energy usage in smart grid using a metaheuristic approach. Int. J. Intell. Syst. Appl. (IJISA) **6**(10), 30–39 (2018)
20. Polatgil, M.: Outlier detection algorithm based on fuzzy c-means and self-organizing maps clustering methods. Int. J. Math. Sci. Comput. (IJMSC) **8**(8), 21–29 (2022)
21. Abu Samra, A.A., Qunoo, H.N., Al Salehi, A.M.: Distributed malware detection algorithm (DMDA). Int. J. Comput. Netw. Inform. Secur. **9**(8), 48–53 (2017). https://doi.org/10.5815/ijcnis.2017.08.07

Mathematical Model of the Process of Production of Mineral Fertilizers in a Fluidized Bed Granulator

Bogdan Korniyenko[✉] and Andrii Nesteruk

National Technical University of Ukraine "Igor Sikorsky Kyiv Polytechnic Institute",
Kyiv 03056, Ukraine
bogdanko@i.ua

Abstract. An analysis of approaches to mathematical modeling of the process of manufacturing mineral fertilizers in a fluidized bed granulator was carried out. A mathematical model of the granulation process in a fluidized bed has been developed, which considers the process as heterogeneous and three-phase, during which three separate phases interact with each other: particles – granulation centers, starting material – ammonium sulfate in the form of drops, and coolant – air. The mathematical model takes into account the hydrodynamics of the fluidized bed, the transfer of kinetic energy, the dissipation of energy, the compression of droplets with particles, their adhesion to the surface, the kinetics of drying the solution on the surface of the particles. The proposed mathematical model can be used to build information technology for managing the granulation process in a fluidized bed.

Keywords: Mathematical Model · Mineral Fertilizers · Fluidized Bed · Granulation

1 Introduction

The industrial application of the fluidization method is caused by a significant list of its advantages. Active mixing of the solid phase takes place in the fluidized bed, the quality of the processing of which directly affects the quality of the finished product. Also, fluidized bed granulators have a relatively simple design, and are quite amenable to mechanization and automation.

With modern world trends towards the growth of consumption of products of various industries and the environmental situation, the problem of rational use of energy and raw materials in industrial production in order to obtain the maximum amount of a finished product of a given quality is acute. An important step in solving this problem is the creation of an adequate mathematical model of the process.

The mathematical model should correctly reflect the technological process, its characteristic features, but also should not be overcomplicated with details that have an insignificant effect on the solution of the given task. Obtaining an adequate model of

© The Author(s), under exclusive license to Springer Nature Switzerland AG 2023
Z. Hu et al. (Eds.): ICAILE 2023, LNDECT 180, pp. 55–64, 2023.
https://doi.org/10.1007/978-3-031-36115-9_6

the device allows you to correctly develop a real strategy for the implementation of the control of the technological process.

In devices with a fluidized bed, granulation is carried out by spraying pulps, solutions or melts on the surface of liquefied particles. At the same time, thin films of granular matter are deposited on the particles, where they are dried and crystallized, thereby increasing the size of the granules to the required size. Another (insignificant) part, which is introduced into the solution layer, dries in the spray zone, forming small dry particles, some of which are carried by the gas flow from the apparatus (carrying out), others remain in the apparatus as an internal return (new granulation centers).

The main advantages of this method include: small dimensions and high productivity of the installation; spherical shape of granules of the finished product; obtaining a product of the required chemical and granulometric composition; intensification of heat and mass exchange processes due to the maximum degree of contact between the solid particles of the suspended layer and the gas coolant; automated control of the installation. Disadvantages of the method include: dust removal and thorough cleaning of the gas leaving the device; different residence time of particles in the apparatus.

In connection with the constant growth of the scale of production and, accordingly, devices with a fluidized bed, significant difficulties arise taking into account the phenomena of gas-particle interaction (resistance force) and particle-particle interaction (collision force). Solving these problems through long-term studies on pilot plants is quite expensive.

To facilitate the design process of devices with a fluidized bed, computer modeling can be a useful tool. The main difficulties in modeling devices with a fluidized bed in natural size are associated with a large difference in scales: the largest flow structures can be of the order of several meters, some structures can directly depend on particle-particle collisions and particle-gas interactions that occur at the millimeter level.

When granulating solutions by dehydration, the requirement to obtain a product with a certain granulometric composition comes to the fore. In general, the mechanism of granule growth depends on the properties of the substances used, the process regime and other factors that determine the nature of the interaction between the dispersed liquid and solid phases.

In our case, numerous studies have confirmed that the condition of the finished product, the main quality indicator of which is the equivalent diameter of the particles, is most affected by the temperature of the fluidized bed in which they are formed. That is why it was chosen as the main controlled value in the device under consideration. Consider the construction of an apparatus for granulating substances in a fluidized bed (Fig. 1). The solution is fed into the fluidized bed granulator 1 with the help of the executive device 2, and the granulation centers are fed with the executive device 3. The heated coolant – air – is supplied from the bottom to the top. The finished product – granules are unloaded using the executive device 4.

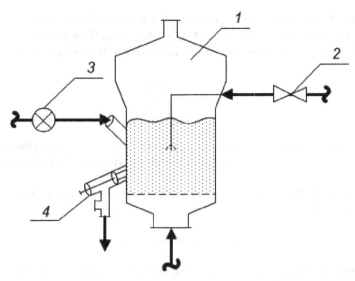

Fig. 1. Apparatus for granulating thermolabile substances in a fluidized bed: 1 – granulator, 2 – device for introducing the initial solution, 3 – device for introducing particles; 4 – a device for unloading the finished product.

2 Review of Mathematical Modelling Methods of Processes of Dehydration and Granulation in the Fluidized Bed

The balance model is used to describe the change in particle size distribution of the granulation process in a fluidized bed [1, 2]. A general packet-mode length-based balance equation that describes the rate of change of the particle number density function $n(t,L)$ is given as follows [3]:

$$\frac{\partial n(t, L)}{\partial t} = B(t, L) \tag{1}$$

$$-\frac{\partial}{\partial L}(G(t, L)n(t, L)) \tag{2}$$

$$+\frac{L^2}{2}\int_0^L \frac{\beta(t, (L^3 - \lambda^3)^{\frac{1}{3}}, \lambda)}{(L^3 - \lambda^3)^{\frac{2}{3}}} n(t, (L^3 - \lambda^3)^{\frac{1}{3}}) n(t, \lambda) d\lambda - n(t, L) \int_0^\infty \beta(t, L, \lambda) n(t, \lambda) d\lambda \tag{3}$$

$$+\int_L^\infty S(t, \lambda) b(t, L|\lambda) n(t, \lambda) d\lambda - S(t, L)n(t, L) \tag{4}$$

The growth of granules in the apparatus with a fluidized bed occurs mainly due to layering. At the same time, it is necessary to take into account the aggregation of particles. Only one kernel of granulation growth in a fluidized bed is used to model the balance [2]:

$$G = \frac{2(1-b)\Phi}{\pi \int\limits_{0}^{\infty} nL^2 dL} \tag{5}$$

In the discrete method, the balance model equation is solved at each size interval. Therefore, the advantage of this method is that as a result we will get a distribution of particles by size.

The moment method is a variant of solving the balance model under moment conditions, which is defined as:

$$m_k(t) = \int\limits_{0}^{+\infty} n(L; t)L^k dL \tag{6}$$

The moments are closely related to important integral quantities of the particle population, such as the average or total surface area and volume of the particles. Compared to the balance model, the numerical solution requires less computing power.

The development of computer technology has made it possible to use the hydrodynamics model, which describes the dynamics of the interaction of gas and solid particles. For modeling in a fluidized bed granulator, there are two different categories of hydrodynamics models: the Euler model and the Lagrangian model [4].

The Euler model allows the inclusion of several secondary phase solids. Conservation of mass and momentum are performed respectively for each phase. Thus, the Euler model solves a set of n continuity and momentum equations, making this approach one of the most complex multiphase models.

The Euler-Euler model, known as the continuum model or the two-fluid model, describes the evolution of solid-gas phase interactions. The interaction between the two phases depends on the hydraulic resistance between the phases, that is, the local relative velocities of the phases and the local volume fractions of the phases. As a result, the simulation of the method of computational hydrodynamics based on the Euler-Euler framework is accepted for the study of the multi-phase flow of gas-solid bodies in a fluidized bed granulator [5–7].

The Lagrangian model solves the equations of motion for each pellet, taking into account particle collisions and forces acting on the particle from the gas side. Therefore, when the number of particles is large, it is better to use Euler-Euler models [8–10].

The Lagrange-Euler model describes gas bubbles as discrete particles that can collide, coalesce, stop, contract, and grow. The Euler model is not suitable for the solid phase, but it is suitable for describing the emulsion of the gas phase and particles [11–15].

3 Mathematical Model of the Process of Production of Mineral Fertilizers in a Fluidized Bed Granulator

During the creation of the model, the process of dehydration and granulation in a fluidized bed was considered as a three-phase heterogeneous process in which three components interact: particles, initial solution and coolant [16–18]. During modeling, it was assumed

Mathematical Model of the Process of Production of Mineral Fertilizers 59

that the fluidized bed parameters change over time without taking into account the change in height and the radial component, the heat exchange between all process components is convective, the particles are monodisperse, there is no porosity and agglomeration, the droplets have a narrow size distribution, between drops do not collide, stick together and stick to the walls of the case [19, 20].

The following system of equations was used to describe the heat exchange process in the granulator:

– Particles temperature change:

$$M_p C_p \frac{d\Theta_p}{dt} = a_p S_p (\Theta_a - \Theta_p) - M_p v_{dry} Q_{dry} + G_d x_d Q_{cryst}$$
$$+ M_p C_p r_p \Theta_{p0} + R_{ad} M_p C_d (\Theta_d - \Theta_p) - \gamma_\Theta - \phi_{ps} \qquad (7)$$

where a_p – heat transfer coefficient of particles, $W/(m^2*K)$; S_p – particles surface area, m^2; Θ_a – heat carrier temperature, K; Θ_p – particles temperature, K; M_p – mass of particles, kg; v_{dry} – the specific speed of drying moisture on particles or in drops, $kg/(kg*s)$; Q_{dry} – specific heat of drops drying, J/kg; G_d – mass flow of drops, kg/s; x_d – concentration of solution drops, Q_{cryst} – specific heat of drops crystallization, J/kg; M_p – mass of particles, kg; C_p – specific heat capacity of particles, $J/(kg*K)$; r_p – coefficient of axial dispersion of particles, $1/s$; Θ_{p0} – initial temperature of particles, K; R_{ad} – specific speed flowing of drops with particles, $kg/(kg*s)$; C_d – specific heat capacity of drops, $J/(kg*K)$; Θ_d – drops temperature, K; γ_Θ – energy dissipation during collision, J/s; φ_{ps} – transfer of kinetic energy, J/s.

– Drops temperature change:

$$M_d C_d \frac{d\Theta_d}{dt} = M_d C_d \Theta_{d0} - R_{ad} M_p C_d (\Theta_d - \Theta_p) +$$
$$+ M_p v_{dry} Q_{dry} + a_d S_d (\Theta_a - \Theta_d) \qquad (8)$$

where M_d – mass of drops, kg; C_d – specific heat capacity of drops, $J/(kg*K)$; Θ_{d0} – initial temperature of drops, K; R_{ad} – specific speed flowing of drops with particles, $kg/(kg*s)$; M_p – mass of particles, kg; Θ_d – drops temperature, K; Θ_p – particles temperature, K; v_{dry} – the specific speed of drying moisture on particles or in drops, $kg/(kg*s)$; Q_{dry} – specific heat of drops drying, J/kg; a_d – heat transfer coefficient of drops, $W/(m^2*K)$; S_d – drops surface area, m^2; Θ_a – heat carrier temperature, K.

– Heat carrier temperature change:

$$M_a C_a \frac{d\Theta_a}{dt} = G_a (C_a \Theta_{a0} - C_a \Theta_a) - a_p S_p (\Theta_a - \Theta_p) - a_d S_d (\Theta_a - \Theta_d) \qquad (9)$$

where G_a – mass flow of heat carrier, kg/s; C_a – specific heat capacity of heat carrier, $J/(kg*K)$; Θ_{a0} – initial temperature of heat carrier, K; Θ_a – heat carrier temperature, K; a_p – heat transfer coefficient of particles, $W/(m^2*K)$; S_p – particles surface area, m^2; Θ_p – particles temperature, K; a_d – heat transfer coefficient of drops, $W/(m^2*K)$; S_d – drops surface area, m^2; Θ_d – drops temperature, K; M_a – mass of heat carrier, kg.

The developed mathematical model takes into account fluidized bed hydrodynamics, kinetic energy transfer, energy dissipation, compression of droplets with particles, their adhesion to the surface, kinetics of solution drying on the surface of particles. The system of equations was developed taking into account empirical ratios for calculating the specific rate of drying, the specific rate of deposition of droplets on particles as a result of adhesion, the coefficient of axial dispersion of particles, heat transfer coefficients, the ratio for calculating the loss of material and the thickness of the coating layer, as well as initial conditions.

The growth of granules is more likely, the greater the adhesion forces of liquid droplets with solid particles and its speed. The adhesive properties of the drop, in turn, depend on the hardness of the surface of the granules and the properties of the sprayed substance.

Specific speed flowing of drops with particles can be calculated by the following formula:

$$R_{ad} = \frac{G_d}{M_{p0}} x_d = \frac{G_d}{M_{p0}} \left(\frac{St_d}{St_d + 0.35} \right) \tag{10}$$

where M_{p0} – initial mass of particles, kg; G_d – mass flow of drops, kg/s; St_d – Stokes number for drops, which we can calculate by the following formula:

$$St_d = \frac{\rho_d v_a d_d^2}{\mu_a d_p} \tag{11}$$

where ρ_d – drops density, kg/m^3; v_a – heat carrier speed of movement, m/s; d_d – drops diameter, m; μ_a – dynamic viscosity of heat carrier, $Pa*s$; d_p – particles diameter, m.

Collision energy dissipation is the rate of energy dissipation within the solid phase due to collisions between particles. This phenomenon is modeled using the Lun correlation:

$$\gamma \Theta = \frac{12(1 - e_p^2) g_0}{d_p \sqrt{\pi} S_p} \rho_p a_p^2 \Theta_p^{1.5} \tag{12}$$

where e_p – coefficient of restoration of collision of particles with other particles, g_0 – radial distribution function, d_p – particles diameter, m; S_p – particles surface area, m^2; ρ_p – particles density, kg/m^3; a_p – heat transfer coefficient of particles, $W/(m^2*K)$; Θ_p – particles temperature, K.

The transfer of kinetic energy of random particles from the solid phase to the liquid phase can be calculated by the formula:

$$\phi_{ps} = \frac{-3K_{ps}\Theta_p}{S_p} \tag{13}$$

where K_{ps} – coefficient of the force of interaction between a liquid and a solid body, S_p – particles surface area, m^2; Θ_p – particles temperature, K.

The amount of heat released when moisture is removed from the surface of the droplets can be rewritten as follows:

$$M_a v_{dry} Q_{dry} = \beta \frac{M_{h2o} S_p}{R \Theta_p} \Delta P Q_{dry} \tag{14}$$

where β – mass transfer coefficient, m/s; M_{h2o} – molecular weight of water, g/mol; R – universal gas constant, $(m^2*kg)/(s^2*K*mol)$; ΔP – partial pressure difference, Pa.

4 The Numerical Results Analysis of the Mathematical Model

To obtain dynamic results of the developed mathematical model, a program was created to calculate this mathematical model using the Runge-Kutt method of the 4th order in the Python programming language. With the help of built-in mathematical libraries, calculations were carried out and the temperature behavior of particles, drops and coolant was visualized from the moment the installation was turned on until the end of the process.

The following values given in Table 1 were used for modeling.

Table 1 Values of the main process parameters.

Name	Marking	Value
Mass of particles	M_p	1.5
Mass of drops	M_d	2
Mass of heat carrier	M_a	3.5
Initial mass of particles	M_{p0}	1.53
Specific heat capacity of drops	C_d	1590
Specific heat capacity of particles	C_p	1420
Specific heat capacity of heat carrier	C_a	1011
Initial heat capacity of heat carrier	C_{a0}	1015
Initial temperature of particles	Θ_{p0}	300
Initial temperature of drops	Θ_{d0}	293
Initial temperature of heat carrier	Θ_{a0}	393
Coefficient of axial dispersion of particles	rp	$1*10^{-5}$
The specific speed of drying moisture on particles or in drops	$vdry$	0.47
Specific heat of drops crystallization	$Qcryst$	82300
Specific heat of drops drying	Q_{dry}	5000
Heat transfer coefficient of particles	ap	5.57
Heat transfer coefficient of drops	ad	5.535
Particles surface area	Sp	$7.07*10^{-7}$
Drops surface area	Sd	$6.07*10^{-6}$
Mass flow of drops	Gd	0.7
Mass flow of heat carrier	Ga	1
Concentration of solution drops	xd	0.37

After the calculations, we get a graph of the dependence of the temperatures of drops, drops, and time on time, which is shown in Fig. 2.

In order to maintain the stable operation of the apparatus with a fluidized bed in the necessary hydrodynamic mode inside them, it is necessary to develop an effective control system for the process of dehydration and granulation. The quality of this control system is closely related to the accuracy of the mathematical model of the object for which this control system is being developed. The obtained results from the calculation

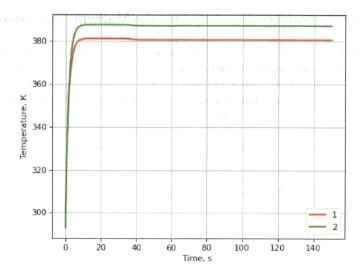

Fig. 2. Graph of dependence of particle, droplet and air temperatures on time (1 – temperature of particles, 2 – temperature of drops).

of the mathematical model prove that it can serve as a basis for the development of an effective control system.

5 Conclusion

A mathematical model of the mineral fertilizer production process in a fluidized bed granulator is proposed, which can serve as a basis for creating a control system for the granulation and dehydration process in a pseudo-fluidized bed. The presented mathematical model expresses the heat exchange between drops, particles and the coolant taking into account such parameters as fluidized bed hydrodynamics, kinetic energy transfer, energy dissipation, compression of drops with particles, their adhesion to the surface, kinetics of solution drying on the surface of particles.

The mathematical model of the mineral fertilizer production process in the fluidized bed granulator was calculated by numerically and it was established that the system needs 110 s to stabilize the temperature, the temperature of the particles stabilizes at 367 K, and the temperature of the particles at 384 K.

References

1. Adetayo, A.A., et al.: Population balance modelling of drum granulation of materials with wide size distribution. Powder Technol. **82**(1), 37–49 (1995)
2. Vreman, A.W., van Lare, C.E., Hounslow, M.J.: A basic population balance model for fluid bed spray granulation. Chem. Eng. Sci. **64**(21), 4389–4398 (2009)
3. Syamlal, M., Rogers, W., O.Brien, T.J.: MFIX Documentation: Volume1, Theory Guide. National Technical Information Service, Springfield, VA (1993). DOE/METC9411004, NTIS/DE94000871993

4. Gidaspow, D., Bezburuah, R., Ding, J.: Hydrodynamics of circulating fluidized beds, kinetic theory approach. in fluidization VII. In: Proceeding of the 7th Engineering Foundation Conference on Fluidization, pp. 75–82 (1992)
5. Lun, C.K.K., et al.: Kinetic theories for granular flow: inelastic particles in couette flow and slightly inelastic particles in a general flow field. J. Fluid Mech. **140**, 223–256 (1984)
6. Kornienko, Y.M., Liubeka, A.M., Sachok, R.V., Korniyenko, B.Y.: Modeling of heat exchangement in fluidized bed with mechanical liquid distribution. ARPN J. Eng. Appl. Sci. **14**(12), 2203–2210 (2019)
7. Kornienko, Y.M., Haidai, S.S., Sachok, R.V., Liubeka, A.M., Korniyenko, B.Y.: Increasing of the heat and mass transfer processes efficiency with the application of non-uniform fluidization. ARPN J. Eng. Appl. Sci. **15**(7), 890–900 (2020)
8. Korniyenko B., Galata L., Ladieva L.: Mathematical model of threats resistance in the critical information resources protection system. Paper presented at the CEUR Workshop Proceedings, vol. 2577, pp. 281–291 (2019)
9. Korniyenko, B., Kornienko, Y., Haidai, S., Liubeka, A., Huliienko, S.: Conditions of non-uniform fluidization in an auto-oscillating mode. In: Hu, Z., Petoukhov, S., Yanovsky, F., He, M. (eds.) ISEM 2021. LNNS, vol. 463, pp. 14–27. Springer, Cham (2022). https://doi.org/10.1007/978-3-031-03877-8_2
10. Korniyenko, B., Kornienko, Y., Haidai, S., Liubeka, A.: The heat exchange in the process of granulation with non-uniform fluidization. In: Hu, Z., Petoukhov, S., Yanovsky, F., He, M. (eds.) ISEM 2021. LNNS, vol. 463, pp. 28–37. Springer, Cham (2022). https://doi.org/10.1007/978-3-031-03877-8_3
11. Korniyenko, B., Ladieva, L.: Mathematical modeling dynamics of the process dehydration and granulation in the fluidized bed. In: Hu, Z., Petoukhov, S., Dychka, I., He, M. (eds.) ICCSEEA 2020. AISC, vol. 1247, pp. 18–30. Springer, Cham (2021). https://doi.org/10.1007/978-3-030-55506-1_2
12. Korniyenko, B., Ladieva, L., Galata, L.: Control system for the production of mineral fertilizers in a granulator with a fluidized bed. In: ATIT 2020 – Proceedings: 2020 2nd IEEE International Conference on Advanced Trends in Information Theory, № 9349344, pp. 307–310 (2020). https://doi.org/10.1109/ATIT50783.2020.9349344
13. Korniyenko, B., Galata, L., Ladieva, L.: Research of information protection system of corporate network based on GNS3. In: 2019 IEEE International Conference on Advanced Trends in Information Theory, ATIT 2019 – Proceedings, № 9030472, pp. 244–248 (2019). https://doi.org/10.1109/ATIT49449.2019.9030472
14. Korniyenko, B.Y., Ladieva, L.R., Galata, L.P.: Mathematical model of heat transfer process of production of granulated fertilizers in fluidized bed. ARPN J. Eng. Appl. Sci. **16**(20), 2126–2131 (2021)
15. Korniyenko, B., Zabolotnyi, V., Galata, L.: The optimization of the critical resource protection system of a mineral fertilizers manufacturing facility. In: Proceedings of the 11th IEEE International Conference on Intelligent Data Acquisition and Advanced Computing Systems: Technology and Applications, IDAACS 2021, vol. 1, pp. 172–178 (2021)
16. Babak, V.P., Babak, S.V., Myslovych, M.V., Zaporozhets, A.O., Zvaritch, V.M.: Methods and models for information data analysis. In: Diagnostic Systems For Energy Equipments. SSDC, vol. 281, pp. 23–70. Springer, Cham (2020). https://doi.org/10.1007/978-3-030-44443-3_2
17. Polishchuk, M., Tkach, M., Parkhomey, I., Boiko, J., Eromenko, O.: Experimental studies on the reactive thrust of the mobile robot of arbitrary orientation. Indonesian J. Electr. Eng. Inform. **8**(2), 340–352 (2020)
18. Karthika, B.S., Deka, P.C.: Modeling of air temperature using ANFIS by wavelet refined parameters. Int. J. Intell. Syst. Appl. **8**(1), 25–34 (2016). https://doi.org/10.5815/ijisa.2016.01.04

19. Ghiasi-Freez, J., Hatampour, A., Parvasi, P.: Application of optimized neural network models for prediction of nuclear magnetic resonance parameters in carbonate reservoir rocks. Int. J. Intell. Syst. Appl. (IJISA) **7**(6), 21–32 (2015). https://doi.org/10.5815/ijisa.2015.06.02
20. Malekzadeh, M., Khosravi, A., Noei, A.R., Ghaderi, R.: Application of adaptive neural network observer in chaotic systems. Int. J. Intell. Syst. Appl. **6**(2), 37–43 (2014). https://doi.org/10.5815/ijisa.2014.02.05

Simulation Study on Optimization of Passenger Flow Transfer Organization at Nanning Jinhu Square Metro Station Based on Anylogic

Yan Chen[1], Chenyu Zhang[1]([✉]), Xiaoling Xie[1], Zhicheng Huang[1], and Jinshan Dai[2,3]

[1] College of Traffic and Transportation, Nanning University, Nanning 530200, China
2676501161@qq.com
[2] Center for Port and Logistics, School of Transportation and Logistics Engineering, Wuhan University of Technology, Wuhan 430072, China
[3] Department of Industrial Systems Engineering and Management, National University of Singapore, Singapore 117576, Singapore

> **Abstract.** Nanning Jinhu square metro station is the transfer station of Nanning metro Line 1 and Line 3. Its passenger flow transfer organization efficiency greatly affects the operation efficiency of the urban rail transit system. This paper collects the station operation data and analyzes the transfer evaluation parameters, builds the station simulation model using the Anylogic software, sets the relevant scenarios to conduct the operation simulation analysis of the station, looks for the flow bottleneck of the passenger flow in the transfer organization, and takes measures such as iron fence diversion to optimize the passenger flow transfer organization, so as to improve the passenger flow transfer efficiency on the basis of reducing the passenger flow congestion. The paper further determines the effectiveness of the optimization measures by comparing the operation results of the simulation model before and after the optimization, and provides support for the design of transfer facilities and the formulation of operation strategies at the transfer station.
>
> **Keywords:** Anylogic simulation research · Passenger flow transfer at metro station · Organizational optimization

1 Introduction

Nanning's rail transit is in a rapid development stage. In 2019, Nanning's subway travel accounted for 41% of the city's total bus travel. The subway transfer station is the hub part of the urban rail transit system and an important distribution point of passenger flow. It undertakes the transfer and connection of passenger flow between lines. The efficiency of passenger flow transfer organization of the subway transfer station affects the operation efficiency of multiple lines. Nanning Jinhu Square Metro Station, an important passenger transfer hub, Nanning Rail Transit Line 3 and Line 1 transfer here. The passenger flow at Jinhu Square Station is relatively large and complex, and the transfer channel will be congested at the morning and evening peak hours, which gradually cannot meet the growing passenger flow transportation demand. Therefore, the optimization

© The Author(s), under exclusive license to Springer Nature Switzerland AG 2023
Z. Hu et al. (Eds.): ICAILE 2023, LNDECT 180, pp. 65–78, 2023.
https://doi.org/10.1007/978-3-031-36115-9_7

of passenger flow transfer organization at Jinhu Square Station is extremely urgent. Due to the scale, uninterrupted operation and passenger flow volatility of passenger transport hub, simulation modeling can be used to explore the bottleneck of passenger flow and provide support for formulating optimization measures.

Among many simulation software, Anylogic software is a widely used simulation modeling tool that is applicable to discrete, system dynamics, multi-agent and hybrid systems [1]. It is mostly used for traffic simulation, pedestrian evacuation, production simulation, etc. Simulation modeling is conducted through Anylogic software to analyze and optimize the passenger organization, passenger flow dispersion, passenger flow conflict, etc. of metro stations, which is a current application hotspot. For example, V M. Antonova; N. A. Grechishkina; N. A. Kuznetsov and others used AnyLogic software to analyze the inbound passenger flow of typical subway stations to check the traffic congestion points [2]. Another scholar, Zhao Min published his master's thesis "Research on Passenger Flow Control Method and Simulation of Urban Rail Transit Transfer Station" in April 2020, and verified the simulation model and method with Chengdu New South Gate Station as an example. According to the passenger flow data and the bottleneck type discrimination table, the identification results of the primary bottleneck are obtained, and then the bottleneck is relieved by adjusting the service time of the gate, setting the one-way passage of stairs, and arranging the fence for batch release. The rationality of passenger flow control measures is verified by simulation results [3].

The scholars' research has laid a good foundation for the simulation study of the transfer passenger flow analysis and evacuation of the Jinhu Square metro station in Nanning City, and provided a good reference case.

2 Analysis of Transfer Facilities in Nanning Jinhu Square Metro Station

2.1 General Layout

Jinhu Square Station is the transfer station between Nanning Metro Line 1 and Line 3. The main station of Line 1 is located at Minzu Avenue, which is arranged along the east-west direction of the road. The main station of Line 3 is located at Jinhu South Road, which is arranged along the north-south direction of the road. In terms of layout location, Nanning Department Store Wuxiang Square Shopping Center is set underground in Wuxiang Square to the west; In the south, there are mainly cultural centers such as Guangxi Press and Publication General Office and Nanning Book City; To the east, there are business office areas such as Modern International Building and East Manhattan Building, and to the north, there are Diwang Building and Agricultural Bank of China Guangxi Branch. It is adjacent to Nanhu Station in the west and Convention and Exhibition Center Station in the east.

According to the field observation, the passenger flow characteristics of Jinhu Square Station have very obvious characteristics of the transfer station: the majority of passengers are transfer passenger flow, and the transfer facilities in the station become the main facilities to bear the passenger flow through, with only a small amount of inbound and outbound passenger flow, so the inbound and outbound facilities only bear a small amount of passenger flow; Such characteristics of passenger flow make it rare that the bottleneck of subway passenger flow such as entrance security check and gate is fully loaded, and most passenger flow is concentrated in the transfer channel.

In terms of transfer mode, Jinhu Square Station belongs to the "L" shape configuration and uses channel transfer to meet the requirements of line interchange. Line 1 is relatively independent from the station located in Line 3. The probability of cross interference of passenger flow between the two stations is small, and the impact of inbound passengers outside the station on station efficiency is small. However, the passenger flow is all transferred through the passage, and the pressure of passenger flow inside and at both ends of the passage is high. The escalator near the passage will bear most of the passenger flow.

2.2 Analysis of Advantages and Disadvantages of Channel Transfer

Channel transfer is generally applied to the relatively independent station structure at the intersection of the two lines. When the station hall cannot be directly used for transfer due to the distance between the platforms or the topographic conditions, separate transfer channels and stairs are set between the two stations for passengers to transfer.

The layout of the channel transfer mode is flexible, which has great adaptability to the intersection of the two lines and the location of the station layout, and is conducive to the phased implementation of the two line projects. Generally, the transfer channel should be set in the middle of the station as far as possible to avoid the cross interference between the passenger flow of both sides and the passenger flow of the entrance and exit of the station, to a certain extent, reducing the concentration of passenger flow in the station hall.

But at the same time, the disadvantages of channel transfer are also obvious. The transfer distance between two stations will increase the length of the passage, and the transfer distance is longer. All passenger flows are transferred through one or two channels. The channel bears greater pressure on passenger flow, and bears less impact of short-term large passenger flow. The transfer stations with height difference between two stations often arrange the vertically moving facilities in the transfer channel in the form of transfer escalator to realize the transfer between the two stations. The speed of passenger flow drops sharply at the transfer escalator, which is easy to cause the aggregation of the escalator ports.

2.3 Analysis of Transfer Facilities in Jinhu Square Station

The station is divided into five parts according to the height setting: Line 1 station hall layer, Line 1 platform layer, transfer channel layer, Line 3 station hall layer, and Line 3 platform layer. A transfer channel is set in the west of the station of Line 1, which is connected with the transfer channel of Line 3 by escalator. There is a transfer node intersecting the transfer channel of Line 1 in the south of the station of Line 3.

Therefore, the facilities at the transfer passage of Jinhu Square Station are arranged as follows: two ascending escalators, one descending escalator and one two-way staircase. Among them, the ascending escalator bears the passenger flow from Line 3 to Line 1, the descending escalator bears the passenger flow from Line 1 to Line 3, and the staircase is used as a diversion measure during peak hours.

It can be seen from the quantity configuration that more passengers transfer from Line 3 to Line 1 than from Line 1 to Line 3. Two ascending escalators and one descending escalator are arranged here. It can be seen from the actual observation that there are different levels of queuing at the entrance of escalators on both sides during the peak period, but passengers at the ascending escalator tend to wait for the escalator. Passengers at the descending escalator tend to change stairs. Therefore, in fact, the queue length at the entrance of the ascending escalator is longer than that of the descending escalator. There are two load-bearing columns at the exit of the descending escalator, which are easy to cause the passenger flow to slow down and small-scale aggregation. According to the field observation, at the transfer escalator, the passenger flow from Line 3 to Line 1 escalator gathered at the entrance. The passenger flow of the escalator from Line 1 to Line 3 gathers at the exit.

3 Results and Analysis of Passenger Flow Transfer at Nanning Jinhu Square Metro Station

3.1 Passenger Flow Data Analysis

Table 1. PassengerFlow Statistics of Jinhu Square Station

Statistical Items	Flat Peak Period (person/hour)	Peak Period (person/hour)
Get off at Line 1	3396	3828
Transfer to Line 3	2630	2804
Get off at Line 3	3126	3556
Transfer to Line 1	2772	3482
Entrance/Exit G (Line 3)	90	125
Entrance/Exit F (Line 3)	48	64
Entrance/Exit D (Line 3)	18	22
Entrance/Exit A (Line 1)	162	201
Entrance/Exit B (Line 1)	258	331
Entrance/Exit C (Line 1)	300	402

In order to have a more comprehensive understanding of the transfer characteristics of passenger flow in Jinhu Square Metro Station and provide enough experimental data for the simulation model, a field survey of Jinhu Square Metro Station was conducted on weekends and holidays. The survey data obtained are shown in Table 1.

3.2 Station Simulation Modeling

The basic simulation of the station model needs to determine the parameters first.

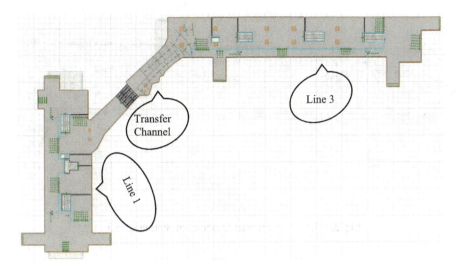

Fig. 1. Plan Structure of Station

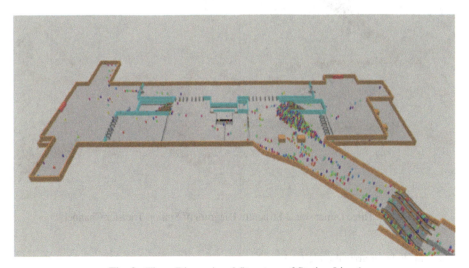

Fig. 2. Three Dimensional Structure of Station Line 1

The basic simulation parameters of the station model include the station plane structure diagram (as shown in Fig. 1), the station three-dimensional structure diagram (as shown in Figs. 2, 3, and 4), and the pedestrian logical flow line. The plane structure of the station shows the horizontal movement ability of the pedestrian, the three-dimensional structure of the station shows the vertical movement ability of the pedestrian, and the logical flow line of the pedestrian constructs the pedestrian action plan.

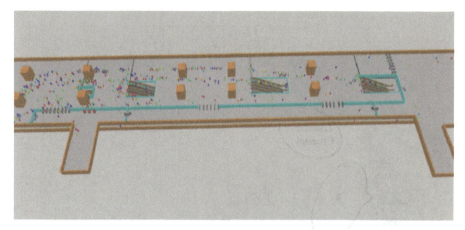

Fig. 3. Three Dimensional Structure of Station Line 3

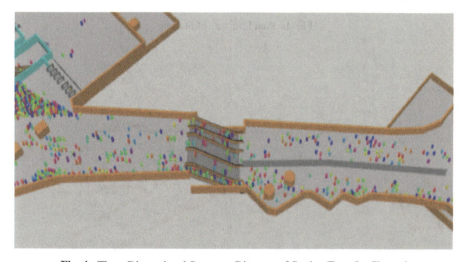

Fig. 4. Three Dimensional Structure Diagram of Station Transfer Channel

Compile train arrival code such as.

"pedSource6.inject(300); pedWait.freeAll(); pedWait6.freeAll();"

Set the target line for passengers to enter and leave the station in a fixed time and quantity, and complete the simulation of train entering and leaving the station.

3.3 Station Simulation Modeling

3.3.1 Station Passenger Flow Density

It refers to the density of passengers gathered in the effective service area of the station. The passenger aggregation density is inversely proportional to the station service level [4]. The dynamic value of passenger flow density at the platform with time can be observed through simulation. The passenger distribution density diagram of different time periods is also presented. According to the density map obtained, the passenger flow density of each region can be roughly obtained according to the passenger flow density comparison bar of the thermal map, and the bottleneck and congestion situation can be analyzed according to this data. According to the simulation of Anylogic, the thermal diagram is shown in Fig. 5.

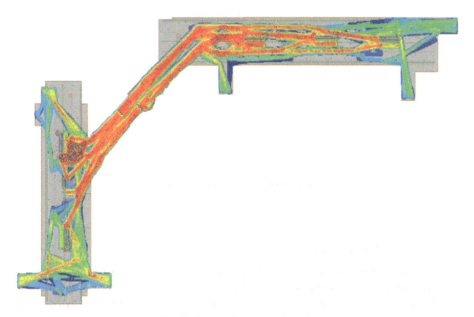

Fig. 5. Station Passenger Density

3.3.2 Passenger Flow Density at the End of the Facility

It refers to the density of passengers gathered at the upper or lower ports of the escalator. Due to the unbalanced characteristics of passenger arrival [5]. The escalator, gate and security check machine are easy to form passenger flow bottlenecks, resulting in congestion and reducing the safety and operation efficiency of the station. It has a great impact on the station service level. Therefore, whether the passenger density gathered at the end of the facility matches the arrival level of passenger flow is selected as the service level evaluation index. According to the Anylogic simulation, the passenger flow density of the facility section is shown in Fig. 6.

72   Y. Chen et al.

By the 180th second of operation, the data had begun to be distorted, and the flow at each port was still fluctuating normally, but the flow of the upper escalator of Line 1 was kept at about 75 people/s, which did not change with the change of passenger flow. It was obvious that there had been irresolvable congestion, leading to the inability of pedestrians to use the escalator normally.

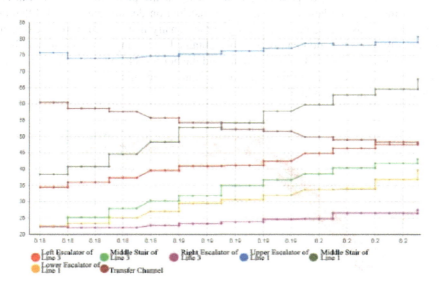

Fig. 6. Flow Diagram of Each Port

3.3.3 Average Transfer Time of Passengers

Average transfer time of passengers refers to the average time it takes for passengers to get off from one line to another. It directly reflects the convenience of the transfer facilities in the subway station. According to the simulation results of Anylogic, the average transfer time of passengers is 11.7 min.

3.4 Bottleneck Analysis of Passenger Flow Transfer at Stations Based on Simulation Model

3.4.1 Analysis of Simulation Operation Data

When the train runs for about 180 s, the model is distorted and there are large congestion points that affect the operation of the station.

1) Through traffic statistics and analysis

Except for the upper escalator of Line 1, all escalators show fluctuations in passenger flow as the train enters and leaves the station, indicating that the passenger flow through the escalator is still within its capacity and can be used normally. The escalator at the upper part of Line 1 fluctuates around 75 people per second, with no downward trend, and is not affected by the train entering and leaving the station, indicating that there has been a large congestion point near it, which is beyond its capacity, and cannot be cleared, and the channel function is lost.

2) Through passenger boarding/departure time analysis

According to statistics, the time for passengers to get into the station from this station is much shorter than the time for passengers to get off and transfer, indicating that there are fewer inbound passengers outside the station compared with transfer passengers, each escalator is within the capacity range, and can enter the platform smoothly, and the passenger flow density at the platform is moderate, which does not affect the passengers to get on the train. Transfer passengers are the main passenger flow of the station. Most passengers enter the station and choose to transfer. Since each station does not affect the passengers entering the station, it indicates that the channel of each platform itself is smooth, and congestion may occur near the transfer channel.

3.4.2 Station Passenger Flow Density/Bottleneck Analysis

As can be seen from Fig. 5, the escalator on the left side of Line 3, the transfer channel and the escalator on the upper side of Line 1 bear most of the passenger flow, with the density of about 1.5 people per square meter. Among them, although the escalator and transfer channel on the left side of Line 3 bear most of the passenger flow, and there is a certain degree of congestion in the simulation, it does not exceed the bearing range of the traffic capacity, and there will be bottlenecks affecting the efficiency, but it will not affect the overall operation of the station [6, 7]. On the contrary, the escalator at the upper part of Line 1 has a large passenger flow, which may be blocked in the transfer channel, blocking the outbound passenger flow and affecting the station function. Therefore, it can be determined that the key optimization point is located at the upper escalator of Line 1, and the secondary optimization point is the left escalator and transfer channel of Line 3.

3.4.3 Analysis of Passenger Flow Line Conflict Points

According to the layout of the facilities in the station, passengers will present different action logic, and streamline is the intuitive expression of action logic. The passenger flow line conflict caused by the passenger facilities arrangement may become one of the reasons for congestion [9, 10]. The paper will draw the passenger flow lines and the conflict points generated by the intersection of different flow lines, so as to observe and determine the improvement plan.

4 Optimization and Simulation of Passenger Flow Transfer Organization at Jinhu Square Metro Station

4.1 Optimization Scheme

After considering the implementation cost and practical operability, combined with the operation situation of the station which mainly focuses on passenger flow, the paper determines to use the combination of iron barrier and manual guidance in the area near the transfer channel, and try to avoid the conflict point by changing the passenger flow line, so as to achieve the purpose of optimizing the transfer organization. The specific optimization measures are as follows.

1) Partial optimization analysis of Line 1

The iron barrier shall be arranged at the upper escalator of Line 1, and the transfer/exit passenger flow line of Line 1 shall be moved down to avoid the passenger flow at the lower platform, and more passenger flow shall be diverted to the upper escalator and the middle staircase to avoid the conflict point. The escalator in the middle of Line 1 is equipped with two layers of iron horses, one is to separate the passenger flow on and off the platform, and the other is to separate the passenger flow from the lower escalator, so that the passenger flow confluence is no longer located at the stair port, and moves to a relatively open area. The right passage of the transfer passage is equipped with iron horses to divert passengers in advance.

2) Partial optimization analysis of Line 3

The iron barrier is set at the entrance of the transfer passage of Line 3 to separate the passenger flow, widen the passage for the passengers who transfer from Line 1 to Line 3, guide the passengers who transfer from Line 3 to Line 1, and avoid passenger flow conflicts.

4.2 Operation Results of Optimized Simulation Model

4.2.1 Data Analysis of Average Transfer Time

The transfer time between different lines is analyzed as follows.

1) Comparison of average transfer time between Line 3 and Line 1

According to the simulation operation statistics, the average transfer time between Line 3 and Line 1 remains unchanged. Compared with that before optimization, the number of passengers with the longest transfer time decreases by 3%, while the proportion of other passengers remains unchanged.

For specific analysis:

After the use of the iron barrier to separate the inbound/outbound passenger flow from the transfer passenger flow, the conflict point disappeared, and the large congestion that caused the escalator facilities to lose their service function disappeared during the peak hours. However, the transfer passenger flow from Line 3 to Line 1 accounted for a high proportion of the passenger flow at the station, and the goal was to ensure the

Simulation Study on Optimization of Passenger Flow Transfer Organization   75

passenger flow under the peak conditions, so the overall transfer time did not change significantly, but the maximum transfer time decreased. The optimization results retained the basic capacity for the peak hours.

2) Comparison of average transfer time between Line 1 and Line 3

The average transfer time between Line 1 and Line 3 is reduced by 5%, the proportion of passengers with medium transfer time is reduced by 7%, and the proportion of passengers with short transfer time is increased by 7%. Compared with the other direction, the optimization effect is better.

For specific analysis:

Before optimization, the transfer of passenger flow from Line 1 to Line 3 is mainly through the narrow passage between the column and the exit gate partition. Although the space above is large, according to the social force model, people like to take short cuts, and often walk by sticking to the column or wall, which is easy to waste space, squeezing the space of the opposite passenger flow, and the insufficient space of the opposite passenger flow will encroach on the passage, and the collision between the two waves of passenger flow will affect the efficiency. The paper uses the iron barrier to separate a new travel channel for the passenger flow in this direction. The passenger flow in this direction can add a new passenger flow channel on the basis of the previous channel. The proportion of passenger flow in this direction is less than that of the target passenger flow, and the passenger flow in the opposite direction is less affected because of the large space. The optimization effect achieved on this basis is better.

4.2.2 Port Traffic Data Analysis

Comparison with previous, the escalator at the upper part of Line 1 has a significant fluctuation of passenger flow caused by the train entering and leaving the station. The passenger flow of the middle staircase of Line 1 increased significantly. The passenger flow of the escalator on the left side of Line 3 has increased significantly. The overall passenger flow of escalators has increased.

For specific analysis:

1) The passenger flow in and out of the station accounts for a very small proportion of the total passenger flow. The escalator that undertakes the passenger flow in and out of the station is always idle, and the optimization effect has no obvious change.
2) The escalator presents normal passenger flow fluctuations, indicating that after the passenger flow is separated from the passenger flow at the iron barrier and the ascending passenger flow is combined with the inbound passenger flow, several nearby congestion points/conflict points are relieved. The upper escalator of Line 1 is within the normal capacity range. The escalator is the key gathering place of passenger flow, with important points, large passenger flow, many conflict points, easy congestion, difficult to optimize, and difficult to reflect the effect. Therefore, the optimization goal is to retain the basic traffic capacity at peak hours and achieve the optimization goal.
3) In order to dissipate the passenger flow diversion results of the upper escalator congestion point measures and relieve the pressure of the upper escalator passenger

flow, there is normal passenger flow fluctuation after the new passenger flow, which is within the range of traffic capacity, and the diversion measures are successful.

4) This point is similar to the escalator point on the upper part of Line 1, and both are important distribution points for transfer passenger flow. However, when the passenger flow is divided from the opposite passenger flow under the condition of deviation to ensure the transfer of Line 1 to Line 3, the passage increases and the arrival efficiency improves significantly.

5) The passenger flow of escalator transfer has increased to a certain extent, indicating that after the use of iron barrier to divide the passenger flow and delay the impact time of large passenger flow, several conflict points/congestion points near some main escalators/stairs of Line 1 have been alleviated, and the efficiency of passenger flow arrival has been improved.

4.2.3 Thermodynamic Diagram Analysis

Comparison with previous, although there are still some conflict points and congestion points after the diversion, there is no large congestion after the optimization of the diversion. The operation of the station during peak hours is guaranteed.

For specific analysis:

The main objectives of the paper are two key points, namely, the upper escalator of Line 1 and the left escalator of Line 3. The specific objectives are to ensure the basic traffic capacity of the upper escalator of Line 1 and improve the traffic efficiency of the left escalator of Line 3 as a whole. According to the thermal diagram, the passenger flow aggregation form of the upper escalator of Line 1 is no longer a block aggregation centered on a conflict point, but a strip aggregation with a certain queuing form. As a diversion measure, the passenger flow to the escalator in the middle of Line 1 also increases orderly. The optimization effect of this point is achieved.

In the passage between the two columns in front of the escalator on the left of Line 3, the blocky passenger flow gathered and disappeared, which was expressed as a strip queue. There were also passengers traveling in the free space at the upper part of the station, no longer wasting space, and the optimization effect was achieved.

5 Conclusion

According to the analysis of simulation model operation data, it can draw the following conclusions.

1) This paper establishes an agent-based simulation model of Jinhu Square transfer station. Integrate the physical configuration of the station, passenger flow, social force performance of passenger flow, train inbound operation and other factors into one system, and conduct unified parameter adjustment. The simulation model can be quickly modified under different passenger flow conditions to adapt to different scheme comparison and selection, and facilitate the adjustment of optimization measures

2) The paper uses the actual collected passenger flow data, through the use of agent-based Anylogic software for simulation, uses the thermal diagram and port flow monitoring to determine the location of passenger flow congestion points and conflict points. The simulation results show that under the operation of the social force model, the passenger flow performance selects the shortest path, which will lead to a large number of conflict points under this condition. This logical choice also causes congestion at some corners. Therefore, certain measures are needed to force the passenger flow to follow the established route to avoid unnecessary conflict points

3) The passenger flow transfer optimization mainly uses measures such as passenger flow organization and facility separation to control the passenger flow density at the bottleneck at a low level. However, through simulation, it is found that measures such as separation of organization and facilities can only disperse passengers to various points and then merge them in a staggered or delayed manner to reduce the impact on facilities in a short time, so as to reduce the density of passenger flow at or within the congestion points. The number of congestion points is reduced, but the main conflict points still exist. The passenger flow at the transfer channel is still saturated at peak hours, and the transfer time of passengers is only slightly increased. When transferring large passenger flow, other auxiliary means should also be used to help improve

4) The simulation model is built by using the Anylogic software, and the passenger flow parameters are substituted and sampled. The fuzzy station operation can be converted into specific visual data, and these data can be used as the station service level evaluation, which is conducive to the improvement of station facilities and management

Acknowledgment. This project is supported by the second batch of school-level teaching team in 2019 "Applied Effective Teaching Design Teaching Team" (2019XJJXTD10) and Nanning University's second batch of teaching reform project of specialized innovation integration curriculum (2020XJZC04).

References

1. Zhu, W.: Analysis of high-speed railway passengers' queuing at the station based on anylogic simulation. Urban Rail Transit Res. **7**, 133–137 (2020). (in Chinese)
2. Liu, W., Wang, F., Zhang, C., et al.: A simulation study of urban public transport transfer station based on anylogic. KSII Trans. Internet Inform. Syst. (TIIS) **15**(4), 1216–1231 (2021)
3. Zhao, M.: Research on Passenger Flow Control Method and Simulation of Urban Rail Transit Transfer Station, pp. 5–6. Southwest Jiaotong University, Chengdu (2020). (in Chinese)
4. He, Y.: Research on the Design of Evacuation Stairs in Underground Commercial Buildings Based on Personnel Evacuation Simulation, pp. 10–12. Chongqing University, Chongqing (2018). (in Chinese)
5. Zhan, Y.: Real Time Task Evaluation of General Aviation Emergency Medical Transfer Based on Hybrid Simulation, p. 8. Nanjing University of Aeronautics and Astronautics, Nanjing (2019). (in Chinese)
6. He, Y.: Control of Transfer Passenger Flow in High-Speed Rail Metro Hub, p. 10. Southwest Jiaotong University, Chengdu (2018). (in Chinese)

7. Shi, H, Yang, Z.: Research on streamline organization optimization of urban rail transit transfer station based on anylogic. Comp. Transport. **42**(06): 69–75+81 (2020). (in Chinese)
8. Zhao, J., Tyler, D.C.: Quantifying the impact of classification track length constraints on railway gravity hump marshalling yard performance with anylogic simulation. Int. J. Comput. Methods Exp. Meas. **10**(4), 345–358 (2022)
9. Liu, Y., Song, Y.: Research on simulation and optimization of road traffic flow based on anylogic. E3S Web Conf. **360**, 01070 (2022)
10. Yang, Y., Chen, J., Du, Z.: Analysis of the passenger flow transfer capacity of a bus-subway transfer hub in an urban multi-mode transportation network. Sustainability **12**(6), 2435 (2020)
11. Omri, A., Omri, M.N.: Towards an efficient big data indexing approach under an uncertain environment. Int. J. Intell. Syst. Appl. (IJISA) **14**(2), 1–13 (2022)
12. Aliyev, A.G.: Technologies ensuring the sustainability of information security of the formation of the digital economy and their perspective development directions. Int. J. Inform. Eng. Electr. Bus. (IJIEEB) **14**(5), 1–14 (2022)
13. Saida, A., Yadav, R.K.: Review on: analysis of an IoT based blockchain technology. Int. J. Educ. Manag. Eng. (IJEME) **12**(2), 30–37 (2022)
14. Samiul Islam, M., Islam, F., Ahsan Habib, M.: Feasibility analysis and simulation of the solar photovoltaic rooftop system using PVsyst software. Int. J. Educ. Manag. Eng. (IJEME) **2**(6), 21–32 (2022)

Engine Speed Measurement and Control System Design Based on LabVIEW

Chengwei Ju[1], Geng E. Zhang[2,3(✉)], and Rengshang Su[1]

[1] Guangxi Special Equipment Inspection and Research Institute, Nanning 530219, China
[2] College of Intelligent Manufacturing, Nanning University, Nanning 530000, China
78704108@qq.com
[3] Faculty of Engineering, University Malaysia Sabah, 88400 Kota Kinabalu, Sabah, Malaysia

Abstract. The engine speed signal is the main control signal of the engine electronic control fuel injection system, which plays a crucial role in the engine power, fuel economy and other performance. In this paper, the principle of engine speed measurement is analyzed, and a new engine speed measurement and control system is designed. The system is based on LabVIEW virtual instrument, using NI USB-6216 data acquisition card as the hardware foundation, collecting the speed signal output by magnetoelectric speed sensor, and using LabVIEW software to calculate and process the measured signal, then the actual speed is obtained. The experimental results show that the system has the characteristics of stable performance, clear data display, high real-time performance, high sensitivity, good precision, simple operation, simple program and so on. It plays a good role in monitoring the engine speed, and is conducive to improving the engine power and fuel economy. This system also provides a reference for other vehicles to design speed measurement and control system. In addition, man-machine interface operation is obvious humanized.

Keywords: LabVIEW · Speed sensor · Measurement and control system

1 Instruction

With the development of automobile industry, automobile performance has attracted more and more people's attention, among which engine speed is an important indicator reflecting automobile power and economy [1]. RPM measurement technology has always been the basis of engine measurement technology, with the continuous efforts of predecessors, a variety of measurement methods have appeared [2]. However, there are some defects in the practical application of these methods. For example, most of the methods require the measurement sensor to be installed on the crankshaft, these methods are direct and effective, but they also bring a lot of inconvenience to engine manufacturing and maintenance [1]. In order to monitor the engine speed signal, many researches have been carried out on the monitoring of engine speed signal in the domestic and overseas. The sensors usually used to measure the rotate speed include Hall type, magnetoelectric type, capacitive type, photoelectric type and photoelectric encoder [3, 4],

---

© The Author(s), under exclusive license to Springer Nature Switzerland AG 2023
Z. Hu et al. (Eds.): ICAILE 2023, LNDECT 180, pp. 79–89, 2023.
https://doi.org/10.1007/978-3-031-36115-9_8

and the corresponding measurement and control systems are also various. Measurement and control systems has the following kinds: C8051F340 MCU as the control center, the use of Fourier transform to realize the system signal analysis and processing, in order to measure the real-time rotate speed; With DSPF2812 as the control core, LabVIEW software realizes signal monitoring and data storage; With the embedded system as the core, the rotate speed value of the starter is calculated indirectly by analyzing the frequency information of vibration [2, 5, 6]. The design results of each system have achieved very good results in practical tests, with relatively high measurement accuracy [7–9].

For the above problems, according to the characteristics of the commonly used car magnetoelectric speed sensor design engine speed measurement and control system based on virtual instrument technology, in the case of without changing the hardware can add new features to the equipment, such not only can save expensive hardware, but also can further improve and perfect the automation of testing instruments. This design takes the engine speed measurement as the goal, LabVIEW as the software development platform, modularity as the design idea, chooses the Baojun 510 vehicle to do the experiment, the magnetoelectric crankshaft position sensor and PFI-6216 data acquisition card as the hardware basis, and finally realizes the real-time data acquisition, display and processing.

The engine speed measurement and control system based on LabVIEW mainly take ordinary computers as the technology foundation. Software and hardware have adopted the modular design concept and ideas, data collection, display, storage and computing information processing are all realized through the computer. Such a design greatly improves efficiency, makes data processing more flexible, and has more powerful application functions. It not only has friendly interface but also has simple user operation, takes into account convenience in security and maintenance, has easy expansion in requirements, and significantly reduces manufacturing costs [2, 10].

2 Hardware System Design

In order to complete the data acquisition and calculation of engine speed measurement and control system, a complete hardware system is needed. The hardware of the measurement and control system mainly consists of the following three parts:

1) Magnetoelectric speed sensor, which is mainly used to convert various physical signals into electronic pulse signals that can be received by the data acquisition card [11].
2) Data set card, which is mainly used to connect the computer with external equipment and working environment, automatically collect non-power or power signal from the analog and digital unit of the sensor and other equipment to be tested, and send it to the host computer for analysis and processing.
3) Auxiliary equipment, including computers, keyboards, monitors, etc. They are used to realize online monitoring and online modification of parameters of measurement and control system, data storage, etc. [12].

2.1 Rotate Speed Sensor

The current rotate speed sensor types include Hall rotate speed sensor, photoelectric rotate speed sensor and magnetoelectric rotate speed sensor. At present, magnetoelectric rotate speed sensor is the most used rotate speed sensor, and its structure is shown in Fig. 1. It is composed of permanent magnet, coil, disk and so on.

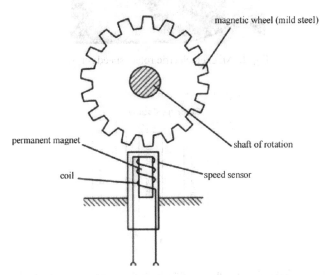

Fig. 1. Measurement principle of magnetoelectric rotate speed sensor

When the rotating shaft rotates, the induction coil with magnetic circuit passing through will induce a pulse potential of a certain amplitude when the magnetic flux changes abruptly. Its frequency is:

$$f = \frac{z * n}{60} \qquad (1)$$

In the formula, z is the number of teeth of the magnetic wheel, n is the number of revolutions of the magnetic wheel in r/min.

The crankshaft position sensor used in the experiment is electromagnetic pulse signal sensor, which is installed on the flywheel. The pulse signal is used to sense the position of the missing two teeth of the 58 teeth of the crankshaft timing disk to determine the speed of the engine and the relative position of the piston when the crankshaft rotates. The engine control unit uses the information from the crankshaft position sensor to generate timing ignition signals and injection pulses, which are then sent to the ignition coil and fuel injector, respectively. The actual rotate speed sensor of the experimental vehicle is shown in Fig. 2.

The sensor is three-wire type, including one signal anti-interference shielding wire, one signal ground wire and one sensor signal wire. The connection between the sensor of the experimental vehicle and the car computer is shown in Fig. 3.

Fig. 2. Magnetoelectric rotate speed sensor

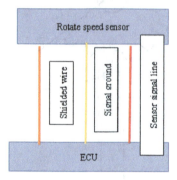

Fig. 3. Connection diagram of magnetoelectric speed sensor and ECU

Fig. 4. NI USB 6216 Data acquisition card

2.2 Data Acquisition Card

Rotate speed sensor measurement and control system requirements: more measurement parameters, high sampling rate, strong compatibility requirements, digital input and output [13]. According to the design requirements, the USB-6216 model data acquisition card is selected, and the communication interface of the data acquisition card is selected, that is, the interface mode. The communication mode between data acquisition card and computer mainly includes serial port (485 module, RS232), parallel port (PCI interface), USB and Ethernet. USB data acquisition card is characterized by plug and play, support hot swap, easy to carry. It is transmitted via USB bus data pass. USB is an external bus standard, conveniently used to regulate the connection and communication between computers and external devices. The appearance of the data acquisition card is shown in Fig. 4, and the main performance indicators of the card are shown in Table 1.

Table 1. Data acquisition card parameters

Sampling frequency	400 k/s	Digital input/output	24 digital I/O lines (5V TTL signal)
Input range	± 0.2V, ± 10 V	Analog output	Two-way
Input accuracy	16-bit	Output accuracy	16-bit
Analog input	16	Counter/timer	2

3 Software System Design

The software design is mainly composed of LabVIEW block diagram, front panel design, data acquisition card and communication module. Data acquisition is the core part, its principle is to first sense the physical signal to be collected by the sensor, and then transmit the sensed signal to the converter, which converts the physical signal into the voltage signal that can be collected by the acquisition card, and then regulate and transmit it to the acquisition card. After the acquisition card goes through the process of amplification, sampling preservation, A/D conversion, etc., LabVIEW program reads the data through the DAQ acquisition channel, filters and calculates through function controls, and the computer displays the collected signals on the front panel of the system after software programming of the virtual instrument [14].

3.1 Program Block Diagram Design

The program of the detection system includes three modules: data acquisition, signal filtering, data calculation and display [15].

1) Data acquisition module

2) It is composed of creating array virtual channels (functions). The sensor pulse signal is collected by DAQ data acquisition channel, and the collected pulse frequency is processed by array size (functions). The corresponding block diagram program design is shown in Fig. 5.
3) Signal filtering module
   The high and low frequency signals of the sensor are sorted according to the value, and the average value is calculated according to the highest and lowest frequency continuously. The sensor with interference signals is effectively filtered. The corresponding block diagram programming is shown in Fig. 6.

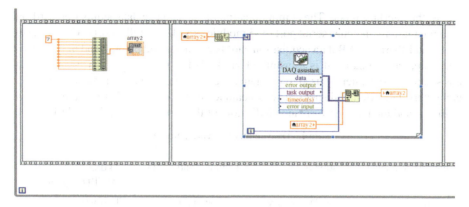

Fig. 5. Data acquisition module

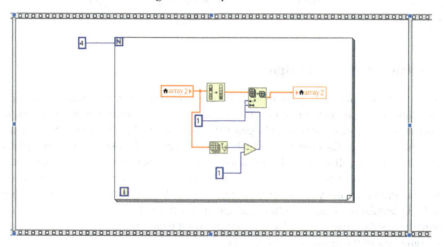

Fig. 6. Signal filtering module

4) Parameter setting module

The sampling period can be set in this module. The quotient of input is calculated, and then the total frequency collected is divided by the number of pulses generated during a rotation of the crankshaft.

Parameters can be flexibly set according to the different pulse numbers of a rotation of different sensors for convenient measurement, with good compatibility, as shown in Fig. 7.

5) Measurement and display module

By creating the pulse frequency signal collected in the virtual channel, and manually configuring the function to output the specified type of data, so that when wiring an array to the function, this function can be automatically resized to display the index input of each dimension in an N-dimensional array, or by adjusting the node

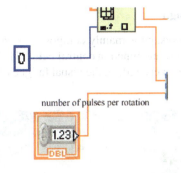

Fig. 7. Parameter setting module

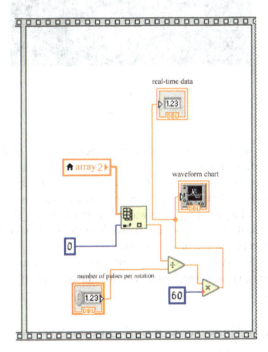

Fig. 8. Measurement and display module

size, adding elements or subarrays, the wire board displays the default data type of the polymorphic function [16]. Each time data is entered from the wire, LabVIEW feeds the data from the wire into a buffer and draws a graph with all the data in the buffer. When the second data input, the input data will clear the data left in the buffer last time, and then send the data into the buffer for plotting. The data will be displayed in real time through the shape chart and the value display control, as shown in Fig. 8.

3.2 Front Panel Design

The front panel of the system is mainly composed of waveform display window, real-time data display window and function control button [17]. The background beautifying is filled with green, which can reduce the visual fatigue of operators, as shown in Fig. 9.

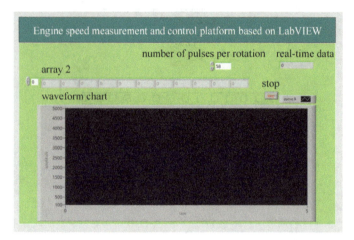

Fig. 9. Front panel

The data display window is arranged on the upper right side of the user interface, so that users can observe the changes of relevant data in real time at a glance. The parameter setting window is placed in the middle position to facilitate parameter setting. The waveform display window is arranged below, which can correspond to the real-time data and observe the changes of the data waveform in real time. The control switch is placed at the top of the data display window, and the eye-catching red button is adopted to facilitate data collection while observing data.

In order to make the system easy to operate and observe, every parameter setting window, waveform display window, function button and other functions are marked with Chinese characters.

4 Experiment of Measurement and Control System

4.1 Experimental Conditions

In the test, a Baojun 510 car was selected for the test, and it was found that the number of teeth of the reluctance ring of the crankshaft position sensor was 58 teeth, which was converted into a pulse number of 58 during one turn of the crankshaft rotation. Before the test of the measurement and control system, it was necessary to know the pulse number of one turn of the crankshaft rotation, and the pulse signal of the engine speed was measured by the data acquisition system during the test. The accurate real-time speed can be obtained by calculating the pulse times of one rotation. The system can adapt to the speed measurement of multiple different frequencies.

4.2 Experimental Procedure

1) Start the vehicle and check whether the experimental vehicle runs normally;
2) Open the computer, start the test program, data acquisition card USB interface connected to the computer;
3) Connect the crankshaft position sensor signal wire and earth wire to the USB - 6216 type of data acquisition card;
4) Select the appropriate measurement channel in the software program;
5) Slowly increase the speed from the idle state, gradually increase the speed to 5000 RPM from 500 to one unit, record the measured speed value, stop the test, and the engine stops running. Click the "Stop" button.

The operation display of the measurement and control system in the experiment is shown in Fig. 10.

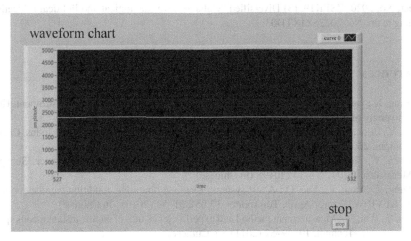

Fig. 10. Display of running pulse signals in the experiment

4.3 Analysis of Experimental Results

It can be seen from the experimental operation that the output pulse signal of the engine revolution sensor is stable. After comparing and analyzing the data of the engine instrument speed record and the measurement and control system, the measured real-time data is basically consistent with the actual speed, which accords with the working principle of the system.

5 Conclusion

This paper analyzes the principle of engine speed test, selected the sensor, data acquisition card, calibrate the sensors, designed the front panel and program speed test system with LabVIEW software, and completed the measurement and control system real vehicle experiment. Through the engine speed test results show that the system is stable performance, high sensitivity, good accuracy, simple operation, simple, program, to achieve the requirements of the design target.

In this paper, there are still many deficiencies in the further research and design of the speed measurement and control system to be improved, such as the multi-functional software measurement and control system, beautiful and simplified operation interface, etc. The experiment in this paper is done with the whole vehicle, the observed accuracy of the automobile instrument speed is 100r/min, this is in order to better reflect the accuracy of the measurement and control system, which can be tested on an engine bench with a digital tachometer.

Acknowledgment. This paper is supported by: (1) Sub-project of Construction of China-ASEAN International Joint Laboratory for Comprehensive Transportation (Phase I), No. GuiKeAA21077011–7(2) Automobile Electrical Appliances, The second batch of Nanning University "Curriculum Ideological and Political Education" demonstration course construction Project, No. 2020SZSFK19. (3) Diversified "Collaborative Ideological and Political Education" teaching team, No. 2022SZJXTD03.

References

1. Yang, X.: Development of Portable Automobile Engine Speed Measuring Instrument. Hefei University of Technology, Hefei (2014). (in Chinese)
2. Li, Y., Zhou, L.: Research on automobile engine speed measurement system. Indus. Control Comput. **23**(04), 34–35 (2010). (in Chinese)
3. Chen, Y., Zhao, Y., Dai, J.: Motor vehicle engine tachometer calibration device. Shanghai Meas. Test **46**(04), 11–14 (2019). (in Chinese)
4. Yang, C., Yang, L., Yao, Q.: A test system of vehicle operating condition parameters based on labVIEW. Tractor Agric. Transporter **37**(05), 51–54 (2010). (in Chinese)
5. Xu, S., Wei, M.: Measurement method and experimental study of engine instantaneous speed. Automobile Technol. **10**, 49–52 (2011). (in Chinese)
6. Song, G.: Design of speed measurement system based on 89C51. J. Weifang Univ. **6**, 32–34 (2008). (in Chinese)
7. Huang, X.: Research and Development of engine Data Acquisition and Display System Based on LabVIEW. Tianjin Vocational and Technical Normal University (2017). (in Chinese)
8. Zhang, X.: Research and analysis of instantaneous speed measurement of vibration and noise engine. Sci. Technol. Innov. Appl. (01), 16 (2017). (in Chinese)
9. Cui, H., Zhu, L.: Research on automobile engine speed measurement and single cylinder power balance test. Sci. Technol. Innov. (07), 9–10 (2019). (in Chinese)
10. Wang, X.: Research on virtual instantaneous speed test system of engine based on LabVIEW. Agr. Equipment Veh. Eng. **51**(11), 64–67 (2013). (in Chinese)
11. Xie, D., Chu, H.: Application analysis of magneto electric and hall type engine speed sensors. Intern. Combust. Engine Accessories (24), 60–61 (2020). (in Chinese)
12. Yue, Y., Zhao, Z.: Design of portable automobile engine speed measuring instrument. Instrum. Tech. Sensor (02), 39–42 (2019). (in Chinese)
13. Cheng, D., Jiang, W., Huang, Z., Zhang, K., Wang, W.: Design of measurement and control platform for engine speed sensor. J. Hubei Univ. Automotive Technol. **27**(04), 14–17+23 (2013). (in Chinese)
14. Li, H., Guo, L., Fu, H.: Design of engine speed measurement system based on single chip microcomputer. J. Yunnan Agric. Univ. (Nat. Sci.) **30**(02), 294–297 (2015). (in Chinese)

15. Patel, D.M., Shah, A.K.: LabView based control system design for water tank heater system. Trends Electr. Eng. **7**(3), 31–40 (2017)
16. Yang, H., Zhou, Q.: Pressure data acquisition and processing system based on LabVIEW and AVR microcontroller. Control Instrum. Chem. Ind. **37**(11), 92–94 (2010)
17. Bo, L., Liu, X., He, X.: Measurement system for wind turbines noises assessment based on LabVIEW. Measurement **44**(2), 445–453 (2011)

Research and Design of Personalized Learning Resources Precise Recommendation System Based on User Profile

Tingting Liang, Zhaomin Liang[✉], and Suzhen Qiu

College of Artificial Intelligence, Nanning University, Nanning 541699, China
minzaa2000@qq.com

Abstract. Online education is developing rapidly, and the scale of digital learning resources is increasing sharply. In the process of learning, learners are faced with relatively redundant resources and information drift, which easily leads to blindness and ineffectiveness in learning. For this reason, from the perspective of students' individualized learning needs, using artificial intelligence and data analysis and mining technology, the corresponding portraits are constructed according to students' basic information, online learning behavior, resource information, domain knowledge, etc. This paper puts forward a mixed recommendation mechanism for different learning stages, which takes the learner as the center, combines four methods of recommendation based on learning style, knowledge map, resource preference and behavior sequence of learning partners, establishes an online precise recommendation system for learning resources, matches the different learning needs of learners, provides precise recommendation-related services for learners and managers, and facilitates the personalized learning and growth of learners.

Keywords: User portrait · Learning resources · Recommended algorithm · Recommendation system · Personalized Recommendation

1 Introduction

In the large data environment, artificial intelligence, cloud computing technology and other rapid development, the field of education has gradually entered the intelligent stage. This puts forward new requirements for meeting the personalization of user learning, guaranteeing collaborative learning, and personalization of recommendation content [1]. Digital resources and platforms are the products of the deep integration of "Internet + Education". At the present stage, more emphasis is placed on the informationization and networking of learning resources, often ignoring the construction of the logical connection of related learning resources. A large number of learning resources are relatively scattered in different system platforms, and the resources of related learning content cannot form a synergy. This often results in the fragmentation and fragmentation of the resources that learners acquire. As a result, learners are prone to overload information or get lost in the process of learning. The learners' systematic learning and growth are facing many challenges.

© The Author(s), under exclusive license to Springer Nature Switzerland AG 2023
Z. Hu et al. (Eds.): ICAILE 2023, LNDECT 180, pp. 90–100, 2023.
https://doi.org/10.1007/978-3-031-36115-9_9

The way learners acquire learning resources has gradually changed from active retrieval to automatic recommendation by the learning system. In the past, digital learning resources were mainly pushed in three ways: Top-N recommendation, keyword query and the latest resource recommendation, which improved the hit rate of knowledge entity search and the efficiency of learning resources, but the accuracy of content-based recommendation is still insufficient. In addition, after the completion of the current stage of learning, there will be difficulties in the correct orientation of learning and systematic learning of courses.

In actual teaching, there are three types of online education in colleges and universities: MOOC platform, practical teaching platform and teaching process management platform. Table 1 is a simple survey of the platforms widely used in the three types of courses.

Table 1. Broadly used platforms and surveys

Type	Name	Description	Search results with "program design" as the keyword
MOOC Platform	China University MOOC	Undertake the task of national high-quality open courses of the Ministry of Education, and provide MOOC courses of well-known universities in China to the public	1517 related courses
MOOC Platform	Superstar Learning	It is the teaching software that many universities choose to cooperate with for online teaching	90 courses related to the demonstration teaching package
Practice teaching platform	EduCoder	It is the official cooperation platform of MOOC Alliance Practical Teaching Working Committee of China's University Computer Education	61 related practical courses
Teaching process management	Rain class	The teaching tools are skillfully integrated into PowerPoint and WeChat, and the hybrid teaching is simple and convenient	Other course resources cannot be retrieved

92   T. Liang et al.

Learners can retrieve multiple courses on the first two types of platforms. Each course contains videos, documents, pictures and other types of resources. Each course contains dozens or even hundreds of resources. The platform is rich in resources. Learners need to click into the curriculum to further search for resources. It is difficult to directly transfer big data in the teaching process between different courses. The sharing and utilization rate of teaching resources is low, and there is a lack of personalized learning resource recommendation services.

How to integrate learners' personalized characteristics, degree of knowledge construction, learning needs and scientific rationality of technology, incorporate some scientific, applicable and innovative technologies to improve the accuracy and efficiency of recommendation, realize learners' individualized acquisition of demand resources and professional systematic learning, in order to reduce learning blindness and improve the logic and relevance of learning process, is still a problem worth further discussion.

2   Research on Recommendation of Educational Resources

2.1   User Portrait

User portraits are an important tool for accurate service recommendation. User portraits are descriptions of the principal features and needs. They are authentic, dynamic, interactive and clustered, which can personalize services and achieve the purpose of successful recommendation.

In order to better achieve personalized recommendations for learners, user portraits are constructed from two aspects: user data collection and behavior data mining, combining learner characteristics. Wang Shu based on the "Chinese Bridge" learner data proves that using user data to build user portraits can improve the learner's experience in learning recommendation, which has practical significance [2]; L Qin By analyzing the user behavior of the resource platform and mining rules from it, combining with the user personality characteristics, dynamic and static data, it can improve the understanding of user behavior and provide personalized service quality [3]. S Shrestha uses the learner's interactive data to determine the learning style and then constructs a picture of the learner to support personalized teaching [4]. T Xian builds user portraits by analyzing logs to get users' acceptance and preferences for content [5]. Teaching designers design content based on portraits to improve the efficiency and effectiveness of the course.

2.2   Research on Personalized Recommendation of Educational Resources

By sorting out the relevant research on Learning Resource Recommendation in the field of education, it can be found that it is mainly divided into three levels: recommendation algorithm, recommendation model and recommendation system. These three are inseparable. Recommendation algorithm is the basis of the latter two, and its recommendation effect and evaluation are directly or indirectly affected by the latter two.

The three commonly used recommendation algorithms for learning resources in the field of education are content-based, collaborative filtering and knowledge-based recommendation algorithms. A single recommendation algorithm often fails to achieve a comprehensive and accurate recommendation for learning content, so a hybrid push of multiple recommendation algorithms is often used in both business and education fields. Multidimensional integration of technologies and algorithms based on tag recommendation, social network, clustering, neural network, in-depth learning, knowledge mapping, hybrid recommendation is a hot spot for scholars [6]. Combining the three technologies of association rule-based recommendation, content filtering recommendation and collaborative filtering recommendation, we can actively recommend resources and information suitable for users' learning needs according to their personalized characteristics, such as learning interests, preferences and learning needs.

Whether the construction of the recommendation model meets the requirements requires selecting the algorithm according to the recommendation mechanism. The construction of Learning Resource Recommendation Model and the design of recommendation strategy focus on the aspects of user attributes and user preference mining, learner characteristics analysis, intelligent recommendation and multivariate model building [8–10]. The recommendation of learning resources, learning partners and learning paths are merged from several dimensions such as recommendation algorithm, learning scenarios and recommendation content, which helps to promote learning resources through multiple channels.

The recommendation system can realize and test the rationality of the recommendation model and the accuracy of the recommendation algorithm. At present, most of the applications of learning resource recommendation system are mainly concentrated in the vertical subdivisions of Library materials, an online learning platform, a course, etc. Few studies are based on cross-domain and multi-source data [11]. It mainly considers the combination of multiple dimensions, such as algorithm recommendation strategy, recommendation content and learning real situation, to better improve the recommendation effect and the level of multiple learning services.

From the perspective of students' individualized learning needs, this paper collects resources construction and electronic imprint of students' learning, uses artificial intelligence and big data technology to analyze and portray learning style and preference data and form a picture of students, designs a recommendation system, realizes the orientation of learning direction and methods of personalized recommendation, and accurately recommends learning resources for them, which has a strong epochal and development value.

3 Key Links to Building a Learner Portrait Model

Student portrait technology is built on the basis of large educational data. Due to the multidimensional and cross-fusion of data information, in order to achieve the efficiency of resource utilization, it is necessary to overcome the problems of data redundancy, heterogeneity and missing [12]. Filter information according to different needs, and use cluster coupling analysis, deep mining and other related technologies to complete the analysis and processing of learner data information. The main steps of building a learner data model include data collection, analysis, presentation and application (evaluation, prediction, intervention). In data fusion, the learner portrait technology needs to obtain the learner's dynamic feature information at multiple levels, and use the multidimensional fusion technology to quantify all aspects of the information of the students, analyze and identify the information with objective data, ensure that all kinds of data complement each other, and achieve the fusion of heterogeneous data from multiple sources [13], so as to ensure the comprehensiveness and integrity of the learner's information. In terms of technical implementation, the natural language processing (NLP) algorithm is generally used to synthesize the various characteristics of the learner to form a full-cycle, full-process portrait data of the learner. Among them, learning behavior feature information is usually extracted through behavior labels, and the establishment of labels mainly through three methods: meta-analysis-based methods, statistics-based methods, and correlation-based analysis methods.

4 Precise Recommendation of Personalized Learning Resources Based on User Portraits

4.1 Technical Framework

The main way to realize the personalized learning resource recommendation service is to collect the specific behavior data generated by the learner using the learning interaction system during the improvement of the learner portrait model, compare the similar learner attribute characteristics, recommend the learning resources that may be needed, and use the Apriori association algorithm to find the frequent itemset and association rules [14]. The online teaching platform covers both learning resources and students' data information. By digging and analyzing these information in depth, it can effectively compare the learning and resource needs of the learners, and then realize the labeling and unit processing of learning resources. At the same time, it systematically compares the learner's characteristic data, extracts the learner's portrait data, and establishes the learner's tag to mark the specific individual attributes of the learner. Then, it digs deeply into the learning resources of interest to the learners and their groups, matches the data collection of the learners and the learning resources, establishes the association rules between the learner label system and the learning resources, improves the recommendation accuracy of the course resources by continuously optimizing the existing learning resources, and combines the collaborative filtering algorithm of the Association rules, so as to help the students obtain the learning resources better, thus improving the learning efficiency of the students. In the system, learners can further improve their own information, take the initiative to assess learning style, pre-test course knowledge, and so on. They can view their own level of development, and compare the level of professional needs, and achieve their own growth based on the personalized learning resources recommended by the system.

4.2 Recommendation Policies

The main basis for the accurate recommendation of personalized learning resources is the learner's portrait. Therefore, when constructing the learner's portrait, the learner's personal characteristics, interests, and dynamic information of learning activities should be fully acquired, the points of interest and learning direction should be analyzed and identified, combined with their existing knowledge level, the learner should be scientifically recommended for learning resources, and even the learning path planning. There are four main strategies for precise recommendation of personalized learning resources [15].

Recommendation based on learning style: When the learner completes the task of assessing learning style with good reliability, the four dimensions of Felder-Silverman model are used to divide the learning style of the learner, which are information processing, information perception, information input and information understanding, and then the system makes corresponding recommendation for learning resources according to the learner's learning style.

Recommendation based on knowledge map: first, build a learning resource classifier to classify resources. When new learning resources enter the resource pool, the classifier classifies them and puts them into the corresponding learning resource pool [16]. When making a recommendation, search the knowledge labels of the students' professional information or currently viewed resources in the feature library of the knowledge map, use the similarity algorithm to extract the knowledge points related to the resource knowledge labels, and then find the learning resources matching the recommended knowledge points in the learning resource database.

Recommendation based on resource preference: After the learner has a certain learning record, the weight of each kind of label value under the learner can be calculated by using $TF = IDF$ algorithm, thus the learner's resource preference can be obtained, and the system can recommend the corresponding type of resources according to the learner's resource preference [17].

Recommendation based on learning partner behavior sequence: record the sequence of learners' behaviors in the set and form a status sequence, use clustering method to calculate similar user (learning partner) groups, and then calculate the similarity of behavior sequence in the cluster space, including state value similarity, state transition similarity and state order similarity [18]. It then filters the learning resources used by the learning partners in their subsequent States and recommends them to the current users.

Because each recommendation method has its advantages and disadvantages and is suitable for a specific scenario, and the process of online learning is also a dynamic and changing process, it is not appropriate to consider only one recommendation method for learning resource recommendation. Therefore, a mixed recommendation mechanism is necessary.

5 Precision Recommendation System Architecture for Online Learning Resources

Aiming at precise recommendation of learning resources, the system is divided into four layers from bottom to top: data layer, data analysis layer, recommendation calculation layer, and human-computer interaction layer. The overall system architecture design is shown in Fig. 1.

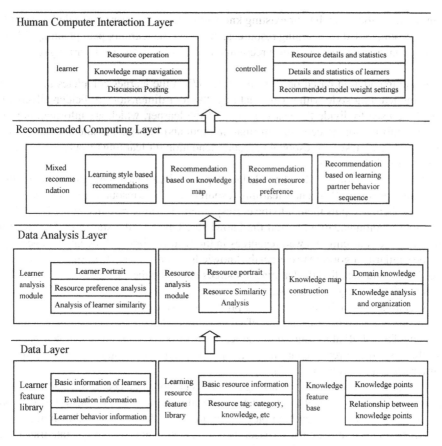

Fig. 1. Architecture of personalized learning resources precise recommendation system

5.1 Data Layer

The data layer is used for portrait modeling and portrays the characteristics of learners, resources and knowledge. It consists of three modules: the learner feature library, the learning resource feature library and the knowledge feature library.

5.1.1 Learner Feature Library

Stores the learner's characteristic information, including the basic information of the learner, the assessment information, and the learner's behavior information. Basic information is inherent to each learner and does not change over time, such as student number, gender, specialty, etc. The assessment information does not change much in a short time, but there are also situations in which the learners do not complete it carefully, resulting in large differences between the data before and after, such as the assessment of learning style, the pre-test of curriculum knowledge, etc. The characteristics of learners' behavior change significantly over time, such as browsing, collecting, commenting, downloading and posting resources.

5.1.2 Learning Resource Feature Base

It consists of the basic information of the resource and the label information of the resource. The basic information of the resource includes subject, resource type (such as video, audio, document, PPT, picture, etc.), storage location (local path, web address), reading amount, favorites, creator, source, language (Chinese, English), etc. Resource label information is a general description of the content and form of learning resources, including categories (such as theory, practice, scenario cases, etc.), owning knowledge (knowledge blocks or points), and other (such as formula derivation, programming language, commentary vocabulary).

5.1.3 Knowledge Feature Base

Is the representation of the knowledge map stored in a computer-understandable way, usually with a head, relation, tail tuple representing the data of the knowledge map. Includes knowledge points (head entities, tail entities) and the relationship between knowledge points (e.g., contained, contained, homologous).

5.2 Data Analysis Layer

This layer is mainly used to process the data of the data layer. It quantifies, statistics and models the three feature libraries of the data layer, corresponding to three modules: learner analysis, resource analysis and knowledge map building.

5.3 Learner Analysis Module

This includes student portraits, resource preference analysis and similarity analysis between learners.

The portrait of the learner describes the learning characteristics of the learner from various perspectives, and constructs the personalized portrait of the learner quantitatively and qualitatively [19]. It includes learning style, learning blocks, points of knowledge already learned, difficulties in knowledge, sequence of commonly used modules in learning, grade of achievement, etc. The learner can understand and master his own learning situation through portraits so as to adjust learning strategies.

Resource preference analysis, through quantitative and statistical learning style test data and learner behavior data, tends to choose the content and form of learning resources preference, preferences and other preferences. For example, users A like learning resources with video as the carrier, while users B prefer learning resources with theoretical type. The calculation formula of resource preference similarity is:

$$Sim\_Area(a, b) = con(X, Y) = \frac{N(X) \cap N(Y)}{\sqrt{N(X) \times N(Y)}} \tag{1}$$

where, $N(X)$ and $N(Y)$ are the number of keywords in the resource type vector browsed by users a and b.

The similarity analysis between learners is the basis for subsequent modeling of learning partner recommendation. The degree of correlation between learners is calculated based on the learner's characteristic information to determine the similarity between learners. Similar learners are referred to as "learning partners" [20]. Finally, the learning resources selected by "learning partners" can be recommended to the current user. The calculation formula of learner similarity is:

$$Sim\_Info(a, b) = \text{con}(A, B) = \frac{\sum_1^n (A_i \times B_i)}{\sqrt{\sum_1^n (A_i)^2} \times \sqrt{\sum_1^n (B_i)^2}} \tag{2}$$

where, A and B are the user attribute vectors of users a and b, and Ai and Bi are the attribute values of each dimension of the two users.

5.4 Resource Analysis Module Includes Resource Portrait and Resource Similarity Analysis

A resource portrait, which describes the content and form of a resource in a hierarchical and refined vocabulary (tags). The tags come from keywords added by the resource creator, keywords extracted from comments, and tags extracted from the tag library by applying semantic calculations.

Resource similarity analysis is the basis for subsequent learning resource recommendation modeling. Compute the degree of correlation of resources based on the tag characteristics of resources to determine the similarity between resources and recommend similar resources to current users.

The construction of knowledge map mainly aims at domain knowledge. The main steps are: defining domain category, extracting terminology, extracting entity and relationship, classifying entity and relationship types, organizing knowledge framework, and completing knowledge attributes.

5.4.1 Recommended Computing Layer

At the beginning of platform learning, learning resources are recommended using knowledge map-based or learning style-based methods, and the weights of these two methods on the reverse order of proposed resources are adjusted according to whether the learner has completed a reliable learning style test or not. With the increase of the learning time and interaction of the learners, the mining analysis of the learners' learning behavior is carried out. The collaborative filtering recommendation is mainly based on the sequence of learning partner behaviors and supplemented by the recommendation based on resource preferences.The recommended system calculation formula is:

$$\forall \in C, S^* = \arg\ max_{s \in S} u(c, s) \tag{3}$$

where, C is the collection of all users, S * is the final collection of recommended items, S is the collection of items to be recommended, and u (c, s) is the recommendation of item s to user c.

5.4.2 Human-Computer Interaction Layer

The application layer is the outermost layer of the platform and the interactive interface between the learner and the platform. In the application layer, the learner can browse the recommended resources, collect, download and browse the resources, select the corresponding resource classifications of knowledge elements according to the navigation of the knowledge map, and exchange discussions with teachers and peers to express their own views. Resource managers can view resource details and learner details, make intelligent assessments and scientific management based on statistical analysis results, and adjust the weight of objects in the recommended model.

6 Conclusion

Accurate recommendation system for personalized learning resources based on user portraits can help students' personalized learning and growth. This paper combs the characteristics of learners, resources and knowledge from the perspective of learners' personalized learning needs, and constructs a portrait. On this basis, in light of the different learning stages of learners, this paper combines four personalized recommendation strategies to design a mixed recommendation model and a precise recommendation system for personalized learning resources. By implementing the "personalized" and "accurate" recommendation of learning resources, we can serve the orientation and personalized learning of learners. With the rapid development of personalized recommendation system and user portrait theory, the recommendation of learning resources will receive more and more attention in the future online learning. In terms of the theoretical model, recommendation mechanism and practical application of the personalized recommendation system for learning resources, there are still many areas that need to be improved continuously and need to be improved and further explored. In the follow-up work, the recommendation system will also be further improved and optimized to make it have better recommendation effect.

Acknowledgment. This project is supported by the Basic Ability Improvement Project for Young and Middle-Aged Teachers in Guangxi Colleges and Universities (2021KY1804, 2021KY1800).

References

1. Gu, X., Li, S., Li, R.: International vision of artificial intelligence innovation application – Prospective progress and future education prospects of NSF Institute of Artificial Intelligence. China Distance Educ. (12), 1–9+76 (2021)
2. Wang, S.: Study on Recommendation of Personalized Learning Path for Portraits of Users of Chinese Bridge. Yunnan Normal University (2021)
3. Qiu, L., Leung, K.Y., Jun Hao, H.O., et al.: Understanding the psychological motives behind microblogging. Stud. Health Technol. Inform. **154**, 140–144 (2010)
4. Shrestha, S., Pokharel, M.: Determining learning style preferences of learners. J. Comput. Sci. Res. **3**(1), 33–43 (2021)
5. Xian, T.: Practice of sandbox game in higher education based on graphic and game programming environment. In: Stephanidis, C., Antona, M. (eds.) HCI International 2020 - Posters: 22nd International Conference, HCII 2020, Copenhagen, Denmark, July 19–24, 2020, Proceedings, Part II, pp. 356–364. Springer International Publishing, Cham (2020). https://doi.org/10.1007/978-3-030-50729-9_51

6. Yinli, S., Sun, Y.: Hot spots, trends and Inspirations of domestic digital learning resource recommendation algorithms. J. Yunnan Normal Univ. (Nat. Sci. Ed.) **42**(03), 60–66 (2022)
7. Zhu, H., Liu, Y., Tian, F., et al.: Across-curriculum video recommendation algorithm based on a video-associated knowledge map. IEEE Access **6**, 57562–5757 (2018)
8. Lifeng, H., Li, C.: User cold start recommendation model combining user attributes with project popularity. Comput. Sci. **48**(2), 114–120 (2021)
9. Mou, Z., Wufati: Study on recommendation of personalized learning resources based on Learner Model in e-book bag. Audio-visual Educ. Res. **36**(1), 69–76 (2015)
10. Lina, Y., Yonghong, W.: Contextualized intelligent recommendation for universal learning resources. Res. Audio-visual Educ. **35**(10), 103–109 (2014)
11. Peng, R.: Intelligent portrait construction system and application for college students based on multi-source data. Compu. Programm. Skills Maintenance (09): 165–168 (2022)
12. Wang, Y., Yang, L., Wu, J., et al.: Mining multi-source campus data: an empirical analysis of student portrait using clustering method. In: 2022 5th International Conference on Data Science and Information Technology (DSIT), pp. 01–06. IEEE (2022)
13. Zhang, L., Xie, Y., Xidao, L., et al.: Multi-source heterogeneous data fusion. In: 2018 International conference on artificial intelligence and big data (ICAIBD), pp. 47–51. IEEE (2018)
14. Wang, H., Fu, W.: Personalized learning resource recommendation method based on dynamic collaborative filtering. Mobile Netw. Appl. **26**, 473–487 (2021)
15. Almu, A., Ahmad, A., Roko, A., et al.: Incorporating preference Changes through users' input in collaborative filtering movie recommender system. Int. J. Inform. Technol. Comput. Sci. (IJITCS) **14**(4), 48–56 (2022)
16. Geng, X., Deng, T.: Research on intelligent recommendation model based on knowledge map. J. Phys.: Conf. Ser. IOP Publishing **1915**(3), 032006 (2021)
17. Qaiser, S., Ali, R.: Text mining: use of TF-IDF to examine the relevance of words to documents. Int. J. Comput. Appl. **181**(1), 25–29 (2018)
18. Chen, M., Yang, X.P., Liu, T.: A research on user behavior sequence analysis based on social networking service use-case model. Int. J. u-and e-Serv., Sci. Technol. **7**(2), 1–14 (2014)
19. Vandewaetere, M., Desmet, P., Clarebout, G.: The contribution of learner characteristics in the development of computer-based adaptive learning environments. Comput. Hum. Behav. **27**(1), 118–130 (2011)
20. Erkens, M., Bodemer, D.: Improving collaborative learning: guiding knowledge exchange through the provision of information about learning partners and learning contents. Comput. Educ. **128**, 452–472 (2019)

Mixed Parametric and Auto-oscillations at Nonlinear Parametric Excitation

Alishir A. Alifov[(✉)]

Mechanical Engineering Research Institute of the Russian Academy of Sciences, Moscow 101990, Russia
a.alifov@yandex.ru

Abstract. The development of the theory of oscillatory processes, taking into account the properties of the energy source, contributes to improving the accuracy of calculating real objects for various purposes, the optimal choice of the power of the energy source to save it. Mixed parametric and self-oscillations under nonlinear parametric excitation in the interaction of an oscillatory system with an energy source are considered. The solution of the equations of motion is constructed using the method of direct linearization of nonlinearity. The conditions of stability of stationary modes of motion are derived. To study the effect of nonlinear parametric excitation on the properties of vibrations, calculations were carried out, the results of which were compared with the results of linear excitation. The results obtained show the difference between the dynamics of systems with linear and nonlinear parametric excitations, which has a quantitative and qualitative character. At the same time, quantitative differences are more significant.

Keywords: nonlinearity · parametric oscillations · self-oscillations · direct linearization · limited excitation · energy source

1 Introduction

Human civilization is inexorably moving towards an environmental crisis on a global scale, which, along with other factors, is promoted by the growth in the amount of energy consumed, and its reduction can make a certain contribution to solving environmental problems. In this context, the systematic theory of oscillatory systems with energy sources, created by V.O. Kononenko [1] and further developed by his followers [2–7, etc.], has become relevant again. It served as the basis for the emergence of a new direction in the theory of oscillations. In the work [8], the connection of environmental problems with metrology, etc. is noted.

During the operation of a number of objects for various purposes, parametrically excited oscillations occur. Among them, one can indicate, for example, pendulums, rods, helical springs, rotating shafts, measuring instruments, cardan transmission, drive systems, gears, railway bridges, ship masts, etc. Therefore, the study of parametric oscillations is important for solving applied problems of solid mechanics, theory of mechanisms and machines, structural mechanics, electrodynamics, optics (in particular,

© The Author(s), under exclusive license to Springer Nature Switzerland AG 2023
Z. Hu et al. (Eds.): ICAILE 2023, LNDECT 180, pp. 101–108, 2023.
https://doi.org/10.1007/978-3-031-36115-9_10

102   A. A. Alifov

lasers), etc. In this context, we note the monograph [9], in which parametric oscillations with linear and nonlinear parametric excitation and an ideal energy source are considered. Oscillations under nonlinear parametric excitation, taking into account the properties of the energy source, are considered only in [10], at least, other works are unknown to the author. According to the classification described in [2], in [10] the parametric type of oscillations was studied, and below the class of mixed (interacting types) parametric and self-oscillations is considered - a self-oscillating system with nonlinear parametric excitation and an energy source of limited power. The purpose of this work is to develop the theory of mixed oscillatory processes taking into account the properties of an energy source for calculating real objects for various purposes.

As is known, a number of approximate methods (averaging, energy balance, harmonic linearization, etc.) of nonlinear mechanics are used to solve nonlinear equations [11–17, etc.]. They are characterized by labor intensity, increasing with increasing degree of nonlinearity, and the presence of this problem for the analysis of coupled oscillatory networks is indicated in [18]. Among them, the asymptotic method of averaging has received wide application [12]. The direct linearization method (DLM) differs significantly from them, which is characterized by rather small labor and time costs and ease of use. Such features are very important in the design and calculation of real technical devices. And also, according to the DLM, one can obtain the final calculated ratios, regardless of the specific type of nonlinear characteristic [20–26, etc.]. For example, if we take y^n, then according to known methods it is impossible to obtain any final calculated ratios for an unknown value of n, a specific value of $n = 2,3$, is necessary… This drawback is absent in the DLM. Comparison of DLM and known methods is given in [20, 24, etc.]. Therefore, using it, solutions of the nonlinear equations considered in this paper are constructed.

2   System Model and Equations of Motion

Let us consider a model of a mechanical self-oscillatory system, which is widely used to analyze self-oscillations in a mechanical system [1–3, 11, etc.]. It well describes self-oscillations under the action of friction, which can occur during the operation of various engineering objects (textile equipment, guides for metal-cutting machines, brakes, etc. [27, 28, etc.]). Taking into account the nonlinear parametric excitation $bx^3 \cos vt$, $b = const$, $v = const$, the equations of motion of the system are:

$$m\ddot{x} + k_0\dot{x} + c_0 x = T(U) - F(x) - bx^3 \cos vt$$
$$J\ddot{\varphi} = M(\dot{\varphi}) - r_0 T(U) \tag{1}$$

where $c_0 = const$ and $k_0 = const$ are, respectively, the spring stiffness and damper resistance coefficients, $T(U)$ is the self-oscillating friction force, $U = V - \dot{x}$, $V = r_0\dot{\varphi}$, $r_0 = const$, $F(x)$ is the nonlinear part of elasticity, $J = const$ is the total moment of inertia of the rotating parts, $M(\dot{\varphi})$ is the torque of the energy source, $\dot{\varphi}$ is the rotation speed.

In practice, the friction force $T(U)$ is widely distributed in a non-linear form

$$T(U) = T_0 (\text{sgn} U - \alpha_1 U + \alpha_3 U^3) \tag{2}$$

which was also observed under space conditions [29].

Here α_1, α_3 are positive constants, T_0 is the normal reaction force, $\mathrm{sgn}U = 1$ for $U > 0$ and $\mathrm{sgn}U = -1$ for $U < 0$. In the case of relative rest $U = 0$, $-T_0 \leq T(0) \leq T_0$ takes place.

Let's nonlinear elasticity in the form of a polynomial $F(x) = \sum_s \gamma_s\, x^s$, $s = 2,3,4,\ldots$.

Using DLM, we replace this polynomial with a linear function

$$F_*(x) = B_F + k_F\, x \tag{3}$$

where B_F and k_F are the linearization coefficients defined by the expressions

$$B_F = \sum_s N_s\, \gamma_s a^s, \qquad s = 2, 4, 6, \ldots\ (s \text{ is even number})$$

$$k_F = \sum_s \overline{N}_s\, \gamma_s a^{s-1}, \qquad s = 3, 5, 7, \ldots\ (s \text{ is odd number})$$

Here $a = \max |x|$, $N_s = (2i + 1)/(2i + 1 + s)$, $\overline{N}_s = (2i + 3)/(2i + 2 + s)$, i is a linearization accuracy parameter.

We also use DLM for the friction force $T(U)$, part of which we represent in the form

$$T_*(\dot{x}) = B_T + k_T\, \dot{x} \tag{4}$$

where

$$B_T = -\alpha_1 V + \alpha_3 V^3 + 3\alpha_3 N_2 V\, \upsilon^2, \quad k_T = \alpha_1 - 3\alpha_3 V^2 - \alpha_3 \overline{N}_3 \upsilon^2, \quad \upsilon = \max |\dot{x}|$$
$$N_2 = (2i + 1)/(2i + 3), \quad \overline{N}_3 = (2i + 3)/(2i + 5)$$

Equations (1) taking into account (3) and (4) take the form

$$\begin{aligned} m\ddot{x} + k\,\dot{x} + cx &= B + T_0\, \mathrm{sgn}U - b\, x^3 \cos \nu t \\ J\ddot{\varphi} &= M(\dot{\varphi}) - r_0 T_0(\mathrm{sgn}U + B_T + k_T\,\dot{x}) \end{aligned} \tag{5}$$

where

$$c = c_0 + k_F, \quad k = k_0 - T_0 k_T, \quad B = T_0 B_T - B_F$$

3 Construction of Solutions of Equations

The solution of the linearized DLM equation can be constructed by two methods [20], one of which is *the method of replacing variables with averaging*. Of these, the method of replacing variables with averaging makes it possible to study stationary and non-stationary processes. We use the *standard form relations* obtained for *a linearized nonlinear equation of a fairly general form*. One of the special cases of such a general type of equation is the first (5) and the standard form allows you to write out its solution. And to solve the second (5), we apply the averaging procedure [22], in which the velocity $\dot{\varphi}$ is replaced by its averaged value Ω. The sign of the relative velocity U in the characteristic of the friction force $T(U)$, which can be either positive or negative, determines the nature of the oscillations [2]. Therefore, we will consider these two cases separately.

104   A. A. Alifov

As a rule, the main interest in the analysis of oscillation dynamics is the main resonance, which is called the main parametric resonance under parametric influence. The solutions in this case have the form

$$x = a \cos \psi, \quad \dot{x} = -ap \sin \psi, \quad \psi = pt + \xi, \quad p = v/2, \quad \dot{\varphi} = \Omega \tag{6}$$

and in this regard, expressions (4) will have $\upsilon = ap$, and instead of V, the expression $u = r_0 \Omega$.

Using the noted standard form of the DLM relations, we have the following equations for (5) in the case of $u \geq ap$:

$$\frac{da}{dt} = -\frac{ak}{2m} + \frac{ba^3}{8pm} \sin 2\xi$$

$$\frac{d\xi}{dt} = \frac{\omega^2 - p^2}{2p} + \frac{ba^2}{4pm} \cos 2\xi \tag{7'a}$$

$$\frac{du}{dt} = \frac{r_0}{J} \left[M\left(\frac{u}{r_0}\right) - r_0 T_0 (1 + B_T) \right]$$

At speeds of $u < ap$, the second equation (phases) in (7'a) remains unchanged, and the other two have the form

$$\frac{da}{dt} = -\frac{a}{2m} \left[k + \frac{4T_0}{\pi a^2 p^2} \sqrt{a^2 p^2 - u^2} \right] + \frac{ba^3}{8pm} \sin 2\xi \tag{7'b}$$

$$\frac{du}{dt} = \frac{r_0}{J} \left[M\left(\frac{u}{r_0}\right) - r_0 T_0 (1 + B_T) - \frac{r_0 T_0}{\pi} (3\pi - 2\psi_*) \right]$$

where $\omega^2 = \omega_0^2 + (k_f/m)$, $\omega_0^2 = c_0/m$, $\psi_* = 2\pi - \arcsin(u/ap)$.

The frequency difference $\omega_0 - p$ in the resonance region is quite small, which is why it is possible to accept $(\omega_0^2 - p^2)/2p \approx \omega_0 - p$.

The conditions $\dot{a} = 0$, $\dot{\xi} = 0$ and $\dot{u} = 0$ give relations for determining the stationary values of the amplitude, phase of oscillations and velocity of the energy source. At $u \geq ap$, these ratios are as follows:

$$16p^2 k^2 + 4m^2(\omega^2 - p^2)^2 = b^2 a^4$$

$$tg 2\xi = -\frac{4pk}{2m\ (\omega^2 - p^2)} \tag{8}$$

$$M(u/r_0) - S(u) = 0, \quad S(u) = r_0 T_0 (1 + B_T)$$

where $S(u)$ represents the load on the energy source, taking into account that in the expression B_T in (4), instead of V, there will be $u = r_0 \Omega$.

In the case of $u < ap$ and small external forces, an approximate equality $ap \approx u$ takes place for the magnitude of the stationary oscillation amplitude, and the load $S(u)$ on its basis can be determined by the expression

$$S(u) \approx r_0 T_0 \left[1 - \alpha_1 u + \alpha_3 (1 + 3N_2) u^3 \right] \tag{9}$$

The stationary values of the velocity u are determined by the intersection points of the curves $M(u/r_0)$ and $S(u)$.

Mixed Parametric and Auto-oscillations   105

4 Stability of Stationary Movements

Having composed the equations in variations for (7'a) and (7'b) and using the Routh-Hurwitz criteria, we have the following conditions for the stability of stationary motions

$$D_1 > 0, \quad D_3 > 0, \quad D_1 D_2 - D_3 > 0, \tag{10}$$

where

$$D_1 = -(b_{11} + b_{22} + b_{33}), \quad D_2 = b_{11}b_{33} + b_{11}b_{22} + b_{22}b_{33} - b_{12}b_{21} - b_{23}b_{32} - b_{13}b_{31}$$
$$D_3 = b_{11}b_{23}b_{32} + b_{12}b_{21}b_{33} - b_{11}b_{22}b_{33} - b_{12}b_{23}b_{31} - b_{13}b_{21}b_{32}$$

For $u \geq ap$ speeds we have

$$b_{11} = \frac{r_0}{J}(Q - r_0 T_0 \frac{\partial B_T}{\partial u}), \quad b_{12} = -\frac{r_0^2 T_0}{J}\frac{\partial B_T}{\partial a}, \quad b_{13} = 0$$

$$b_{21} = \frac{T_0 a}{2m}\frac{\partial k_T}{\partial u}, \quad b_{22} = -\frac{1}{2m}(k - aT_0\frac{\partial k_T}{\partial a}) + \frac{3ba^2}{8pm}\sin 2\xi$$

$$b_{23} = \frac{ba^3}{4pm}\cos 2\xi, \quad b_{31} = 0, \quad b_{32} = \frac{ba}{2pm}\cos 2\xi$$

$$b_{33} = -\frac{ba^2}{2pm}\sin 2\xi$$

and with $u < ap$ undergo changes only

$$b_{11} = \frac{r_0}{I}\left[Q - r_0 T_0\frac{\partial B_T}{\partial u} - \frac{2r_0 T_0}{\pi\sqrt{a^2 p^2 - u^2}}\right], \quad b_{12} = -\frac{T_0 r_0^2}{J}\left[\frac{\partial B_T}{\partial a} + \frac{2u}{\pi a\sqrt{a^2 p^2 - u^2}}\right]$$

$$b_{21} = \frac{a}{2m}\left[T_0\frac{\partial k_T}{\partial u} + \frac{4u T_0}{\pi a^2 p^2\sqrt{a^2 p^2 - u^2}}\right]$$

$$b_{22} = -\frac{1}{2m}(k - aT_0\frac{\partial k_T}{\partial a} + \frac{4T_0 u^2}{\pi a^2 p^2\sqrt{a^2 p^2 - u^2}}) + \frac{3ba^2}{8pm}\sin 2\xi$$

where $Q = \frac{d}{du}M(\frac{u}{r_0})$ represents the steepness of the energy source characteristic.

5 Calculations

The calculation parameters are: $\omega_0 = 1$ s^{-1}, $m = 1$ kgf $\cdot$ s$^2 \cdot$ cm^{-1}, $k_0 = 0.02$ kgf $\cdot$ s $\cdot$ cm^{-1}, $b = 0.07$ kgf $\cdot$ cm^{-1}, $T_0 = 0.5$ kgf, $\alpha_1 = 0.84$ s $\cdot$ cm^{-1}, $\alpha_3 = 0.18$ s$^3 \cdot$ cm^{-3}, $r_0 = 1$ cm, $J = 1$ kgf $\cdot$ s $\cdot$ cm^2. Depending on the linearization accuracy parameter i, the values of the coefficients N_2 and $\overline{N}_3$ are taken as follows: $N_2 = 0.6$ ($i = 1$), $\overline{N}_3 = 0.75$ ($i = 1.5$). The nonlinear elasticity is chosen in the form $f(x) = \gamma_3 x^3$ and, in accordance with (3), is replaced by $k_f = 0.75\gamma_3 a^2$ ($\overline{N}_3 = 0.75$ at $i = 1.5$), where $\gamma_3 = \pm 0.2$ kgf $\cdot$ cm^{-3}. Note that the number 0.75 also occurs if using, for example, the widespread averaging method is used [12].

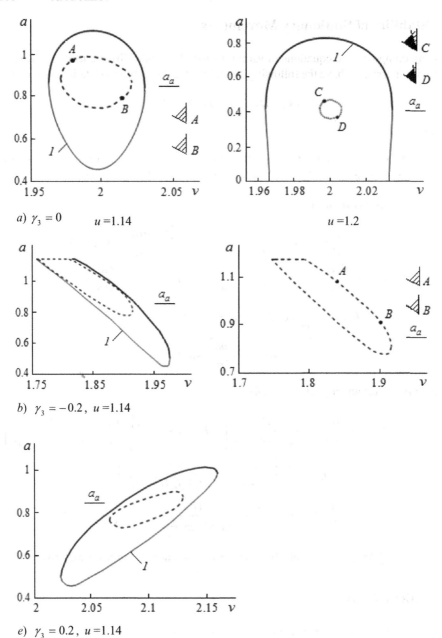

Fig. 1. Amplitude-frequency curves

Figure 1 shows the amplitude-frequency curves for the velocity values $u = 1.14$ cm·s^{-1}, $u = 1.2$ cm·s^{-1}. They completely coincide with the results of the well-known

Bogolyubov-Mitropolskii averaging method. Solid line *1* corresponds to the linear parametric excitation $x \cos \nu t$ in eq. (1), is shown for comparison, and a_a denotes the self-oscillation amplitudes. The upper branch of curve *1*, shown by the thick line, corresponds to stable oscillations, while the lower branch (thin line) corresponds to unstable ones. The dashed ($u = 1.14$) and dotted ($u = 1.2$) curves correspond to nonlinear parametric excitation.

Figure 1,*a* takes place for a linear characteristic of elasticity ($\gamma_3 = 0$), Fig. 1,*b* is soft ($\gamma_3 < 0$), and Fig. 1,*e* is rigid ($\gamma_3 > 0$). In the case of a nonlinear elastic force, there are no amplitude-frequency curves at $u = 1.2$, i.e. there are no real solutions to Eq. (8). The horizontal line in Fig. 1,*b* corresponds to $ap \approx u = 1.14$. As can be seen from the graphs, in the case of $\gamma_3 = 0$, the nonlinear parametric excitation reduces the amplitude, at $u = 1.2$ it significantly narrows the frequency range and changes the shape of the amplitude curve.

Within the shaded sector, for the steepness of the characteristics of the energy source, fluctuations with amplitude are stable. Note that these sectors should actually be indicated on the load curve $S(u)$. Sectors with black color reflect rather weak stability. In these sectors, the stability criteria (criterion) (10) are met with $0.000Y > 0$ values, where $Y \le 9$.

6 Discussion and Conclusion

The above calculation results show that nonlinear parametric excitation narrows the range of amplitude curves and can change their shape. The dynamics of the system under nonlinear parametric excitation differs from linear excitation, the differences are both quantitative and qualitative. At the same time, quantitative differences are more significant. However, the possibility is not ruled out that for other parameters of the system, these differences will change in the other direction.

References

1. Kononenko, V.O.: Vibrating Systems with Limited Power-Supply. Iliffe, London (1969)
2. Alifov, A.A., Frolov, K.V.: Interaction of Nonlinear Oscillatory Systems with Energy Sources. Hemisphere Publishing Corporation, New York, Washington, Philadelphia, London (1990)
3. Frolov, K.V.: Selected Works: in 2 vol. Nauka, Moscow (2007). (in Russian)
4. Krasnopolskaya, T.S., Shvets, A.Y.: Regular and Chaotic Dynamics of Systems with Limited Excitation. Regular and chaotic dynamics. M.-Izhevsk (2008)
5. Samantaray, A.K.: On the non-linear phenomena due to source loading in rotor-motor systems. Proc. Instit. Mech. Eng. J. Mech. Eng. Sci. **223**(4), 809–818 (2008)
6. Cveticanin, L., Zukovic, M., Cveticanin, D.: Non-ideal source and energy harvesting. Acta Mech. **228**(10), 3369–3379 (2017). https://doi.org/10.1007/s00707-017-1878-4
7. Pust, L.: Electro-mechanical impact system excited by a source of limited power. Eng. Mech. **15**(6), 391–400 (2008)
8. Alifov, A.A.: About calculation of self-oscillatory system delayed and limited excitation. In: Proceedings of the International Conference on "Measurement and quality: problems, perspectives", pp. 289–293. AzTU, Baku (2018)
9. Schmidt, G.: Parametric Oscillations. Mir, Moscow (1978). (in Russian)

10. Alifov, A.A.: Oscillations at nonlinear parametric and limited excitation. Dyn. Syst., Mech. Mach. **10**(1), 2–6 (2022)
11. Vibrations in the Technique, vol. 2. Mashinostroyeniye, Moscow (1979). (in Russian)
12. Bogolyubov, N.N., Mitropolskii, Y.: Asymptotic Methods in Theory of Nonlinear Oscillations. Nauka, Moscow (1974). (in Russian)
13. Moiseev, N.N.: Asymptotic Methods of Nonlinear Mechanics. Nauka, Moscow (1981). (in Russian)
14. Butenin, N.V., Neymark, Y., Fufaev, N.A.: Introduction to the theory of Nonlinear Oscillations. Nauka, Moscow (1976). (in Russian)
15. Hayashi, C.: Nonlinear Oscillations in Physical Systems. Princeton University Press, New Jersey (2014)
16. Wang, Q., Fu, F.: Variational iteration method for solving differential equations with piecewise constant arguments. Int. J. Eng. Manuf. **2**(2), 36–43 (2012)
17. Karabutov, N.: Structural identification of nonlinear dynamic systems. Int. J. Intell. Syst. Appl. **09**, 1–11 (2015)
18. Gourary, M.M., Rusakov, S.G.: Analysis of oscillator ensemble with dynamic couplings. In: AIMEE 2018. The Second International Conference of Artificial Intelligence, Medical Engineering, Education, pp. 150–160 (2018)
19. Ziabari, M.T., Sahab, A.R., Fakha-ri, S.N.S.: Synchronization new 3D chaotic system using brain emotional learning based intelligent controller. Int. J. Inform. Technol. Comput. Sci. **7**(2), 80–87 (2015)
20. Alifov, A.A.: Methods of Direct Linearization for Calculation of Nonlinear Systems. Regular and chaotic dynamics. M.-Izhevsk 2015. (in Russian)
21. Alifov, A.A.: Method of the direct linearization of mixed nonlinearities. J. Mach. Manuf. Reliab. **46**(2), 128–131 (2017)
22. Alifov, A.A.: About calculation of oscillatory systems with limited excitement by methods of direct linearization. Eng. Autom. Problems **4**, 92–97 (2017)
23. Alifov, A.A.: About some methods of calculation nonlinear oscillations in machines. In: Proceedings of the International Symposium of Mechanism and Machine Science, pp. 378–381. Izmir (2010)
24. Alifov, A.A.: About direct linearization methods for nonlinearity. In: Advances in Artificial Systems for Medicine and Education III. Advances in Intelligent Systems and Computing, 1126, pp. 105–114. Springer, Cham (2020)
25. Alifov, A.A.: On the calculation by the method of direct linearization of mixed oscillations in a system with limited power-supply. In: Advances in Computer Science for Engineering and Education II. Advances in Intelligent Systems and Computing, 938, pp. 23–31. Springer, Cham (2020)
26. Alifov, A.A.: On Mixed Forced and Self-oscillations with Delays in Elasticity and Friction. In: Zhengbing, H., Petoukhov, S., He, M. (eds.) CSDEIS 2020. AISC, vol. 1402, pp. 1–9. Springer, Cham (2021). https://doi.org/10.1007/978-3-030-80478-7_1
27. Murashkin, L.S., Murashkin, S.L.: Applied nonlinear mechanics of machine tools. Mashinostroenie, Leningrad (1977). (in Russian)
28. Ponomarev, A.S., et al.: Transverse self-oscillations of power tables caused by friction forces. Bull. Kharkov Polytech. Inst., Mashinostroenie **130**(8), 67–69 (1977)
29. Bronovec, M.A., Zhuravljov, V.F.: On self-oscillations in systems for measuring friction forces. Izv. RAN, Mekh. Tverd. Tela **3**, 3–11 (2012). (in Russian)

Spectrum Analysis on Electricity Consumption Periods by Industry in Fujian Province

Huawei Hong, Lingling Zhu, Gang Tong, Peng Lv, Xiangpeng Zhan[✉], Xiaorui Qian, and Kai Xiao

State Grid Fujian Electric Power Co., Ltd., Fuzhou 350003, China
xiangpengzhan@whu.edu.cn

Abstract. Electricity is the basic energy for urban economic development and people's daily life. Exploring periods of electricity consumption has a great significance in improving quality of life, promoting industry development and completing electric power planning. This paper applies the single spectral analysis method to explore the periods of electricity consumption in nine cities at Fujian Province, i.e. Fuzhou, Putian, Quanzhou, Xiamen, Zhangzhou, Longyan, Sanming, Nanping and Ningde, for six industries, i.e. big industry, on-general industry, residential living, non-residential lighting, commercial electricity and agricultural electricity. The data used is the daily electricity consumption data of Fujian Province from January 1, 2019 to April 17, 2022 for each pair of city and industry. It is found that the electricity consumption of most industries has an obvious annual period, except for residential living. In addition, the electricity consumption of some industries in some cities also have week period, quarter period, four-month period or semi-annual period, respectively.

Keywords: Spectrum analysis · Electricity consumption · Periods · Different cities · Industry classified

1 Introduction

Electricity is an indispensable energy source in today's production and life, and electricity consumption is a direct manifestation of electricity consumption. As the "barometer" [1] and "vane" [2] of the national economy, electricity consumption is an important indicator of the city's economic activities. Generally speaking, increasement of electricity consumption in industry means increasement of output value [3, 4]. Therefore, it is of great economic and social significance to study the electricity consumption periods in Fujian Province by industry, which leads to a more deeply understanding of electricity consumption period pattern.

There have been some researches focusing on electricity consumption. [5] made a medium and long-term forecast of China's electricity demand by using gray system theory and analyzed the fluctuation period by using autoregressive model; [6] predicted the annual household electricity consumption by calculation experiment; [7] used classical time series ARIMA method, ETS model and neural network autoregressive model to

© The Author(s), under exclusive license to Springer Nature Switzerland AG 2023
Z. Hu et al. (Eds.): ICAILE 2023, LNDECT 180, pp. 109–119, 2023.
https://doi.org/10.1007/978-3-031-36115-9_11

forecast electricity consumption in Jiangsu Province, and finally found that the combined model was better than the non-combined model for electricity consumption; [8] based on feature selection, clustering and Markov process techniques to structure model and predict consumption. The simulation results show the proposed model over other similar algorithms such as LASSO-QRNN and HyFIS. Although electricity consumption has been studied by many scholars, there is no detailed and in-depth study on electricity consumption periods by industry so far.

Time domain has always been the focus of various fields, but time domain is to study the time series as a whole. Spectral analysis can make up for the shortcomings of time domain, and many foreign scholars have introduced Spectral analysis into the research field and achieved important research results [9–11]. Spectral analysis is one of the main methods used in Statistics to find hidden periods in time series data and complete period analysis in various industries [12–15]. Besides, [16] used singular spectrum analysis (SSA) to detect the hidden periodicity information in the data and compared the forecasting performance with method combining linear recurrent formulas and artificial neural networks (ANNs), getting the conclusion that SSA-ANN model is more accurate; [17] extracted periodic component by SSA, and concluded that the performance of SSA was best, compared with the classical methods; [18] extracted a financial period by using multi-channel SSA to obtain more robust and conclusive results of the global financial period based on time-series data from 1970 to 2018; [19] employed a wavelet spectrum analysis to study globalization and business periods in China and G7 countries, found that the overall co-movements and trade tended not to be significantly related.

In summary, most of the studies on electricity consumption stayed at the level of constructing models to forecast electricity consumption [20, 21], while the studies on electricity consumption periods are insufficient, and the studies on electricity consumption periods by industries have not been reported. Spectral analysis, as one of the most excellent methods for period theory research, has been widely used in period research. In this paper, we will explore the period pattern of electricity consumption by industry in different cities of Fujian Province through single-spectrum analysis, and provide new ideas for the in-depth study of electricity consumption.

2 Basic Idea of Spectral Analysis

According to the Fourier transform theory, a smooth time series can be regarded as a superposition of several regular sine or cosine waves with appropriate amplitude, frequency and phase. Spectral analysis is to convert the sequence from the time domain to the frequency domain with the help of Fourier transform, so as to decompose the time series into the superposition of different harmonics, and analyze the periodic characteristics of the time series by comparing the characteristics of different frequency harmonics. The basic principle of single-spectrum analysis is to find out the main frequency components according to the spectral density function. There will be obvious spikes in the frequency-spectral density diagram, and the periods corresponding to the spikes are the main periods of the time series.

First, the expression for the spectral density function [22, 23] is given.

For the time series $\{Y_t\}$, there is a self-covariance function $\{\gamma(h)\}$, a monotonically nondecreasing right continuous function $F(\omega)$ at the frequency $\omega \in (-1/2, 1/2)$ satisfying $F(-1/2) = 0$, such that

$$\gamma(h) = \int_{1/2}^{-1/2} e^{2\pi i \omega h} dF(\omega), h = 0, \pm 1, \pm 2, \cdots . \tag{1}$$

$F(\omega)$ is said to be the spectral distribution function of $\{Y_t\}$, and i i is imaginary unit root. If there is a non-negative function $f(\omega)$ with $\omega \in (-1/2, 1/2)$ such that

$$\gamma(h) = \int_{1/2}^{-1/2} e^{2\pi i \omega h} f(\omega) d\omega, h = 0, \pm 1, \pm 2, \cdots , \tag{2}$$

then $f(\omega)$ is said to be the spectral density function of $\{Y_t\}$, referred to as the spectral density.

Reversing Eq. (2) gives the spectral density as

$$f(\omega) = \sum_{h=-\infty}^{\infty} \gamma(h) e^{-2\pi i \omega h}. \tag{3}$$

Define $d(\omega_j)$ as the Discrete Fourier Transformation (DFT) of Y_t with the following expression:

$$d(\omega_j) = n^{-1/2} \sum_{t=1}^{n} Y_t e^{-2\pi i \omega_j t}, j = 0, 1, \cdots n - 1, \tag{4}$$

where n is the number of samples. Due to $\sum_{t=1}^{n} e^{-2\pi i \omega_j t} = 0, j \neq 0$, the DFT can be rewritten as

$$d(\omega_j) = n^{-1/2} \sum_{t=1}^{n} (Y_t - \overline{Y}) e^{-2\pi i \omega_j t}. \tag{5}$$

It can be shown that $f(\omega_j)$ has the following relationship with $d(\omega_j)$

$$\begin{aligned}
\tfrac{n}{4} \left| d(\omega_j) \right|^2 &= 4n^{-2} \sum_{t=1}^{n} \sum_{s=1}^{n} (Y_t - \overline{Y})(Y_s - \overline{Y}) e^{-2\pi i \omega_j (t-s)} \\
&= 4n^{-2} \sum_{h=-(n-1)}^{n-1} \sum_{t=1}^{n-|h|} (Y_{t+|h|} - \overline{Y})(Y_t - \overline{Y}) e^{-2\pi i \omega_j} \\
&= 4n^{-2} \sum_{h=-(n-1)}^{n-1} \hat{\gamma}(h) e_j^{-2\pi i \omega_j h} = 4n^{-1} \hat{f}(\omega_j),
\end{aligned} \tag{6}$$

where $j \neq 0$, $\hat{\gamma}(h) = n^{-1} \sum_{t=1}^{n-|h|} (Y_{t+|h|} - \overline{Y})(Y_t - \overline{Y})$, $h = t - s$ is an estimator of $\gamma(h)$ and $\hat{f}(\omega_j)$ is an estimator of $f(\omega)$.

Second, prove the relationship between periodogram and the spectral density.

Time series $Y_t, t = 1, \cdots n$ can be expressed as a linear combination of sine and cosine components of different frequencies according to Fourier transform. For $m = (n-1)/2$, Y_t at $\omega_j = j/n$, the expression is as follows:

$$Y_t = A_0 + \sum_{j=1}^{m} [A_j \cos(2\pi \omega_j t) + B_j \sin(2\pi \omega_j t)] \tag{7}$$

where

$$A_0 = \overline{Y} \tag{8}$$

$$A_j = \frac{2}{n} \sum_{t=1}^{n} Y_t \cos(2\pi t \omega_j) \tag{9}$$

$$B_j = \frac{2}{n} \sum_{t=1}^{n} Y_t \sin(2\pi t \omega_j) \tag{10}$$

Periodic graph I is defined as

$$I(\omega_j) = A_j^2 + B_j^2 \tag{11}$$

Next, prove that

$$I(\omega_j) = A_j^2 + B_j^2 = \frac{4}{n} |d(\omega_j)|^2 \tag{12}$$

By Euler's formula, we have

$$
\begin{aligned}
d(\omega) &= n^{-1/2} \sum_{t=1}^{n} Y_t e^{-2\pi i \omega_j t} \\
&= n^{-1/2} \left(\sum_{t=1}^{n} Y_t \cos(2\pi \omega_j t) - i \sum_{t=1}^{n} Y_t \sin(2\pi \omega_j t) \right).
\end{aligned} \tag{13}
$$

Thereby,

$$|d(\omega_j)|^2 = n^{-1} \left(\sum_{t=1}^{n} Y_t \cos(2\pi i \omega_j t) \right)^2 + n^{-1} \left(\sum_{t=1}^{n} Y_t \sin(2\pi i \omega_j t) \right)^2 \tag{14}$$

Combining Eqs. (9), (10), (11) and (14) yields that

$$
\begin{aligned}
I(\omega_j) &= A_j^2 + B_j^2 \\
&= 4n^{-2} \left(\sum_{t=1}^{n} Y_t \cos(2\pi \omega_j t) \right)^2 + 4n^{-2} \left(\sum_{t=1}^{n} Y_t \sin(2\pi \omega_j t) \right)^2 \\
&= 4n^{-1} |d(\omega_j)|^2
\end{aligned} \tag{15}
$$

and Eq. (12) is proved.

Spectrum Analysis on Electricity Consumption Periods   113

That is, we can analyze the sample spectral density $f(\omega_j)$ of Y_t by study the periodogram $I(\omega_j)$. Hence, we often use the periodogram method instead of estimating the spectral density in practice. According to Eq. (11), we can quickly calculate the periodogram and obtain the spectral density estimator.

3 Data Preprocessing

The daily electricity consumption data of six types of industries in nine cities at Fujian Province from January 1, 2019 to April 17, 2022 were collected as raw data. In total, the data used are 54 time series data for six types of industries in nine cities. Daily trend will introduce very low frequency components in the periodogram, which often masks the higher frequency. Preprocessing is completed by fitting the raw data to a linear regression over time, i.e. $y_t = \alpha + \beta t$, and the residuals, i.e. $\epsilon_t = y_t - \hat{y}_t$, are used as data for the subsequent spectral analysis.

The F-test is used to test whether the linear regression is significant. Under null hypothesis $H_0 : \beta = 0$, the test statistic $F = \frac{SSR/1}{SSE/(n-2)}$ follows $F(1, n-2)$ distribution, where:

$$SSR = \sum_{t=1}^{n} (\hat{y}_t - \bar{y})^2, \ SSE = \sum_{t=1}^{n} (y_t - \hat{y}_t)^2.$$

Then P-value is calculated as $P(F(1, n-2) > F)$. Finally, all the P-values of 54 linear regressions are smaller than 0.05, which means the null hypothesis should be rejected and the regressions are significant. Hence, the linear trend of raw data should be removed by using the residuals instead.

4 Study on Periods by Single-Spectrum Analysis

In the following, single-spectrum analysis will be applied on the time series data. The periods corresponding to the spikes on the frequency-spectral density plot are the main period of the time series. To facilitate the conclusion, the horizontal coordinates of the frequency-spectral density plot are transformed to $T = 1/\omega$, so that the single-spectrum analysis plot with the horizontal coordinate as period and the vertical coordinate as spectrum is obtained.

4.1 Big Industry

The spectrum-period curves of big industry are shown in Fig. 1.

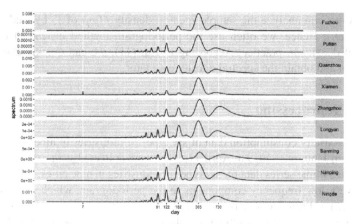

Fig. 1. Spectrum analysis of big industry daily electricity consumption in 9 cities.

From Fig. 1, the periods of electricity consumption of big industry in nine cities are concluded as follows:

1) All nine cities have annual period, but Sanming has a weaker one;
2) All nine cities have semi-annual period, four-month period, and quarter period, but the periods are more obvious in northwestern cities of Longyan, Sanming, Nanping and Ningde than that in southeastern cities;
3) Only Xiamen has a weak week period.

4.2 Non-general Industry

The spectrum-period curves of non-general industry are shown in Fig. 2.

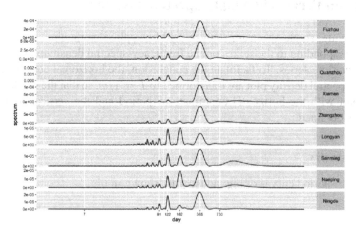

Fig. 2. Spectrum analysis of non-general industry daily electricity consumption in 9 cities

From Fig. 2, the periods of electricity consumption of non-general industry in nine cities are concluded as follows:

1) There is an obvious annual period in all nine cities;
2) Longyan, Sanming and Nanping have an obvious semi-annual period, while Putian, Quanzhou, Zhangzhou and Ningde have a weak semi-annual period;
3) Except Xiamen, other cities have a weak quarter period.

4.3 Residential Living

The spectrum-period curves of residential living are shown in Fig. 3.

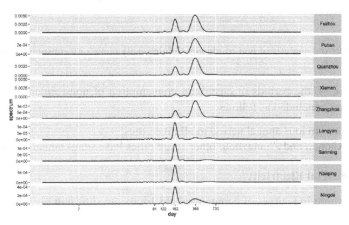

Fig. 3. Spectrum analysis of residential living daily electricity consumption in 9 cities

From Fig. 3, the periods of electricity consumption of residential living in nine cities are concluded as follows:

1) Southern cities, including Fuzhou, Putian, Quanzhou, Xiamen and Zhangzhou, have an obvious annual period, and Ningde has a weak annual period;
2) Besides Fuzhou and Putian, northwestern cities, including Longyan, Sanming, Nanping and Ningde, have an obvious semi-annual period, and Quanzhou and Zhangzhou have a weak one, while Xiamen's semi-annual period is the weakest.

According to the above results, we can conclude that the main period of residential living electricity consumption in southeastern Fujian Province is annual period, while the main period in northwestern Fujian Province is semi-annual period. This phenomenon may be derived from climate [24]. Since winter temperature in the northwest is lower than that in the southeast, northwestern cities have a higher electricity consumption in winter. Hence, the electricity consumptions in summer and winter of northwestern cities are approximate, which leads to a semi-annual period.

4.4 Non-residential Lighting

The spectrum-period curves of non-residential lighting are shown in Fig. 4.

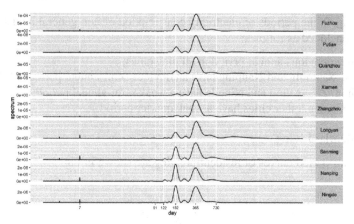

Fig. 4. Spectrum analysis of non-residential lighting daily electricity consumption in 9 cities

From Fig. 4, the periods of electricity consumption of non-residential lighting in nine cities are concluded as follows:

1) There is an obvious annual period in all nine cities;
2) Except Xiamen and Zhangzhou, other seven cities have semi-annual period.

4.5 Commercial Electricity

The spectrum-period curves of commercial electricity are shown in Fig. 5.

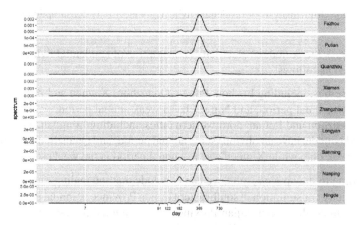

Fig. 5. Spectrum analysis of commercial electricity daily electricity consumption in 9 cities

From Fig. 5, the periods of electricity consumption of commercial electricity in nine cities are concluded as follows:

1) There is an obvious annual period in all nine cities;
2) Four northwestern cities, including Longyan, Sanming, Nanping and Ningde, have a weak semi-annual period.

4.6 Agricultural Electricity

The spectrum-period curves of agricultural electricity are shown in Fig. 6.

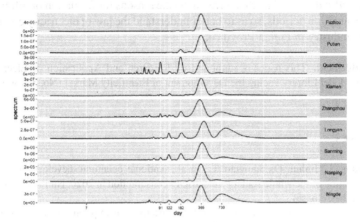

Fig. 6. Spectrum analysis of agricultural daily electricity consumption in 9 cities

From Fig. 6, the periods of agricultural electricity consumption in nine cities is concluded as follows:

1) There is an obvious annual period in all nine cities;
2) Quanzhou has an obvious semi-annual period, and Putian, Zhangzhou, Longyan, Sanming and Ningde have a weaker one;
3) Longyan, Sanming and Ningde have a weak four-month period;
4) Only Quanzhou has an obvious quarter period.

Quanzhou has high quality vegetables industry, and Longyan, Sanming and Ningde have well developed animal husbandry. This may be the reason that these cities have special agricultural electricity consumption period.

5 Conclusion

In this paper, the periods of electricity consumption by industry in Fujian Province are studied by using single-spectrum analysis, based on daily electricity consumption data of six types of industries in nine cities in Fujian Province.

In general, first, except residential living, other five types of industries in all nine cities have an obvious annual period; second, the industries with four-month period and quarter period concentrates in big industry and non-general industry; third, northwestern cities, including Longyan, Sanming, Nanping and Ningde, have similar daily electricity consumption periods, and their quarter, four-month and semi-annual periods are more obvious than southeastern cities; finally, Xiamen has the most simple period pattern, where only annual period is obvious.

Although this paper has studied the period of electricity consumption by industry in Fujian Province through single-spectrum analysis, further analysis is required to find the reasons of above results. Next, we plan to first extend our research to study influencing factors of electricity consumption periods in different cities and industries, such as geographic location, climate, electricity consumption habits of users, industry development policies, GDP and so on. And then investigate the reasons for formation of different electricity consumption periods, so as to better understand the law of electricity consumption periods by industry in Fujian Province.

Acknowledgement. This paper is supported by Research and Application of Electric Power Forecasting Model Library Facing the Development of Electricity Market (No. 52130X230008).

References

1. Guoqiang, J., Jinhui, D., Jiangtao, L., et al.: National economic development barometer-interpretation of the whole society's electricity consumption in 2020. National Grid **2**, 50–53 (2021). (in Chinese)
2. Yang, L.: Stable economic recovery: July electricity consumption up 12.8% year-on-year. China Securities Journal, 2021-08-12 (A02) (in Chinese)
3. Zufei, X., Meng, W., Wen, H., et al.: An empirical study on the relationship between industrial structure, electricity consumption and economic growth in Shanghai. China Business Journal **03**, 42–47 (2021). (in Chinese)
4. Limin, J., Lei, X.: Research on the relationship between electricity consumption and economic growth in China. Management Observation **30**, 43–46 (2016). (in Chinese)
5. Wenxia, Z.: Economy forecasting and periodicity analysis for power demand. Information on Electric Power **04**, 14–16 (2001). (in Chinese)
6. Sidorov, A.I., Tavarov, S.S.: Method for forecasting electric consumption for household users in the conditions of the Republic of Tajikistan. Int. J. Sustain. Dev. Plan. **15**(04), 569–574 (2020)
7. Qi, W.: Analysis of residential electricity consumption forecasting in Jiangsu Province based on combined model. Jiangsu Business Theory **01**, 11–14 (2022). (in Chinese)
8. Dalkani, H., Mojarad, M., Arfaeinia, H.: Modelling electricity consumption forecasting using the Markov process and hybrid features selection. Int. J. Intell. Sys. Appli. (IJISA) **13**(5), 14–23 (2021)
9. Yasir, M., Shah, Z.S., Memon, S.A., et al.: Machine learning based analysis of cellular spectrum. Int. J. Wireless and Microwave Technolog. (IJWMT) **11**(2), 24–31 (2021)
10. Zerihun, B.M., Olwal, T.O., Hassen, M.R.: Spectrum sharing technologies for cognitive iot networks: Challenges and future directions. Int. J. Wireless and Microwave Technolog. (IJWMT) **10**(01), 17–25 (2020)
11. Joshi, D., Sharma, N., Singh, J.: Spectrum sensing for cognitive radio using hybrid matched filter single cycle cyclostationary feature detector. Int. J. Info. Eng. Elect. Bus. (IJIEEB) **5**, 13–19 (2015)
12. Elsner, J.B., Tsonis, A.A.: Singular. Spectral analysis: a new tool in time series analysis. Plenum Press, New York and London (1996)
13. Hassani, H., Zhigljavsky, A.: Singular spectrum analysis: methodology and application to economics data. J. Syst. Sci. Complexity **22**, 372–394 (2009)
14. Dabbakuti, J.R.K.K., Gundapaneni, B.L.: Application of singular spectrum analysis using artificial neural networks in TEC predictions for ionospheric space weather. IEEE J. Selec. Top. Appl. Earth Observ. Remo. Sens. **12**(12), 5101–5107 (2019)

15. Andi, Z., Lele, H., Huewen, W.: The implicit period of non-stationary time series based on spectral analysis. Math. Practice Theory **46**(18), 197–203 (2016). (in Chinese)
16. Sun, M., Li, X., Kim, G.: Precipitation analysis and forecasting using singular spectrum analysis with artificial neural networks. Clust. Comput. **22**(S12), 633–640 (2019)
17. Coussin, M.: Singular spectrum analysis for real-time financial cycles measurement. J. Int. Money Financ. **120**, 102532 (2022)
18. Škare, M., Porada-Rochoń, M.: Multi-channel singular-spectrum analysis of financial cycles in ten developed economies for 1970–2018. J. Bus. Res. **112**, 567–575 (2020)
19. Poměnková, J., Fidrmuc, J., Korhonen, I.: China and the world economy: wavelet spectrum analysis of business cycles. Appl. Econ. Lett. **21**(18), 1309–1313 (2014)
20. Lee, Y.W., Tay, K.G., Choy, Y.Y.: Forecasting electricity consumption using time series model. Int. J. Eng. Technol. **07**(04), 218–223 (2018)
21. Guofeng, F., Xiao, W., Yating, L., Weichiang, H.: Forecasting electricity consumption using a novel hybrid model. Sustain. Cities Soc. **61**, 102320 (2020)
22. Shumway, H., Robert, S.S.D.: Time Series Analysis and Applications. Springer International, USA, pp. 165–203 (2016)
23. Jonathan, C.D., Chan, K.-S.: Time series analysis and applications: R language, 2nd edition. In: Pan, H., et al. (eds.) Translation, pp. 229–238. Machinery Industry Press, Beijing (2011)
24. Chen, Y.: Research on the influence of temperature change on urban residential electricity consumption. Nanjing University of Information Engineering (Nanjing), 42–49 (2022). (in Chinese)

Determination of the Form of Vibrations in Vibratory Plow's Moldboard with Piezoceramic Actuator for Maximum Vibration Effect

Sergey Filimonov, Sergei Yashchenko, Constantine Bazilo$^{(\boxtimes)}$, and Nadiia Filimonova

Cherkasy State Technological University, Shevchenko blvd., 460, Cherkasy 18006, Ukraine
{s.filimonov,n.filimonova}@chdtu.edu.ua, b_constantine@ukr.net

> **Abstract.** Agriculture is the practice of growing plants and livestock. The task of agriculture is to supply raw materials for industry and provide the population with food. One of the main problems of agriculture is the complicacy and efficiency of soil cultivation. The plow is the main instrument for soil's tillage in agriculture. The analysis of present-day plows is conducted and their characteristics are analyzed. The method of reducing the friction of the cultivation unit with the help of ultrasonic vibrations has been experimentally tested. For the first time, the oscillation forms of the vibratory plow's moldboard with a piezoceramic actuator at ultrasonic frequencies are obtained. According to the results of the research, it is found that the frequencies of 36 kHz and 37.4 kHz can be chosen as the resonant ones. The influence of negative factors, such as soil sticking during its cultivation, is reduced, as well the resistance created by the soil during its cultivation is reduced. Thus, the time of plow operation is increased. When using a piezoceramic actuator, more flexible possibilities open up in the control of the vibratory plow.
>
> **Keywords:** Vibratory plow · Vibration technologies · Chladni figures · Friction reduction · Prototype · Ultrasonic transducer

1 Introduction

One of the most important sectors of the economy is agriculture, because it provides the population with food, other sectors with raw materials. The economic condition and food security of the country depends on the state of agriculture. The International Food Policy Research Institute says that agricultural technologies will affect food production if they are applied jointly. Technological advances are helping to provide farmers with the tools and resources to make agriculture more sustainable [1, 2].

© The Author(s), under exclusive license to Springer Nature Switzerland AG 2023
Z. Hu et al. (Eds.): ICAILE 2023, LNDECT 180, pp. 120–128, 2023.
https://doi.org/10.1007/978-3-031-36115-9_12

The main tool in agriculture for tillage is the plow. Most of the agricultural enterprises are sure that it is impossible to completely abandon the plow, since it is necessary for loosening the soil, controlling weeds and improving the quality of the soil in general, and is also one of the well-proven methods of tillage [3, 4]. Plowing is a method of tillage, which consists in cutting the treated layer, raising it with loosening and turning by 130–180° and laying a previously open furrow on the bottom [5, 6].

In the technological process of growing crops, tillage is the most energy intensive operation. The main problem of agriculture is the labour intensity and efficiency of tillage. One of the most common tillage methods is loosening the soil, so improving its properties is one of the most effective solutions of the problem of excessive energy consumption [7]. When plowing, the soil layer is deformed, destroyed and displaced [8]. Loosening the ground is created through the use of conventional plows. One of the main disadvantages of this plow is the high resistance (friction) that occurs when plowing the soil, which results in high energy consumption, faster wear of the material from which the plow is made and its failure [9].

The effect of friction during plowing is shown schematically in Fig. 1.

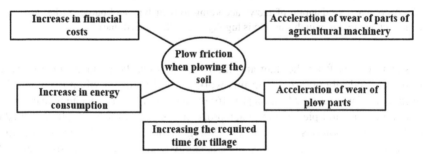

Fig. 1. Friction and its negative factors in tillage

As can be seen from Fig. 1, the effect of friction is of great importance in tillage. Therefore, an important task is to reduce its influence.

2 Literature Review

The plow consists of several components, two of which (a plowshare and a moldboard) are the main ones [10, 11]. Plow moldboard's main task is to cut from the wall of the soil layer, crumble it and turn it over. That is, in many respects, how well the soil will be prepared depends on the quality of the moldboard [12, 13].

The ease of manufacture and relatively low price are plow's main advantages. The main limitation is the high resistance (friction) that occurs when plowing the soil, while energy costs increase significantly.

Figure 2 shows a comparative diagram of plows by their type and purpose.

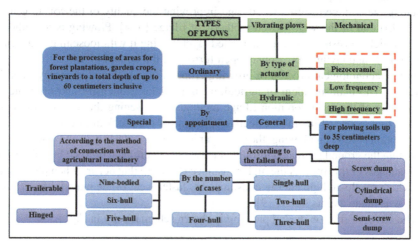

Fig. 2. A comparative diagram of plows according to their type and purpose (the plow under development and research in this article is highlighted with a dotted line)

As can be seen from the diagram, modern plows can be divided into two types: conventional and vibratory ones. In this work we will focus on vibratory plows. The main elements for creating vibration [14–16] in vibratory plow are mechanical elements, hydraulic systems and piezoceramic actuators. It has been experimentally established that when using vibrations in a plow, the sliding friction of the soil, which is the main component in the total traction resistance, is significantly reduced. The sticking of the working bodies is also reduced [17].

It is presented and experimentally proved in [18] the decrease in friction between two surfaces of bodies when exposed to ultrasonic vibrations. Thereby, many constructions of plows with vibrating operating units have recently appeared [19]. Figure 3 shows a diagram of a vibratory plow.

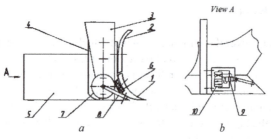

Fig. 3. Functional diagram of a hydraulic vibratory plow: a) general view; b) view A; 1 – plowshare, 2 – shelves, 3 – rack, 4 – shoe, 5 – field board, 6 – rubber-metal absorbers, 7 – hydraulic vibrator, 8 – lever, 9 – piston, 10 – spool valve.

The traction resistance of the plow is reduced due to the pseudo liquefaction of the soil, which occurs under the action of a vibrating share 1. Rubber-metal absorbers 6 must allow vibrations of the share 1 of the vibrating plow body with an amplitude of 3–4 mm. Forced oscillations transmitted to the plowshare 1 by the vibrator 7 make changes to the natural vibratory process of soil destruction. Plowshare 1 oscillation frequency is 300 oscillations per minute. They contribute to this destruction, that is, the traction resistance during forced vibrations of the tool decreases in comparison with the traction resistance without vibrations and reduces the overall energy intensity of the process [20]. However, such improvements have a number of limitations: an increase in the weight of the working body and the complexity of maintenance and manufacturing of the product, as well as decrease in the operating time of the vibration unit, etc.

In paper [21] the construction of a vibratory plow is discussed. To create vibrations, a mechanical system with an offset eccentric is used. The limitation of such plow is the complexity of the design and the limited vibrations' power. Another example is a vibratory plow model using a piezoceramic actuator [22], which is given in the work [23] (Fig. 4).

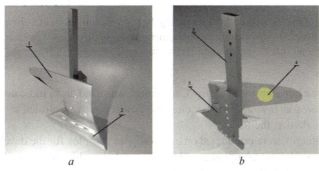

Fig. 4. 3D Model of vibratory plow with piezoelectric actuator: a) front view; b) back view; 1 – moldboard, 2 – plowshare, 3 – body stand, 4 – piezoelectric actuator in the form of a disk, 5 – shoe

In the works cited above, it is not experimentally given how the oscillations will propagate in the plow moldboard when they are created. Determining the shape of the oscillations in a working vibratory plow will allow to select the operating mode more effectively and correctly distribute the load.

The aim of the work is reducing the plow's friction during soil plowing as well as reducing the sticking of the operating units by determining the form of vibration of the vibratory plow based on a piezoceramic actuator.

3 Materials and Methods

In order to choose the correct operating mode of the vibratory plow, it is necessary to have an idea about the form of oscillations occurring in it, which directly depends on the frequency. To visually determine the shape of vibrations of a plow with piezoelectric actuator, the method of Chladni figures for the main resonant frequencies was used.

The principle of the method is as follows. First, the current-frequency characteristic of a vibratory plow with a piezoceramic actuator is measured to determine the main resonant frequencies. To study and determine the Chladni figures, the installation shown in Fig. 5 is used.

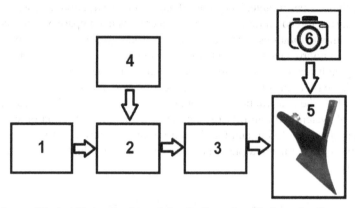

Fig. 5. Scheme of the installation for determining the form of oscillations of a vibratory plow with piezoelectric actuator: 1 – alternating voltage generator, 2 – high-voltage amplifier, 3 – ammeter, 4 – voltmeter, 5 – plow with piezoceramic actuator, 6 – video camera

Activated carbon in powder form "Carbolong" was used as an indicator. In the case of standing waves at resonant frequencies, the powder is concentrated at the vibration nodes and thus detects them.

The experiments were conducted on an intelligent complex for the development and research of piezoelectric components [24, 25], created within the scientific work (the number of state registration 0117U000936).

The installation works as follows. A rectangular voltage of 300 V is supplied from the generator to the piezoceramic actuator. Due to the action of the reverse piezoelectric effect, oscillations occur in the piezoceramic actuator. Since the piezoceramic actuator is rigidly connected to the plow blade, and "Carbolong" is poured on its surface, during oscillations, part of the powder falls off, and part remains in the oscillation nodes, showing what oscillations occur.

4 Experiments and Results

A prototype of a vibratory plow with piezoelectric actuator is shown in Fig. 6.

Determination of the Form of Vibrations

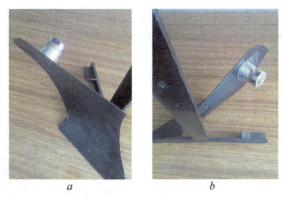

a *b*

Fig. 6. Prototype of vibratory plow with a piezoceramic actuator: a) front view; b) back view

The plow "Lux 200" of Ukrainian production was used for research. The piezoceramic actuator was connected to the plow moldboard with epoxy adhesive. The parameters of the ultrasonic piezoelectric actuator are presented in Table 1.

Table 1. Main parameters of the ultrasonic piezoelectric actuator

Parameter of ultrasonic actuator	Value
Waveguide's diameter maximum	45 mm
Waveguide's diameter minimum	38 mm
Waveguide's length	25 mm
Reflector's thickness	14 mm
Piezoelectric elements' diameter	38 mm
Total length	49 mm
Power	60 W
Resonant frequency	40 kHz

Figure 7 shows the initial state of the experimental study (part of the plow moldboard, piezoelectric actuator and "Carbolong" powder).

Fig. 7. The initial state of the experimental study of a vibratory plow with a piezoceramic actuator

Figure 8 shows some experimental results of the vibration form of the moldboard of a vibratory plow with piezoelectric actuator, obtained using the Chladni figures.

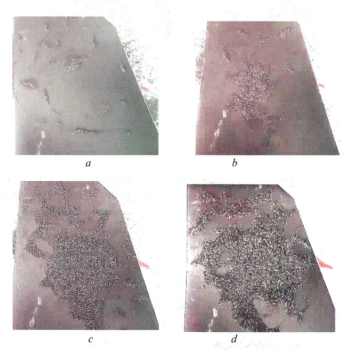

Fig. 8. The experimental results of Chladni figures of a vibratory plow with a piezoceramic actuator: a) 36 kHz; b) 37.1 kHz; c) 39.1 kHz; d) 40 kHz

From Fig. 8 it can be seen that the maximum vibration area of the moldboard of a plow with piezoceramic actuator is observed at a frequency of 36 kHz.

Separately, it should be noted the obtained Chladni figures at a frequency of 37.4 kHz (Fig. 9).

Fig. 9. The experimental results of Chladni figures of a vibratory plow with piezoelectric actuator at a frequency of 37.4 kHz

From Fig. 7 it can be seen that the generated vibrations occupy a large part of the plow moldboard. Thus, a decrease in friction in the moldboard occurs due to a decrease in friction contact, which is indicated by the Chladni figures.

Further research by the authors can be aimed at studying the possibility of alternately using a piezoceramic transducer (vibrator) as an actuator and a sensor for adjusting the oscillation amplitude, as well as modeling the developed design of a vibratory plow in the COMSOL Multiphysics software package.

5 Conclusions

The advantages and disadvantages of modern models of vibratory plows were revealed. The method of reducing the friction of the cultivation unit with the help of ultrasonic vibrations has been experimentally tested.

For the first time, the oscillation forms of the moldboard of a vibratory plow with piezoelectric actuator at ultrasonic frequencies were obtained.

According to the results of the research, it was found that the frequencies of 36 kHz and 37.4 kHz can be chosen as the resonant ones.

When using a piezoceramic actuator, more flexible possibilities for controlling the vibratory plow open up, and the weight and dimensions of the plow practically do not increase.

The use of a vibratory plow based on a piezoelectric actuator will reduce the effect of soil sticking during its cultivation, reduce the resistance created by the soil, and increase the operating time of the plow.

References

1. Kalogiannidis, S., Kalfas, D., Chatzitheodoridis, F., Papaevangelou, O.: Role of Crop-Protection Technologies in Sustainable Agricultural Productivity and Management. Land **11**, 1680 (2022). https://doi.org/10.3390/land11101680
2. Rosegrant, M., Ringler, C., Cenacchi, N., Cindy C.: Food security in a world of natural resource scarcity: The role of agricultural technologies. International Food Policy Research Institute (IFPRI), Washington, D.C. (2014). https://doi.org/10.2499/9780896298477
3. Safety and health in agriculture. International Labour Organization, Report VI (1) (2000). https://www.ilo.org/public/english/standards/relm/ilc/ilc88/rep-vi-1.htm
4. United Nations Environment Programme. Making Peace with Nature: A scientific blueprint to tackle the climate, biodiversity and pollution emergencies. Nairobi (2021). https://www.unep.org/resources/making-peace-nature
5. Weil, R.C., Brady, N.C.: Elements of Nature and Properties of Soil. Pearson (2009)
6. Acquaah, J.: Principles of Crop Production: Theory, Techniques, and Technology. Prentice-Hall (2011)
7. Hukov, Ya.S.: Soil cultivation. Technology and technique. Mechanical and technological substantiation of energy-saving means for the mechanization of soil cultivation in the conditions of Ukraine. Kyiv, Nora-print (1999)
8. Lurie, A.B., Grombchevskii, A.A.: Calculation and design of agricultural machinery Leningrad, Mashinostroenie (1977)

9. Filimonov, S.O., Yashchenko, S.S., Batrachenko, A.V., Filimonova, N.V.: Using smart piezoceramics for soil cultivation in agriculture. Bulletin of Cherkasy State Technological University **2**, 30–36 (2019). https://doi.org/10.24025/2306-4412.2.2019.169817
10. Chen, P., Tao, W., Zhu, L., Wu, Q.-M.: Effect of varying remote cylinder speeds on plough-breast performances in alternative shifting tillage. Computers and Electronics in Agriculture **181**, 105963 (2021). https://doi.org/10.1016/j.compag.2020.105963
11. Lim, Y., Zhai, Z., Cong, P., Zhang, Y.: Effect of plough pan thickness on crop growth parameters, nitrogen uptake and greenhouse gas (CO_2 and N_2O) emissions in a wheat-maize double-crop rotation in the Northern China Plain: A one-year study. Agricultural Water Management **213**, 100081 (2019). https://doi.org/10.1016/j.agwat.2018.10.044
12. Hrabě, P., Müller, M., Hadač, V.: Evaluation of techniques for ploughshare lifetime increase **61**, 72–79 (2015). https://doi.org/10.17221/73/2013-RAE
13. Ansorge, D., Godwin, R.J.: The effect of tyres and a rubber track at high axle loads on soil compaction: part 3: comparison of virgin compression line approaches. Biosystems Engineering **104**(2), 278–287 (2009). https://doi.org/10.1016/j.biosystemseng.2009.06.024
14. Li, J.: Design of active vibration control system for piezoelectric intelligent structures. Int. J. Edu. Manage. Eng. (IJEME) **2**(7), 22–28 (2012). https://doi.org/10.5815/ijeme.2012.07.04
15. Tomaneng, S., Docdoc, J.A.P., Hierl, S.A., Cerna, P.D.: Towards the development a cost-effective earthquake monitoring system and vibration detector with SMS notification using IOT. Int. J. Eng. Manuf. (IJEM) **12**(6), 22–31 (2022). https://doi.org/10.5815/ijem.2022.06.03
16. Hu, Z., Legeza, V., Dychka, I., Legeza, D.: Mathematical model of the damping process in a one system with a ball vibration absorber. Int. J. Intelli. Sys. Appli. (IJISA) **10**(1), 24–33 (2018). https://doi.org/10.5815/ijisa.2018.01.04
17. Bulgakov, V.M., Sviren, M.O., Palamarchuk, I.P., Dryga, V.V., Chernysh, O.M., Yaremenko, V.V.: Agricultural Vibrating Machines. Kirovograd, KOD (2012)
18. Dong, S., Dapino, M.: Experiments on ultrasonic lubrication using a piezoelectrically-assisted tribometer and optical profilometer. J. Vis. Exp. **103**, 52931 (2015). https://doi.org/10.3791/52931
19. Bulgakov, V.M., Sviren, M.O., Kisiliov, R.V., Oryshchenko, S.B., Lisovyi, I.O.: Research of vibration processes at the main processing of the soil. Scientific Bulletin of the Tavriya State Agrotechnological University. Melitopol **1**(5), 3–13 (2015)
20. Loveikin, V.S., Yaroshenko, V.F., Bychenko, L.A.: Vibratory plough body. Patent 19528 UA. Publ. 15.12.06, Bul. 12
21. Doering, D., Campbell, T.: Vibratory plow. Patent US 7,546,883 B1 (2009)
22. Sharapov, V.M., et al.: Improvement of piezoceramic scanners. In: IEEE XXXIII International Scientific Conference Electronics and Nanotechnology (ELNANO), Kiev, pp. 144–146 (2013). https://doi.org/10.1109/ELNANO.2013.6552063
23. Bazilo, C., Filimonov, S., Filimonova, N., Yashchenko, S.: Method of reducing friction in the plow moldboard with soil during cultivation due to the implementation of ultrasonic vibrations. In: Hu, Z., Petoukhov, S., Yanovsky, F., He, M. (eds.) Advances in Computer Science for Engineering and Manufacturing. ISEM 2021. Lecture Notes in Networks and Systems, vol 463, pp. 281–289 (2022). https://doi.org/10.1007/978-3-031-03877-8_25
24. Filimonov, S.A., Bazilo, C.V., Bondarenko, Y., Filimonova, N.V., Batrachenko, A.V.: Creation of a highly effective intellectual complex for the development and research of piezoelectric components in instrument engineering, medicine and robotics. Bulletin of Cherkasy State Technological University **3**, 33–43 (2017)
25. Bazilo, C., Filimonov, S., Filimonova, N., Bacherikov, D.: Determination of Geometric Parameters of Piezoceramic Plates of Bimorph Screw Linear Piezo Motor for Liquid Fertilizer Dispenser. In: Hu, Z., Petoukhov, S., Yanovsky, F., He, M. (eds.) ISEM 2021. LNNS, vol. 463, pp. 84–94. Springer, Cham (2022). https://doi.org/10.1007/978-3-031-03877-8_8

Evaluating Usability of E-Learning Applications in Bangladesh: A Semiotic and Heuristic Evaluation Approach

Samrat Kumar Dey[1], Khandaker Mohammad Mohi Uddin[2(✉)], Dola Saha[2],
Lubana Akter[2], and Mshura Akhter[2]

[1] School of Science and Technology, Bangladesh Open University, Gazipur 1705, Bangladesh
[2] Department of Computer Science and Engineering, Dhaka International University,
Dhaka 1205, Bangladesh
jilanicsejnu@gmail.com

Abstract. In this age of technology, the popularity, and necessity of mobile applications in the field of education are increasing day by day. Many students in Bangladesh are not able to develop as qualified due to a lack of a suitable environment and skilled teachers. This problem has been solved to a large extent through various E-Learning applications. But no research has yet been done on exactly how efficient, intuitive, and usable these applications are. Therefore, this study's objectives are to assess these E-Learning apps' usability in the context of Bangladesh and to offer potential architectural recommendations for improving their usability and acceptance. To accomplish the objectives, the usability of five E-Learning applications (Bacbon, Prottoy, Shikho, Classroom, and Muktopaath) developed in the context of Bangladesh was evaluated through Heuristic Evaluation and Semiotic Evaluation process. The study showed that all selected applications have several usability problems. The heuristic evaluation is based on the principle of efficiency, consistency, flexibility, conventions, and error management of each application while the semiotic evaluation upheld the intuitiveness of the interface elements of the selected applications. Finally, to enhance the usability of these applications, some sets of recommendations and implications have been presented for practitioners.

Keywords: Usability · Heuristic Evaluation · Semiotic Evaluation · E-Learning · Usability Problems · HCI · Mobile Applications

1 Introduction

In the field of education, mobile technology has taken a big place for itself. Education is gaining popularity through mobile apps in Bangladesh like other developed countries for the advancement of Information and Communication Technology (ICT). Although the rate of education has increased in Bangladesh, in many cases due to lack of a proper efficacy education system, many children drop out prematurely. The Teacher-Student Ratio acts as a big influencer in this case. According to the Bangladesh Bureau of

© The Author(s), under exclusive license to Springer Nature Switzerland AG 2023
Z. Hu et al. (Eds.): ICAILE 2023, LNDECT 180, pp. 129–140, 2023.
https://doi.org/10.1007/978-3-031-36115-9_13

Educational Information and Statistics (BANBEIS), the Teacher-Student Ratio (TSR) in primary and secondary schools is 1:37 and 1:45 respectively [1, 2]. Some private education organizations maintain a proper teacher-student ratio. But it is quite hard to maintain the ratio in government schools. In this case, mobile technology can be a ground breaking tool. Through a website or app full of numerous educational resources, everyone in every region of the country may get the proper education. It will be easy for everyone to get education through this. It is possible to teach that subject to all the students of the country through the teaching material of one teacher.

According to a UNESCO report [3], more than 5,050 million children and young people, due to the epidemic, could not attend classes in almost half of the world's student population. In such difficult times, mobile learning apps have smoothed the path of learning for students. The use of mobile learning apps for epidemics in the education sector has increased a lot. But the question is, how useful are these learning apps? Can students really gain accurate knowledge through this app? How much will learn through the app benefit students?

Usability refers to the quality of a user's experience when interacting with the products or systems, including websites, software, devices, or applications [4]. Usability is about effectiveness, efficiency, and the overall satisfaction of the user. Applications should be designed in a way that makes them comprehensible (so that it is easily understood, comfortable by using, required jobs can be done easily and faster) for the benefit of users. High usability is required for these applications in order to increase students' knowledge bases.

These apps have been evaluated through Heuristic Evaluation and Semiotic Evaluation and also give priority to customer experience. Based on their user interface, decoration some app has positive user rating some has negative. The entire app has some similar problem that was lack of proper content. It is quite easy because the maximum app is new and most of these are in developing mode. At last, by following all the problems like user experience and app Efficacy the necessary steps have been highlighted to increase the effectiveness, also for growing the acceptance and usage of these five apps in Bangladesh.

Therefore, the aim of this study is to investigate and assess the usability of learning applications created in Bangladesh, to identify the features that make them less user-friendly for students, and to suggest potential design changes that may be made to these apps to increase their usability. In this research paper, to attain these objectives, from Google play store five learning apps namely Bacbon School [5], Classroom [6], Prottoy [7], Shikho [8], and Muktopaath [9] have been selected based on ratings, the number of downloads and features.

The rest of the articles are organized as follows: in Sect. 2, related works along with theoretical background are presented, in Sect. 3, the research methodology with an overview. In Sect. 4, the data analysis and findings of this research are presented, and in Sect. 5, this study concluded with the outcomes and the discussion of the highlighted recommendations.

2 Related Works

In this pandemic time, the world runs with its way to make our locked life easy. Today's world is dependent on the internet, and also dependent on different kinds of mobile application systems. People are showing an imperative interest in different kinds of mobile application systems. People are showing an imperative interest in different kinds of mobile based interactive solution. These applications are made up by applying various usability heuristic and semiotic heuristic rules. There has been a lot of research based on a semiotic and heuristic evaluation for many years now. Over the past few years, researchers have been focusing more on evaluation of mobile applications. Muaz *et al.* [10] have experimented with the truck hiring mobile application in Bangladesh. In the paper, researchers worked on 3 different mobile applications. Researchers were attempting to discover usability and semiotic problems from these three applications. Authors have find out usability heuristic problems and intuitiveness scores of the interface of the selected applications. In Digi truck mobile application authors have identified 20 usability heuristic problems, in truck Lagbe, 21 usability heuristic problems, and in Trux24, 27 usability heuristic problems were discovered.

A study based on Pakistan by Bibi *et al.* [11] have evaluated different Islamic learning mobile applications. Author have evaluated around 50 Islamic mobile learning applications as the cornerstone to effectiveness, learnability, and user satisfaction which was selected based on popularity at the play store. Evaluators have categorized those fifty mobile applications into 3 catalogs, each application containing five applications. These categories were Duas and Kalimas, How to Learn Namaz, and99 Names of Allah. About 75 participants were participated in the study and all were children who are aged between five to twelve.

In another study, Mubeen *et al.* [12] considered the available healthcare based mobile application and those application were chosen from two different digital platform (App Store and Play Store). All the mobile applications were selected based on the five features: the first one was assessment, the second was a questionnaire, the third one was privacy statement, the fourth one was Precautionary measures if test positive for COVID-19, and last one was plans of pandemic and region.

In January 2020, some students at a university in Malaysia created their own TARC [13] mobile application to establish their communication and daily resource. TARC mobile applications are quite exoteric applications that helped their daily basis to access necessary resources comfortably. Researchers have attempted to evaluate the usability of the application by balancing with the Concordia mobile application from the aspects of effectiveness, efficiency, and satisfaction. It is observed that completion rate for the TARCApp was 86% and the Concordia application was 96%. The mean completion times for the TARCApp mobile application ranged from 1.88s to 4.66s and the Concordia mobile application ranged from 4.17s to 7.11s.

In another work, the authors have evaluated the usability of the pregnancy tracker application in Bangladesh [14]. In the article, it applied heuristic rules on the two Bangladeshi mobile applications (MAA and Aponjon). Researchers have analyzed the usability of these two applications and applied them to maternity applications that was developed in Bangladesh and took steps to possible design solicitation for improving the overall usability of such applications. The findings of the study also discovered usability heuristic and semiotic heuristic and compared their number of heuristics. It concluded that the semiotic heuristic rule 17 was mostly violated in MAA apps for 6 times. In MAA application, a total of 23 usability heuristic and 45 semiotic heuristic problems were identified. However, in Aponjon, a total of 105 usability heuristics and 50 semiotic heuristic problems were discovered. The initiative score of Aponjon apps was lower than MAA apps, respective values were 2.7612 and 4.038.

3 Methodology

This section explains the study's methodology. First, we go into great depth on the applicant selection procedure. Next, the heuristic assessment strategy is explained, and after that, the semiotic evaluation method is covered.

3.1 Application Selection

We have selected 5 applications from among many renowned learning applications in Bangladesh. These 5 applications are selected for Heuristic Evaluation (HE) and Semiotic Evaluation (SE) based on ratings, the number of installations, and the date of release. These 5 apps are, 1. "Bacbon" (rating: 4.5, number of installations: 10k+ and the release date: 5th August 2019, updated on: 26th July 2021), 2. "Prottoy-Visual Learning App" (rating: 4.7, number of installations: 1k+ and the release date: 21st February 2021, updated on: 12th July 2021), 3. "Shikho Learning App" (rating: 4.5, number of installations: 50k+ and the release date:27th October 2020, updated on: 6th June 2021), 4. "Classroom" (rating: 3.9, number of installations:100k+ and the release date:12th July 2019, updated on: 27 May 2021), 5. "Muktopaath" (rating: 3.3, number of installations: 100k+ and the release date: 17th June 2019, updated on: 22nd June 2021).

The first one the "Bacbon School" is developed for those students who are making preparations for every public examination such as JSC, SSC, HSC, and admission tests. This app provides the opportunity for students for testing themselves after every video and provides a descriptive file about the video content so it becomes easier to learn. The second one is "Prottoy- Visual Learning App" which is a one-stop learning app, where learners would find complete course materials as per the NCTB curriculum. The third one is the "Shikho Learning App" for the Bangladeshi National Curriculum; SSC exam. Though this app is specifically designed for the NCTB and eventually covers a wide range of subjects for the JSC, SSC, and HSC curriculum and exams, the initial release of the app features a full exam preparation course for the SSC General Math Syllabus. The fourth one is "Classroom" which provides a comprehensive E-Learning platform for the students of SSC, HSC, and University Admission. Another related app to "Classroom" is "Classroom Parent" for parents which is the monitoring app to check the progress of a student's education. Parents can check their children's educational progress and can easily track the activity related to their children's learning. And the last one is "Muktopaath" which is an e-learning platform for education, skills, and professional development. This app also offers courses and tutorials for the unemployed and underemployed youths in Bangladesh intending to encourage them in self-employment. And also this app provides digital certificates after completing each course. The authors of this article selected these 5 apps based on the given information.

3.2 Heuristic Evaluation

Firstly, these five selected applications have been evaluated using heuristic evaluation based on [15, 16] heuristic and the followings are: "Visibility of system status and find ability of the mobile device (heuristic 1/ H1)", "Match between system and the real world (H2)", "User control and freedom (H3)", "Consistency and standards (H4)", "Flexibility and efficiency use (H5)", "Recognition rather than recall (H6)", "Aesthetic and minimalist design (H7)", "Error prevention (H8)". After that these problems were categorized based on severity ratings (0-4) proposed by Neilson [17]. In this concept where 0 indicates not a usability problem at all, 1 indicates a cosmetic problem, 2 indicates a minor usability problem, 3 indicates a major usability problem, and 4 indicates a catastrophic usability problem. The findings from the independent analysis of the 3 authors of this article have been aggregated.

3.3 Semiotic Evaluation

Subsequently the obtained problems were analyzed through Semiotic Evaluation following the semiotic heuristic based on the proposal of Islam *et al.* [18, 19] (Table 1).

The evaluation is based on the interface signs of the selected applications including signs, buttons, navigation bars, symbols, and other visual directives [20]. These signs are common to all selected applications for semiotic evaluation. Again, M. Speroni [21] proposed the W-SIDE (Web Semiotic Interface Design Evaluation) framework that pays explicit attention to the intuitiveness of user interface signs (even if it is the tiniest element of the UI) to design and assess web interfaces. Authors of this article have brought these interface elements under intuitiveness score (1 to 9) where 1-3 is recognized as low intuitiveness, 4-6 is recognized as middle intuitiveness and 7-9 is recognized as high intuitiveness score (1 is less and 9 is highly intuitive). And after that, again Neilson's [17] severity rating was adopted.

4 Results

Table 2 and Fig. 2 showed the result of the semiotic evaluation. Backbone has high intuitiveness (7-9) in all selected interface signs. Prottoy and Shikho both have seven high intuitiveness (7-9) and low intuitiveness for the remaining 1 icon. Classroom displays 4 high intuitiveness (7-9) for the selected interfaces, 1 medium intuitiveness (4-6), and 2 low intuitiveness for the remaining interfaces. Similarly, Muktopaath has 5 high intuitiveness (7-9) except 3 low intuitiveness (1-3) for the selected icons. The av-erages of the five apps are, Bacbon is 7.625, Prottoy is 6.875, Shikho is 7.125, Class-room is 5.625 and Muktopaath is 5.00.

Table 1. Semiotic Heuristics of the Side Framework

Level	Semiotic Heuristics
Syntactic	SH1. Clearly present the purpose of interactivity
	SH2. Make the effective use of color to design an interface sign
	SH3. Make the representamen readable and clearly noticeable
	SH4. Make a sign presentation clear and concise
	SH5. Create the representamen context appropriately
	SH6. Follow a consistent interface sign design strategy
Pragmatic	SH7. Place the interface sign in the proper position in a UI
	SH8. Make effective use of amplification features
	SH9. Create good relations among the interface signs of a UI
	SH10. Retain logical coherence in interface sign design
Social	SH11. Design interface signs to be culturally sensitive or reactive
	SH12. Match with the reality, conventions, or real-world objects
	SH13. Make effective use of organizational features
	SH14. Map with metaphorical and attributing properties
Environment	SH15. Model the profiles of the focused end-users SH16. Make effective use of ontological guidelines
Semantic	SH17. Realize a match between a designer's encoded and a user's decoded meaning.

Table 2. Intuitiveness score of the selected interface sign of 5 apps

Interface Sign	Bacbon	Prottoy	Shikho	Classroom	Muktopath
Logo	8	8	8	8	8
Notifications	6	8	8	1	7
Log out	8	8	8	8	7
Query	8	7	1	8	7
Buying course	8	7	8	6	1
Learning progress	7	8	8	6	1
Downloads	8	1	8	1	8
Recent activity	8	8	8	7	1
Average	7.625	6.875	7.125	5.625	5.00

The result showed that the most intuitive application from a semiotic perspective is Bacbon compared to the other 4 applications. It is clear from Fig. 1, that none of these applications have any catastrophic (severity rating 4) violation in semiotic evaluation. Bacbon and Shikho both have seven cosmetics (severity rating1) violations and one minor (severity rating 2) violation which occurred in Bacbon for notification symbols and Shikho for a query.

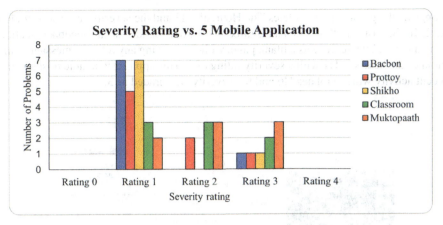

Fig. 1. Number of problems to each Severity Rating

Prottoy violates five cosmetic (severity rating 1) problems, two minor (severity rating 2), and one major (severity rating 3) violations that occurred in the download option (no download option in this application). The classroom application has three cosmetic (severity rating 1) and three minor (severity rating 2) violations except for two major (severity rating 3) violations for notifications and downloads icons. The last one Muktopaath has two cosmetic problems, three minor problems, and three major problems. The result shows that maximum major (severity rating 3) violations have occurred in Muktopaath. Moreover, Bacbon and Shikho both have the maximum cosmetic problem (severity rating 1), and Muktopaath and Classroom both have the highest number of minor (severity rating 2) problems from all the selected applications.

The Heuristic evaluation has been done on these five selected apps based on Bertini*et al.* [15, 16] heuristics. Some examples are given to show how the evaluation was conducted. For "Bacbon School" in the profile section (Not Applicable) shown in Fig. 2(a) is not clear which violates the Heuristic H4. The severity of this violation has been graded as 2 by the evaluators. Another violation is that the app shows only the last action, it is unable to show all history, shown in Fig. 2(b). This violates the heuristic H6 and H3 with severity rating 3. In the second app, "Prottoy-Visual Learning App" it was important to keep the download option because without an internet connection the videos of this app cannot be viewed. Also, the method of subscription is not flexible, there are so many processes to buy a course. Figure 2(c) shows that nothing was said clearly about the "promo code". Both violations violate The Heuristic H5 with a severity rating of 3. In "Shikho Learning App" there is no way to contact the tutor for any study-related problems or any queries which violate both The Heuristic H1 and H2 with severity rating 2. Another violation is the terms and conditions are not given which again violates The Heuristic H1 and alongside H3, was given a severity score 2 by all evaluators. In "Classroom" there are so many subject options with no videos, basic and concept papers, shown in Fig. 2(d) and Fig. 2(e). This violates both the Heuristics H1 and H2 with a severity score of 3. In another related app "Classroom Parents", it would be easier for the parents to understand their child's progress, if it had a brief description alongside the graph which violates The Heuristic H5 and the severity score was graded as 2. In the last one "Muktopaath", three evaluators found that the feedback section shown in Fig. 2(f), represents a blank page, without showing any relevant message. This violates the heuristic H1 with a severity rating of 2. This app is also unable to show any recent activity which violates H6 and the severity was graded as 3.

Fig. 2. Some screenshots of bacbon (a, b), Prottoy (c), Classroom (d, e), and Muktopaath (f)

Table 3 showed that the number of violated heuristics for all eight heuristics are maximum (n = 10) in both Bacbon and Classroom apps on the other hand minimum (n = 4) in Shikho. Besides, the second maximum (n = 8) violations found in Muktopaath app and the second minimum (n = 5) on Prottoy. The table also showed that Bacbon is the app that violates all 8 heuristics, on the other hand just 3 heuristics violations occurred in Shikho app.

Table 3. Usability problem to each heuristic

Heuristics	Bacbon	Prottoy	Shikho	Classroom	Muktopaath
H1	1	1	2	2	1
H2	1	0	1	1	1
H3	2	1	1	0	1
H4	2	0	0	2	1
H5	1	2	0	3	2
H6	1	0	0	1	1
H7	1	1	0	0	1
H8	1	0	0	1	0
Total	10	5	4	10	8

Here are some examples of semiotic evaluation in Fig. 3. The intuitiveness scores of the notification icons of the five apps which are Bacbon (Fig. 3(a)), Prottoy (Fig. 3(b)), Shikho (Fig. 3(c)), Classroom (not available) and Muktopaath (Fig. 3(d)) were 6,8,8,1 and 7 respectively. The notification icon of Bacbon is placed within the list of the dashboard with other buttons which violate a pragmatic heuristic (SH7) and there is no notification icon in the classroom app, for this reason, its intuitiveness score is very low. On the other side, icons of buying courses of these 5 apps are holding the score 8,7,7,6 and 1 respectively.

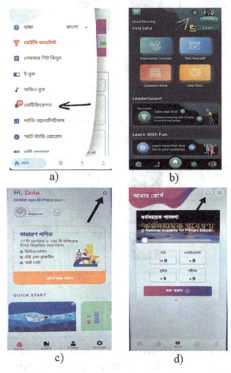

Fig. 3. Notification sign (a)Bacbon, (b)Prottoy, (c)Shikho, (d)Classroom (no notification option), and Muktopaath

In semiotic evaluation there have also some rules on the based on graphical interface. Those are "logo, notification, exit, query, buying course, learning progress, download, recently activity". For the experiment on the semiotic evaluation the interface signs in Table 4 have been considered and selected.

138   S. K. Dey et al.

Table 4. Interface sign selected for Semiotic Evaluation

Inter-facesign	Bacbon	Prottoy	Shikho	Classroom	Muktopath
Logo					
Notification				Not Available	
Logout/Exit	সাইন আউট	Sign Out	Logout	Logout If you need a break	Exit
Query	সাধারন জিজ্ঞাসা		Not Available	Chat With Us Let's Talk	সচরাচর জিজ্ঞাসা
Course Buy	কোর্স ক্রয় করুন	My Courses	শায়াকত্র কিনুন		Doesn't show properly
Learning progress		Progress	LEARNING ANALYSIS		Not Available
Downloads	ডাউনলোডস	Not Available		Not Available	
Recent activity	সর্বশেষ যা পড়েছিলেন	Recent	RECENTLY VIEWED		Not Available

5 Conclusion

This study makes these five programs more accessible to users, mostly students, by giving a comprehensive understanding of their present usability status and prevalent problems. The information obtained from the study will help to further develop and improve application performance significantly so many more users are encouraged to use these applications. The results of the study would be extremely helpful to practitioners in creating study-related applications in Bangladesh with enhanced usability and user experience so that these apps may offer users the most advantages. Students should be provided an easier way for data input, self-error control, the flexibility of buying courses, and the video download process. It needs to be ensured for novice users. The meaning of the interface signs should be more consistent with their convention. The symbol used as an interface sign should be more compatible with their elaborate meaning of having a constant phase relationship with users. The evaluations are made separately by each evaluator without including any common users. Again, this research only focused on E-Learning applications in the context of Bangladesh to evaluate by principles of heuristic and semiotic evaluations. Future studies may use different applications to explore deeper usability issues.

References

1. Teacher shortage makes education recovery a stiff job, https://www.tbsnews.net/bangla-desh/education/teacher-shortage-makes-education-recovery-stiff-job-206482. Accessed 30 May 2022
2. Teacher-student ratio not ideal at many public universities | Daily Sun |, https://www.daily-sun.com/printversion/details/331748/Teacherstudent-ratio-not-ideal-at-many-public-uni-ver sities. Accessed 30 May 2022
3. Half of world's student population not attending school: UNESCO launches global coalition to accelerate deployment of remote learning solutions, https://en.unesco.org/news/half-wor lds-student-population-not-attending-school-unesco-launches-global-coalition-acceler-ate. Accessed 30 May 2022
4. Moumane, K., Idri, A., Abran, A.: Usability evaluation of mobile applications using ISO 9241 and ISO 25062 standards. Springerplus **5**(1), 1–15 (2016). https://doi.org/10.1186/s40064-016-2171-z
5. BacBon School, https://bacbonschool.com/. Accessed 30 May 2022
6. Classroom: Crunchbase Company Profile & Funding, https://www.crunchbase.com/organization/. Accessed 30 May 2022
7. Prottoy | Click. Learn. Excel., https://www.prottoy.com.bd/. Accessed 30 May 2022
8. Shikho Academic Program, https://shikho.tech/. Accessed 30 May 2022
9. Muktopaath- Bangla eLearning Platform, https://www.muktopaath.gov.bd/. Accessed 30 May 2022
10. Muaz, M.H., Islam, K.A., Islam, M.N.: Assessing the usability of truck hiring mobile applications in Bangladesh using heuristic and semiotic evaluation. In: Advances in Design and Digital Communication: Proceedings of the 4th International Conference on Design and Digital Communication, Digicom 2020, November 5–7, 2020, Barcelos, Portugal, pp. 90–101 (2020)
11. Bibi, S.H., Munaf, R.M., Bawany, N.Z., Shamim, A., Saleem, Z.: Usability evaluation of Islamic learning mobile applications. J. Islamic Sci. Technol. **6**(1), 1–12 (2020)
12. Mubeen, M., Iqbal, M.W., Junaid, M., Sajjad, M.H., Naqvi, M.R., Khan, B.A., Tahir, M.U.: Usability evaluation of pandemic health care mobile applications. In: IOP Conference Series: Earth and Environmental Science, Vol. 704, No. 1, p. 012041. IOP Publishing (2021)
13. Tunku Abdul Rahman University College - ICDXA, https://www.tarc.edu.my/icdxa/icdxa-2020/. Accessed 30 May 2022
14. Kundu, S., Kabir, A., Islam, M.N.: Evaluating usability of pregnancy tracker applications in Bangladesh: a heuristic and semiotic evaluation. In: 2020 IEEE 8th R10 Humanitarian Tech-nology Conference (R10-HTC), pp. 1–6. IEEE (2020)
15. Bertini, E., Gabrielli, S., Kimani, S., Catarci, T., Santucci, G.: Appropriating and assessing heuristics for mobile computing. In: Proceedings of the Working Conference on Advanced Visual Interfaces, 119–126 (2006)
16. Bertini, E., Catarci, T., Dix, A., Gabrielli, S., Kimani, S., Santucci, G.: Appropriating heuristic evaluation for mobile computing. Int. J. Mob. Hum. Comput. Interact. (IJMHCI) **1**(1), 20–41 (2009)
17. Severity Ratings for Usability Problems: Article by Jakob Nielsen, https://www.nngroup.com/articles/how-to-rate-the-severity-of-usability-problems/. Accessed 30 May 2022
18. Islam, M.N., Bouwman, H.: Towards user–intuitive web interface sign design and evaluation: a semiotic framework. Int. J. Hum. Comput. Stud. **86**, 121–137 (2016)
19. Islam, M.N., Bouwman, H.: An assessment of a semiotic framework for evaluating user-intuitive Web interface signs. Univ. Access Inf. Soc. **14**(4), 563–582 (2015). https://doi.org/10.1007/s10209-015-0403-6

20. Islam, M.N.: A systematic literature review of semiotics perception in user interfaces. J. Syst. Inf. Technol. **15**(1), 45–77 (2013)
21. Speroni, M., Paolini, P.: Mastering the Semiotics of Information-Intensive Web Interfaces 212

Perceptual Computing Based Framework for Assessing Organizational Performance According to Industry 5.0 Paradigm

Danylo Tavrov[1(✉)], Volodymyr Temnikov[2], Olena Temnikova[1], and Andrii Temnikov[2]

[1] National Technical University of Ukraine "Igor Sikorsky Kyiv Polytechnic Institute", 37 Peremohy Ave., Kyiv 03056, Ukraine
d.tavrov@kpi.ua

[2] National Aviation University, 1 Kosmonavta Komarova Ave., Kyiv 03680, Ukraine

Abstract. The new economic paradigm of Industry 5.0, unlike Industry 4.0, puts greater emphasis on fostering human-centric, sustainable, and resilient industrial practices, and is focused on human well-being and shifting towards new forms of sustainable economics. However, the concept is still at a nascent stage and remains largely ill-defined. Existing organizational performance approaches do not take into account the new goals of Industry 5.0, largely because of their subjective and imprecise nature.

In this paper, we propose a formalized framework for evaluating performance of an organization in terms of compliance to subjective and ill-defined Industry 5.0 criteria. Our framework is based on perceptual computing and enables the regulators to express their subjective opinions using words.

We illustrate applicability of our framework to a specific task of evaluating the functional state of a person as one of the core aspects of human well-being. Given uncertainties involved in measuring appropriate physiological, psychological, mental, and physical characteristics, we show that perceptual computing can allow us to assess such criteria using words, which makes the whole process easily adoptable by regulators.

Keywords: Perceptual computing · Type-2 fuzzy set · Industry 5.0 · Functional state of a person

1 Introduction

Extensive literature on evaluating organizational performance focuses on defining various measures and metrics, which can help ensure high overall performance of an organization. It has long been recognized that financial and economic indicators such as return on assets, return on capital, economic value added, and others, when taking alone, lead to [1, p. 689] short-term performance at the expense of long-term performance, by reducing expenditure on new product development, human resources, customer development, and so on.

© The Author(s), under exclusive license to Springer Nature Switzerland AG 2023
Z. Hu et al. (Eds.): ICAILE 2023, LNDECT 180, pp. 141–151, 2023.
https://doi.org/10.1007/978-3-031-36115-9_14

142   D. Tavrov et al.

Multidimensional performance measure approaches try to consider many aspects of organizational performance at once. For instance, a widely used performance prism [2] aims at measuring different facets, including defining stakeholder satisfaction, laying out strategies and processes needed to achieve it, defining the capabilities needed to achieve it, and involving contributions of stakeholders. Another popular approach called balanced scorecard [3] suggests a wide range of metrics to be tracked to assess organization's success, including operational improvement metrics, metrics for innovation, employee capabilities, and others. Some authors argue [4] that performance management practices work better when combined with human resource management techniques, as maintaining organizational social climate is key to improving efforts of the employees.

However, these and other approaches presented in the literature are not very formalized and often represent broad and generic recommendations.

New challenges such as climate change, biodiversity deterioration, and sustainability issues require new approaches to measuring organizational performance. European Commission proposes a new concept of Industry 5.0 [5] in place of Industry 4.0, which was profit-driven and focused on enhanced efficiency by incorporating technological advancements, artificial intelligence, and digitalization into production processes [6]. The three pillars of Industry 5.0 are [5]:

- human-centric approach, whereby technological advances are used to adapt production to workers' needs, with an emphasis on training and well-being;
- sustainability, which means that circular processes are needed to decouple resource and material use from negative impact on the environment;
- resilience, which means that industrial production must become more robust against disruptions and crises.

Thus, a new framework for evaluating organizational performance is needed. Industry 5.0 paradigm requires to re-imagine existing metrics and indicators, to allow [7] measurement of the above criteria and guide accelerated compliance to them. Many of new metrics and indicators will be qualitative. It is also recognized [7] that evaluation of compliance should be more "user-friendly." In our opinion, a holistic approach, when objective parameters are supplemented with linguistic and qualitative data, is essential.

The concept of Industry 5.0 is significantly understudied. Sustainability dimension of Industry 5.0 was analyzed in [8] using interpretive structural modeling. Authors of [9] propose a reference model for the human-centric dimension. However, no comprehensive approach exists for evaluating organizational performance in terms of Industry 5.0, taking into consideration its imprecise and ill-defined criteria. In our opinion, fuzzy methods should be applied, which are tailored to subjective and imprecise data. Using subjective measures for organizational performance is not new [10], and methods such as fuzzy TOPSIS [11], fuzzy AHP [12], and other fuzzy methods for multicriteria decision making [13, 14] were successfully used.

In this work, we propose a novel framework for assessing organizational performance according to Industry 5.0 criteria. We formalize the whole process by offering an hierarchical structure of organizational performance, where each specific indicator describing the organization is evaluated using words, which are then processed using computing with words (CWW) methodology to arrive at the overall assessment. According to [15],

CWW is a way of "dealing with real-world problems [by means of] exploiting the tolerance for imprecision," which is superior to applying ordinary fuzzy sets. Application of CWW can enable regulators to perform evaluation using words, treat subjective judgments in a mathematically coherent way, and reach easily interpreted decisions that can be used to formulate recommendations.

The rest of the paper is organized as follows. Section 2 describes a proposed framework for assessing compliance to Industry 5.0 criteria. In Sect. 3, we discuss one specific area related to functional states of employees. We detail the inner workings of a CWW procedure and offer a model exercise that illustrates the proposed approach. We conclude in Sect. 4 and indicate directions for further research.

2 Framework for Assessing Organizational Performance

2.1 Outline of Assessment Process

In the framework we propose, assessment of organizational performance for compliance with the criteria of Industry 5.0 is done through an hierarchical and distributed process [16] according to Fig. 1, which is based on [5, 17]. Evaluations at lower levels are aggregated and propagated forward, level by level, until a final decision is reached.

Indicators numbered from 1 to N in Fig. 1, which are fed into the lowest-level nodes, can be evaluated as any of the following:

- binary values corresponding to either absence or presence of a given characteristic (e.g., whether certain recycling practices are in place or not);
- numbers, obtained by objective measurements (e.g., levels of carbon emissions). Every number needs to be converted to a [0; 1] scale, so that higher numbers contribute more to the overall performance level;
- words, which correspond to subjective, ill-defined, or immeasurable indicators (e.g., employee satisfaction or engagement).

Arcs connecting indicators to each other are weighted, to allow for possibility of some indicators having more influence on the overall decision. Weights are expressed as words, which are customary in practice of perceptual computing and reflects imprecise nature of interrelations between different indicators. Some weights can be set to 0, if lower-level indicators have no effect on a given middle-level indicator.

Level of compliance of each middle-level indicator in Fig. 1 such as "Human-technology collaboration" or "Supply chains" is obtained as a weighted average of all lower-level indicators it is connected to (not shown), weighted accordingly. Such indicators can be interesting in themselves, as corresponding performance levels can be used to identify weak links, or to make recommendations for improvement.

The general architecture of the decision-making process shown in Fig. 1 captures all the main aspects of Industry 5.0 paradigm. Of course, for each specific industry and even individual enterprise, indicators can be added, removed, or modified.

2.2 Representation of Words as Type-2 Fuzzy Sets

To process indicators expressed as words, we need to represent them in a computer. Often words are represented as fuzzy sets [18]. A fuzzy set A is characterized by a membership function $\mu_A(x): X \to [0; 1]$. Unlike an ordinary set, to which any element either belongs or does not, any element x can belong to a fuzzy set A with some membership degree $\mu_A(x)$.

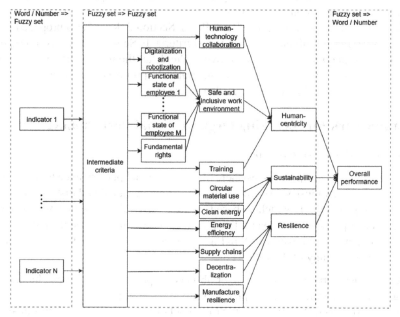

Fig. 1. Framework for assessing organizational performance according to Industry 5.0 criteria

However, type-1 fuzzy sets are not sufficient to model words, and instead we use type-2 fuzzy sets [19], in particular the so-called interval type-2 fuzzy sets (IT2FS). Each IT2FS $\tilde{A}$ is characterized by a lower membership function (LMF), $\underline{\mu}_{\tilde{A}}(x)$, and an upper membership function (UMF), $\overline{\mu}_{\tilde{A}}(x)$. Membership of each element x in an IT2FS is described by an interval $[\underline{\mu}_{\tilde{A}}(x); \overline{\mu}_{\tilde{A}}(x)]$, which captures the uncertainty associated with a linguistic concept being modeled.

When LMF and UMF are trapezoidal, i.e. when their functional form is

$$\mu(x; a, b, c, d) = \begin{cases} \frac{x-a}{b-a}, & a \leq x \leq b \\ 1, & b \leq x \leq c \\ \frac{d-x}{d-c}, & c \leq x \leq d \\ 0, & \text{otherwise} \end{cases} \quad (1)$$

we obtain what is called a trapezoidal IT2FS (Fig. 2). The shaded area is called the footprint of uncertainty (FOU). Thus, each trapezoidal IT2FS $\tilde{A}$ used in this paper can be uniquely represented as a tuple $(a_l, a_r, o_l, o_r, b_l, b_r)$.

2.3 Computing with Words in the Perceptual Computer

As noted above, in this work, we propose to use one of the existing methodologies for CWW called perceptual computing, which is described in its entirety in [16]. A perceptual computer (Per-C) is a specific architecture for CWW, in which words are used in decision making in three stages:

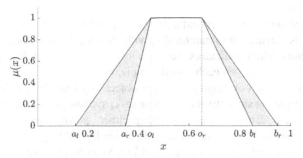

Fig. 2. Trapezoidal interval type-2 fuzzy set

- input words are mapped to their IT2FS models according to a codebook, which maps each word to an appropriate IT2FS;
- at each node in Fig. 1, input IT2FSs are processed using linguistic weighted average (LWA):

$$Y_{LWA} = \frac{\sum\limits_{i=1}^{n} X_i W_i}{\sum\limits_{i=1}^{n} W_i} \qquad (2)$$

where X_i are IT2FSs representing evaluations, and W_i are IT2FSs representing weights on corresponding arcs. At each node we obtain output IT2FSs, which are then fed as inputs to upper-level nodes, and so on until the overall performance is determined. Discussion of LWA lies beyond the scope of our paper;

- the IT2FS corresponding to "Overall performance" can be used to rank different companies, or it can be converted into an aggregate compliance level, expressed either as a word from the codebook or as a number from [0; 1].

To condense an IT2FS $\tilde{A}$ to a number, average centroids can be applied:

$$c(\tilde{A}) = \frac{c_l(\tilde{A}) + c_r(\tilde{A})}{2} \qquad (3)$$

where $c_l(\tilde{A})$ and $c_r(\tilde{A})$ can be computed, e.g., using an algorithm described in [20].

3 Subdomain of Assessing Functional State of Employees

3.1 Assessing Functional State as Part of Human-Centric Approach

In this section, we reuse and modify our previous approach in [21] and fully specify application of Per-C to a subdomain of assessing functional states (FuncS) of employees. According to Fig. 1, it is part of the human-centricity criterion, which is aligned with recommendations of [5], where it is stated that under Industry 5.0 guidelines, the well-being of employees should encompass not only physical health in the workplace, but also mental health, especially in the current era of digital technologies that can lead to frequent burnouts.

By functional state we understand a complex of indicators shown in Fig. 3. Assessment of FuncS is carried out hierarchically, with each level corresponding to a certain group of indicators. These indicators are by no means exhaustive and/or obligatory, and can be modified depending on each specific application.

Experts assess FuncS of employees either objectively (e.g., by conducting medical tests or examinations) or subjectively (e.g., to assess psychological state of an employee). Different indicators can be evaluated using either words (from a corresponding codebook) or numbers. Each indicator is evaluated independently.

For the sake of concreteness, indicators in Fig. 3 can be evaluated as follows:

- indicators A, B, E, F, L are evaluated using objective tests and protocols, and then mapped into [0; 1], where higher numbers stand for higher compliance;
- indicators C, D, G, H, I, J, K are evaluated by experts using words, because they are highly subjective and do not lend themselves to objective testing;
- indicators M–Q are computed using LWA of respective lower-level indicators, as shown by connecting lines in Fig. 3.

We can use numbers instead of words because each number x can be considered an IT2FS with parameters $a_l = a_r = o_l = o_r = b_l = b_r = x$.

3.2 Components of the Perceptual Computer

In the Per-C discussed in this paper, words belong to two different categories:

- words used by experts to set levels of compliance of individual indicators;
- words assigned by (a different set of) experts to each arc in Fig. 3 as weights.

All the words are represented as IT2FS with the same universe of discourse, [0; 1]. Adhering to recommendations from [21, 22], we chose the following words:

- for assessment: good (G), normal (N), satisfactory (S), acceptable (A), unsatisfactory (U);
- for weights: essential effect (Es), significant effect (Sg), average effect (Av), insignificant effect (In), little or no effect (Lt).

The same words can have different IT2FS representation. Experts from different domains can have different levels of uncertainty associated with the same word. We propose to create two codebooks: one for evaluating indicators C–D and K from Fig. 3, and another one with the same words but used for evaluating indicators G–J. We will index these words by 1 or 2 to distinguish the codebooks they belong to.

IT2FS models for each word should be created using, for instance, the Hao-Mendel approach [23] or similar. As we did not have access to many experts needed for this approach, we modeled words using a person FOU approach [24], which involves interviewing only one expert. The third author of this paper, based on her expertise, provided data to create trapezoidal IT2FS models (1) of all the words. Corresponding parameters are given in Table 1 in reference to Fig. 2 (all the numbers in this paper are shown up to two significant digits).

Weights corresponding to arcs from Fig. 3 were assigned as words from the codebook by the second author of this paper based on his expertise (Table 2).

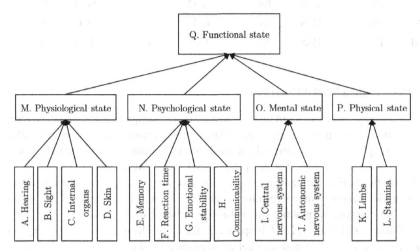

Fig. 3. Graph model of the functional state of enterprise employees

Table 1. Parameters of IT2FSs used to represent words from the codebook

Word	Parameters					
	a_l	a_r	o_l	o_r	b_l	b_r
G_1	0.92	0.93	0.95	1.00	1.00	1.00
N_1	0.79	0.81	0.87	0.93	0.95	0.97
S_1	0.68	0.70	0.74	0.82	0.87	0.89
A_1	0.60	0.60	0.64	0.70	0.74	0.76
U_1	0.00	0.00	0.00	0.60	0.64	0.65
G_2	0.83	0.86	0.92	1.00	1.00	1.00
N_2	0.73	0.74	0.78	0.87	0.92	0.94
S_2	0.62	0.63	0.63	0.74	0.78	0.79
A_2	0.54	0.55	0.55	0.63	0.65	0.66
U_2	0.00	0.00	0.00	0.54	0.57	0.59
Es	0.86	0.88	0.92	1.00	1.00	1.00
Sg	0.63	0.66	0.75	0.87	0.91	0.94
Av	0.37	0.40	0.50	0.67	0.76	0.78
In	0.19	0.25	0.35	0.43	0.52	0.55
Lt	0.00	0.00	0.00	0.25	0.37	0.39

Table 2. Linguistic weights of arcs in the Per-C

Arc	Weight	Arc	Weight	Arc	Weight	Arc	Weight
A–M	Sg	E–N	Sg	I–O	Es	M–Q	Sg
B–M	Es	F–N	Es	J–O	Sg	N–Q	Es
C–M	Av	G–N	Es	K–P	In	O–Q	Sg
D–M	Lt	H–N	Av	L–P	Av	P–Q	Av

3.3 Results for a Model Exercise

To illustrate applicability of our approach to assessing subjective and imprecise criteria such as functional state, we created model data for five synthetic employees with different levels of compliance to specified indicators (Table 3). Evaluations for each indicator were assigned by the fourth author of this paper based on his experience.

Resulting IT2FSs for each employee are given in Fig. 4. To complement visual representation, Table 3 shows resulting centroids (3), which can be used, for instance, to rank the employees, making this submodule useful in its own right. As part of a bigger framework in Fig. 1, LWA is applied to the IT2FS from Fig. 4 and corresponding IT2FS calculated at other nodes to get an aggregate level of "Safe and inclusive work environment," and so on up the hierarchy until the overall performance is determined.

Results for employees 2 and 5 are especially interesting. Their IT2FSs and centroids are very close, even though they differ in indicators B and F, which are "essential" for estimating the overall state. Thus, the Per-C takes into account the whole multitude of parameters and reaches conclusions that are robust to imprecise and subjective data.

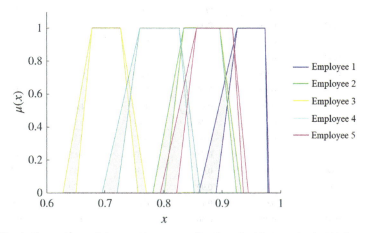

Fig. 4. Interval type-2 fuzzy sets representing FuncS of five synthetic employees

Table 3. Evaluations of indicators from Fig. 3 for five synthetic employees and centroids of the corresponding output IT2FSs

#	Evaluation												Centroid	
	A	B	C	D	E	F	G	H	I	J	K	L	c_l	c_r
1	0.95	0.97	N_1	N_1	0.93	0.95	G_2	G_2	G_2	G_2	N_1	0.98	0.93	0.94
2	0.88	0.96	N_1	S_1	0.92	0.96	G_2	N_2	N_2	N_2	S_1	0.75	0.86	0.87
3	0.70	0.73	A_1	S_1	0.82	0.85	S_2	A_2	A_2	A_2	S_1	0.72	0.70	0.71
4	0.94	0.97	N_1	N_1	0.71	0.62	S_2	A_2	S_2	S_2	N_1	0.98	0.78	0.79
5	0.89	0.87	N_1	S_1	0.99	0.87	G_2	N_2	G_2	G_2	S_1	0.72	0.87	0.88

4 Conclusions and Further Research

In this work, we introduced a new framework for assessing organizational performance in accordance with the concept of Industry 5.0. It captures three main pillars of Industry 5.0—human-centric view of production, sustainability, and resilience. The overall performance level is determined using perceptual computing that allows us to evaluate adherence to subjective and ill-defined criteria of Industry 5.0 using words.

The proposed framework enables the regulators to compare different enterprise by ranking them according to the centroids of the overall assessment levels, as well as track the dynamics over some period of time, e.g. from year to year.

We treated in more detail a subdomain of human-centricity related to assessment of functional states of employees and discussed how the obtained results fit in the general framework. To assess the FuncS, one needs to combine objective measurements and subjective evaluations of psychological and mental state of employees, which is why we decided to use perceptual computing. We described in detail a codebook for our perceptual computer and a method of converting the output IT2FS into a number. Model exercise related to assessing FuncS of synthetic employees showed that the perceptual computer is user-friendly, and the results produced by it are easily interpreted.

Further research can be led in the direction of full specification of a working prototype of a perceptual computer that can be used to evaluate organizational performance using Industry 5.0 criteria. Work towards implementation of this approach in practice could help guide the process of defining appropriate metrics and measures to be used as inputs to such a system.

References

1. Langfield-Smith, K., Smith, D., Andon, P., et al.: Management Accounting: Information for Creating and Managing Value, 8th edn. McGraw-Hill, North Ryde (2017)
2. Neely, A., Adams, C., Crowe, P.: The performance prism in practice. Meas. Bus. Excell. **5**(2), 6–12 (2001)
3. Kaplan, R.S.: Conceptual foundations of the balanced scorecard. In: Chapman, C.S., Hopwood, A.G., Shields, M.D. (eds.) Handbook of management accounting research, Vol. 3, pp. 1253–1269 (2009)

4. Bourne, M., Pavlov, A., Franco-Santos, M., Lucianetti, L., Mura, M.: Generating organisational performance: the contributing effects of performance measurement and human resource management practices. Int. J. Oper. Prod. Manag. **33**(11/12), 1599–1622 (2013)
5. Breque, M., De Nul, L., Petridis, A.: European Commission, Directorate-General for Research and Innovation. Industry 5.0: towards a sustainable, human-centric and resilient European industry, Publications Office (2021)
6. Aliyev, A.G., Shahverdiyeva, R.O.: Scientific and methodological bases of complex assessment of threats and damage to information systems of the digital economy. IJIEEB **14**(2), 23–38 (2022)
7. Renda, A., Schwaag, S.S., Tataj, D., et al.: European Commission, Directorate-General for Research and Innovation. Industry 5.0, a transformative vision for Europe: governing systemic transformations towards a sustainable industry. Publications Office of the European Union (2022)
8. Ghobakhloo, M., Iranmanesh, M., Mubarak, M.F., et al.: Identifying industry 5.0 contributions to sustainable development: a strategy roadmap for delivering sustainability values. Sustain. Prod. Consumption **33**, 716–737 (2022)
9. Lu, Y., Zheng, H., Chand, S., et al.: Outlook on human-centric manufacturing towards Industry 5.0. J. Manuf. Syst. **62**, 612–627 (2022)
10. Singh, S., Darwish, T.K., Potocnik, K.: Measuring organizational performance: a case for subjective measures. Br. J. Manag. **27**, 214–224 (2016)
11. Gupta, H.: Assessing organizations performance on the basis of GHRM practices using BWM and Fuzzy TOPSIS. J. Environ. Manage. **226**, 201–216 (2018)
12. Modak, M., Pathak, K., Ghosh, K.K.: Performance evaluation of outsourcing decision using a BSC and Fuzzy AHP approach: a case of the Indian coal mining organization. Resour. Policy **52**, 181–191 (2017)
13. Verma, R., Sharma, B.D.: A new inaccuracy measure for fuzzy sets and its application in multi-criteria decision making. IJISA **6**(5), 62–69 (2014)
14. Dey, P.K., Ghosh, D.N., Mondal, A.C.: IPL team performance analysis: a multi-criteria group decision approach in fuzzy environment. IJITCS **7**(8), 8–15 (2015)
15. Zadeh, L.A.: Fuzzy logic = computing with words. IEEE Trans. Fuzzy Syst. **4**(2), 103–111 (1996)
16. Mendel, J.M., Wu, D.: Perceptual Computing. Aiding People in Making Subjective Judgments. John Wiley & Sons Inc, Hoboken, New Jersey (2010)
17. European Commission, Directorate-General for Communication, Towards a sustainable Europe by 2030: reflection paper, Publications Office (2019)
18. Zadeh, L.A.: Fuzzy sets. Inf. Control **8**(3), 338–353 (1965)
19. Mendel, J.M.: Type-2 fuzzy sets and systems: an overview. IEEE Comput. Intell. Mag. **2**(1), 20–29 (2007)
20. Wu, D., Nie, M.: Comparison and practical implementations of type-reduction algorithms for type-2 fuzzy sets and systems. In: FUZZ-IEEE 2011, pp. 2131–2138 (2011)
21. Tavrov, D., Temnikova, O., Temnikov, V.: Perceptual computing based method for assessing functional state of aviation enterprise employees. In: Chertov, O., Mylovanov, T., Kondratenko, Y., Kacprzyk, J., Kreinovich, V., Stefanuk, V. (eds.) ICDSIAI 2018. AISC, vol. 836, pp. 61–70. Springer, Cham (2019). https://doi.org/10.1007/978-3-319-97885-7_7
22. Pavlenko, P., Tavrov, D., Temnikov, V., Zavgorodniy, S., Temnikov, A.: The method of expert evaluation of airports aviation security using perceptual calculations. In: 2018 IEEE 9th International Conference on Dependable Systems, Services and Technologies (DESSERT), pp. 406–410 (2018)

23. Hao, M., Mendel, J.M.: Encoding words into normal interval type-2 fuzzy sets: HM approach. IEEE Trans. on Fuzzy Systems **24**(4), 865–879 (2016)
24. Mendel, J.M., Wu, D.: Determining interval type-2 fuzzy set models for words using data collected from one subject: Person FOUs. In: 2014 IEEE International Conference on Fuzzy Systems (FUZZ-IEEE), pp. 768–775 (2014)

Petroleum Drilling Monitoring and Optimization: Ranking the Rate of Penetration Using Machine Learning Algorithms

Ijegwa David Acheme[1], Wilson Nwankwo[3]([⊠]) [iD], Akinola S. Olayinka[2], Ayodeji S. Makinde[1], and Chukwuemeka P. Nwankwo[1]

[1] Department of Computer Science, Edo State University, Uzairue, Nigeria
[2] Department of Physics, Edo State University, Uzairue, Nigeria
[3] Department of Cyber Security, Delta State University of Science and Technology, Ozoro, Nigeria
wnwankwo@dsust.edu.ng

Abstract. Oil prospection and exploration is a major economic venture requiring enormous expertise and technologies in developing countries. In this study, we set out to examine an important issue that affects the efficiency of oil drilling operations that is, the rate of penetration (ROP) during oil exploration. We attempt to apply different machine learning algorithms in order to discover the most efficient algorithm that could be used for the prediction of the ROP. We use the North Sea oil field data set while eighteen intelligent models were developed. On comparative analysis of the models, it is shown that the random forest algorithm exhibits the best efficiency in the prediction workflow having an RMSE value of 0 0010, and an R2 score of 0 891. Findings also showed that significant parameters of concern during drilling operations were the measured depth, bit rotation per minute, formation porosity, shale volume, water saturation, and log permeability respectively. We conclude that the result of this study provides a good template to guide the selection of machine learning algorithms to drive solutions for the optimization of oil drilling operations.

Keywords: Petroleum drilling · Drilling operations · Rate of Penetration · Machine learning · Optimization

1 Introduction

In different jurisdictions including developed countries oil and mineral drilling operations had been designated as very costly ventures in the oil and mineral production workflow [1–5]. Hence continuous research efforts, targeted towards the optimization of the drilling process so as to ensure a reduction in the total costs associated with the drilling process has been reported. However, despite the constant improvement in tools, products, and processes, machine learning methods have only very recently began to play

© The Author(s), under exclusive license to Springer Nature Switzerland AG 2023
Z. Hu et al. (Eds.): ICAILE 2023, LNDECT 180, pp. 152–164, 2023.
https://doi.org/10.1007/978-3-031-36115-9_15

Petroleum Drilling Monitoring and Optimization: Ranking the Rate of Penetration   153

significant roles in oil drill optimization, this has been mostly enhanced by the current availability of massive datasets [6]. Machine learning (ML) and Artificial Intelligence (AI) models have been recognized as pervasive tools that find interesting applications in the various sectors where intelligent computerization is required [7–19]. In the domain of oil exploration and drilling, ML and AI also hold promising expectations. This is because, drilling and oil exploration operations are data and knowledge-driven, hence, efficient processing of the massive amounts of data generated from several sensor networks at the oil rigs to aid cannot be overemphasized. Presently, there are implementations of Real-Time Operation Centers (RTOC) with major oil companies spending thousands of dollars, in the IT infrastructure to read drilling data from rigs in real-time. These readings enable experts to analyze real-time data in the centers, leading to faster decision making, reduction in stuck pipe incidents, less hole cleaning issues and fluid losses events, while also increasing the number of wells that can be monitored with the same number of personnel [20]. The availability of this data has also provided the necessary foundation for the implementation of artificial intelligence and machine learning techniques for the development of smart models for more accurate and robust real-time drilling performance monitoring and optimization.

With current drilling rigs holding massive amounts of instrumentation for the collection of parameters from almost every equipment installed in the drilling rig, using sensors measuring their states, and allowing for remote and safe operations, there has been an exponential growth of data generated at oil rigs which has laid the foundation for the development of machine learning models for predictive analytics and development of decision support systems.

As researchers continue to explore these datasets generated at oil rigs, one challenge comes to bear, selecting the right machine learning algorithms and features for the effective prediction of ROP because the ROP is influenced by several factors which have complex relationships, the extent of the influence of these factors also vary as some are more relevant than others. A machine learning model built with many of the lowly influential factors or with a less efficient algorithm is not likely to give satisfactory results, hence this research work sets out to investigate and rank the factors that have been reported in literature for ROP prediction as well as implement and compare several ML algorithms. Identifying the most important factors will improve prediction performance, computation speed and training time while providing better explain ability for these models [21].

2 Related Literature

The rate of penetration (ROP) can simply be defined as the speed at which a wellbore is being drilled. This can be manually calculated by measuring the depth per time intervals in units of feet or meters per hour. The high ROP values imply fast drilling which in turn emphasize an increased productivity of the drilling process. Because ROP is such a direct measurement of the overall time required to drill an oil well, reducing this time in order to achieve higher ROP is a key optimization strategy for oil companies. In this section, different strategies that have been employed in optimizing ROP are presented. These strategies have focused on modelling and predicting the ROP using selected drilling parameters which can be controlled on the surface such as weight-on-bit (WOB), rotary speed (RPM) etc., and they can be broadly classified into traditional and data-driven models Traditional models are mathematical equations that have been formulated from experiments and field experience, while data- driven models refer to machine learning algorithms for the prediction of ROP.

2.1 Traditional ROP Models

According to [22], one of the earliest mathematical models for ROP prediction was established by Maurer in 1962 who used a rock cratering technique to create a formula with the parameters as; bit diameter, rock strength, weight on bit (WOB), and rotations per minute (RPM). The Bingham model, reported in [23] is another early mathematical equation-based ROP prediction model, it uses comparable input parameters with an additional empirical constant, "k," which represents a parameter that was dependent on formation. Eckel presented another early traditional model in [24] this model investigated the effects of mud on ROP. Among these early models, Bourgoyne and Young's (BY) model [25] has received the most attention and been widely reported. It included additional geological and physical factors which included; the formation strength, normal compaction trend, under compaction, differential pressure, bit diameter and bit weight, rotational speed, tooth wear, and bit hydraulics.

2.2 Data Science Models

The application of data science and machine learning techniques for the prediction and optimization of ROP seeks to build predictive models of ROP using data collected during drilling. Such models make use of parameters measured at the surface like weight on bit, rotations per minute, and flow rate as input variables in order to predict the rate of penetration. Generally, the process of oil drilling involves massive collection of data from both surface and sub-surface areas using IOT sensors. These sensors are able to collect large volumes of data about the state of the bit underground. The data collected are used to plot, analyze and control bit performance in this case the ROP. The availability of these datasets have laid the foundation for the building of ML models for the prediction of ROP because ML models the relationship between input variables in order to predict an output (target) variable. In these section we review different studies that have presented ML models for ROP prediction.

Research works on the implementation of neural networks machine learning algorithm have been mostly reported. These neural network models have utilized different input parameters for ROP prediction [26]. [27] presented a hybrid neural network model that utilized Savitzky–Golay (SG) smoothing filter for removing noise from extracted data for the prediction of rate of penetration. [5] Presented a feedforward neural network model for the prediction of rate of penetration. The work utilized selected features such as; bit rotations per minute, mud flow, weight on bit, and differential pressures. These features comprised the input variables. Datasets utilized for the development of their model were obtained from laboratory simulations as well an oil field. The predicted ROP values are then used to identify bit malfunction or failure by comparison of the

Petroleum Drilling Monitoring and Optimization: Ranking the Rate of Penetration 155

predicted values with the actual measured value. Any observed deviation indicates an underperformance of the bit and this can serve as a warning flag.

In the study by [28], an artificial neural network (ANN) model was used to predict hydraulics pump pressure and to provide early warnings, the model was implemented through the fitting tool of MATLAB, the sensitivity of the selected input parameters were analyzed by using a forward regression method. Data sets were obtained from selected well samples and were used for the validation of the model. The model predicted pump pressure against well depth in similar formations. While neural networks are powerful tools which have been very efficient in handling high dimensional modelling. The works in [29–35] argue that they generally underperform in comparison to simpler machine learning models such as random forest when applied to low dimensional problems, as such these simpler models have reported higher prediction accuracies.

Traditionally, ROP is measured in real-time by instruments which carryout Measurement-While-Drilling (MWD). Higher ROP values means that faster distance is being achieved in drilling, and it is the desire of oil drilling companies to achieve deeper depths in shorter time intervals in order to save time and money, hence the need for the optimization of the rate of penetration. Several variables influence the rate ROP, some of them are WOB and RPM which can be controlled on the surface. The other variables (PHIF, VSH, SW, and KLOGH) are a function of the soil formation. Unfortunately, ROP does not always increase proportionally with increasing or decreasing values of these variables, as observation has shown that it increases initially until a point called the founder point or the sweet spot (optimum point) from where it begins to wane, The values of the outside variables must be increased from this point in order to maintain the optimum performance.

3 Proposed Methodology

This work adopts the standard procedure for developing machine learning and data science models. The methodology as shown in Fig. 1 comprises of five (5) phases, these are:

(1) Understanding the Business problem
(2) Data cleaning and Preparation
(3) Data modelling
(4) Model Evaluation
(5) Model Operationalization

3.1 Maintaining the Integrity of the Specifications

The template is used to format your paper and style the text. All margins, column widths, line spaces, and text fonts are prescribed; please do not alter them. You may note peculiarities. For example, the head margin in this template measures proportionately more than is customary. This measurement and others are deliberate, using specifications that anticipate your paper as one part of the entire proceedings, and not as an independent document. Please do not revise any of the current designations.

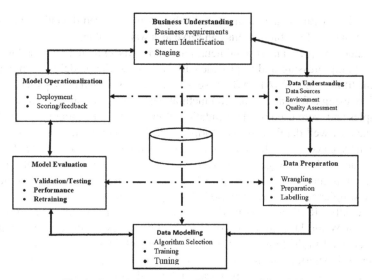

Fig. 1. Data Science Methodology (Nwankwo, 2020)

3.2 Understanding the Problem and Data Collection

The dataset utilized for this research is the open source volve data. This is a complete set of data from a North Sea oil field available for research, study and development purposes arranged from two original data; real-time drilling data and a Computed Petrophysical Output (CPO) log data from well number 15/9-F-15 in the Volve Oil Field in the North Sea [32] This dataset comprises of seven (7) input variables and one (1) target variable. These are.

i. Depth (measured depth)
ii. WOB (weight on bit)
iii. SURF_RPM (rotation per minute at the surface)
iv. PHIF (formation porosity)
v. VSH (shale volume)
vi. SW (water saturation)
vii. KLOGH (log permeability)
viii. ROP_AVG (rate of penetration average TARGET VARIABLE).

The dataset contained a total of one hundred and fifty (150) records with eight (8) features.

Table 1. Snapshot of the Dataset

	Depth	WOB	SURF_RPM	ROP_AVG	PHIF	VSH	SW	KLOGH
0	3305	26217.864	1.314720	0.004088	0.086711	0.071719	1.000000	0.001000
1	3310	83492.293	1.328674	0.005159	0.095208	0.116548	1.000000	0.001000

(*continued*)

Table 1. (*continued*)

	Depth	WOB	SURF_RPM	ROP_AVG	PHIF	VSH	SW	KLOGH
2	3315	97087.882	1.420116	0.005971	0.061636	0.104283	1.000000	0.001000
3	3320	54793.206	1.593931	0.005419	0.043498	0.110040	1.000000	0.001000
4	3325	50301.579	1.653262	0.005435	0.035252	0.120808	1.000000	0.001000
...	...	...	...	...	...	...	...	...
146	4065	71081.752	2.104258	0.008808	0.087738	0.291586	1.000000	0.162925
147	4070	72756.626	2.333038	0.008824	0.019424	0.503175	1.000000	−0.001124
148	4075	83526.789	2.333326	0.008799	0.054683	0.689640	1.000098	0.002261
149	4080	84496.549	2.334673	0.008375	0.022857	0.640100	1.000000	0.001000
150	4085	86658.559	2.331339	0.008454	0.022857	0.640100	1.000000	0.001000

3.3 Methodology

In order to achieve the objectives of this research, the selected machine learning algorithms were built using the dataset described in Table 1. The steps of the entire process are shown in Fig. 2.

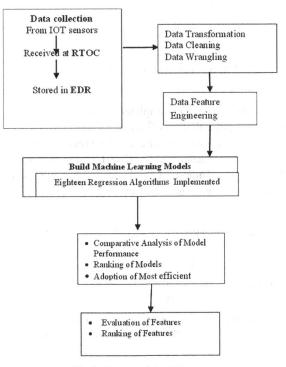

Fig.2. Proposed Architecture

3.4 Data Modelling and Evaluation

Exploratory data analysis was then performed to obtain additional knowledge from the data and uncover hidden patterns. With the selected features, the data is then divided in the ratio of 70:30 for training and testing purposes. This is then passed on to machine learning algorithms. Table one shows the machine learning algorithms used and their performance comparison.

In order to estimate outliers in the dataset, the cooks distance outlier detection was used, (Fig. 3). The Cook's Distance is an estimate of a data point's influence. It considers both the leverage and residual of each observation. Cook's Distance is a calculation that summarizes how much a regression model changes when the ith observation is removed.

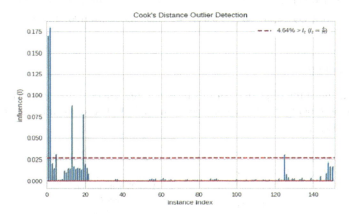

Fig. 3. Cook's distance outlier detection

The models shown in Table 2. were implemented in order to compare the prediction accuracy of the target variable, ROP_AVG using the selected features. The metrics for comparison are the established metrics for evaluating regression models such as MAE, MSE, RMSE, R2 etc. From our results, the random forest regressor model outperformed all the others and it is the ranked number 1, while the passive aggressive regressor showed the least prediction accuracy and ranks number 18.

Table 2. Selected Regression Models and their Performances

	Model	MAE	MSE	RMSE	R2	RMSLE	MAPE	TT (Sec)
rf	Random Forest Regressor	0.0006	0.0000	0.0010	−0.0891	0.0010	0.1082	0.407
gbr	Gradient Boosting Regressor	0.0006	0.0000	0.0010	−0.1076	0.0010	0.1133	0.045

(*continued*)

Petroleum Drilling Monitoring and Optimization: Ranking the Rate of Penetration   159

Table 2. (*continued*)

	Model	MAE	MSE	RMSE	R2	RMSLE	MAPE	TT (Sec)
et	Extra Trees Regressor	0.00 06	0.00 00	0.00 09	−0.180 5	0.00 09	0.1064	0.363
huber	Huber Regressor	0.00 07	0.00 00	0.00 10	−0.185 6	0.00 10	0.1213	0.029
dt	Decision Tree Regressor	0.00 07	0.00 00	0.00 12	−0.361 7	0.00 12	0.1298	0.014
knn	K Neighbors Regressor	0.00 09	0.00 00	0.00 13	−0.536 4	0.00 12	0.1572	0.059
ridge	Ridge Regression	0.00 07	0.00 00	0.00 10	−0.551 6	0.00 10	0.1219	0.012
br	Bayesian Ridge	0.00 07	0.00 00	0.00 10	−0.553 0	0.00 10	0.1224	0.014
en	Elastic Net	0.00 09	0.00 00	0.00 12	−0.572 9	0.00 12	0.1531	0.013
lightg bm	Light Gradient Boosting Machine	0.00 07	0.00 00	0.00 11	−0.577 1	0.00 11	0.1281	0.046
lr	Linear Regression	0.00 07	0.00 00	0.00 10	−0.586 1	0.00 10	0.1227	0.304
lar	Least Angle Regression	0.00 07	0.00 00	0.00 10	−0.586 1	0.00 10	0.1227	0.013
lasso	Lasso Regression	0.00 09	0.00 00	0.00 12	−0.604 0	0.00 12	0.1532	0.014
llar	Lasso Least Angle Regression	0.00 09	0.00 00	0.00 12	−0.692 9	0.00 12	0.1531	0.014
dum my	Dummy Regressor	0.00 09	0.00 00	0.00 12	−0.692 9	0.00 12	0.1531	0.013
omp	Orthogonal Matching Pursuit	0.00 07	0.00 00	0.00 11	−0.719 7	0.00 11	0.1262	0.012
ada	AdaBoost Regressor	0.00 07	0.00 00	0.00 11	−0.784 5	0.00 11	0.1267	0.067
par	Passive Aggressive Regressor	0.00 79	0.00 01	0.00 80	−138.4 456	0.00 80	1.0000	0.013

3.5 Evaluation of the Random Forest Regressor

Having established the random forest (rf) algorithm as the most efficient among the chosen eighteen (18) machine learning algorithms tested with the dataset, further evaluation analysis of the algorithm was carried out, this includes the residual plot, error plot, learning and validation curves. Furthermore, feature importance ranking was also carried out to rank the most important features for predicting ROP. These results are shown in Fig. 4, 5 and 6

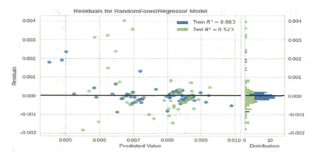

Fig. 4. Residuals for the random forest algorithm

The residual plot was used to confirm the model's fit with assumptions of constant variance, normality and independence of errors. The plot reveals the difference between observation and fitted values. The plot (Fig. 4) shows randomly scattered points which are maintain an approximately constant width about the line of identity, this is indicative of a good model as it is close to a null residual plot.

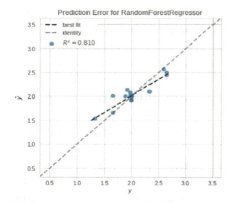

Fig. 5. Prediction error for the random forest algorithm

The model's performance was further investigated with the learning and validation curves (Fig. 6). These plots show a model's performance as the set of training data increases or over a defined period of time. They are useful for models built with datasets that are incremental. The training curves shows how well the model learns, while the validation curve shows how well the model with generalize with yet to be seen values.

3.6 Feature Importance and Ranking

The decrease in node impurity weighted by the probability of reaching that node is used to calculate feature importance. The node probability can be calculated by the number of

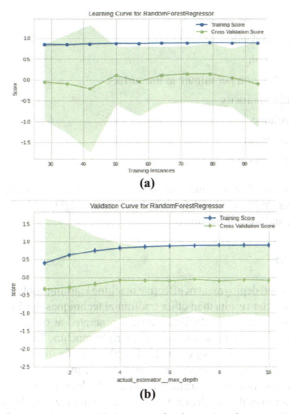

Fig. 6. Learning and validation curve for the random forest regressor

samples that reach the node, divided by the total number of samples. The more important features have higher values (see Fig. 7).

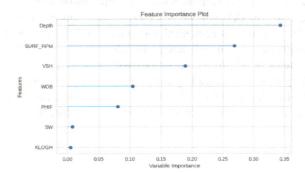

Fig. 7. Feature importance

KEY:

Depth : Measured depth

WOB: Weight on bit

SURF_RPM: Rotation per minute at the surface

PHIF: Formation porosity

VSH: Shale volume

SW: Water saturation

KLOGH: Log permeability.

As shown in Fig. 7, the features in increasing order of importance are; measured depth, rotation per minute on the surface, shale volume, weight on bit, formation porosity, water saturation and log permeability.

4 Conclusion

The development and deployment of effective machine learning application for ROP prediction promises better results than other traditional techniques that have been reported. This is so because of more availability datasets generated at oil rigs, however, a clear challenge comes to bear, which is selecting the right machine learning algorithms and features. A model built with many of the lowly influential factors or with a less efficient algorithm is not likely to give satisfactory results. Hence in this research work, we investigated eighteen (18) machine learning algorithms by building these models from the Volve drilling dataset of the North Sea in order to comparatively study their performance and rank them. The result of this work provides a template for the choice of algorithms as well as selection of features for implementing ML solutions for oil drilling optimization. This can be helpful in hybridization and development of the real-time ROP prediction models.

Future research will emphasize the development of innovative embedded systems that incorporate use specific optimized machine learning models for controlling drilling operations. These ML models will be designed to take into account various drilling conditions and factors such as geological formations, well depth, and drilling speed, among others, to improve drilling efficiency and reduce downtime. Additionally, the use of embedded systems in oil drilling operations will provide real-time monitoring and analysis of drilling parameters, enabling early detection of potential problems and minimizing the risk of equipment failure. Ultimately, this collaboration has the potential to revolutionize the oil drilling industry in Nigeria and significantly enhance its productivity and profitability.

References

1. Cao, J., Gao, J., Jiao, T., Xiang, R., Pang, Y., Li, T.: Feature investigation on the ROP machine learning model using realtime drilling data. J. Phys. Conf. Ser. **2024**(1), (2021). IOP Publishing
2. Sircar, A., Yadav, K., Rayavarapu, K., Bist, N., Oza, H.: Application of machine learning and artificial intelligence in oil and gas industry. Petroleum Res. (2021)

3. Darwesh, A.K., Rasmussen, T.M., Al-Ansari, N.: Controllable drilling parameter optimization for roller cone and polycrystalline diamond bits. J. Petroleum Explor. Prod. Technol. **10**(4), 1657–1674 (2019). https://doi.org/10.1007/s13202-019-00823-1
4. Ameloko, A.A., Uhegbu, G.C., Bolujo, E.: Evaluation of seismic and petrophysical parameters for hydrocarbon prospecting of G- field, Niger Delta, Nigeria. J. Petroleum Explor. Prod. Technol. **9**(4), 2531–2542 (2019)
5. Takbiri-Borujeni, A., Fathi, E., Sun, T., Rahmani, R., Khazaeli, M.: Drilling performance monitoring and optimization: a data- driven approach. J. Petroleum Explor. Prod. Technol. **9**(4), 2747–2756 (2019)
6. Braga, D.C.: Field Drilling Data Cleaning and Preparation for Data Analytics Applications. Louisiana State University and Agricultural & Mechanical College (2019)
7. Olayinka, A.S., Adetunji, C.O., Nwankwo, W., Olugbemi, O.T., Olayinka, T.C.: A study on the application of bayesian learning and decision trees IoT-enabled system in postharvest storage. In: Artificial Intelligence-based Internet of Things Systems, pp. 467–491. Springer, Cham (2022)
8. Nwankwo, W., Adetunji, C.O., Olayinka, A.S.: IoT-driven bayesian learning: a case study of reducing road accidents of commercial vehicles on highways. In: Pal, S., De, D., Buyya, R. (eds.) Artificial Intelligence-based Internet of Things Systems. Internet of Things. Springer, Cham (2022). https://doi.org/10.1007/978-3-030-87059-1_15
9. Adetunji, C.O., et al.: Machine learning and behaviour modification for COVID-19. Med. Biotechnol. Biopharmaceutics, Forensic Sci. Bioinform. 271–87
10. Adetunji, C.O., et al.: The role of an intelligent feedback control system in the standardization of bio-fermented food products. Fermentation and Algal Biotechnologies for the Food, Beverage and Other Bioproduct Industries 143–62
11. Nwankwo, W., Adetunji, C.O., Ukhurebor, K.E., Makinde, A.S.: Artificial intelligence-aided bioengineering of eco-friendly microbes for food production: policy and security issues in a developing society. In: Agricultural Biotechnology, pp. 301–313. CRC Press, Dec 21 2022
12. Nwankwo, W., Nwankwo, C.P., Wilfred, A.: Leveraging on artificial intelligence to accelerate sustainable bioeconomy. IUP J. Knowledge Manag. **20**(2), 1 (2022)
13. Chinedu, P.U., Nwankwo, W., Masajuwa, F.U., Imoisi, S.: Cybercrime detection and prevention efforts in the last decade: an overview of the possibilities of machine learning models. Rev. Int. Geographical Educ. **11**(7), 1 (2021)
14. Olayinka, T.C., Olayinka, A.S., Nwankwo, W.: Evolving feed-forward artificial neural networks using binary and denary dataset. SAU Sci. Tech J. **6**(1), 96–108 (2021)
15. Nwankwo, W., et al.: The adoption of AI and IoT technologies: socio-psychological implications in the production environment. IUP J. Knowl. Manag. **19**(1), 1 (2021)
16. Acheme, I.D., Makinde, A.S., Udinn, O., Nwankwo, W.: An intelligent agent-based stock market decision support system using fuzzy logic. IUP J. Inf. Technol. **16**(4), 1 (2020)
17. Nwankwo, W., Umezuruike, C., Njoku, C.C.: Enhancing learning systems using interactive intelligent components. Int. J. **9**(3) (2020)
18. Olayinka, A.S., Nwankwo, W., Olayinka, T.C.: Model based machine learning approach to predict thermoelectric figure of merit. Archive of Science and Technology. 1(1) (2020)
19. Nwankwo, W., Ukhurebor, K.E.: Web forum and social media: a model for automatic removal of fake media using multilayered neural networks. Int. J. Sci. Technol. Res. **9**(1), 4371–4377 (2020)
20. Al-Khudiri, M.M., et al.: Application Suite for 24/7 Real Time Operation Centers. InSPE Saudi Arabia Section Annual Technical Symposium and Exhibition 2015 Apr 21. OnePetro
21. Acheme, I.D., Vincent, O.R., Olayiwola, O.M.: Data science models for short-term forecast of COVID-19 spread in Nigeria. In: Hassan, S.A., Mohamed, A.W., Alnowibet, K.A. (eds.) Decision Sciences for COVID-19. International Series in Operations Research & Management Science, vol. 320. Springer, Cham (2022). https://doi.org/10.1007/978-3-030-87019-5_20

22. Alsaihati, A., Elkatatny, S., Gamal, H.: Rate of penetration prediction while drilling vertical complex lithology using an ensemble learning model. J. Petrol. Sci. Eng. 1(208), 109335 (2022)
23. Hegde, C., Soares, C., Gray, K.: Rate of penetration (ROP) modeling using hybrid models: deterministic and machine learning. InUnconventional Resources Technology Conference, Houston, Texas, 23-25 July 2018 2018 Sep 28 (pp. 3220-3238). Society of Exploration Geophysicists, American Association of Petroleum Geologists, Society of Petroleum Engineers
24. Eckel, J.R.: Microbit studies of the effect of fluid properties and hydraulics on drilling rate. J. Petrol. Technol. 19(04), 541–546 (1967)
25. Bourgoyne, A.T., Young, F.S.: A multiple regression approach to optimal drilling and abnormal pressure detection. Soc. Petrol. Eng. J. 14(04), 371–384 (1974)
26. Jahanbakhshi, R., Keshavarzi, R., Jafarnezhad, A.: Real-time prediction of rate of penetration during drilling operation in oil and gas wells. In: 46th US Rock Mechanics/Geomechanics Symposium 2012 Jun 24. OnePetro
27. Ashrafi, S.B., Anemangely, M., Sabah, M., Ameri, M.J.: Application of hybrid artificial neural networks for predicting rate of penetration (ROP): a case study from Marun oil field. J. Petrol. Sci. Eng. 1(175), 604–623 (2019)
28. Wang, Y., Salehi, S.: Application of real-time field data to optimize drilling hydraulics using neural network approach. J. Energy Resour. Technol. 137(6), 1 (2015)
29. Hinton, G.E., Srivastava, N., Krizhevsky, A., Sutskever, I., Salakhutdinov, R.R.: Improving neural networks by preventing co- adaptation of feature detectors. arXiv preprint arXiv:1207. 0580. 3 Jul 2012
30. Hegde C, Wallace S, Gray K. Using trees, bagging, and random forests to predict rate of penetration during drilling. InSPE Middle East Intelligent Oil and Gas Conference and Exhibition 2015 Sep 15. OnePetro
31. Equinor. Volve data village dataset: released under a license based on CC BY 4.0
32. Makinde, A.S., Agbeyangi, A.O., Nwankwo, W.: Predicting mobile portability across telecommunication networks using the integrated-KLR. Int. J. Intell. Inf. Technol. (IJIIT) 17(3), 50–62 (2021). https://doi.org/10.4018/IJIIT.2021070104
33. Rajasekar, M., Geetha, A.: Comparison of machine learning algorithms in domain specific information extraction. Int. J. Math. Sci. Comput. (IJMSC) 9(1), 13–22 (2023). https://doi.org/10.5815/ijmsc.2023.01.02
34. Joseph, I., Imoize, A.L., Ojo, S., Risi, I.: Optimal call failure rates modelling with joint support vector machine and discrete wavelet transform. Int. J. Image Graph. Signal Process. (IJIGSP), 14(4), 46–57 (2022). https://doi.org/10.5815/ijigsp.2022.04.04
35. Abd El-Latif, E.I., Khalifa, N.E.: A model based on deep learning for COVID-19 X-rays classification. Int. J. Image graph. Signal Process. (IJIGSP), 15(1), 36–46 (2023). https://doi.org/10.5815/ijigsp.2023.01.04

Application of Support Vector Machine to Lassa Fever Diagnosis

Wilson Nwankwo[5]([⊠]) [iD], Wilfred Adigwe[2], Chinecherem Umezuruike[3],
Ijegwa D. Acheme[1], Chukwuemeka Pascal Nwankwo[1], Emmanuel Ojei[4],
and Duke Oghorodi[2]

[1] Department of Computer Science, Edo State University, Uzairue, Nigeria
[2] Department of Computer Science, Delta State University of Science and Technology,
Ozoro, Nigeria
[3] College of Computing and Communication Studies, Bowen University, Iwo, Nigeria
[4] Department of Software Engineering, Delta State University of Science and Technology,
Ozoro, Nigeria
[5] Department of Cyber Security, Delta State University of Science and Technology,
Ozoro, Nigeria
wnwankwo@dsust.edu.ng

Abstract. Lassa fever is a type of viral hemorrhagic fever with high fatality rate. Precise and prompt diagnosis and successful treatment of this disease is very important in the control and prevention of the spread as unrestricted spread can easily erupt into an epidemic. An insightful decision support system can help make quick and more exact findings on this disease. In this paper, an intelligent model is proposed for predicting Lassa fever is proposed. The Lassa fever dataset was extracted from the clinical records of patients from a known Specialist Teaching Hospital in Southern Nigeria. The relevant feature variables were identified and encoded in a manner appropriate for use in algorithmic analysis. The support vector classifier was used to develop a characterization model for the disease. Assessment and evaluation were based on exactness, accuracy, and affectability. Findings showed that the proposed system could improve and encourage quick decision-making as it utilizes real historical data to effectively characterize Lassa fever spread as well as promote the timely treatment and management of patients.

Keywords: Diagnostics · Lassa fever · Machine learning · Support vector machine · Epidemic

1 Introduction

The healthcare delivery sector has continued to suffer setbacks in providing quality healthcare services, especially in developing countries [1–7]. Quality connotes fitness for purpose hence, quality healthcare entailed correct and accurate diagnosis and the subsequent prompt treatment of diagnosed cases. Quality healthcare is largely a function of right and accurate clinical decision-making. Poor clinical decisions can lead to disaster and loss of lives in the healthcare space. Medical diagnosis is a subjective task;

© The Author(s), under exclusive license to Springer Nature Switzerland AG 2023
Z. Hu et al. (Eds.): ICAILE 2023, LNDECT 180, pp. 165–177, 2023.
https://doi.org/10.1007/978-3-031-36115-9_16

that is, it depends on the physician making the diagnosis [8]. Furthermore, the amount of data (signs, symptoms, laboratory tests, etc.) to be analyzed before arriving at the correct diagnosis may be enormous and, at times, unmanageable. Machine learning (ML) and deep learning have proved reliable and efficient in most socioeconomic spheres, especially in the provision of quick and high-precision prediction rules from historical data [9–19]. ML could help healthcare specialists make the diagnostic process more objective and more reliable [20].

Lassa fever (LF) is an acute and viral hemorrhagic disease with a cycle of between two and twenty-one days. The disease was first discovered in the North-East of Nigeria in 1969 and the virus (an RNA Virus) was named the lassa virus after a town where it was discovered in Nigeria. The LF virus has been isolated in other West African countries such as Sierra Leone, Liberia, Mali, Cote d'Ivoire, and Guinea. Similarly, the virus has also been isolated in some countries in eastern Africa. As a disease with high mortality rates, effective and efficient approaches need be explored in controlling the spread of the disease [21]. Usually, human infections is through direct contact with items such as food, water, etc. contaminated with the urine or feces of the multimammate rat, a rodent from Mastomys family [22, 23]. LF may be transmitted from human to human which may often result to epidemics with case fatality rates as high as 60%. LF may be difficult to diagnose in its early stages in hospitals [21]. The common symptoms include pharyngitis, cough, and gastrointestinal symptoms [22]. It has been noted that the early symptoms of the disease are not specific but include fever, sore throat, malaise, myalgia, cough, nausea, headache, chest pain, abdominal pain, diarrhea, and vomiting. Late signs of infection or disease are facial oedema, bleeding, effusions and convulsion, coma, pleural, and pericardial. With these known symptoms, it is possible to develop and deploy a system that is capable of predicting the spread of the disease so that Government agencies, medical experts, and individuals can take precautionary measures in the fight of the disease. A successful prediction would help healthcare practitioners and other stakeholders to manage emergencies such as outbreaks of the disease. Current researches focus on intelligent insights drawn through spatiotemporal infectious diseases outbreak modeling and prediction [24, 25]. This study aims to develop an intelligent diagnostic model for Lassa fever disease. The specific objectives are to: (1) design a machine learning model that can serve as a decision support system for health experts in the diagnosis of Lassa fever; (2) implement the design in (i) for the diagnosis of Lassa fever using the python programming language.

2 Related Works

Owing to the fatality of the Lassa fever disease, research is ongoing in the area of developing sophisticated tools and technologies such as machine learning applications and mathematical models for managing Lassa fever outbreaks. However, due to the limited localization of the disease, not much studies have been documented especially in the domain of machine learning applications and algorithmic modeling. Some of the related works in the domain of disease management using intelligent and smart approaches are briefly discussed herein. Madueme and Chirove [22] developed a mathematical model for the study of LF epidemiology. Their model comprises a system of nonlinear ordinary differential equations (ODE) which were aimed at determining the effect of the various LF transmission pathways as well as the progression of the infection in humans and rodents. Their model analysis and numerical simulations showed that the rate of LF infection increases if there are different transmission routes.

Bakere et al. [26] developed a mathematical model which they used to analyse the transmission of Lassa fever. They employed a periodically forced seasonal nonautonomous system of a nonlinear ordinary differential equation to capture the dynamics of Lassa fever transmission as well as the seasonal variation in the birth of *Mastomys* rodents which are the main vectors of the Lassa fever virus. The researchers showed that their model could offer epidemiological insights into the control of Lassa fever spread.

Ossai et al. [27]. Examined the preventive measures against Lassa fever among heads of family units in Abakaliki city, Southeast Nigeria using an expressive cross-sectional study. A four-phase examining configuration was utilized to choose 420 respondents from Abakaliki city. They concluded that though most of their respondents exhibit a good understanding of preventive measures against Lassa fever, there is a need to involve more health workers in community awareness programmes to educate the population on appropriate measures.

Alabdulkarim et al. [28] proposed a privacy-policy single decision tree model for clinical decision support which could help healthcare officers to protect patients' data from privacy breaches. The system is secure and the patients' data are protected through encryption using homomorphic encryption code in such a way that only authorized organizations utilizing the Internet of Things could access the patients' information.

Oonsivilai et al. [29] applied ML algorithms to guide the empirical antibiotic prescription process in children. They deduced that the intelligent model could provide exceptionally instructive predictions on anti-microbial susceptibilities using patients' data thereby aiding in controlling experimental anti-toxin therapy.

Osaseri and Osaseri [30]. Proposed an Adaptive Neuro-Fuzzy Inference System (ANFIS) for use in Lassa Fever prediction among patients. They adopted four input parameters: White Blood Count, Temperature on Admission, Abdominal Pain, and Proteinuria respectively. They appropriated their LF dataset into a 61: 59 ratio for training and testing of their model respectively.

Roosan et al. [31] examined the multifaceted nature of clinical thinking in infections utilizing subjective topical examination, that is a meeting of coauthors to autonomously note pertinent ideas joined with multifaceted nature, psychological objectives, versatile systems, and sense-production.

Shoaib et al. [32] discussed a predictive computational network that is aimed at predicting the transmission of Lassa fever in Nigeria. They designed a mathematical model comprising an artificial neural networks-based hybrid algorithm of Genetics Optimization and Sequential Quadratic Programming to predict the dynamics of Lassa fever. The model emphasized the transmission patterns of Lassa fever.

3 Methodology

The Lassa fever clinical dataset was obtained from the epidemiological unit of a specialist hospital designated by the Nigerian Government for the management of and research on Lassa fever infections. The dataset contains a thousand seven hundred and twenty-eight records. Following data cleaning, the dataset was split into 80: 20 ratios for training, testing, and evaluation purposes. Figure 1 shows the methodology adopted for this study.

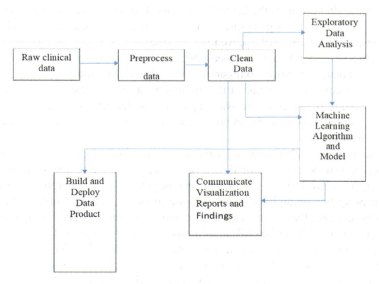

Fig. 1. Methodology for developing the prediction model

3.1 Data Collection

The dataset collected was unstructured with some columns not required for training and testing; such columns were the patients' name ID and home address. As part of that data preparation process, these columns were removed from the dataset. Table 1 presents the useful features selected for the machine learning model. From the raw dataset, essential features relevant to the prediction/diagnosis of Lassa fever disease were identified (Table 1).

Application of Support Vector Machine to Lassa Fever Diagnosis

Table 1. Description of selected variables

S/N	Condition
1	Lassa Fever (Target Variable: Yes or No)
2	Date/Time
3	Fever (Body Temperature)
4	Body Pain
5	Headache
6	Vomiting
7	Jaundice
8	Bleeding
9	Age
10	Gender
11	Underlying health condition (Diabetes/Hepatitis/High blood pressure)
12	Body Mass Index
13	Alcohol or cigarette intake

3.2 Data Preprocessing

During this stage, the raw data was transformed into a machine-usable format since the raw data contained incomplete inconsistent values and outliers. The dataset was also labeled and encoded appropriately. Hence the resulting dataset was more reliable and fit for knowledge discovery. Furthermore, in a randomized manner, the entire dataset was split into a train and test set.

3.3 Input Variable Transformation and Target Variable

The symptoms manifested by patients are the input variables. These are transformed into two classes (yes or no) and three classes depending on the nature of the manifestation, while the target/predicted variable is a diagnosis of suspected Lassa fever which is a binary classification). Table 2 presents the transformed dataset for the input variables.

The data transformation and encoding, as shown in Table 2, are essential for the dataset used for training machine learning models. This is because machine learning algorithms require that input and output variables are numeric values. As a result, categorical data such as is in this study must be encoded to numbers before we can use it to train and test a model.

Table 2. Input Variables Transformation

S/N	Variable definition	Domain 1	Domain 2	Domain 3
1	Lassa Fever (Target Variable: Yes or No)	Yes = 1	No = 0	NA
2	Date/Time	NA	NA	NA
3	Fever (Body Temperature)	High = 1	Medium = 0.5	Low = 0
4	Body Pain	Yes = 1	No = 0	NA
5	Headache	High = 1	Moderate = 0.5	Low = 0
6	Vomiting	Yes = 1	No = 0	NA
7	Jaundice	Yes = 1	No = 0	NA
8	Haemorrhage	Yes = 1	No = 0	NA
9	Age	NA	NA	NA
10	Gender	NA	NA	NA
11	Underlying Health condition (Diabetes/Hepatitis/High blood pressure)	Yes = 1	No = 0	NA
12	Body Mass Index	NA	NA	NA
13	Alcohol or cigarette intake	Yes = 1	No = 0	NA

3.4 Use Case Diagram of the System

The use case diagram is used for representing a user's interaction with the system, in this case, medical personnel since this system is built to aid in the diagnosis of Lassa fever disease. The use case diagram helps in understanding and identifying the operation and users' role in the entire system as well as showing the relationship between the user and designated functions. Figure 2 is the use case diagram of the proposed method.

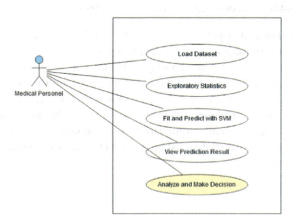

Fig. 2. Use case Diagram

3.5 The Support Vector Machine Algorithm

The Support Vector Machine (SVM) belongs to the family of supervised machine learning algorithms used for regression and classification problems [33–35]. The SVM algorithm plots each data item as a point in n-dimensional space with the value of each feature represented by the value of a coordinate and n refers to the number of features in the dataset. Thus, classification is performed by finding the hyper-plane that most distinguishes the two classes. The study utilized the Lassa fever patients' dataset in the form:

$$\{(x_k, y_k)\}_{k=1}^{n} \tag{1}$$

where x is the features, and y is a 1x n vector; n is the training dataset. Given that training variable x and output variable y are defined by Eq. (2) and (3).

$$\{x_1, x_2, x_3 \ldots\} \in R^N \tag{2}$$

$$\{y, y_2, y_3 \ldots\} \in R \tag{3}$$

Equation (1) represents the training variable while Eq. (2) is the output variable. The SVM algorithm is the function y(x) that minimizes the error for all the learning features x_i, hence the objective is to:

$$minimize: \frac{1}{2}||w||^2 + c\sum_{l=1}^{\ddots l}(\alpha_l + \overline{\alpha}) \tag{4}$$

Subject to: $w.x_i + b\text{-}y \leq \mathcal{E} + \alpha_i$, $y_i\text{-}w.x_i\text{-}b < \mathcal{E} + \alpha_i$.

Where w is an n-dimensional vector representing the weight $i = 1, \ldots l$; $c > 0$ determines the trade-off between the differences in the decision function and α, $\overline{\alpha} \geq 1$.

3.6 Proposed Evaluation

The results of the SVM model are evaluated by the prediction accuracy, represented by a confusion matrix. Other evaluation metrics used include AUC-ROC, F_1 Score, precision, and recall though these are more suitable for binary classification problems, unlike a multiclass problem discussed in the present study.

3.7 Exploratory Analysis

The dataset comprises over one thousand seven hundred records of patients admitted for treatment of Lassa fever. Figure 3–4 show the summary and description of the dataset in relation to the member variables: 'Lasssa_Fever_Case', and 'Not_Lasssa_Fever_Case' respectively.

```
df.describe()
```

	Fever_and_HeadAche	Vomiting	Alcohol_and_Cigaret	Age	Bleeding	Underlyning_Illness	Lassa_or_Not
count	1728	1728	1728	1728	1728	1728	1728
unique	4	4	4	3	3	3	2
top	vhigh	vhigh	4	adult	big	low	Lasssa_Fever_Case
freq	432	432	432	713	576	576	1210

```
df['Lassa_or_Not'].describe()
count                    1728
unique                      2
top         Lasssa_Fever_Case
freq                     1210
Name: Lassa_or_Not, dtype: object
```

Fig. 3. Description of dataset

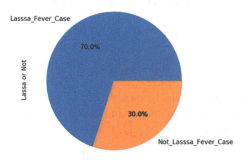

Fig. 4. Pie Chart showing the percentage distribution of Lassa case

3.8 Evaluation Metrics

To establish the efficiency of this model, well-known and generally acceptable evaluation metrics are used. First, the prediction accuracy and confusion matrix are considered, as shown in Fig. 5. The "Area Under the Curve (AUC)" of the "Receiver Operating Characteristics(ROC)" i.e. AUC-ROC curve that evaluates the number of true positives against false positives is then plotted. This model is also evaluated with the precision and recall plots. Figure 6 shows the AUC-ROC and Precision/Recall plots.

Confusion Matrix
It is a productivity metric for the problem of classification of machine learning where two or more classes can be tested. It is a table comprising four different expected and actual value combinations. Figure 5 and Table 3 show the confusion matrix of the support vector machine of the test data.

Application of Support Vector Machine to Lassa Fever Diagnosis 173

```
print("Accuracy:",accuracy_score(Y_test, y_pred))
plot_confusion_matrix(svm_clf, X_test, Y_test, display_labels=ylabels)

Accuracy: 0.7283236994219653

<sklearn.metrics._plot.confusion_matrix.ConfusionMatrixDisplay at 0x1164c529788>
```

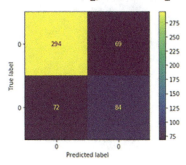

Fig. 5. Prediction Accuracy and Confusion Matrix

Table 3. Confusion Matrix computation

N = 519	Predicted: 0	Predicted: 1	
Actual: 0	TN = 294	FP = 69	363
Actual: 1	FN = 72	TP = 84	98
	366	153	

The rates computed from the confusion matrix for the classifier model are presented in Table 4. The plots of AUC-ROC, precision, and recall are shown in Fig. 6.

Table 4. Rates Computed from the confusion matrix

Rates	Values
Accuracy	0.73
Misclassification Rate	0.27
True Positive Rate	0.86
False Positive Rate	0.19
True Negative Rate	0.81
Precision	0.55
Prevalence	0.19
Recall	0.80

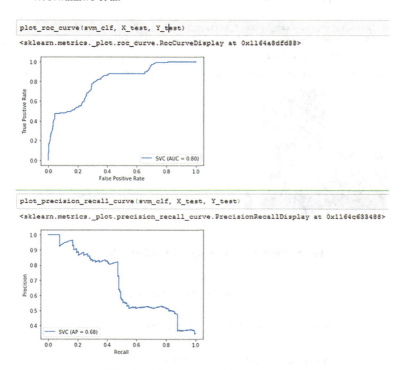

Fig. 6. AUC-ROC; Precision and Recall

One of the major advantages of SVM is relatively high accuracy and speed (efficient use of processing resources) compared to many fundamental machine learning algorithms. This constitutes the bedrock for the choice of the SVM. However, we are aware that different SVM kernels may perform differently on same dataset. In the above computation, the linear kernel was used. From Table 4, the accuracy of predictions on cases of Lassa fever is placed at 73% however other metrics such as the true positive and true negative predictions are placed at 86% and 81% respectively. The model in its present state predicts 80% of the true Lassa fever cases (recall = 0.80). The recall is considered very important to us considering the fact that we are more interested in identifying and managing a Lassa fever case. This shows that the model exhibits the potential for improved accuracy and precision which might be attained on carrying out model tuning by way of kernel modification, regularization, modifying the gamma value(increasing it). It is also instructive to note that though model tuning is ideal, it does not necessarily guarantee much significant performance improvements considering the small dataset. Furthermore, in Fig. 6, we got 0.80 as the AUC, and we considered this a pretty good score in the sense that the model could differentiate those patients with Lassa fever and those patients without Lassa fever 80% of the time. This percentage/score may be improved by applying different hyperparameter values.

4 Conclusion

Clinical decision support and intelligent systems have a history of enhancing the efficiency of valuable socio-economic workflows. The application of AI-based systems in medical diagnosis and disease management cannot be overemphasized. In this paper, we set out to examine the applicability of support vector machine algorithms in the detection, diagnosis, and profiling of Lassa fever in patients using historical clinical data from a specialist hospital. The developed prediction model shows that valuable insights could be drawn from historical data and used to drive new diagnoses. The proposed system improves and encourages decision-making in this regard and would be a valuable tool in the hands of medics who are saddled with the task of providing care for the helpless during Lassa fever outbreaks.

The findings in this study notwithstanding, we recognize the limitation in the sense that the dataset available was not detailed enough and also fall short of satisfying the adequacy metric in terms of the volume of data available. However, we believe that this study would constitute a good cornerstone for conducting future research in the construction of hybrid models that could provide better results for Lassa disease diagnosis and management.

References

1. Victor-Ikoh, M.I., Moko, A., Nwankwo, W.: Towards the Implementation of a Versatile Mobile Health Solutions for the Management of Immunization Against Infectious Diseases in Nigeria. In: Salvendy, G., Wei, J. (eds.) Design, Operation and Evaluation of Mobile Communications. HCII 2022. Lecture Notes in Computer Science. Springer, Cham (2022)
2. Umezuruike, C., Nwankwo, W., Tibenderana, P., John, P.A., Muhirwa, R.: Corona Virus Disease (COVID 19): analysis and design of an alert and real-time tracking system. Int. J. Emerg. Trends Eng. Res. **8**(5), 1743–1748 (2020)
3. Umezuruike, C., Nwankwo, W., Okolie, S.O., Adebayo, A.O., Jonah, J.V., Ngugi, H.: Health informatics system for screening arboviral infections in adults. Int. J. Inf. Technol. Comput. Sci. (IJITCS), **11**(3), 10–22 (2019)
4. Nwankwo, W., Umezuruike, C.: An object-based analysis of an informatics model for Zika virus detection in adults. Comput. Biol. Bioinform. **6**(1), 1–20 (2018)
5. Nwankwo, W.: Harnessing e-healthcare technologies for equitable healthcare delivery in Nigeria: the way forward. Int. J. Sci. Res. **6**(3), 1875–1880 (2017)
6. Umezurike, C., Nwankwo, W., Kareyo, M.: Implementation challenges of health management information systems in Uganda: a review. J. Multidisciplinary Eng. Sci. Technol. **4**(7), 7726–7731 (2017)
7. Umezurike, C., Nwankwo, W., Okolie, S.O., Adebayo, A.: Developing an informatics model for effective healthcare in military health facilities in Nigeria. World J. Eng. Res. Technol. **3**(4), 69–99 (2017)
8. Das, R., Turkoglu, I., Sengur, A.: Effective diagnosis of heart disease through neural networks ensembles. Expert Syst. Appl. **36**(4), 7675–7680 (2009)
9. Nwankwo, W., Chinedu, U.P., Aliu, D., et al.: Integrated FinTech solutions in learning environments in the post-COVID-19 era. IUP J. Knowl. Manag. **20**(3), 1–22 (2022)
10. Nwankwo, W., Nwankwo, C.P., Wilfred, A.: Leveraging on artificial intelligence to accelerate sustainable bioeconomy. IUP J. Knowl. Manag. **20**(2), 38–59 (2022)

11. Acheme, D.I., Makinde, A.S., Osemengbe, U., Nwankwo, W.: An intelligent agent-based stock market decision support system using fuzzy logic. IUP J. Inf. Technol. **16**(4), 1–20 (2020)
12. Nwankwo, W., Adetunji, C.O., Olayinka, A.S.: IoT-driven bayesian learning: a case study of reducing road accidents of commercial vehicles on highways. In: Pal, S., De, D., Buyya, R. (eds.) Artificial Intelligence-based Internet of Things Systems. Internet of Things. Cham: Springer, pp. 391–418 (2022)
13. Chinedu, P.U., Nwankwo, W., Masajuwa, F.U., Imoisi, S.: Cybercrime detection and prevention efforts in the last decade: an overview of the possibilities of machine learning models. Rev. Int. Geographical Educ. (RIGEO) **11**(7), 956–974 (2021)
14. Nwankwo, W., Adetunji, C.O., Olayinka, A.S., et al.: The Adoption of AI and IoT technologies: socio-psychological implications in the production environment. IUP J. Knowl. Manag. **19**(1), 50–75 (2021)
15. Olayinka, A.S., Adetunji, C.O., Nwankwo, W., et al.: A study on the application of bayesian learning and decision trees IoT-enabled system in postharvest storage. In: Pal, S., De, D., Buyya, R. (eds.) Artificial Intelligence-based Internet of Things Systems. Internet of Things. Cham: Springer, 467–491 (2022)
16. Osikemekha, A.A., Adetunji, C.O., Olaniyan, T.O., Hefft, D.I., Nwankwo, W., Olayinka, A.S.: IoT-based monitoring system for freshwater fish farming: analysis and design. In: Abraham, A., Dash, S., Rodrigues, J.J.P.C., Acharya, B., Pani, S.K. (eds.) Intelligent Data-Centric Systems: AI, Edge and IoT-based Smart Agriculture, pp. 505–515. Academic Press, Amsterdam (2022)
17. Adetunji, C.O., Osikemekha, A.A., Olaniyan, T.O.: Toward the design of an intelligent system for enhancing salt water shrimp production using fuzzy logic. In: Abraham, A., Dash, S., Rodrigues, J.J.P.C., Acharya, B., Pani, S.K. (eds.) Intelligent Data-Centric Systems: AI, Edge and IoT-based Smart Agriculture, pp. 533–541. Academic Press, Amsterdam (2022)
18. Nwankwo, W., Ukhurebor, K.E.: Big data analytics: a single window IoT-enabled climate variability system for all-year-round vegetable cultivation. IOP Conference Series: Earth and Environmental Science, 655, 012030 (2021)
19. Nwankwo, W., Ukhurebor, K.E., Ukaoha, K.C.: Knowledge discovery and analytics in process re-engineering: a study of port clearance processes. In: International Conference in Mathematics, Computer Engineering and Computer Science (ICMCECS). Lagos: IEEE Explore, pp. 1–7
20. Adetunji, C.O., Nwankwo, W., Olayinka, A.S., Olaniyan, O.T., et al.: Machine learning and behaviour modification for COVID-19. In: Inuwa, H.M., Ezeonu, I.M., Adetunji, C.O., Ekundayo, E.O., Gidado, A., Ibrahim, A.B., Ubi, B.E. (eds.) Medical Biotechnology, Biopharmaceutics, Forensic Science and Bioinformatics. Florida: CRC Press, pp. 271–87 (2020)
21. Asogun, D.A., Günther, S., Akpede, G.O., Ihekweazu, C., Zumla, A.: Lassa fever: epidemiology, clinical features, diagnosis, management and prevention. Infect. Dis. Clin. North Am. **33**(4), 933–951 (2019)
22. Madueme, P.U., Chirove, F.: Understanding the transmission pathways of Lassa fever: a mathematical modeling approach. Infectious Dis. Model. **8**(1), 27–57 (2023)
23. Musa, S.S., Zhao, S., Abdullahi, Z.U.: COVID-19 and Lassa fever in Nigeria: a deadly alliance? Int. J. of Infectious Diseases **117**, 45–47 (2022)
24. Li, Y.: Genetic basis underlying Lassa fever endemics in the Mano River region, West Africa. Virology **579**, 128–136 (2023)
25. Okokhere, P., Colubri, A., Azubike, C., et al.: Clinical and laboratory predictors of Lassa fever outcome in a dedicated treatment facility in Nigeria: a retrospective, observational cohort study. Lancet. Infect. Dis **18**(6), 684–695 (2018)

26. Bakare, E.A., Are, E.B., Abolarin, O.E., et al.: Mathematical modelling and analysis of transmission dynamics of Lassa Fever. J. Appl. Math. **2020**, 1–18 (2020)
27. Ossai, E.N., Onwe, O.E., Okeagu, N.P., et al.: Knowledge and preventive practices against Lassa fever among heads of households in Abakaliki metropolis, Southeast Nigeria: a cross-sectional study. Proc. Singapore Healthc. **29**(2), 73–80 (2020)
28. Alabdulkarim, A., Al-Rodhaan, M., Ma, T., Tian, Y.: PPSDT: a novel privacy-preserving single decision tree algorithm for clinical decision-support systems using IoT devices. Sensors **19**(1), 142 (2019)
29. Oonsivilai, M., et al.: Using machine learning to guide targeted and locally-tailored empiric antibiotic prescribing in a children's hospital in Cambodia. Wellcome Open Res. **3**(131), 1–18 (2018)
30. Osaseri, R.O., Osaseri, E.I.: Soft computing approach for diagnosis of Lassa fever. Int. J. Comput. Inf. Eng. **3**(11) (2016)
31. Islam, R., Weir, C.R., Jones, M., Del Fiol, G., Samore, M.H.: Understanding complex clinical reasoning in infectious diseases for improving clinical decision support design. BMC Med. Inform. Decis Making **15**(101), 2–12 (2015)
32. Shoaib, M., Tabassum, R., Raja, M.A.Z., Nisar, K.S., Alqahtani, M.S., Abbas, M.: A design of predictive computational network for transmission model of Lassa fever in Nigeria. Results Phys. 39 (2022)
33. Rajasekar, M., Geetha, A.: Comparison of machine learning algorithms in domain specific information extraction. Int. J. Math. Sci. Comput. (IJMSC) **9**(1), 13–22 (2023)
34. Joseph, I., Imoize, A.L., Ojo, S., Risi, I.: Optimal call failure rates modelling with joint support vector machine and discrete wavelet transform. Int. J. Image Graph. Signal Process. (IJIGSP), **14**(4), 46–57 (2022)
35. Abd El-Latif, E.I., Khalifa, N.E.: A model based on deep learning for COVID-19 x-rays classification. Int. J. Image Graph. Signal Process. (IJIGSP), **15**(1), 36–46 (2023)

A Novel Approach to Bat Protection IoT-Based Ultrasound System of Smart Farming

Md. Hafizur Rahman[1] , S. M. Noman[2] , Imrus Salehin[3(✉)] ,
and Tajim Md. Niamat Ullah Akhund[4]

[1] Department of Digital Anti-Aging Healthcare, Inje University, Gimhae, Republic of Korea
[2] Faculty of Computer Science and Engineering, Frankfurt University of Applied Sciences, Frankfurt, Germany
[3] Department of Computer Engineering, Dongseo University, Busan, Republic of Korea
deeplab43@gmail.com
[4] Department of Computer Science and Engineering, Daffodil International University, Dhaka, Bangladesh

Abstract. In past decades, scientific research has explored various bat detection systems methods, a common trend to identify heterogeneous species of bats. In this article, we represented an ultrasonic bat detection and protection method by interfacing an IoT automated device for the first time in agriculture development to introduce steady security in a preventive way. The automated bat banisher system is an effectual model for farmers to protect their land at a very affordable cost from nuisance bats. Moreover, the highest-quality fruits are the ones that bats choose to eat when they attack. This prevents the farmer from getting the best possible yield, Farmers have tried everything from handcrafted sound systems and medication to nets and traps to get rid of the bats, but they have yet to be effective. We created an Internet of Things-based automated bat banisher to combat bats invading farms. In this research, our High-Frequency Ultrasound (IoT automated) System has the potential to save farmers money, time, and effort. Farmers may increase agricultural output to provide the highest suitable fruit supply. Therefore, an ultrasonic IoT-based bat Banisher device is the most straightforward and functional means of dealing with nuisance bats.

Keywords: Ultrasonic Waves · High-Frequency Ultrasounds · IoT · Smart Farming · Agriculture Development · Bat Banisher

1 Introduction

The Bat discharges ultrasonic waves with high frequencies. Its calls are pitched at 20–200 Kilohertz (kHz), usually absurdly sharp for people to hear [1]. Their sounds are reflected in the earth, hitting different articles, and returning to the Bat as echoes. The resounding sign empowers the Bat to lay out a psychological guide to its condition. Sound is prompted through a medium as the particles are vibrated, making weight waves with the locale of weight and rarefaction [2]. The waves have trademark highlights of

© The Author(s), under exclusive license to Springer Nature Switzerland AG 2023
Z. Hu et al. (Eds.): ICAILE 2023, LNDECT 180, pp. 178–186, 2023.
https://doi.org/10.1007/978-3-031-36115-9_17

wavelength, rehash, and ample. Wavelength (λ) separates two zones of maximal weight (or rarefaction). The immensity of wavelength is that the attack of the ultrasound wave is as for wavelength, and picture targets are close to 1–2 wavelengths [3]. Rehash (f) is the number of wavelengths that appreciate relief. It is reliably evaluated as cycles (or wavelengths); the unit is hertz (Hz). It is a particular portion of the critical stone utilized in the ultrasound transducer. It may be moved by the administrator inside the set reasons for constraint – the higher the rehash, the better the targets, yet the lower the entrance. The level of weight change is given by the ample. It is passed on in decibels on a logarithmic scale [21]. Farmers can employ ultrasonic bat banisher devices to regulate bat communities because they emit sound waves that interfere with echolocation, making the region inhospitable to bats and creating a friendly environment for the farming system. The ultrasonic sound waves produced by bats are protected barriers that do not harm people [22]. Farmers have suffered from their traditional method of protecting their crops for a few decades, but they may significantly improve by adopting a bat banisher device. The bat colony will look for a new home in unremarkable structures or areas.

This paper presents the technique and structures for ousting bats from agriculture. In one putting, the procedure that may accomplish on a system fabricated perceiving from an ultrasound structure for Bat removing to guarantee agriculture field. Much more research has to be done on the pervasive use of ultrasonic sound as a monitoring technique. Many bats, cetaceans, certain rodents, insectivores, birds, and insects are among the numerous animals that create ultrasound [4]. O'Donnell et al. described their papers BATBOX III bat detector worked in two different types of bats, short-tailed and long-tailed bats, with detection frequency levels 27 kHz and 40 kHz [5]. In contrast, our bat banisher devices are more effective in the 20 kHz to 60 kHz range in recognizing ultrasound frequencies that accelerate from various bat categories.

This article suggests an IoT-based methodology to guard against possible harm to agriculture from wild animals and weather. Using IoT (Internet of Things) agricultural applications, the economy could improve crop quality, save expenses, and raise operational efficiency [6]. Various methods are used in bat detection systems of ultrasound conversion techniques, such as heterodyning, frequency division, and time expansion [19]. Through ultrasonic recording technology, analysis software, and conservation, the UK's iBat program has devised a system for monitoring ultrasound biodiversity in three bat species. According to their research, agricultural land contains 42.4% of Pipistrellus species and 21.7% of Myotis species [20]. This paper addresses a prototype ultrsound IoT device system for banishing bats and a procedure for growing the agricultural rate that leads to our country's economic growth.

- To reduce time, cost, and human resources
- To increase agricultural production
- To mitigate many types of virus attacks, including Nipah virus (NIV) from bats.

2 Related Work

Bats pose a significant risk, harm to fruit growth and infecting viruses. The technology is advertised as simple, affordable, and autonomous. A watertight container houses a Batbox III bat detector, a tiny voice-activated tape recorder, a talking clock, and an optional long-life battery are a bat detecting methods [5]. Oisin Mac Aodha et al. identified echolocating bats' ultrasonic, full-spectrum sounds during the search phase and created an open-source pipeline based on convolutional neural networks [7]. Ultrasonic signal transmission via a wireless connection in the commercial radio frequency range is the subject of this paper's early investigation (UHF). Herman et al. used in the building of massive wireless sensor networks for the study of bat behaviour [8]. Tomas et al. described in their paper superheterodyne QMC Mini Bat Detector inaccuracies action due to an adequate calibration system. The echolocation call features of bats cannot be used to identify them as a result [9]. Using bat detectors to record bat behavior, such as flight patterns and hunting strategy, the researchers provided examples of their research [17]. According to the study by Griffin et al., heterodyne bat detectors have had trouble detecting some species of bats since the 19th century because of their poor performance at 12 kHz, even though bat noises have a frequency range of up to 120 kHz. They implemented a bat detector at frequencies between 35 kHz and 45 kHz audible [18].

S. Noman et al. designed a prototype IoT device to detect insects from farmer's crop fields. In their study, they simulated agriculture automation systems to dispose of conventional farming through the blessing of modern technology [16]. Besides, bats employ an echo they make to forage for food at night. The primary diet of bats consists of insects and a variety of fruits, including mango, litchi, guava, plum, banana, white Jamun, Ziziphus mauritiana, tamarind, olive, and betel nuts. IoT-based bat detection and protection device is a unique method to redress conventional man-made systems to enlarge fruit production in farming areas. According to S. Sofana Reka et al., who utilized an IoT-Based Smart Greenhouse Farming System, food production may be boosted by using even dry and infertile terrain, improving output. Higher quality yields are produced by eliminating wastage and keeping greater control over crops' operations and environment [11]. Sensor technology and wireless IOT network integration have been researched and examined in the context of the current state of smart agricultural farming systems using the internet and cellular communications. The Remote Monitoring System (RMS) is suggested [12]. I. Salehin et al. proposed an intelligent device, interfacing between Arduino UNO and ultrasonic sensors to get an overload water alert in the drainage system. Their motive was to reduce the time and cost of innovating IoT to enable a model for a smart city [10]. Ryu et al. demonstrate an IoT-based linked farm that seeks to provide consumers with smart agricultural solutions. The benefits of linked farms over earlier smart farms are described with service scenarios and a complete design and execution for connected farms [13]. The proposed Farm as a Service (FaaS) integrated system offers high-level application services by managing connected equipment, data, and models and running and monitoring farms. This system analyses environmental and growth data as well as registers, connects and controls Internet of Things (IoT) devices [14]. An in-depth analysis of many techniques, including drip irrigation, greenhouse cultivation, IoT-based monitoring systems, wireless networks, smart agriculture, and precision farming, is addressed in this research [15].

3 Methodology of Circuit Implementation

We used an Arduino UNO R3, a speaker 8Ω, a Voice Recorder Module ISD1820, connection cables, an IR sensor, a motion sensor PIR, red and green LED lights, a breadboard, and a 5-V Lithium-ion battery, power source to build the system. The thickness and compressibility of the medium affect the propagation velocity (v) of the sound waves. The link between these variables is transmitted via the Wave Eq. 1. The bat behavior expressed in terms of frequency of "f" is caused by wavelength λ (because λf = v, v is the velocity of sound waves in the air) in the Bat algorithm. The magnitude of the wavelength denoted the shape of the aliment that bats are hunting to their focusing target [21].

$$v = \lambda f \tag{1}$$

3.1 Model Selection Diagram

Figure 1 illustrates the activity completed from beginning to end. At that time, we need first to tap the switch start. The PIR sensor is quickly approached. Whether the IR sensor receives signals from PIR sensor to differentiate an object, the recording of the article's voice transfers to the ultrasound sensor, the ultrasound sensor expands the recorded sound by voice recorder module between 20 kHz up to 60 kHz, and the ultrasound sensor moves to play, after which the recording stops, and the process ends. An algorithm for global streamlining is called the Bat calculation. It was propelled by the echolocation of microbats, whose heartbeats fluctuated and produced noise and a lot of noise. [22].

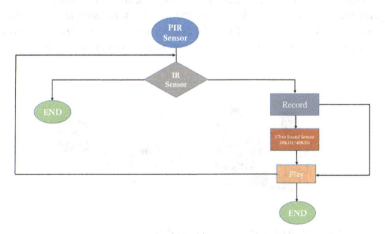

Fig. 1. Flow chart diagram of system activities.

The activation process shows in Fig. 2, after activated the device first emits an IR signal and, if an ultrasonic echo is detected, PIR sensor sends signal in Arduino device and receives an IR signal from the opposite side. If echoes are found, it will work red light in system and record then play back to transfer at the ending of the framer's field.

Fig. 2. Block diagram of system activities.

3.2 System Circuit Implementation

We integrated the ultrasonic sound speaker terminals connected to the SP+ and SP-pins of the ISD1820 Voice Recorder Module. The module's VCC and GND are linked to its +5V and GND. Arduino's Computerized IO Pins 2 and 3 are connected to the REC and PLAYE pins. Here, two advanced infrared sensors are used, PIR and IR sensors. Its driven output is connected to Arduino Sticks 0 and 4. In this experiment, we used a 5-V Lithium-ion battery connected to the battery connection port. We create the connection as the circuit schematic specifies, then turn on the circuit. The PIR Sensor output is LOW, and the Arduino is idle when there is no object in front of it. The IR Sensor's output rapidly rises when there is an object in front of it, and Arduino then begins recording a message by setting the REC Pin HIGH for around 5 s. The LED connected to Pin 13 is now lit RED to show that the module is currently encoding a message. Following the strategy, the message is replayed by raising the play stick for around five seconds. The LED connected to Pin 12 is now lit GREEN after exterminating the detected object. Figure 3 shows the circuit diagram of bat detection and protection system.

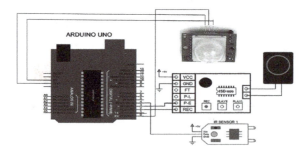

Fig. 3. Circuit diagram of bat detection and protection system.

3.3 Architecture Model Implementation

A particular form of a sound wave is sent into the air as a bat moves. The sound is reflected off buildings, trees, mountains, and other large obstructions and returns to the bat's ears. The bat's brain knows the distance between the obstacle and the reflected sound. When they join our agricultural fields, the bat will make high-frequency sounds ranging from 40 to 200 kHz [2]. The Voice Recorder Module in our system will capture the high-frequency sound made by bats. The IR Sensor and Motion Sensor detect the sound when bats come within range of the device up to 110 m. The Voice Recorder Module starts recording the sound once the sensor sends a signal to the Red LED. The Red LED will then flash in a millisecond, and the speaker will begin producing the same sound frequency. Bats will assume this passage is dangerous if they hear the same echo frequency. So, they'll stay away from this route at this time, and LED Green will blink when an object is finished off. Our agricultural land will be secure at the same time. The following Fig. 4 shows architecture model of Bat detection and protection system in real image.

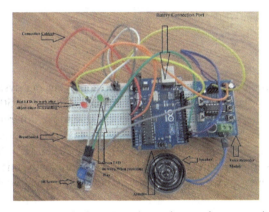

Fig. 4. Architecture model of Bat detection and protection system in real image.

4 Experimental Result and Discussion

In our experiment, the PIR sensor's maximum range outside is from 10 to 150 m. The PIR sensor will signal the voice recorder module when the bats come into the range. The red light will be ON, meaning the sound is recorded with the help of an IR sensor. Then within a few seconds green lights is ON when the tweeter speaker (high-frequency speaker) emits the sound into the environment and blows away the object. After bat banishes, our agricultural fruit will be safe, and we will get the proper output from future utilization. Olive, litchi, Ziziphus Mauritiana, and plum trees are used in our investigation of farming land. After conducting fieldwork, we discovered that out of 68 bat detection alerts, 54 bats flew alerts using red and green lights (on/off), whereas device accuracy was an average of 79.41%. The sensing number of bats depended on various time durations in different locations in tree fields. As we see, Litchi tree has the highest number of detection alerts producing 84% accuracy. In the morning, to verify the field status, we haven't seen damaged fruits in the ground as a before, comparatively past time.

184 Md. H. Rahman et al.

We set up the device in a 100-m farmers field for the experiment and observed the bat's activity by seeing the green and red lights on and off. Table 1 shows the experimental results of four variant trees on farming land.

Table 1. Experimental results of IoT Bat Banisher

Tree Name	Time (hour)	RED Light on/off (time)	GREEN Light on/off (time)	IoT Device Accuracy (%)
Olive	9pm–10pm	12	9	75
Litchi	10pm–11pm	25	21	84
Ziziphus Mauritiana	11pm–12am	13	10	76.92
Plum	12am–1am	18	14	77.77
Total	**4 h**	**68**	**54**	**79.41**

We had a pleasant, successful outcome after deploying our system in a field of agricultural trees. For better accuracy, more than one device above 50 m of land space is suggested depending on the farmer's land space. Due to the different genetics of bat species and producing high frequency, some bat was unaffected by the bat banisher. A man can hear the range of frequencies between 20 Hz and 20 kHz [22]. As a result, people won't be harmed by this system of ultrasonic sound. Additionally, this technology does not affect the biosphere, as we saw when we employed it in a real-world setting.

5 Conclusion

Innovative tools are essential to the success of today's agricultural methods. As a result, our Bat Banisher may be a useful tool for farmers. From this experiment, we might infer that protecting agricultural land is the farmer's responsibility and that doing so is essential to successfully implementing the novel approach. Bat detection and protection via the Internet of Things is a versatile and valuable addition. The bats are driven away by the ultrasonic barrier created by the continuous transmission of ultrasonic sound waves from the ultrasonic bat banisher. The Internet of Things Bat Banisher is the most recent method developed to remove bats from agricultural locations. Adopting this strategy will boost the agricultural sector's gross domestic product (GDP), allowing us to raise the country's total GDP each year. This proposed device would also be beneficial in eliminating bat-borne viruses like Covid-19 and Nipah.

Acknowledgments. The Daffodil International University Innovation Lab and Dongseo University Machine Learning /Deep Learning lab assisted in the development of the suggested framework.

References

1. Brigham, R., Elizabeth, K., Gareth, J., Stuart, P., Herman, L.: Bat Echolocation Research. Tools, Techniques, and Analysis (2004)
2. Dekker, J., Steen, W., Bouman, H.B., van der Vliet, R.E.: Differences in acoustic detectibility of bat species hamper environmental impact assessment studies. Eur. J. Wildl. Res. **68**(2), 1–8 (2022). https://doi.org/10.1007/s10344-022-01562-1
3. Fenton, M.B., Bouchard, S., Vonhof, M.J., Zigouris, J.: Time-expansion and zero-crossing period meter systems present significantly different views of echolocation calls of bats. J. Mammal. **82**(3), 721–727 (2001)
4. Pettorelli, N., Baillie, J.E., Durant, S.M.: Indicator bats program: a system for the global acoustic monitoring of bats. Biodiversity monitoring and conservation: bridging the gap between global commitment and local action, 211–247 (2013)
5. O'Donnell, C.F.J., Sedgeley, J.: An automatic monitoring system for recording bat activity, 1–16 (1994)
6. Haque, M.A., Haque, S., Sonal, D., Kumar, K., Shakeb, E.:. Security enhancement for IoT enabled agriculture. Materials Today: Proceedings (2021)
7. Mac Aodha, O., et al.: Bat detective—Deep learning tools for bat acoustic signal detection. PLoS Comput. Biol. **14**(3), e1005995 (2018)
8. Herman, K., Debski, M., Furmankiewicz, J.: A preliminary analysis of wireless link for bats signals acquisition system. Acta Phys. Pol. A **120**(4), 635–638 (2011)
9. Thomas, D.W., West, S.D.: On the use of ultrasonic detectors for bat species identification and the calibration of QMC Mini Bat Detectors. Can. J. Zool. **62**(12), 2677–2679 (1984)
10. Salehin, I., Baki-Ul-Islam, S.M., Noman, M.M., Hasan, S.T., Hasan, M.: A smart polluted water overload drainage detection and alert system: based on IoT. In: 2021 International Mobile, Intelligent, and Ubiquitous Computing Conference (MIUCC), pp. 37–41 (2021). https://doi.org/10.1109/MIUCC52538.2021.9447664
11. Reka, S.S., Chezian, B.K., Chandra, S.S.: A novel approach of IoT-based smart greenhouse farming system. In: Green Buildings and Sustainable Engineering, pp. 227–235. Springer, Singapore (2019)
12. Patil, K.A., Kale, N.R.: A model for smart agriculture using IoT. In: 2016 International Conference on Global Trends in Signal Processing, Information Computing and Communication (ICGTSPICC). IEEE (2016)
13. Ryu, M., Yun, J., Miao, T., Il-Yeup, A., Choi, S.H., Kim, J.: Design and implementation of a connected farm for smart farming system. In: 2015 IEEE SENSORS, pp. 1–4. IEEE (2015)
14. Kim, S., Lee, M., Shin, C.: IoT-based strawberry disease prediction system for smart farming. Sensors **18**(11), 4051 (2018)
15. Prasad, G., Vanathi, A., Devi, B.S.: A Review on IoT Applications in Smart Agriculture (2023). https://doi.org/10.3233/ATDE221332
16. Salehin, I., et al.: IFSG: intelligence agriculture crop-pest detection system using IoT automation system. Indonesian J. Electr. Eng. Comput. Sci. **24**(2), 1091 (2021). https://doi.org/10.11591/ijeecs.v24.i2.pp1091-1099
17. Limpens, H.J.G.A.: Field identification: using bat detectors to identify species. Bat Echolocation Res. Tools, Tech. Anal. 46 (2004)
18. Griffin, D.R.: The past and future history of bat detectors. Bat echolocation research: tools, techniques, and analysis. Bat Conservation International, Austin, Texas, USA, 6–9 (2004)
19. Pettersson, L.: The properties of sound and bat detectors. Bat Echolocation Research: tools, techniques and analysis, 9 (2004)
20. Linton, D.M.: Bat ecology & conservation in lowland farmland. Diss. University of Oxford (2009)

21. Yong, J., He, F., Li, H., Zhou, W.: A novel bat algorithm based on collaborative and dynamic learning of opposite population. In: 2018 IEEE 22nd International Conference on Computer Supported Cooperative Work in Design ((CSCWD)), pp. 541–546. IEEE (2018)
22. Britzke, E.R., Gillam, E.H., Murray, K.L.: Current state of understanding of ultrasonic detectors for the study of bat ecology. Acta Theriologica **58**, 109–117 (2013)

Synthesis and Modeling of Systems with Combined Fuzzy P + I - Regulators

Bohdan Durnyak[1], Mikola Lutskiv[1], Petro Shepita[1], Vasyl Sheketa[2], Nadiia Pasieka[2(✉)], and Mykola Pasieka[2]

[1] Ukrainian Academy of Printing, Pid Goloskom Str., 19, Lviv 79020, Ukraine
[2] National Technical University of Oil and Gas, 76068 Ivano-Frankivsk, Ukraine
pasyekanm@gmail.com

Abstract. A mathematical model of the automatic control system with a combined fuzzy P + I - controller with an unclouded I - component, which allows eliminating the static error of regulation of static objects, has been developed. The method of determining the parameters of the controller adjustment based on the identification of the object by the transient characteristic, under the condition when the transfer coefficient of the object is equal to one by conducting an extended tangent through the point of intersection determined the parameters of adjustment of the controller. The blurring of the P - component of control has been carried out based on three linguistic variables of the normalized error signal. The fuzzy control described by the fuzzy base of rules has been given, three triangular membership functions have been chosen, and their parameters have been adjusted. MatLab Simulink package on the basis of which the structural scheme of the system was developed was used for modelling and research of fuzzy automatic control systems. The combined fuzzy P + I - regulator consists of two main blocks: the algorithm of normalized control, fusion units, logic output and denormalization, on the basis of which the regulated action on the object has been formed. For convenience of adjustment and research of separate blocks and system visualization blocks have been provided. An example of synthesis and modelling of a combined fuzzy system for inertial objects of the third and fourth order has been considered. The settings for adjusting the controller have been defined and adjusted. The results of simulation modelling in the form of transient characteristics for objects of different dimensions that satisfy the specified parameters of control quality and provide better quality indicators than with a traditional controller have been given. The maximum value of the regulatory action is small, so it is not physically limited, which is an advantage.

Keywords: Synthesis · Modelling · Fuzzification · Fuzzy controller · Parameters · Inference · Accuracy

1 Problem Setting

Growing demands on the quality of manufactured products place new demands on systems of automatic control of technological processes and objects in the presence of various influences. The main disadvantage of traditional standard industrial regulators is

© The Author(s), under exclusive license to Springer Nature Switzerland AG 2023
Z. Hu et al. (Eds.): ICAILE 2023, LNDECT 180, pp. 187–196, 2023.
https://doi.org/10.1007/978-3-031-36115-9_18

that they do not provide quality control when changing the parameters of the object and limiting the regulatory effect on the object. Instead, fuzzy regulators control processes and objects much better with incomplete information about the object, changes in its parameters and the action of various influences. Today there is a development of various algorithms and fuzzy automatic control systems in various industries [1–5, 7, 8, 10].

In the available sources, little attention is paid to the synthesis of systems with fuzzy controllers and the formation of fuzzy control on an object that is fundamentally different from the classical theory of automatic control. Traditional control algorithms are described by an analytical expression and have two or three debug parameters. Instead, fuzzy regulators are described by a base of fuzzy rules, linguistic variables, a series of fuzzy transformations (error normalization, fuzzification, defuzzification, fuzzy inference, which are different and have different implementation schemes), and so on. Therefore, the solution of the synthesis problem and the definition of the parameters of the fuzzy controller is ambiguous and complex. To date, there are no generally accepted sound methods for the synthesis and determination of the parameters of fuzzy regulators, which makes it impossible to optimize them and complicates their development and implementation. Thus, the task of synthesis and modelling of automatic control systems of the combined fuzzy controller is relevant.

2 Analysis of Recent Research and Publications

The demand for fuzzy systems is because they are developed faster, simpler, and cheaper than traditional controllers [6, 9, 11–14]. Works [15–17] are devoted to the synthesis and calculation of fuzzy digital controllers in automatic control systems. The methods of designing fuzzy controllers are based on the analytical expression for the regulatory action at the output of the fuzzy controller for different membership functions and the general functional and structural schemes, based on which it is possible to implement fuzzy program controllers by software and hardware. Schematic diagrams of fuzzy controllers with different membership functions are presented. The study of automatic control systems was conducted by modeling in the MatLab: Simulink package. For the implementation of the given task, numerous complex schemes were used, which additionally contain a scheme for estimating the maximum value of error and its scaling, and the I-component that introduces inertia into the system is not used, but the first and second derivatives of the error signal complicating the system are considered.

The monographs [17, 19–24] present the tasks of designing fuzzy systems in the package MatLab: Simulink using Fuzzy Logic Toolbox. Information on fuzzy sets, fuzzy logic and the application of fuzzy rules and transformations in different systems has been given. Books [18, 25, 26] present different versions of fuzzy controllers, rule bases, block diagrams of different regulators, their analysis, and debugging choices. In the works of the authors [27, 28, 31] the analysis of automatic control systems with simplified versions of fuzzy controllers has been carried out, and the simulation results have been presented. It has been established that the quality of the transient process changes a little when the object transfer coefficient is doubled. Significant fluctuations occur in a system with a traditional regulator, instead [29, 30]. Purpose of the article: To synthesize an automatic control system with a simple version of the combined P + I -

controller for static objects, to determine the debugging parameters, and to investigate its properties.

3 Presentation of the Main Research Material

The problem of synthesis of automatic control systems with the fuzzy controller is complex and multifaceted, and contains a significant number of partial problems and possible ways to solve them, so we have chosen a simple algorithm for fuzzy combined P-controller with I-component that does not blur and eliminates static control error in static control objects. We set the basic requirements for the quality of regulation: to ensure sufficient system speed and 20% overshoot when changing the transfer rate of the object. It is necessary to determine the debugging parameters of the controller, which provide quality indicator tasks for a given object of regulation.

Debugging parameters will be determined based on the identification of the object by the transient characteristic, provided that the transfer factor of the object is equal to one and its parameters will be determined on the basis of tangent to the transition characteristic through the point of intersection, which is shown in Fig. 1.

The simplest substitute model of the object of regulation is the inertial link with delay, which is widely used in determining the parameters of traditional regulators [6, 7].

$$W_0(S) = \frac{k_0 e^{-\tau s}}{T_o S + 1}, \qquad (1)$$

where T0- is the substitution constant of the object time, τ- is the delay time, k0 - is the transmission coefficient of the object.

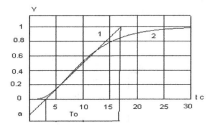

Fig. 1. Transient characteristics of the object

To determine the debugging parameters of the P + I – fuzzy controller, we additionally perform a tangent to the intersection with the axis and determine the parameters a, and τ as shown in Fig. 1.

Based on the identification of the object by the transient characteristic in Fig. 1 for the fuzzy P + I - regulator, the following values of the debugging parameters were obtained: the scale P of the component Mp = 0,17/a Mr = and I component Ki = 0.125/τ. The presented methods of determining the parameters of adjustment of the controller are approximate and therefore require additional experimental tuning or simulation.

For the convenience of presentation, we give the algorithm of normalized control of P + I – fuzzy controller in operator form.

$$U = F(e) + \frac{k_i}{S}e, \text{ if } e = \frac{E}{y_0}, \quad (2)$$

where $E = y_0 - y$ is the control error, k_i is the transmission factor I of the algorithm component, y_0, y is the output and the set point of the controlled value, $F(e)$ is the fuzzy transformations of the normalized e-error signal.

To synthesize the fuzzy P-component of regulation, we accept three linguistic variables of the normalized error signal e: negative B, zero H, positive D, which correspond to three functions of belonging to fuzzy sets of left type - right and triangular symmetric, then control is described by fuzzy rule base.

$$\begin{aligned}R1 &: if(e = B)then(U = B1)\\ R2 &: if(e = H)then(U = B2),\\ R3 &: if(e = D)then(U = B3)\end{aligned} \quad (3)$$

where B_i are blurred controls.

Each rule corresponds to the initial sets Bi, i = 1, 2, 3. To model and study the fuzzy automatic control system, use the package MatLab: Simulink, which developed a scheme of combined P + I-fuzzy controller, given in Fig. 2.

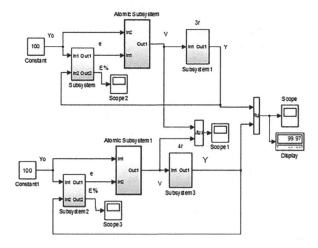

Fig. 2. Block diagram of the model of automatic control system with fuzzy controller

Fuzzy P + I– regulator consists of two main units: the normalized control algorithm developed by expression (2) for competence is located in the Subsystem unit and the fuzzification, logical output and denormalization units located in the Enabled Subsystem unit. The inertial static object of the third dimension is located on the right. At the bottom there is a copy of the fuzzy control system of the fourth-dimension object, necessary for comparing the results of modeling and analysis.

The scheme of the algorithm of normalized control has been developed on the basis of the expression (2) shown in Fig. 3.

To determine the normalized signal of e – error, its scaling is carried out by dividing the adjustment error E by a given value of the input task y0, which ensures its rationing regardless of the value of the task, which may be different, which simplifies rationing. The normalized signal is sent to the input of the next unit for fuzzification and fuzzy transformations. The general scheme of the unit of fuzzification of logical output and denormalization is shown in Fig. 4.

The scheme consists of three triangular functions of accessories such as Triangular MF, the following parameters B [-2000; -1; 1], H [-1; 0; 1], D [0; 1; 2000] have been adjusted.

Conclusion is carried out by optimizing the function of belonging to the Up level, which is set by a normalized e-error signal, and is limited to the Saturation Dynamic unit.

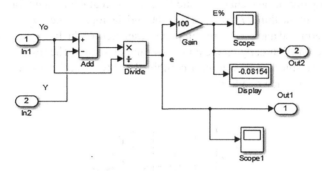

Fig. 3. Scheme of a normalized control algorithm

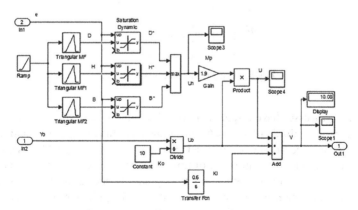

Fig. 4. Scheme of a logical output fuzzification unit

The functions of belonging to B*, H*, D* fuzzy sets are modified in this way at the input of the MAX operator at the input of which the normalized fuzzy control Un is obtained, which is fed to the input of the Gain unit, in which the Mp debugging parameter of the P-component of the regulator is set, the output of which is fed to the first input of the Produkt unit, and a displacement of $U_0 = 1/k_0 y_0$ is submitted to its second input as a result of this, denormalization (scaling) of the U control is carried out. To create an adjusting action V on the object on the ground of the addition unit, three components are supplied: denormal U control, U0 displacement and I-component of Ui control.

$$V = U + U_0 + U_i, \text{ if } U_0 = \frac{1}{k_0} y_0 \text{ then } U = M_p U_n.$$

For the convenience of debugging individual units of the regulator and the automatic control system, scope and display visualization units are provided. An example of synthesis and modeling of a combined fuzzy system with a P + I-regulator for inertial static objects of the third and fourth order with a transmission coefficient $k_0 = 10$ and time levels $T_1 = 5c$, $T_2 = 3c$, $T_3 = 2c$, $T_4 = 1c$. is considered.

The 20% walking, and the $y_0 = 100$ input task has been set. Based on the identification of the object by transitional characteristic (Fig. 1), the regulator debugging parameter x + rs for objects of the third $M_p = 1,9$, $k_i = 0,60$ and the fourth order $M_p = 1,0$, $k_i = 0,25$ have been determined and set. More complete parameters of the fuzzy automatic control system are presented directly on the corresponding diagrams. The results of simulation modeling of transition processes in a combined fuzzy control system for objects of different orders are shown in Fig. 5.

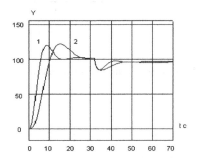

Fig. 5. Transition processes in systems

Transition processes have a specified 20%. Walking the growth time in the system with the object of the third order $t_н = 6c$, the adjustment time is $t_p = 14c$, and $t_н = 10c$, $t_p = 26c$ with the objects of the fourth order. Transition processes quickly disappear, and there is no static error. At the moment of time, $t = 30c$, a degree perturbation $z_0 = 20$ is presented on an object that, with a slight reregulation, goes away. Consequently, the combined fuzzy automatic control system with the P + I regulator provides the specified indicators of the quality of regulation. The results of the study of the regulatory action quality on the object at the step task $y_0 = 100$ are presented in Fig. 6.

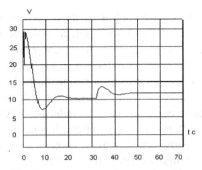

Fig. 6. Regulatory action on the object

When the jumping task $y_0 = 100$, the combined fuzzy regulator forms a regulatory effect that differs significantly from the traditional one. At the initial moment of time, a short-term amplitude pulse of which is 30 is created and gradually approaches the established value. Consequently, the maximum value of the adjustable action is small, so it is not physically limited, which is an advantage.

Studies confirm that automatic control systems with a combined P + I fuzzy regulator with inertial static objects provide better quality indicators of regulation than with traditional PI regulator. In particular, with small transfer coefficients of the object ($k_O \approx 1$), the maximum value of the adjustable action increases significantly ($V_H \geq 250$), which can be physically limited, which impairs achievable quality indicators.

It is established that systems with a traditional PI regulator are quite sensitive to changes in the transfer coefficient of the object. With a twofold increase in the coefficient in the system, there is a significant re-regulation and oscillation. Instead, with a twofold decrease in the transmission coefficient, the transition process becomes periodic and significantly decreases the performance of the system. To eliminate the impact of changes in the transfer coefficient of the object, it is necessary to re-determine and set the parameters for debugging the PI regulator, which is a disadvantage. Instead, systems with a combined fuzzy regulator are little sensitive to changes in the parameters of the object, which is their advantage.

4 Conclusions

Nowadays, there are no generally accepted reasonable methods of synthesis and determination of parameters for setting fuzzy regulators, which makes it impossible to optimize them, complicates their development and implementation. The problem of synthesis of the automatic control system with fuzzy regulators is complex and multifaceted, it contains a significant number of partial problems and possible ways to solve them. A fuzzy combined P-regulator with the I-component is proposed, which does not heat up and eliminates the static error of regulation of static regulatory objects, which simplifies the regulator's system. It is proposed to determine the parameters of setting the regulator on the basis of identification of objects by transitional characteristics, provided that the transfer coefficient of the object is equal to one that ensures the choice of parameters of the regulator, regardless of the transfer factor of the object. A structural diagram of the fuzzy system model in the MatLab: Simulink package has been developed, which makes it possible to calculate and build transitional characteristics of the system to analyze its properties and interactively test the optimal adjustment parameters of the regulator.

The results of simulation modeling in the form of transitional characteristics of the system for objects of different order are presented and the quality indicators of regulation are determined. It is established that automatic control systems with combined fuzzy P + I -regulator with inertial objects provide better indicators of quality of regulation than with traditional regulators. With small transfer coefficients of the object ($k_0 \approx 1$), the maximum value of the adjusting action increases significantly ($V_H \geq 250$), which can be physically limited, which impairs achievable quality indicators.

It has been established that systems with a traditional PI regulator are quite sensitive to changes in the transfer coefficient of the object. With a twofold increase in the coefficient of the system, significant reregulation and oscillation occurs, and with a decrease in the coefficient, the performance decreases significantly. To eliminate the effect of changing the coefficient, it is necessary to re-determine and set the adjustment parameters of the regulator, which is a disadvantage. Instead, systems with a combined fuzzy regulator are little sensitive to changes in object parameters, which is their advantage.

References

1. Durnyak, B., Lutskiv, M., Shepita, P.: Fuzzy model of raster transformation of square elements. Paper presented at the CEUR Workshop Proceedings, 3156, pp. 140–149 (2022)
2. Durnyak, B., Lutskiv, M., Shepita, P., Nechepurenko, V.: Simulation of a Combined Robust System with a P-Fuzzy Controller. Intellectual Systems of Decision Making and Problems of Computational Intelligence: Proceedings of the XV International Scientific Conference. **1020**, 570–580 (2019)
3. Imamović, B., Halilčević, S.S., Georgilakis, P.S.: Comprehensive fuzzy logic coefficient of performance of absorption cooling system. Expert Syst. Appl. **190**, art. no. 116185 (2022). https://doi.org/10.1016/j.eswa.2021.116185
4. Pasieka, N., Sheketa, V., Romanyshyn, Y., Pasieka, M., Domska, U., Struk, A.: Models, methods and algorithms of web system architecture optimization. In: IEEE International Scientific-Practical Conference: Problems of Infocommunications Science and Technology, PIC S&T 2019 - Proceedings, pp. 147–152. https://doi.org/10.1109/PICST47496.2019.906 1539
5. Salem, A., Desouky, A., Alaboudy, A.: New analytical assessment for fast and complete pre-fault restoration of grid-connected FSWTs with fuzzy-logic pitch-angle controller. Int. J. Electr. Power Energy Syst. **136**, art. no. 107745 (2022). https://doi.org/10.1016/j.ijepes.2021. 107745
6. Yoo, J., et al.: RaScaNet: learning tiny models by raster-scanning images. In: Proceedings of the IEEE/CVF Conference on Computer Vision and Pattern Recognition, pp. 13673–13682 (2021)
7. Durnyak, B., Lutskiv, M., Shepita, P., Karpyn, R., Savina, N.: Determination of the optical density of two-parameter tone transfer for a short printing system of the sixth dimension. In: IntelITSIS, pp. 134–140 (2021)
8. Xia, J., Zhang, J., Feng, J., Wang, Z., Zhuang, G.: Command filter-based adaptive fuzzy control for nonlinear systems with unknown control directions. IEEE Trans. Syst. Man Cybern. Syst. **51**(3), 1945–1953 (2019)

9. Kurniawan, E., Widiyatmoko, B., Bayuwati, D., Afandi, M., Suryadi, Rofianingrum, M.: Discrete-time design of model reference learning control system. In: 24th International Conference on Methods and Models in Automation and Robotics (MMAR), pp. 337–341 (2019). https://doi.org/10.1109/MMAR.2019.8864713
10. Brik, A., Labrak, L., O'Connor, I., Saias, D.: Fast hierarchical system synthesis based on predictive models. In: 18th IEEE International New Circuits and Systems Conference (NEWCAS), pp. 70–73 (2020). https://doi.org/10.1109/NEWCAS49341.2020.9159765
11. Nejati, A., Zhong, B., Caccamo, M., Zamani, M.: Controller synthesis for unknown polynomial-type systems: a data-driven approach. In: The 2nd International Workshop on Computation-Aware Algorithmic Design for Cyber-Physical Systems (CAADCPS), pp. 11–12 (2022). https://doi.org/10.1109/CAADCPS56132.2022.00007
12. Allaboyena, G., Yilmaz, M.: Robust power system stabilizer modeling and controller synthesis framework. In: IEEE International Conference on Electro Information Technology (EIT), pp. 207–212 (2019). https://doi.org/10.1109/EIT.2019.8833656
13. Pasyeka, M., Sheketa, V., Pasieka, N., Chupakhina, S., Dronyuk, I.: System analysis of caching requests on network computing nodes. In: The 3rd International Conference on Advanced Information and Communications Technologies, AICT2019 - Proceedings, pp. 216–222 (2019). https://doi.org/10.1109/AIACT.2019.8847909
14. Valenzuela, P., Ebadat, A., Everitt, N., Parisio, A.: Closed-loop identification for model predictive control of hvac systems: from input design to controller synthesis. IEEE Trans. Control Syst. Technol. **28**(5), 1681–1695 (2020)
15. Danilushkin, A.. Closed-loop optimal system :synthesis in the mode of "special" control of a continuous induction heater. In: 2019 XXI International Conference Complex Systems: Control and Modeling Problems (CSCMP), pp. 214–217 (2019). Doi: https://doi.org/10.1109/CSCMP45713.2019.8976505
16. Pasyeka, M., Sviridova, T., Kozak, I.: Mathematical model of adaptive knowledge testing. In: 5th International Conference on Perspective Technologies and Methods in MEMS Design, MEMSTECH 2009, pp. 96–97 (2009)
17. Oleg, B., Marina, T.: Regulator synthesis for the proportional electromagnet control system. In: 2022 International Conference on Industrial Engineering, Applications and Manufacturing (ICIEAM), pp. 610–614 (2022). https://doi.org/10.1109/ICIEAM54945.2022.9787185
18. Loubach, D.: A high-level synthesis approach applicable to autonomous embedded systems. In: 2022 IEEE XXIX International Conference on Electronics, Electrical Engineering and Computing (INTERCON), pp. 1–4 (2022). https://doi.org/10.1109/INTERCON55795.2022.9870103
19. Romanyshyn, Y., Sheketa, V., Poteriailo, L., Pikh, V., Pasieka, N., Kalambet, Y.: Social-communication web technologies in the higher education as means of knowledge transfer. In: IEEE 14th International Scientific and Technical Conference on Computer Sciences and Information Technologies (CSIT). – 2019(3), pp. 35–39 (2019)
20. Shebanin, V., Atamanyuk, I., Kondratenko, Y.: Synthesis of control systems on the basis of a canonical vector decomposition of random sequences. In: 2020 IEEE Problems of Automated Electrodrive. Theory and Practice (PAEP), pp. 1–5 (2020). doi: https://doi.org/10.1109/PAEP49887.2020.9240808
21. Liu, K., Teel, A., Sun, X., Wang, X.: Model-based dynamic event-triggered control for systems with uncertainty: a hybrid system approach. IEEE Trans. Autom. Control **66**(1), 444–451 (2021). https://doi.org/10.1109/TAC.2020.2979788
22. Vodyaho, A., Zhukova, N., Abbas, S., Kulikov, I., Annam, F.: On one approach to the dynamic digital twins models synthesis. In: 2022 XXV International Conference on Soft Computing and Measurements (SCM), pp. 126–128 (2022). doi: https://doi.org/10.1109/SCM55405.2022.9794895

23. Botan, C., Ostafi, F.: Discrete time model-following problem for linear systems with variable disturbances. In: The 24th International Conference on System Theory, Control and Computing (ICSTCC), 2020, pp. 54–59 (2020). https://doi.org/10.1109/ICSTCC50638.2020.925 9778

24. Khalifa, T., El-Nagar, A., El-Brawany, M., El-Araby, E., El-Bardini, M.: A novel hammerstein model for nonlinear networked systems based on an interval type-2 fuzzy takagi–sugeno–kang system. IEEE Trans. Fuzzy Syst. **29**(2), 275–285 (2021). https://doi.org/10.1109/TFUZZ. 2020.3007460

25. Rashkevich, Y., Peleshko, D., Pasyeka, M., Stetsyuk, A.: Design of web-oriented distributed learning systems. Upravlyayushchie Sistemy i Mashiny, (3–4), 72–80

26. Guo, Y., Hou, Z., Liu, S., Jin, S.: Data-driven model-free adaptive predictive control for a class of MIMO nonlinear discrete-time systems with stability analysis. IEEE Access **7**, 102852–102866 (2019)

27. Zeng, Y., Lam, H., Wu, L.: Hankel-norm-based model reduction for stochastic discrete-time nonlinear systems in interval type-2 T-S fuzzy framework. IEEE Trans. Cybern. **51**(10), 4934–4943 (2021). https://doi.org/10.1109/TCYB.2019.2950565

28. Sheketa V., Poteriailo L., Romanyshyn Y., Pikh V., Pasyeka M., Chesanovskyy M.: Case-based notations for technological problems solving in the knowledge-based environment." Paper presented at the International Scientific and Technical Conference on Computer Sciences and Information Technologies, 1, pp. 10–14 (2019). doi:https://doi.org/10.1109/STC-CSIT.2019. 8929784

29. Hussain, N., Ali, S., Ridao, P., Cieslak, P., Al-Saggaf, U.: Implementation of nonlinear adaptive U-model control synthesis using a robot operating system for an unmanned underwater vehicle. IEEE Access **8**, 205685–205695 (2020). Doi:https://doi.org/10.1109/ACCESS.2020. 3037122

30. Pasieka, M., Sheketa, V., Pasieka, N., Chupakhina, S., Dronyuk, I.: System analysis of caching requests on network computing nodes. In: 3rd International Conference on Advanced Information and Communications Technologies, AICT 2019, pp. 216–222 (2019)

31. Mushar, K., Hote, Y., Pillai, G.: New mixed model order reduction approach for linear system. In: 2022 IEEE International Conference on Signal Processing, Informatics, Communication and Energy Systems (SPICES), pp. 343–348 (2022). https://doi.org/10.1109/SPICES52834. 2022.9774083

Protection of a Printing Company with Elements of Artificial Intelligence and IIoT from Cyber Threats

Bohdan Durnyak[1], Tetyana Neroda[1], Petro Shepita[1], Lyubov Tupychak[1], Nadiia Pasieka[2(✉)], and Yulia Romanyshyn[2]

[1] Ukrainian Academy of Printing, Pid Goloskom str., 19, 79020 Lviv, Ukraine
[2] National Tech, University of Oil & Gas, Ivano-Frankivsk 76068, Ukraine
pms.mykola@gmail.com

Abstract. The article considers the actual challenges of cyber security that arise at the current stage of information technology development, with the introduction of artificial intelligence and machine learning. On the basis of the received information, it is proposed to carry out the design of the management system of the printing enterprise with the imitation of an expert. When implementing this function, it was decided to use an element of artificial intelligence, namely neural networks. When designing the proposed analytical apparatus, two alternative options for ANN training were developed for comparison. As a result of the simulation of disturbance perception, the ANN trained using TrustScore showed its effectiveness and superiority.

Keywords: Printing company · Internet of things · Artificial Neural network · Cyber attack · Knowledge base

1 Introduction

The introduction of existing control and data collection systems into the technological process of small printing enterprises, along with the significant cost of highly specialized, mostly built-in equipment of a limited range, provision of equipment for the placement of the control room and server room, and the introduction of separate positions, requires high qualification of operators and free service, which significantly increases the cost of manufactured products. And services provided. The problems of the application of information technologies in printing are focused on the study of materials and equipment, the analysis of individual effects on the quality of printing, methods of rasterization and their influence on the transfer of image color, the development of means of automation and informatization of printing production. In the conditions of innovative reforming of enterprises, there is a growing shortage in the design of technologies to optimize the management of multi-stage information processes, in particular, machine and instrument engineering, in the printing industry, as well as in education computerization systems. Today, the list of industries in which the operator makes decisions either at key stages of

© The Author(s), under exclusive license to Springer Nature Switzerland AG 2023
Z. Hu et al. (Eds.): ICAILE 2023, LNDECT 180, pp. 197–205, 2023.
https://doi.org/10.1007/978-3-031-36115-9_19

the technological process or when coordinating the stages of the technological process among themselves is expanding. Significant progress in the modernization of the enterprise is achieved with the introduction of production complexes, which are integrated into the existing technological process with the gradual displacement of the outdated elemental base in the weakest places, in particular, allowing to carry out an operational analysis of the means of production to make an adequate management decision. Along with the introduction of such necessary and modern information technologies, a number of threats to cyber security are observed. Basically, the development of machine learning and artificial intelligence not only facilitates the mental activity of users within the Internet of Things industrial or Internet of Things but also floods these environments with growing. The most common cause is a malfunctioning machine learning model. An adversarial attack can involve presenting a model with inaccurate or false data during training or injecting maliciously crafted data to trick an already trained model. Thus, there is an urgent need to protect the provision of information systems with elements of artificial intelligence implemented at printing enterprises.

2 Analysis of Recent Research and Publications

One of the important sources about the inadequacy of the protection of artificial intelligence systems can be considered the report of the National Security Commission of the USA [1], which notes the very small number of studies on the protection of artificial intelligence systems. And neglecting the basic principles of cyber security in conducting scientific research. Also, certain systems that have already been implemented in production are not 100% protected from attacks.

In the study [2], the authors placed improvised road markings on the surface of a car with autopilot, which caused the car to cross into the oncoming lane. Work [3–5] proved that small and almost imperceptible changes in the images provided for medical diagnosis can lead to a misdiagnosis and harm to a person. The authors of the study [6–8] gave an example of how a road sign learned by the car control system during machine learning can easily lead to an accident by making corrections to it with improvised means. In the article [9–12], scientists from Google and the University of California proved that even the best forensic classifiers - artificial intelligence systems developed by the US government security departments, trained to distinguish and separate real and synthetic content are vulnerable to attacks. As a VentureBeat contributor points out [13–15], there has been a surge in research on adversarial attacks in recent years. Thus, from 2014 to 2022, the number of papers submitted to the preprint server Arxiv.org on adversarial machine learning increased from 2 to about 1,800.

Competitive attacks on artificial intelligence systems received wide publicity at ICLR, Usenix, and Black Hat international conferences. Therefore, when designing and developing systems for IIoT, protection against adversarial attacks must be provided.

3 Presentation of the Main Research Material

The paper examines the management of a printing enterprise as a whole structure based on elements of artificial intelligence and machine learning. Using the available elements of information technologies, a decision-making system was built based on the training of an artificial neural network and the formation of a knowledge bank. Two implementation options were considered and an analysis of the system's reaction to external threats was carried out.

3.1 Building an Expert Simulator and Training ANNs by Conventional Means

The main element of the enterprise management system is the analytical unit, which is responsible for understanding the problem and solving the tasks and localization of disturbances. When designing it, the parameters and main blocks of the knowledge base are first formed [16, 17]. An operator-expert survey was conducted, which determined the main stages and important points of the operation of the technological equipment (Fig. 1). Since the formed knowledge base is also used at the stages preceding the direct production of products, the survey also takes place among other experts.

To form a knowledge base, a production goal is defined, after which a certain number of expert operators OEn are interviewed, who describes the situation St that arises during the work, and also give a description of the controlling influences Kv that must be implemented to obtain the desired result, the result Rz [18, 19]. These data form the

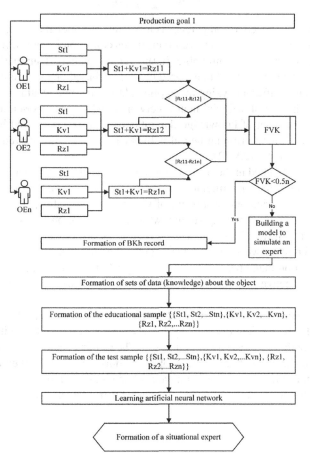

Fig. 1. A model of knowledge base formation and imitation of an expert

properties of experts Fig. 1.

$$OE_n = \{St,\ Kv,\ Rz\} \tag{1}$$

The survey results of each of the expert operators are formed into a mathematical relationship:

$$St_n + Kv_n = Rz_n \tag{2}$$

where at the output, numerical values of expert operator testing are obtained that correspond to the resulting value of the control action.

The results of a mathematical operation are compared:

$$G_n = |Rz_n - Rz_{n+1}| \tag{3}$$

where G is the difference between the resulting scores of expert operators;

Parameter G is sent to the unit for forming the value of knowledge competence, where the values are added up:

$$FVK = \sum_{i=1}^{n} G_i \tag{4}$$

If the value of knowledge competence is less than 0.5n, then the corresponding entry is entered into the knowledge base as correct. If this condition is not fulfilled, then these statements are recognized as unreliable. In this regard, an expert simulation block was introduced into the model. To obtain a predicted result based on the received statements of expert operators, after analyzing the applied algorithms that would satisfy the needs of the system, the use of artificial neural networks (ANN) was chosen as the optimal option. After that, a set of knowledge data about the object is formed in the form of a training sample for an artificial neural network from the data received from experts about production goal 1 [20–22].

The data are tabulated in the form of normalized values, and a test and training sample is formed for training an artificial neural network. Training and testing of an artificial neural network is carried out, as a result of which a situational expert is formed based on the formed neural network apparatus, which, depending on the situation, provides predicted results. The resulting situational expert is used if the interviewed expert operators give different answers about the situation. The situation is considered stochastic and may depend on many parameters. In this regard, it is proposed to use a situational expert for the effective operation of an intelligent system with a knowledge base [11, 12, 23, 24].

For its construction, a neural network of direct propagation and the method of the fastest descent, which is implemented by the Levenberg–Marquardt algorithm, i.e. using the nonlinear method of least squares, are used. The training sample is formed from normalized data on the production situation, obtained during the survey of expert operators. The input vector of the expert's opinion values (3.5) is applied to the input of the neural network, which reflects the parameters necessary for training and further functioning of the system.

$$X = [St1] \tag{5}$$

After processing the data according to the selected algorithm, the output vector of values (6) is obtained at the output of the artificial neural network:

$$Y = \begin{bmatrix} Rz1 \\ Kv1 \end{bmatrix} \tag{6}$$

In this case, the neural network is represented as the expression of its output $Y = Y(X, Q)$ after adjustment by the weights of Q neurons.

The calculation of the network error for one epoch is represented by expression 7:

$$F(Y) = \frac{1}{2}\left((St1 - Rz1)^2 + (Kv1 - Rz1)^2\right) \tag{7}$$

For the selected algorithm, the input sample must be divided into three subsamples (Fig. 3) in the percentage ratio. The largest number of examples required for training the network is 70%, the other two subsamples require a smaller number of examples, then the validation and testing samples each receive 15% of the records of the total sample [13, 14].

3.2 Learning ANNs for Recognition from Adversarial Attacks

After importing the necessary normalized data, we build a CNN and conduct training on clean data without involving additional ones for training. To prevent external interference, the construction model of the autoencoder, TrustScore, was built and trained on the obtained data set. The essence of this procedure is to calculate indicators of confidence in the existing knowledge base and data set for training. First, we preprocess the training data to find the α-high density sample of each class, which is defined as the training samples within that class after filtering out the α-fraction of samples with the lowest density.

Let $0 \leq \alpha < 1$ and let f be a continuous density function with a compact support X $\subseteq$ RD. After that, we define Hα(f), an α-set f with high density, as a set of level $\lambda\alpha$ f defined as the expression:

$$\{x \in X : f(x) \geq \lambda\alpha\}\partial e\ \lambda\alpha := \inf\{\lambda \geq 0 : \int X 1[f(x) \leq \lambda]f(x)dx \geq \alpha\} \tag{8}$$

To approximate the set of α-high density, the α fraction of points is filtered by the smallest empirical density, based on k-nearest neighbors. This data filtering step is independent of the given classifier h. The next stage is to provide a test sample, we determine the confidence estimate as the ratio between the linguistic data from the sample under study and the high-density α-set of the closest class, different from the predicted class. It is assumed that if the classifier h predicts a label that is significantly further than the closest label, then this is a warning that the simulated expert may be wrong. Thus, our procedure can be seen as a comparison with a modified nearest neighbor classifier, where the modification is to initially filter out linguistic variables that are not included in the high-density α-set for each class [16, 26, 27].

3.3 Comparison of System Performance

In the course of testing the traditional model and the autoencoder with TrusctScore on clean data and on disturbed samples with different disturbance indicators, the following results were obtained, shown in Fig. 2 and Fig. 3, which indicate that the expert imitation model with protection against competitive attacks has been developed, using an autoencoder and TsustScore, is significantly more accurate in recognition and resistant to adversarial FGM attacks compared to the traditional model. Despite the rather high efficiency, with powerful disturbances with a high delta, the information is severely damaged, which makes it virtually indeterminate, although even in this case we observe that the accuracy underlying the agreement with TsustScore is higher.

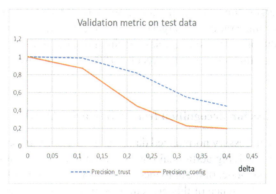

Fig. 2. Dependence of precession on delta.

where precision_config-accuracy is based on confidence at the output of a regular classifier without a corrector in the form of trust score; precision_trust-precision based on agreement with trust score. An analysis of the obtained graphs was performed, which demonstrates that the accuracy of the classification in agreement with the trust score (Precision_trust) is significantly higher than the accuracy of the base classifier at all delta values, even when the considered samples contain external disturbance and/or are strongly distorted.

Where accuracy valuation is the accuracy of the classifier in its classical task without a corrector in the form of a trust score.

Analyzing the resulting graphs, it is obvious that with an increase in delta, which causes an increase in the distortion of samples and their loss of important information, the number of samples for which a decision is made decreases, and the accuracy of recognition of disturbance signals obtained from polygraphic possession decreases somewhat. Given that the samples that have not passed the inspection are not allowed to the decision-making stage about them and do not affect the accuracy of the control process on the production task.

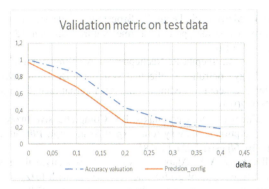

Fig. 3. The dependence of precession on delta

4 Conclusion

Artificial intelligence systems and Industrial Internet of Things tools, like any software tool, require protection against external unauthorized access, reliability, and safe operation in all respects and conditions.

As a result of the study, modern and effective technologies and approaches for the implementation of the analytical block of the management system of a printing company with the ability to recognize various types of cyber-attacks on artificial intelligence systems and protection against them were considered and analyzed.

References

1. Dr. Eric Schmidt and others National security commissionon artificial intelligence interim report november 2019
2. Experimental Security Research of Tesla Autopilot /Tencent Keen Security Lab 2019–03
3. Finlayson, S.G., Bowers, J.D., Ito, J., Zittrain, J.L., Beamand, A.L., Kohane, I.S.: Adversarial attacks on medical machine learning 2019, vol. 363, Issue 6433, pp. 1287–1289 (2019)
4. Eykholt, K., et al.: Robust physical-world attacks on deep learning visual classification. Paper presented at the Proceedings of the IEEE Computer Society Conference on Computer Vision and Pattern Recognition, pp. 1625–1634 (2018). https://doi.org/10.1109/CVPR.2018.00175
5. Carlini, N., Farid, H.: Evading deepfake-image detectors with white-and black-box attacks. Paper presented at the IEEE Computer Society Conference on Computer Vision and Pattern Recognition Workshops, 2020-June 2804–2813 (2020). https://doi.org/10.1109/CVPRW50498.2020.00337
6. Metzen, J.H., Kumar, M.C., Brox, T., Fischer, V.: Universal adversarial perturbations against semantic image segmentation. arXiv preprint arXiv:1704.05712 (2017)
7. Wang, J., Liang, J., Zhang, L., Ding, X.: Non-fragile dynamic output-feedback control for-gain performance of positive FM-II model with PDT switching: an event-triggered mechanism. Int. J. Robust Nonlinear Control
8. Lopes, I.O., Zou, D., Abdulqadder, I.H., Ruambo, F.A., Yuan, B., Jin, H.: Effective network intrusion detection via representation learning: a denoising AutoEncoder approach. Comput. Commun. **194**, 55–65 (2022). https://doi.org/10.1016/j.comcom.2022.07.027
9. Pan, Y., Du, P., Xue, H., Lam, H.K.: Singularity-free fixed-time fuzzy control for robotic systems with user-defined performance. IEEE Trans. Fuzzy Syst. **29**(8), 2388–2398 (2020)

10. Crețu, A., Monti, F., Marrone, S., Dong, X., Bronstein, M., de Montjoye, Y.: Interaction data are identifiable even across long periods of time. Nature Commun. **13**(1). https://doi.org/10.1038/s41467-021-27714-6
11. Durnyak B., Lutskiv M., Shepita P., Nechepurenko V.: Simulation of a Combined Robust System with a P-Fuzzy Controller. Intellectual Systems of Decision Making and Problems of Computational Intelligence: Proceedings of the XV International Scientific Conference, vol. 1020, pp. 570–580 (2019)
12. Imamović, B., Halilčević, S.S., Georgilakis, P.S. Comprehensive fuzzy logic coefficient of performance of absorption cooling system. Expert Syst. Appl. **190**, art. no. 116185 (2022). https://doi.org/10.1016/j.eswa.2021.116185
13. Salem, A.A., ElDesouky, A.A., Alaboudy, A.H.K.: New analytical assessment for fast and complete pre-fault restoration of grid-connected FSWTs with fuzzy-logic pitch-angle controller. Int. J. Electr. Power Energy Syst. **136**, art. no. 107745 (2022). https://doi.org/10.1016/j.ijepes.2021.107745
14. Tang, P., Ma, Y.: Exponential stabilization and non-fragile sampled-date dissipative control for uncertain time-varying delay TS fuzzy systems with state quantization. Inf. Sci. **545**, 513–536 (2021)
15. Lin, J., Njilla, L.L., Xiong, K.: Secure machine learning against adversarial samples at test time. EURASIP J. Inf. Secur. **2022**(1), 1–15 (2022). https://doi.org/10.1186/s13635-021-00125-2
16. Mahmoodabadi, M.J., Andalib Sahnehsaraei, M.: Parametric uncertainty handling of underactuated nonlinear systems using an online optimal input–output feedback linearization controller. Syst. Sci. Control Eng. **9**(1), 209–218 (2021)
17. Yoo, J., et al.: RaScaNet: learning tiny models by raster-scanning images. In: Proceedings of the IEEE/CVF Conference on Computer Vision and Pattern Recognition, pp. 13673–13682 (2021)
18. Durnyak, B., Lutskiv, M., Shepita, P., Karpyn, R., Savina, N.: Determination of the optical density of two-parameter tone transfer for a short printing system of the sixth dimension. In: IntelITSIS, pp. 134–140 (2021)
19. Pasieka, N., Sheketa, V., Romanyshyn, Y., Pasieka, M., Domska, U., Struk, A.: Models, methods and algorithms of web system architecture optimization. In: IEEE International Scientific-Practical Conference: Problems of Infocommunications Science and Technology, PIC S&T 2019 - Proceedings, pp. 147–152. https://doi.org/10.1109/PICST47496.2019.9061539
20. Pasieka, M., Sheketa, V., Pasieka, N., Chupakhina, S., Dronyuk I.: System analysis of caching requests on network computing nodes. In: 3rd International Conference on Advanced Information and Communications Technologies, AICT 2019-pp. 216–222 (2019)
21. Pasyeka, M., Sheketa, V., Pasieka, N., Chupakhina, S., Dronyuk, I.: System analysis of caching requests on network computing nodes. In: 3rd International Conference on Advanced Information and Communications Technologies, AICT2019 - Proceedings, pp. 216–222. https://doi.org/10.1109/AIACT.2019.8847909
22. Pasyeka, M., Sviridova, T., Kozak, I.: Mathematical model of adaptive knowledge testing, 5th International Conference on Perspective Technologies and Methods in MEMS Design, MEMSTECH 2009, pp. 96–97 (2009)
23. Sun, K., Mou, S., Qiu, J., Wang, T., Gao, H.: Adaptive fuzzy control for nontriangular structural stochastic switched nonlinear systems with full state constraints. IEEE Trans. Fuzzy Syst. **27**(8), 1587–1601 (2018)
24. Pasieka, N., Sheketa, V., Romanyshyn, Y., Pasieka, M., Domska, U., Struk, A.: Models, methods and algorithms of web system architecture optimization. In: IEEE International Scientific-Practical Conference: Problems of Infocommunications Science and Technology, PIC S&T 2019 - Proceedings, pp. 147–152. Doi:https://doi.org/10.1109/PICST47496.2019.9061539

25. Bandyra, V., Malitchuk, A., Pasieka, M., Khrabatyn, R.: Evaluation of quality of backup copy systems data in telecommunication systems. In: IEEE International Scientific and Practical Conference Problems of Infocommunications Science and Technology.-PIC S&T'2019.-08–11 October 2019, Ukraine, pp. 329–325, Doi:https://doi.org/10.1109/PICST47496.2019.9061379
26. Yoo, J., et al.: RaScaNet: learning tiny models by raster-scanning images. In: Proceedings of the IEEE/CVF Conference on Computer Vision and Pattern Recognition, pp. 13673–13682 (2021)
27. Romanyshyn, Y., Sheketa, V., Poteriailo, L., Pikh, V., Pasieka, N., Kalambet, Y.: Social-communication web technologies in the higher education as means of knowledge transfer. In: IEEE 14th International Scientific and Technical Conference on Computer Sciences and Information Technologies (CSIT), vol. 3.-2019.-Lviv, Ukraine, pp. 35–39 (2019)

AMGSRAD Optimization Method in Multilayer Neural Networks

S. Sveleba, I. Katerynchuk[✉], I. Kuno, O. Semotiuk, Ya. Shmyhelskyy, S. Velgosh, N. Sveleba, and A. Kopych

Department of Optoelectronics and Information Technologies,
Ivan Franko National University of Lviv, Lviv 79017, Ukraine
{serhiy.sveleba,ivan.katerynchuk,ivan.kuno}@lnu.edu.ua

Abstract. The AMSGrad optimization learning method for multilayer neural network (MNN) was tested using the logistic function and the Fourier spectra of the error function. The logistic function describes the process of doubling the number of local minima. It was established that due to retraining of MNN, the learning error function of each neuron is characterized by a set of wave vectors with different periodicities. The average value of the learning error for all neurons can be considered as the average value for all existing periodicities. At the same time, the wave vector of the total oscillation can take both commensurate and incommensurate values. The optimization method of AMSGrad training leads to a change in the frequency spectrum of the existing periodicities of the error function on each neuron. That is, the speed of learning of each neuron is corrected, which removes the degeneracy of this system by preventing relearning processes of each neuron.

Keywords: Multilayer neural network (MNN) · AMSGrad optimization method · Block structure

1 Introduction

1.1 AMSGrad Optimization Method

The most common way to optimize neural networks is the gradient descent method. Gradient descent is an optimization algorithm that follows the negative gradient of an objective function to find the minimum of the error function. In most cases, gradient methods are based on an iterative procedure implemented according to the formula:

$$w_{k+1} = w_k + \eta_k p(w_k)$$

where w_k, w_{k+1} the current and new approximation of the values of controlled variables to the optimal solution, respectively, η_k is the convergence step, $p(w_k)$ is the search direction in the N-dimensional space of controlled variables. The method of determining $p(w_k)$ and α_k at each iteration depends on the specifics of the method. The most common Cauchy method (the method of fastest descent) consists in implementing the rule:

$$w_{k+1} = w_k - \alpha \frac{\partial J(w, b, x, y)}{\partial w}$$

© The Author(s), under exclusive license to Springer Nature Switzerland AG 2023
Z. Hu et al. (Eds.): ICAILE 2023, LNDECT 180, pp. 206–216, 2023.
https://doi.org/10.1007/978-3-031-36115-9_20

Denoting the current gradient $g = \frac{\partial J}{\partial w}$, we get: $w_{k+1} = w_k - \alpha_k g_k$, where α_k is the learning rate. The complete form of the derivative of the evaluation function $\frac{\partial J(w,b,x,y)}{\partial w_{ik}} = \alpha_k \delta_i$, where $\delta_i - (y_i - h_i)f(z_i)$, $z_i = \sum_i^n w_i x_i$, x_i - input variables, y_i - expected output values, h_i - calculated output values, and node number in the output layer.

A limitation of the gradient descent method is that this method applies a single learning rate for all input variables. An extension of the gradient descent method, such as the Adaptive Movement Estimation (Adam) algorithm, uses a different learning rate for each input variable, but as a result, the learning rate can quickly decrease to very small values [1].

The AMSGrad method is an extended version of the Adam method, which attempts to improve the convergence properties of the algorithm by avoiding large abrupt changes in the learning rate for each input variable. Technically, gradient descent is called a first-order optimization algorithm because it explicitly uses the first-order derivative of the objective function.

It is known [2, 3] that the AMSGrad algorithm updates the exponential moving averages of the gradient (m_t) and the square of the gradient (v_t), where the hyperparameters β_1, β_2 (the value of which varies in the interval [0,1)) control the exponential decay rates of these moving averages. The moving averages themselves are estimates of the 1st moment (mean) and 2nd moment (uncentered variance) of the gradient [4].

Thus, AMSGrad results in no step size increase, which avoids the problems experienced by Adam. For simplicity, the authors of [2] also remove the offset step used in Adam. A full AMSGrad update without corrected grades can be submitted as follows:

$$m_t = \beta_1^{N+1} m_{t-1} + \left(1 - \beta_1^{N+1}\delta_t\right)$$

$$v_t = \beta_2 v_{t-1} + \left(1 - \beta_2\delta_t^2\right)$$

$$\widehat{v}_t = max\left(\widehat{v}_{t-1}, v_t\right)$$

$$w_{t+1} = w_t - \alpha h_t m_t / \left(\sqrt{\widehat{v}_t}\right)$$

where: m_t – is the moving average of the gradient of the error function, v_t – is the moving average of the square of the gradient of the error function, α – is the learning speed, N – is the number of epochs.

Our preliminary studies have shown that, unlike MNN in which no optimization methods were used, in the considered MNN, when the AMSGrad optimization method is applied, with an increase in the learning speed, cascades of transition to a chaotic state and exit from it are traced. Their number increases as learning speed increases. Therefore, the purpose of our work is to conduct a study of chaotic states and the processes of entering and exiting these states, that is, the mechanisms of the cascade learning process when applying the AMSGrad optimization method. The goal of present work is to analyze the appearance of local minima of the error function and the mechanism of influence of the AMSGrad optimization method on the process of blocking their appearance.

208 S. Sveleba et al.

2 Methodology

2.1 Software Implementation

A program for a MNN with hidden layers for recognizing printed digits was written in the Python programming environment. The array of each number consisted of a set of "0" and "1" of size 4x7. The sample of each digit contained a set of 4 possible distortions of the digit and a set of 3 arrays that did not correspond to any of the digits. For example, for the number "0" an array of x values:

Numt1=[0,0,0,0,1,1,1,1,1,1,1,1,0,0,0,0,0,0,0,0,0,0,0,0,0,0,0,0]
Numt2=[1,1,1,1,1,1,1,1,1,1,1,1,1,1,1,1,1,1,1,1,1,1,1,1,1,1,1,1]
Numt3=[0,0,0,0,0,0,0,0,0,0,0,0,0,0,0,0,0,0,0,0,0,0,0,0,0,0,0,0]
Num01=[1,1,1,1,1,0,0,1,1,0,0,1,1,0,0,1,1,0,0,1,1,0,0,1,1,1,1,1]
Num02=[1,1,0,1,1,0,0,1,1,0,0,1,1,0,0,1,1,0,0,1,1,0,0,1,1,1,1,1]
Num03=[1,1,1,1,1,0,0,1,0,0,0,1,1,0,0,1,1,0,0,1,1,0,0,1,1,1,1,1]
Num04=[1,1,1,1,1,0,0,1,1,0,0,1,1,0,0,1,1,0,0,1,0,0,0,1,1,1,1,1]
Num05=[1,1,1,1,1,0,0,1,1,0,1,1,1,0,0,1,1,0,0,1,1,0,0,1,1,1,1,1]

an array of y values:

Num0Y= [[0],[0],[0],[1],[1],[1]]

The neural network described with the help of this program contained 3 hidden layers with 28 neurons in each layer. The choice of the number of hidden layers and neurons in each of them was determined by the smallest error of learning and recognizing numbers. According to [5], this is a three-layer neural network with 28 neurons in each layer. The $\beta_1 = 0.9$ and $\beta_2 = 0.999$ parameter values were chosen as in [4]. According to [6], the sigmoidal function was chosen as the activation function. The implementation of this optimization method was carried out with the help of the following code:

```
for i in range (num -1):
        layer_errors.append(layer_deltas[i].dot(synapse[num - 1 - i].T))
        layer_deltas.append(layer_errors[i + 1] *
sigmoid_output_to_derivative(layers[num - 1 - i]))
        layer_deltass=layer_errors[i + 1] *
sigmoid_output_to_derivative(layers[num - 1 - i])
        # m(t) = beta1(t) * m(t-1) + (1 - beta1(t)) * g(t)
        m = beta1**(age+1) * m[i-1] + (1.0 - beta1**(age+1)) * layer_deltass
        # v(t) = beta2 * v(t-1) + (1 - beta2) * g(t)^2
        v = (beta2 * v[i-1]) + (1.0 - beta2) * layer_deltass**2
        # vhat(t) = max(vhat(t-1), v(t))
        vhat = max(max(vhat.reshape(-1,1)), max(v.reshape(-1,1)))
        dd= m / np.sqrt(vhat+ 1e-8)
        d.append(dd)
    for i in range (num):
        synapse[num -1 -i] -= alpha * (layers[num -1 -i].T.dot(d[i]))
```

where num $= 3$ is number of layers of MNN.

where δ- layer_deltass, h_i-layers, age is the number of epochs, num is the number of hidden layers.

2.2 Logistics Function

To analyze the error function, a logistic function of the following type was used:

$$x_{n+1} = \alpha - x_n - x_n^2$$

where n is a step, alpha is a parameter that determines the learning rate. Its fixed points:

$$x_{1,2} = -1 \pm (\alpha + 1)^{1/2}$$

eigenvalues, which can be calculated as follows:

$$\rho_{1,2} = 1 \pm 2(\alpha + 1)^{1/2}$$

The choice of this logistic mapping is due to the fact that it describes the process of doubling the frequency of oscillations [7]. In our case, this process is due to the occurrence of local minima when approaching the global minimum. For one-dimensional mappings, there are 2 ways to change the stability of a fixed point, when the point multiplier is $\rho = +1$ and $\rho = -1$. However, the number of related bifurcations (doublings) is noticeably larger. This is because they often involve more than one fixed point. 4 variants of bifurcations correspond to such a situation: tangential bifurcation (fold, saddle-nodal); transcritical bifurcation; fork-shaped bifurcation (bifurcation of loss of symmetry); doubling bifurcation.

3 Application of the AMSgrad Optimization Method

3.1 AMSGrad and Different Number of Epochs

Figure 1 shows the result of the developed program. Provided that beta1 = 0.9, beta2 = 0.999 at, Fig. 1 shows the dependence of the value of the logistic error function on the parameter α and Fourier spectra for 100 epochs and $\beta_1 = 0.9$, $\beta_2 = 0.999$. . The resulting branching diagram proves that the entire investigated range of α changes (0.000001 ÷ 0.008) can be divided into 4 parts:

1) the range of a sharp decrease in the value of the error function (α = 0.000001 ÷ 0.00002) – no retraining;
2) the range of slightly variable, monotonous behavior of the error function (α = 0.00002 ÷ 0.00025) is a satisfactory learning process;
3) the bifurcation range on the branching diagram (α = 0.00025 ÷ 0.00047) - the process of relearning (Fig. 1, b);
4) the range of chaotic non-monotonic behavior of the error function from alpha (α = 0.00047 ÷ 0.008) the appearance of chaos.

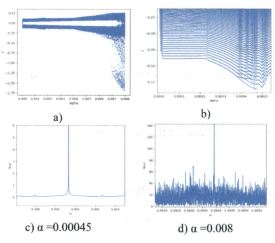

Fig. 1. Branching diagrams (*a, b*) in the range of α = 0 ÷ 0.00055. Fourier spectra (*c*) - during retraining, and (*d*) - in conditions of chaos, 100 epochs, $\beta_1 = 0.9$, $\beta_2 = 0.999$, digit "0", when using the AMSGrad optimization method.

The process of retraining is accompanied by a transition through the global minimum and a doubling of the number of local minima. This process is especially well manifested at small values of the number of epochs (Fig. 2). Although here, too, the process of blocking the doubling of the number of local minima begins to follow, which makes it impossible for the system to transition to a chaotic state (Fig. 2).

With an increase in the number of epochs (*N*), a decrease in the gradient of the error function and an increase in the value of the expression $(1 - \beta_1 N)$, at small values of $(1 - \beta_2)$, lead to the fact that the gradients become rarefied, since the ratio of the vectors of the first and second moments forms block structure that specialize in certain patterns. This is clearly manifested at small epoch values ($N = 5$ at $\alpha > 0.0017$; (Fig. 2)). Therefore, the optimization process especially manifests itself when the number of epochs increases, leading to a decrease in the gradient when approaching the global minimum.

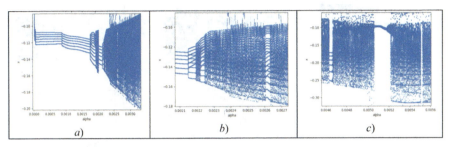

Fig. 2. Branching diagrams for learning speed in the enlarged version, (a, b, c), 5 epochs, $\beta_1 = 0.9$, $\beta_2 = 0.999$, digit '0', when using the AMSGrad optimization method.

3.2 Analysis of the Block Structure in the AMSGrad

To analyze the appearance of the block structure in the branching diagrams, consider the behavior of the learning error function from the learning rate at 10 epochs. Figure 3a shows the branching diagram under the condition of 10 epochs, $\beta_1 = 0.9$, $\beta_2 = 0.999$, , digit "0", and in an enlarged version of the intervals of existence of the block structure (Fig. 3b - m). The entire range of retraining, on the branching diagram, is characterized by the division into blocks (Fig. 3b). Let us consider in more detail each interval of existence of the block structure and conduct an analysis of the Fourier spectra in these ranges. It should be noted that the AMSGrad optimization method is based on the analysis of the objective function based on the values calculated in the previous step, so the Fourier spectra of the error function were calculated for a small interval of the learning rate change in each block. The first interval is the interval $\alpha = 0.0002 \div 0.00075$ of a satisfactory learning process. At the same time, the Fourier spectra show the absence of harmonics. On the interval $\alpha = 0.00175 \div 0.018$, the process of retraining of the neural network with the appearance of a bifurcation on the branching diagram (Fig. 3c) and the second harmonic on the Fourier spectra (Fig. 4a) can be traced. Moving further along the branching diagram as a result of the doubling process, we fall into the interval of the appearance of a chaotic state ($\alpha = 0.001825 \div 0.01914$) (Fig. 3d). The Fourier spectra of this chaotic state, according to Fig. 4c, prove that a chaotic state arises as a result of doubling the number of local minima and not passing through the global minimum once. This interval of the existence of a chaotic state is characterized by areas of "transparency", where there is no chaotic state of the neural network. These areas prove the absence of a chaotic state, and therefore a sharp decrease in the number of local minima. That is, the emerging chaotic state is characterized by the existence of a block structure, which, according to the authors, consists of areas of constant values of the number of local minima, and the averaged value over this block gives a chaotic value. This assumption is supported by the fact that the considered system is dynamic, since the considered optimization method works only if the next value depends on the previous one. Entry into the chaotic state goes through the doubling process, and the appearance

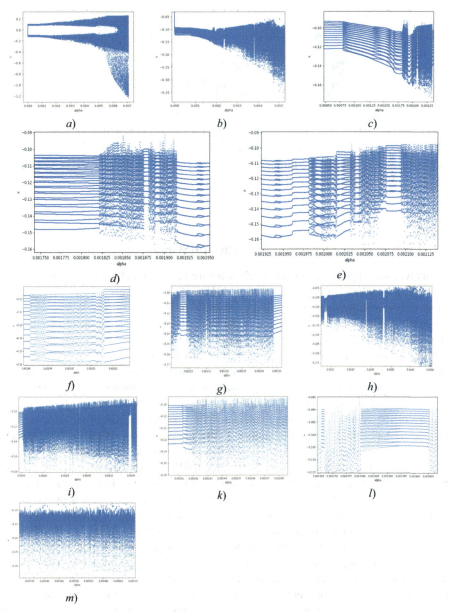

Fig. 3. Branching diagrams (*a*) from the learning speed in the enlarged version, (*b* - *m*)), 10 epochs, $\beta_1 = 0.9$, $\beta_2 = 0.999$, digit '0', when using the AMSGrad optimization method.

of intervals of the absence of chaotic solutions in the chaos domain is accompanied by a sharp transition from chaos to symmetry and vice versa.

Moving on to the next area of change $\alpha = 0.001925 \div 0.001975$ where the neural network is characterized by the absence of a chaotic state (Fig. 3d, f and Fig. 4c).

In this region of α change, the Fourier spectra of the target function are characterized by the presence of harmonics (Fig. 4c), and therefore the learning process of the neural network is accompanied by retraining. According to the branching diagram, this state of the neural network in this area is not chaotic, but is characterized by the processes of branching initiation and its disappearance (Fig. 3d, f). Therefore, the Fourier spectrum

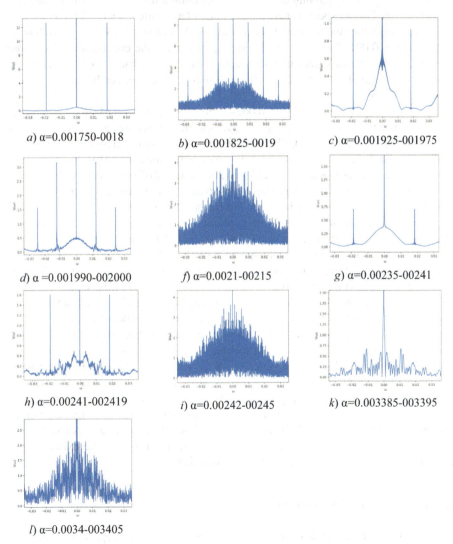

Fig. 4. Fourier spectra in different ranges of learning speed changes, 10 epochs, $\beta_1 = 0.9$, $\beta_2 = 0.999$, digit '0', when using the AMSGrad optimization method.

of the objective function in this area of change in learning speed is characterized by the appearance of harmonics (Fig. 4c), which in terms of signal power is much smaller than in the area of α change, where only the relearning of this network can be traced (Fig. 3c).

The next area of alpha change is the area of cascade transformations ($\alpha = 0.001990–002000$) shown in Fig. 3e. The branching diagram shows several transitions to and from the chaotic state. Under these conditions, the Fourier spectra are characterized by the appearance of higher harmonics and the absence of a chaotic state (Fig. 4d).

Moving to the next interval of change $\alpha = 0.00210 \div 0.00234$, the branching diagram shows a transition to a chaotic state due to the process of doubling the number of local minima and repeatedly passing through the global minimum (Fig. 3g). Under the given conditions, the Fourier spectra prove the existence of a chaotic state, but this chaotic state is different from the chaotic state that exists at $\alpha = 0.001825 \div 0.01914$. The difference is that at $\alpha = 0.001825 \div 0.01914$ Fourier spectra show the coexistence of a chaotic state with a state characterized by the existence of only a few harmonics. Therefore, the interval of change of $\alpha = 0.001825 \div 0.01914$ may be characterized by the existence of a block structure.

The next interval $\alpha = 0.00235 \div 0.00241$ (Fig. 3g-k) is characterized by the absence of doubling on the branching diagram. Fourier spectra in this range also show the almost absence of harmonics (Fig. 4e). Therefore, in this interval of α changes ($\alpha = 0.00235 \div 0.00241$), neural network learning is not characterized by retraining, that is, a satisfactory learning process takes place. Approaching the next interval of existence of a chaotic state ($\alpha = 0.00241 \div 0.002419$), branching is observed on the branching diagram (Fig. 3k) with the appearance of harmonics in the Fourier spectra (Fig. 4g).

A further increase in α leads to the appearance of a chaotic state on the branching diagram (Fig. 3k, l) in the range $\alpha = 0.00242 \div 0.00338$. Fourier spectra at these values of α also confirm the existence of a chaotic state (Fig. 4h, $\alpha = 0.00242–00245$). This chaotic state is no different from other chaotic states that occur at $\alpha = 0.001825–0019$, $\alpha = 0.0021–00215$ (Fig. 4b, f). With a further increase in the learning speed, as a result of a decrease in the number of doublings (according to the branching diagram in Fig. 3l, $\alpha = 0.003385–003395$), possibly due to a decrease in the number of local minima, the absence of a chaotic state can be traced. Fourier spectra (Fig. 4i) of this interval of α change are characterized by a small number of harmonics, which are several times smaller in magnitude than in other intervals of the absence of chaos. A further increase in the learning speed is accompanied by a transition to a chaotic state (Fig. 3m), which is inherent in a larger range of changes in $\alpha > 0.004$ (Fig. 3h). An interesting feature of this chaos is that its Fourier spectra are not characterized by the existence of n-fold harmonics. According to Fig. 4k at $\alpha = 0.0034–003405$, the Fourier spectrum of this chaotic state is characterized by a smaller number of harmonics, which proves the existence of a mechanism for blocking the occurrence of harmonics.

4   Summary and Conclusion

The learning error on a single neuron is described by its functional dependence, therefore the learning error at a given speed and a given step is a symbiosis of all neurons involved in the learning process. Since the learning error is calculated as the root-mean-square deviation, the given functional dependences of the learning error on each neuron are, in the first approximation, periodic functions with different periods. That is, these functional dependencies are characterized by a spectrum of wave vector values. These vectors are characterized by rational values (that is, proportional fluctuations determined by the number of neurons in a layer and the number of hidden layers). The total value of the wave vector for such an ensemble of periodicities can take both commensurate and incommensurate values.

It is known that in MNN, the correction of the weights of one neuron is influenced by all the neurons of the previous layer. Since this influence from the learning speed becomes non-uniform when approaching the global minimum, it can lead to its weakening, and therefore to non-uniform learning. This heterogeneity should manifest itself more when using optimization methods that are based on the use of an algorithm that updates the exponential sliding mean gradients (m_t) and the square of the gradients (v_t) based on previous values. That is, the appearance of a block structure, which is characterized by the coexistence of both a chaotic state and a state characterized by several harmonics, should be inherent to such multilayer neural networks. Based on the above results (Fig. 3), this state may exist in the interval $\alpha = 0.001982$–0.001990.

Therefore, when applying the AMSGrad method to the learning process of MNN, a block structure may arise. This block is characterized by the stochastic learning process of the neural network. Namely, in a certain range of changes in the learning speed, retraining of the neural network is observed, which is accompanied by the appearance of local minima. An increase in the number of local minima leads to the appearance of higher order harmonics. An increase in the number of doublings of existing harmonics causes the system to transition to a chaotic state. This occurs under the condition that the considered neural network training method is related to the correction of the weights of each neuron at a given epoch, and this weight correction is affected by all neurons from the previous layer. So, in this case, the learning error function for each neuron should be considered as a functional dependence described by a set of wave vectors of different periodicities. In this case, the average value of the learning error for all neurons can be considered as the average value for all existing periodicities. And the value of the wave vector of the total oscillation can take both commensurate and incommensurate values. Therefore, AMSGrad algorithm leads to a change frequency spectrum of existing periodicities of functional dependencies of each neuron. That is, the speed of learning of each neuron is corrected, which removes the degeneration of this system by preventing the processes of relearning the neural system.

The results obtained above prove that the appearance of local minima is caused by non-uniform learning of the neural network, which is associated with the retraining of individual neurons. An increase in the number of local minima with an increase in the learning rate indicates an increase in the number of such neurons. The AMSGrad optimization method, due to the control of the exponential rate of decline of the average gradients and the square of the gradient of the target error function, causes a decrease in the number of retraining neurons.

References

1. Kingma, D.P., Ba, J.L.: Adam: a method for stochastic optimization. Published as a conference paper at the 3rd International Conference for Learning Representations, San Diego (2015). https://doi.org/10.48550/arXiv.1412.6980
2. Phuong, T.T., Phong, L.T.: On the Convergence Proof of AMSGrad and a New Version IEEE Access (Volume: 7), 2019, 61706–61716. https://ieeexplore.ieee.org/document/8713445
3. Reddi, S.J., Kale, S., Kumar, S.: On the convergence of adam and beyond computer science. Published as a conference paper at ICLR 2018 [Submitted on 19 Apr 2019] arXiv:1904.09237v1 https://doi.org/10.48550/arXiv.1904.09237
4. Brownlee, J.: Gradient Descent Optimization with AMSGrad From Scratch, 9 June 2021 https://machinelearningmastery.com/gradient-descent-with-adagrad-from-scratch/
5. Sveleba, S., et al.: Chaotic states of multilayer neural network. Electronics and information technologies. Issue 13, pp. 20–35 (2021). https://doi.org/10.30970/eli.16.3
6. Sveleba, S., Katerynchuk, I., Kuno, I., Semotiuk, O., Shmyhelskyy, Ya., Sveleba, N.: Peculiarities of the dependence of the learning error of multilayer neural networks on the activation function in the process of recognizing of printed digits. Electron. Inf. Technol. **17**, 36–53 (2022). https://doi.org/10.30970/eli.17.4
7. Taranenko, Yu.: Information entropy of chaos. https://habr.com/ru/post/447874

Regional Economic Development Indicators Analysis and Forecasting: Panel Data Evidence from Ukraine

Larysa Zomchak[✉], Mariana Vdovyn, and Olha Deresh

Ivan Franko National University of Lviv, Lviv 79000, Ukraine
lzomchak@gmail.com

Abstract. The development of Ukraine's economy as a socio-economic system is determined by the development of its constituent subsystems - regions that function successfully if there are resources necessary for development and their economic evaluation. In this regard, it is important to investigate, due to which factors this economic growth is achieved, what is the contribution of each of these factors to the overall economic success of the country or region. The analysis of the economic development of the regions highlighted the important factors of the regional economic growth: the gross regional product, turnover of retail trade; volume of sold industrial products; capital investments; export volumes. The panel data approach is used in the investigation for regional economic development of Ukraine modelling. Gross regional product as the main economic indicator in the level of region is a dependent variable, independent variables are gross regional product in two previous periods, export, capital investment in previous periods, turnover of retail trade and volume of sold industrial products with two lags. So, with econometric modeling methods the main determinants of the regional economic development of Ukraine were revealed and the level of impact for each of them was estimated. The value of the gross regional product in the regions of Ukraine for the next period is forecasted.

Keywords: Regional development · Economic development · Macroeconomic modelling · Panel model

1 Introduction

Considering the peculiarities of regional development, the strengths and weaknesses of the country's regions functioning, negative and positive trends investigations allows making effective decisions at both the mezzo and macro levels. At the level of the regions socio-economic development projects are implemented, and the indicators of the regions determine the level of country development. It can be argued that the study of the development of the regions of Ukraine is a relevant and timely issue, especially in the conditions of the full-scale invasion of russia on the territory of Ukraine, which will certainly have and will have a negative impact on the development of the regions and Ukraine as a whole, will cause great disparities in the development of the regions,

© The Author(s), under exclusive license to Springer Nature Switzerland AG 2023
Z. Hu et al. (Eds.): ICAILE 2023, LNDECT 180, pp. 217–228, 2023.
https://doi.org/10.1007/978-3-031-36115-9_21

will lead to reduction of all economic indicators due to large expenditures by the state to ensure a peaceful sky over the heads of all Ukrainians, due to the occupation of part of the territory, due to the cessation of business activities, etc.

The decentralization reform, which began in 2014 in Ukraine, was aimed at creating active and capable territorial communities, and not just transferring powers from state bodies to local self-government bodies. In the conditions of war, it is obvious that the emphasis shifts again to a strong central government, while it is important to understand the role, place and prospects of each of the individual regions in the defense of the country.

Due to the special structure of the panel data allow to build more flexible and meaningful models and obtain answers to questions not available within models based, for example, only on spatial data for individual national economies. It becomes possible to take into account and analyze individual differences between economic units, which cannot be done within the framework of standard regression models.

The purpose of the investigation is modeling and analysis of the specifics of economic growth, factors that influence it, in terms of individual regions of Ukraine at a certain point in time, development of such a model that would be suitable for further forecasting. The application of models on panel data makes it possible to obtain regression equations for each region of Ukraine in particular, and thus to investigate the influence and weight of economic impact factors in each region, to compare the economic development of different regions of Ukraine, to identify the strengths and weaknesses of the regional economy.

2 Literature Review

The investigation of the economic development at the level of the region is considered in the articles of many scientists. A review of the main theories of regional economic development was conducted by Coccia M. [1], a review of the latest research on regional development can be founded in the article by Malecki E. [2], Gibbs D. and O'Neill K. [3] reviewed the scientific problems of regional development in the context of the green economy, Smol M. and co-authors[4] – in the circular economy, Aliyev, A. [5] – in the digital economy, and Amin E. [6] – in the context of institutional perspectives. An overview of models of regional economic development can be found in the study of Treys G. [7], and the implementation of models of regional economic development based on US statistical data in the article by Buchholz M., Bathelt H. [8]. More about methods and models, used for investigation on the regional level, in the Table 1.

For modeling regional development, it is advisable to use models based on panel data, because they allow simultaneous consideration of several variables collected over time for the same objects.

The purpose of the study is to model and analyze the specifics of economic growth, factors that influence it, in a section of regions in Ukraine at a certain point in time. On the basis of the selected factors of regional economic growth, the implementation of such a model, which would be suitable for further forecasting.

Regional Economic Development Indicators Analysis and Forecasting 219

Table 1. Methods and models used for regional economic development investigation

Authors	Method and model
Xin, J., & Lai, W. (2021) [9]	multiple statistic analysis
Guangyu, T. (2020) [10]	ARIMA
Hu, Y., & Wang, Z. (2022) [11]	support vector machine
Yu, W., & Huafeng, W. (2019) [12], Matviychuk, A., Lukianenko, O., & Miroshnychenko, I. (2019) [13]	fuzzy logic
Xu, X., & Zeng, Z. (2021) [14], Chagovets, L., Chahovets, V., & Chernova, N. (2020) [15]	machine learning
Li, Z., Cheng, J., & Wu, Q. (2016) [16]	principal component analysis
Zhong, Z., & He, L. (2022) [17]	agent-based model
Emmanuel Halkos, G., & Tzeremes, N. G. (2010) [18]	data envelopment analysis
Santos, L. D., & Vieira, A. C. (2020) [19]	spatial econometrics
Shen, Y. (2020, October) [20]	panel vector autoregressive model
Márquez, M. A., Ramajo, J., & Hewings, G. J. (2013) [21]	spatial vector autoregressive model
Zomchak, L., & Lapinkova, A. (2023) [22]	ARDL
Jiang, X., He, X., Zhang, L., Qin, H., & Shao, F. (2017) [23]	structural equation model
Allo, A. G., Dwiputri, I. N., & Maspaitella, M. (2022) [24]	input-output model
Chen, L., Yu, N. N., & Su, Y. S. (2014, August) [25]	game theory
Dutta, P. K., Mishra, O. P., & Naskar, M. K. (2013) [26], Alhosani, N. (2017) [27], Xu, F., Chen, Z., Li, C., Xu, C., Lu, J., & Ou, Y. (2011) [28]	topography analysis
Koroliuk, Y., & Hryhorenko, V. (2019) [29], Ivan Izonin, et. al.[30], Tkachenko R., et. al.[31]	neural network
Lytvynenko, V., Kryvoruchko, O., Lurie, I., Savina, N., Naumov, O., & Voronenko, M. (2020). [32]	self-organizing algorithm

3 Method of Panel Data Modelling in Economic Research

First of all, the application of models on panel data makes it possible to detect and analyze changes at the individual level, which is impossible neither within the framework of individual time series models nor within the framework of variation series models. Panel data models can be used not only to explain the behavior of different research objects, they can also explain why individual research objects behave differently in different time periods.

The panel data models require compliance with the following assumptions: panel data should be balanced, panels are characterized by short time series; in order to take into account the time effect, additive dummy variables can be used; the possibility of taking into account specific individual effects (disturbances).

In general, the longitudinal data model can be presented in the form:

$$Y_{it} = \alpha + X_{it}\beta_{it} + \varepsilon_{it}$$

where Y_{it} is the value of the indicator for the object i (individual, region, firm, etc.) in the time period t (i = 1,2, 3,..., N; t = 1, 2, 3,..., T); X_{it} is a vector of independent variables; ε_{it}—disturbance for the object i (individual, region, firm, etc.) in the time period t (i = 1,2, 3,..., N; t = 1, 2, 3,..., T); α is a scalar; β_{it} are model parameters that measure the partial effects of a change in X_{it}.

As we can see, such a model is similar to the multiple regression, so the methods of unknown parameters estimating are also similar. In addition, models of longitudinal data make it possible to additionally divide disturbances into several components, and accordingly, according to this feature, all models can be divided into one- (one-way error component model) and two-way error component (two-way error component) errors. In the paper, we consider models with a one-dimensional error component, because they are the most common in practice.

4 Results: Economic Development of the Regions of Ukraine Model Specification

The number of determinants of the regional economic development is rather large, so it is quite difficult to take all of them into account, and even more so to accurately forecast them, especially when the entire economy of the country works in wartime mode. Therefore it is necessary to analyze a certain set of factors and select those that have the most significant influence.

The following determinants of the regional economic development were identified:

- Retail turnover (million UAH);
- Volume of sold industrial products (million UAH);
- Capital investments (million UAH);
- Volume of manufactured construction products (million UAH);
- Export (million USD);
- Employed population (thousands of people).

Gross regional product is the dependent variable as one of the main indicators of the regional economic development.

The model for forecasting the gross regional product, as an indicator of the economic growth of the region, can be written in the form (1):

$$VRP = f(ORT; OR\text{Pr}; KI; OVBP; EXPORT; ZN) \tag{1}$$

where VRP is gross regional product (million UAH); ORT – turnover of retail trade (million UAH); ORPr – volume of sold industrial products (million UAH); KI – capital investments (million UAH); OVBP – volume of manufactured construction products (million UAH); EXPORT – export volumes (million USD); ZN – employed population (thousands of people).

The construction of the model will take place across 24 regions of Ukraine based on retrospective data from 2010 to 2020.

The first important step in the specification of the model is to check the series for stationarity - Panel Unit Root Test. For example, export as a factor of economic growth is not stationary because the probability value for the tests is greater than 5%, that is why we reject the null hypothesis and can say that the original series is not stationary. Next, we have to check whether this series is stationary in the first differences, etc.

So, after checking all the series, we conclude that they are all stationary in the first differences.

The next important step in the study of such a phenomenon as economic development is conducting a test for the causal relationship between the dependent and independent variables - the Granger test (Table 2).

Table 2. Granger 's test for causality

Variable	X^2 - statistics	p- value
ZN	12.60959	0.126
OVBP	9.092023	0.3346
ORT	26.19425	0.001
ORPR	22.84009	0.0036
KI	16.36029	0.0375
EXPORT	18.80705	0.0159

Variables for which the p- value is larger than 5% have no or rather weak effect on the gross regional product. So, we can say that the employed population and the volume of manufactured construction products do not have a significant impact and these variables can be excluded from the model.

Panel autoregressive model in which the previous values of the variables can influence the current values with a certain lag, that is, with a certain "lag" in time. The definition of lags also occupies a very important place in the specification of models. For this, it is necessary to conduct a test - VAR Lag Order Selection test, which is based on different criteria, namely Akaike, Schwartz, Hannan-Quinn. For conclusions, it is necessary to take the smallest informational criterion. In our case, the value of the number of lags for all criteria coincides and is two (Table 3).

Also confirmed by another test - VAR Lag Exclusion Wald Test. The assumption that four lag can be applied on quarterly data was not confirmed.

For choosing the type of model, it is necessary to conduct the Durbin -Vue- Hausman test, which will answer the question of which model to choose. This test also checks whether random effects estimation produces unbiased estimates or not.

After calculating this value, we obtained the following result (Table 4).

222 L. Zomchak et al.

Table 3. Lag determination according to Akaike 's criteria, Schwartz and Hannan-Quinn criterious

Lag	AIC (Akaike)	SC (Schwartz)	HQ (Hannah-Queen)
0	19.9785	20.16548	20.05408
1	19.83488	20.04857	19.92125
2	19.78521*	20.02562*	19.88239*
3	19.80579	20.07291	19.91376
4	19.815	20.10883	19.93377
5	19.8157	20.13624	19.94527
6	19.81178	20.15903	19.95214

Table 4. The Darbin -Vue- Hausman test results

Summary test	X 2 - statistics	Prob
Random period	53.373701	0.0000

As one can see, the calculated value of the X^2 statistic is less than the critical value calculated according to the table, and the p - value is less than 5%. That is, the null hypothesis is not rejected. Therefore, we can conclude that the data will be better described by a model with fixed effects.

The next important test for model specification is the Wald test (Table 5). It gives a clear understanding of whether we really need to use cross-sectional effects.

Table 5. The Wald test results

Summary test	Value	Prob
F-statistics	2.359498	0.0013
X 2 - statistics	58.230242	0.0001

According to the value of the F-statistic, we cannot reject the null hypothesis (p-value less than 0.05) that the model with the common cross-section is worse than the model with fixed effects.

So, as a result of the conducted tests, we obtained a panel autoregression model with fixed cross-sectional effects of the economic development of the regions of Ukraine.

5 Discussion: Panel Data Model of Ukraine Regions Economic Development Application

In order to estimate model parameters, it is necessary to use the generalized least squares method (panel EGLS), because this approach will allow us to avoid problems related to autocorrelation and heteroskedasticity (assumption of constancy of variance) in the future.

That is, in contrast to the least squares method, this method takes into account information about the inequality of variance and therefore makes it possible to obtain the best linear estimates.

The model specified above will look like this (2):

$$VRP_{it} = \alpha_i + \sum_{j=1}^{p} \varphi_{ij} VRP_{i,t-j} + \sum_{j=1}^{q} \sigma_{ij} EXPORT_{i,t-j} + \sum_{j=1}^{q} \beta_{ij} KI_{i,t-j}$$

$$+ \sum_{j=1}^{q} \gamma_{ij} ORT_{i,t-j} + \sum_{j=1}^{q} \theta_{ij} ORPR_{i,t-j} \tag{2}$$

where i is the index of regions; t – time lag; p, q – number of lag values;

α— scalar; VRP – gross regional product (million UAH); ORT – turnover of retail trade (million UAH); ORPr – volume of sold industrial products (million UAH); KI – capital investments (million UAH); EXPORT – export (million USD).

After estimating the parameters of the model, we get the following results, where not all variables are statistically significant. This is evidenced by the p- value. Therefore, we can get rid of the model EXPORT(-1), EXPORT(-2), Kl, Kl (-1), ORT and ORT(-2).

Estimates of the new model are presented in the Table 6. It is obvious that all variables are statistically significant because the p - value is less than 0.05.

Table 6. Results of estimation of model parameters

Variable	Coefficients	Standard error	t-statistics	Prob
C	-5625.650	1256.985	-4.475511	0.0000
VRP (- 1)	0.884471	0.048593	18.20159	0.0000
VRP (- 2)	0.229284	0.028394	8.075005	0.0000
EXPORT	2.656496	0.493614	5.381723	0.0000
Cl (- 2)	-0.767422	0.216006	-3.552778	0.0005
LOCATION (- 1)	-0.239889	0.049652	-4.831429	0.0000
ORPR	0.212883	0.014747	14.43527	0.0000
ORPR (- 1)	-0.237092	0.024662	-9.613807	0.0000
ORPR (- 2)	0.082794	0.023641	3.502198	0.0006

224   L. Zomchak et al.

As in the case of paired regressions, the standard error characterizes the dispersion of the actual values of the resulting variable around the theoretical ones. We can claim that for each variable from Table 7, this value is quite small and close to 0.

Then the model will look like this (3):

$$VRP_{it} = -5625.65 + 0.884471*VRP_{i,t-1} + 0.229284*VRP_{i,t-2} + 2.656496*$$

$$EXPORT_{it} - 0.767422*KI_{i,t-2} - 0.239889*ORT_{i,t-1} + 0.212883*ORPR_{it} -$$

$$0.237092*ORPR_{i,t-1} + 0.082794*ORPR_{i,t-2} \qquad (3)$$

Figure 1 presents actual, calculated values of gross regional product and confidence intervals.

If comparing the results of the values calculated by the panel autoregression model and the actual values, for example, for 2020, one can see that the model is quite accurate, and the difference in values is insignificant (Fig. 2).

If we talk about the evaluation of quality criteria of models of longitudinal data, it is somewhat different from classical multivariate analysis. At the same time, the values of the coefficient of determination are in the interval [0, 1] regardless of which evaluation method was used to obtain the calculated (theoretical) values.

After the calculation, get the following values (Table 7).

Table 7. Evaluation of the quality criterion of logit data models

Mean square deviation	0.997531
d-statistics	2.230461
The relation of determination	0.996968
F-statistics	1772.28
Prob	0.0000

So, according to the results from Table 7, we can conclude about the adequacy of the proposed model. The root mean square deviation becomes rather small. This means that the random variables are normally distributed.

According to the Durbin-Watson test (d-statistic), it follows that the autocorrelation is uncertain, because the d-statistic is greater than 2 and falls into the interval [2.1;3,07]. That is, we cannot assert the presence or absence of autocorrelation.

The determination ratio becomes almost unity. This gives an understanding that such a model almost completely describes the variation of the variables.

So, on average across the regions of Ukraine, 99.69% of changes in the gross regional product are explained by changes in the gross product itself, the volume of capital investments, realized industrial products and exports, as well as retail trade turnover.

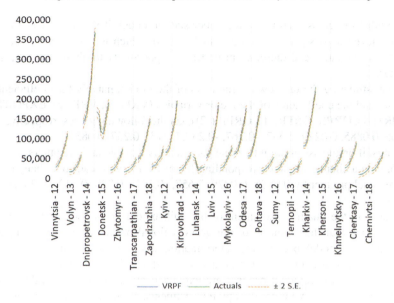

Fig. 1. Actual and calculated GRP values

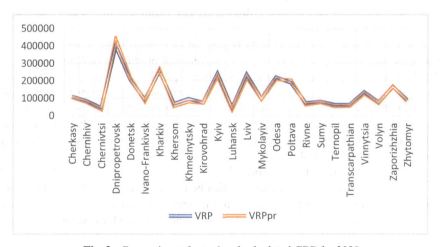

Fig. 2. Comparison of actual and calculated GRP for 2020

6 Summary and Conclusion

Therefore, with the help of the analysis of the economic development of the regions, a number of important factors of influence on the regional economic growth were singled out, namely on such an indicator as the gross regional product, in particular: turnover of retail trade; volume of sold industrial products; capital Investments; export volumes.

After conducting a number of tests (causality test and series stationarity test, Darbin -Wue- Hausman test and Wald test), the model was specified as a panel autoregression model with fixed cross-sectional effects. For such a model, it was necessary to apply the generalized least squares method (panel EGLS), which is based on the assumed inequality of variance and therefore provides the opportunity to obtain the best linear estimates.

After estimating the unknown parameters of the model, namely the coefficients of the scalar and the coefficients of the factor variables (VRP(-1), VRP(-2), EXPORT, Kl (-2), ORT(-1), ORPR, ORPR(-1), ORPR (-2)), get the following values, respectively – - 5625.65, 0.0885, 0.0229, 2.657, -0.767, -0.239, 0.213, -0.237, 0.083.

That is, if on the assumption that one parameter is variable and the others are equal to zero, then a change in the corresponding factor by 1% will lead to a change in the gross regional product by 0.0885, 0.0229, 2.657, -0.767, -0.239, 0.213, -0.237, 0.083, respectively.

In general, the quality assessment of the built model gives good results. We can definitely state that the variation of the gross regional product is almost completely explained by the model, because the determination ratio is close to unity.

Based on this, it is obvious that on average in the regions of Ukraine, 99.69% of changes in the gross regional product are explained by changes in the values of the gross product itself, the volumes of capital investments, realized industrial products and exports, as well as the turnover of retail trade, with a corresponding lag.

And the calculated value of the F-statistic exceeds the empirical value (found according to the Fisher distribution table), that is, we can claim that the model adequately describes such a dependence.

In this study, where the object is the regions of Ukraine, with the help of models on panel data, not only the behavior of the gross regional product in different regions of Ukraine, but also its dynamics over time was investigated.

In general, the proposed model of economic development of regions is suitable and can be used for forecasting. It is worth remembering that the concept of "economic growth" is very complex and multifaceted, it requires a deeper study, coverage and analysis of a greater number of factors affecting it, which will result in a better quality of forecasting at the regional level.

Further improvements to the approach to modeling the economic development of the regions of Ukraine on panel data can also be in the direction of using other types of models, in particular, structural equations on panel data instead of theoretical vector autoregressive equations and justification dependencies between the resulting and factor variables.

Of course, Russian armed aggression against Ukraine will make adjustments to the development of the economy of the country and its regions in the future. Therefore, the model characterizes the pre-war situation and may inadequately describe the development of economic processes after Ukraine's victory in the war.

References

1. Coccia, M.: An introduction to theories of national and regional economic development. Turkish Economic Review 5(4), 350–358 (2019)

2. Malecki, E.J.: Entrepreneurs, networks, and economic development: A review of recent research. Reflections and extensions on key papers of the first twenty-five years of advances (Advances in Entrepreneurship, Firm Emergence and Growth) [M]. Emerald Publishing Limited, Bingley, 71–116 (2018)
3. Gibbs, D., O'Neill, K.: Future green economies and regional development: a research agenda. Transitions in Regional Economic Development, 287–309 (2018)
4. Smol, M., Kulczycka, J., Avdiushchenko, A.: Circular economy indicators in relation to eco-innovation in European regions. Clean Technol. Environ. Policy **19**(3), 669–678 (2017)
5. Aliyev, A.G.: Technologies ensuring the sustainability of information security of the formation of the digital economy and their perspective development directions. Int J Inf Eng Electron Business (IJIEEB) **14**(5), 1–14 (2022)
6. Amin, A.: An institutionalist perspective on regional economic development. Economy, 59–72 (2017)
7. Treyz, G.I.: Regional economic modeling: a systematic approach to economic forecasting and policy analysis. Springer Science & Business Media (2013)
8. Buchholz, M., Bathelt, H.: Models of regional economic development: Illustrations using US data. Zeitschrift für Wirtschaftsgeographie **65**(1), 28–42 (2021)
9. Xin, J., Lai, W.: Research on regional economic development of guangxi based on multiple statistic analysis. Annals of Mathematical Modeling **13**(1), 12–21 (2022)
10. Guangyu, T.: Analysis and Decision-making of Regional Economic Vitality and Its Influencing Factors. The Frontiers of Society, Science and Technology **2**(14) (2020)
11. Hu, Y., Wang, Z.: Support vector machine-based nonlinear model and its application in regional economic forecasting. Wireless Communications and Mobile Computing (2022)
12. Yu, W., Huafeng, W.: Quantitative analysis of regional economic indicators prediction based on grey relevance degree and fuzzy mathematical model. Journal of Intelligent & Fuzzy Systems **37**(1), 467–480 (2019)
13. Matviychuk, A., Lukianenko, O., Miroshnychenko, I.: Neuro-fuzzy model of country's investment potential assessment. Fuzzy economic review **24**(2), 65–88 (2019)
14. Xu, X., Zeng, Z.: Analysis of regional economic evaluation based on machine learning. J. Intell. Fuzzy Sys. **40**(4), 7543–7553 (2022)
15. Chagovets, L., Chahovets, V., Chernova, N.: Machine learning methods applications for estimating unevenness level of regional development. Data-Centric Business and Applications. Springer, Cham., 115–139 (2020)
16. Li, Z., Cheng, J., Wu, Q.: Analyzing regional economic development patterns in a fast developing province of China through geographically weighted principal component analysis. Lett. Spat. Resour. Sci. **9**(3), 233–245 (2016)
17. Zhong, Z., He, L.: Macro-regional economic structural change driven by micro-founded technological innovation diffusion: an agent-based computational economic modeling approach. Comput. Econ. **59**(2), 471–525 (2022)
18. Emmanuel, H.G., Tzeremes, N.G.: Measuring regional economic efficiency: the case of Greek prefectures. Ann. Reg. Sci. **45**(3), 603–632 (2010)
19. Santos, L.D., Vieira, A.C.: Tourism and regional development: a spatial econometric model for Portugal at municipal level. Port. Econ. J. **19**(3), 285–299 (2020)
20. Shen, Y.: The interactive effect of science and technology finance development and regional economic growth in the yangtze river economic belt: analysis based on panel vector autoregressive (PVAR) model of interprovincial data. Journal of Physics: Conference Series **1624**(2), 022026 (2020, October)
21. Márquez, M.A., Ramajo, J., Hewings, G.J.: Assessing regional economic performance: regional competition in Spain under a spatial vector autoregressive approach. Geography, institutions and regional economic performance. Springer, Berlin, Heidelberg, pp. 305–330 (2013)

22. Zomchak, L., Lapinkova, A.: Key interest rate as a central banks tool of the monetary policy influence on inflation: the case of Ukraine. Advances in Intelligent Systems, Computer Science and Digital Economics IV, pp. 369–379. Springer Nature Switzerland, Cham (2023)
23. Jiang, X., He, X., Zhang, L., Qin, H., Shao, F.: Multimodal transportation infrastructure investment and regional economic development: a structural equation modeling empirical analysis in China from 1986 to 2011. Transp. Policy **54**, 43–52 (2017)
24. Allo, A.G., Dwiputri, I.N., Maspaitella, M.: The impact of electricity investment on inter-regional economic development in Indonesia: An Inter-Regional Input-Output (IRIO) approach. Journal of Socioeconomics and Development **5**(1), 1–12 (2022)
25. Chen, L., Yu, N.N., Su, Y.S.: The application of game theory in the Hercynian economic development. International Conference on Management Science & Engineering 21th Annual Conference Proceedings, pp. 794–799 (2014)
26. Dutta, P.K., Mishra, O.P., Naskar, M.K.: A method for post-hazard assessment through topography analysis using regional segmentation for multi-temporal satellite imagery: a case study of 2011 Tohuku earthquake region. Int. J. Ima. Graph. Sig. Proc. **5**(10), 63 (2013)
27. Alhosani, N.: Mapping urban expansion due to special economic zones in the united arab emirates using landsat archival data (Case Study Dubai). Int. J. Ima. Graph. Sign. Proc. **9**(4), 22 (2017)
28. Xu, F., Chen, Z., Li, C., Xu, C., Lu, J., Ou, Y.: A Study on sustainable development of grain production coping with regional drought in China. Int. J. Modern Edu. Comp. Sci. **12**, 80–86 (2011)
29. Koroliuk, Y., Hryhorenko, V.: ANN model of border regions development: approach of closed systems. Int. J. Intell. Sys. Appli. **11**(9), 1 (2019)
30. Izonin, I., et al.: Stacking-based GRNN-SGTM ensemble model for prediction tasks. Proceedings of the DASA, pp. 60–66 (2020)
31. Tkachenko, R., et al.: Piecewise-linear approach for medical insurance costs prediction using SGTM neural-like structure. CEUR Workshop Proceedings **2255**, 170–179 (2018)
32. Lytvynenko, V., Kryvoruchko, O., Lurie, I., Savina, N., Naumov, O., Voronenko, M.: Comparative studies of self-organizing algorithms for forecasting economic parameters. Int. J. Modern Edu. Comp. Sci. **12**(6), 1–15 (2020)

An Enhanced Performance of Minimum Variance Distortionless Response Beamformer Based on Spectral Mask

Quan Trong The[1]( ) and Sergey Perelygin[2]

[1] Digital Agriculture Cooperative, Hanoi, Vietnam
[2] Faculty of Media Technologies, State University of Film and Television St Petersburg, Saint, Petersburg, Russian Federation

Abstract. Minimum Variance Distortionless Response (MVDR) is one of the most attractive techniques commonly used in microphone array beamforming for speech enhancement applications. The conventional MVDR beamformer is very sensitive to the direction-of-arrival of useful signal and microphone gains. Several research has successfully made MVDR beamformer more robust in practice scenario, where exist several types of noise, transport vehicle, third-party speaker, teleconference, communication. MVDR-based equipment is already installed in numerous audio devices: hearing aid, surveillance, telephone, dialogue customer. Nevertheless, MVDR performance is often corrupted in complex surrounding environments, due to imperfect propagation of sound source, phase error or imprecise of the direction-of-arrival of desired speaker. In this paper, the author introduced a solution for enhancing performance of MVDR beamformer in noisy scenario. The method uses a spectral mask, which is based on Differential Microphone Array and an additive function of a priori SNR. Experimental results on real recoding microphone array received data show an increase of the signal-to-noise ratio (SNR) from 5.5 (dB) to 8.0 (dB), and reduce speech distortion to 3.5 (dB) in comparison of conventional MVDR and the suggested method. The effectiveness of the proposed technique shows us that it can be integrated into multi-microphone system for complicated further speech processing.

Keywords: Microphone array · Minimum variance distortionless response · Speech enhancement · Noise reduction · Spectral mask

1 Instruction

Speech corrupted or degraded performance is a complex task in a speech-based system due to interferences and surrounding noise. In a few decades, the demand of speech enhancement for clean - speech use in teleconference system, audio device, hearing aid, communication system has become a vital part. The effectiveness or capability of these systems depends on the quality and intelligibility of processed signal and the amount of suppressed noise from mixture of noisy signal. The purpose of speech enhancement is obtaining desired original signal and alleviating all negative effects caused by unwanted

© The Author(s), under exclusive license to Springer Nature Switzerland AG 2023
Z. Hu et al. (Eds.): ICAILE 2023, LNDECT 180, pp. 229–238, 2023.
https://doi.org/10.1007/978-3-031-36115-9_22

noise in hardy noisy environments. Facing the difficulty for people to understand the speech and words in the situation, where the existence of various types of noise corrupt speech quality, noise reduction is demanded to help avoid this drawback (Fig. 1).

Fig. 1. Separation of speech source is always the most complex task in speech enhancement.

Many single-channel algorithms usually use frequency-domain to process signal. Boll presented spectral subtraction technique, which concerned subtracting spectral noise component from mixture of noisy spectrum. The approach, which is based on statistical model, was introduced in Ephraim and Malah have been evaluated to suppress background at high SNR environment. Method OM-LSA was confirmed its effectiveness in Cohen and Berdugo's work. However, this approach is too sensitive with dependencies on the type of background noise, an imprecise estimation of spectral noise component often leads to distortion of desired speech signal and introduce musical noise at the output. Theoretically, multi-microphone [1–5, 12–14] speech enhancement direction, which allows implementing more less distortion when compared to single-channel methods, has been harshly developed (Fig. 2).

Fig. 2. The use of microphone array beamforming.

Microphone array (MA) beamforming - based method, which has been intended to be the most widely used tool for study the spatial characteristics of acoustic environment.

Beamforming algorithm steer MA's response to a preferred direction of desired source on the interesting region toward target speech, and they provide a spatial distributed beampattern with a certain level pressure. The amount of spectral speech component can be computed by using a priori information of direction of arrival of useful signal, distributed geometry of MA to capture desired signal, an appropriate signal processing algorithm.

Spatial filtering has been vital component in this approach, which helps to boot the useful target speech at certain direction while attenuating annoyed interference from other directions. In real-life situations, the MA's working can be adjusted by using an appropriate beamforming. The first type of beamformer, which has a static frequency response and static constant coefficients, is named fixed beamformer. The simplest one is a delay-and-sum beamformer. The most popular widely used adaptive beamformer is MVDR [6–10]. Adaptive beamformer allow utilize a certain number of microphones, while efficiently mitigating surrounding noise and saving desired speech. Based on these factors, dual-microphone system (MA2) is considered a famous solution to implement numerous signal processing algorithms for improving the quality and intelligibility. There are several methods to improve the performance of MVDR beamformer (Fig. 3).

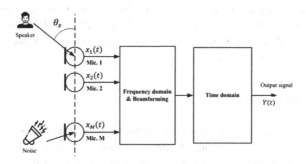

Fig. 3. An implementation of microphone array in frequency domain

Relative transfer function (RTFs) is one of the most essential quantities required for accomplishing speech enhancement by using MVDR beamformer. This vector RTF can be determined a priori or based on assumptions about talker location, acoustic environment, microphone characteristics, position. A precise RTF makes performing MVDR beamformer more robustness to achieve high diversity, high noise reduction.

Diagonal loading becomes a commonly used technique to enhance the robustness of MVDR beamformer. This approach is derived from criteria of the Euclidean norm of the coefficients vector and an additional quadratic constraint. Diagonal loading also alleviates some complicated tasks of using the covariance matrix of microphone array signals and better adjust the peak of sidelobe. However, it is not exactly how to choose an appropriate parameter based on the uncertainty steering vector.

The subspace-based adaptive beamforming technology need the information of the initial noise covariance matrix. Hence, the performance is often degraded in presence of imprecise knowledge of steering vector and matrix.

In low environments with low SNR, a real-valued unitary transformation MVDR algorithm is used to improve the performance of beamformer. Otherwise, the scholar proposed root MVDR algorithm, which overcomes spectral search and decreases the computational complexity.

The traditional adaptive beamformer tends to work on determined acoustical scenario and cannot outperform good performance in the non-stationary or complex environment. In this paper, the author suggested the use of a spectral mask, which based on the principle of Differential Microphone Array (DMA) and a function of a priori SNR for enhancing the performance of MVDR beamformer. Many unwanted reasons: the microphone mismatch, the difference of received data, phase error, the influence of interference or imperfect environment that significantly attenuates target directional useful signal. Illustrated experiments were verified and show the effectiveness of suggested method for speech enhancement while mitigating all non-target directional signals and keeping desired speech source.

This contribution is organized as follows: The next section introduces the model working of MVDR beamformer. In section III, the spectral mask, which is based on DMA and an additional function of a priori SNR, is determined and used as pre-processing step to block the target directional speech component. Experiment with DMA2 in real environment was shown in Section IV, a comparison between the conventional MVDR beamformer (conMVDR) and suggested method (spmMVDR) show the improvement of suggested method in term of the signal-to-noise ratio (SNR). Concluding remark and future work are described in Section V.

2 Mvdr Beamformer

MVDR beamformer is one of the most efficient microphone array beamforming to obtain the desired target speaker by using the priori information of the direction of arrival (DOA) and observed matrix of microphone array signal without speech distortion. In this contribution, the author will illustrate the scheme of MVDR beamformer in dual-microphone system (DMA2), due to its simplicity and convenience to implement almost digital signal processing.

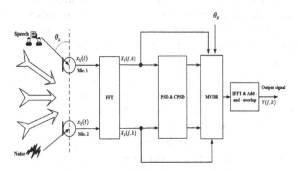

Fig. 4. The scheme of MVDR filter on dual-microphone system.

Consider DMA2 depicted in Fig. 4. The received microphone array signals in the frequency domain $X_1(f, k), X_2(f, k)$ are written as:

$$X_1(f, k) = S(f, k)e^{j\Phi_s} + V_1(f, k) \tag{1}$$

$$X_2(f, k) = S(f, k)e^{-j\Phi_s} + V_2(f, k) \tag{2}$$

where f, k: the index of frequency and frame respectively. $S(f, k)$: the original target direction speech source, $V_1(f, k), V_2(f, k)$ are the obtained noise at DMA2. $\Phi_s = \pi f \tau_0 cos\theta_s$, θ_s: : the direction-of-arrival (DOA) of interest signal relative to MA, $\tau_0 = d/c, d$: the distance between two microphones, $c = 343(m/s)$: the speed of sound in the air, τ_0: the time delay.

If we denote, $X(f, k) = \begin{bmatrix} X_1(f, k) \, X_2(f, k) \end{bmatrix}^T D_s(f, \theta_s) = \begin{bmatrix} e^{j\Phi_s} \, e^{-j\Phi_s} \end{bmatrix}^T$, $V(f, k) = \begin{bmatrix} V_1(f, k) \, V_2(f, k) \end{bmatrix}^T$ so the Eq. (1–2) can be presented as following:

$$X(f, k) = S(f, k)D_s(f, \theta_s) + V(f, k) \tag{3}$$

where $D_s(f, \theta_s)$: steering vector has the information of DOA, which plays a significant role in almost beamforming algorithm.

The optimum constraint of minimum of the power of noise while preserving the original of target speech, leads to the optimum problem, which can be described the following equation:

$$\underset{W}{min} \; W^H(f, k)\Phi_{VV}(f, k)W(f, k)\text{st } W^H(f, k)D_s(f, \theta_s) = 1 \tag{4}$$

where $\Phi_{VV}(f, k) = E\{V^H(f, k)V(f, k)\}$ is the covariance matrix of noise. Using Lagrange method, (4) gives us the optimum weight vector as:

$$W(f, k) = \frac{\Phi_{VV}^{-1}(f, k)D_s(f, \theta_s)}{D_s^H(f, \theta_s)\Phi_{VV}^{-1}(f, k)D_s(f, \theta_s)} \tag{5}$$

In almost speech applications, the information of noise is not always available, so the covariance matrix of observed microphone array signals $\Phi_{XX}(f, k)$ is used instead of. Therefore, the final solution for MVDR beamformer can be yielded that:

$$W(f, k) = \frac{\Phi_{XX}^{-1}(f, k)D_s(f, \theta_s)}{D_s^H(f, \theta_s)\Phi_{XX}^{-1}(f, k)D_s(f, \theta_s)} \tag{6}$$

where:

$$\Phi_{XX}(f, k) = \begin{Bmatrix} E\{X_1^*(f, k)X_1(f, k)\} \; E\{X_1^*(f, k)X_2(f, k)\} \\ E\{X_2^*(f, k)X_1(f, k)\} \; E\{X_2^*(f, k)X_2(f, k)\} \end{Bmatrix}$$

The power spectral density (PSD), $E\{X_i(f, k)X_j^*(f, k)\} = P_{X_iX_j}(f, k)$ is determined as:

$$P_{X_iX_j}(f, k) = \alpha P_{X_iX_j}(f, k - 1) + (1 - \alpha)X_i^*(f, k)X_j(f, k)\forall i, j \in \{1, 2\} \tag{7}$$

where α is the smoothing parameter and often in range $\{0..1\}$.

3 The Proposed Method

The author's ideal of improving the speech enhancement is using a spectral mask for pre-processing alleviate the speech component in the microphone signals $X_1(f, k), X_2(f, k)$ as:

$$\hat{X}_1(f, k) = SM(f, k) \times X_1(f, k) \tag{8}$$

$$\hat{X}_2(f, k) = SM(f, k) \times X_2(f, k) \tag{9}$$

where $SM(f, k)$ is the spectral mask. $\hat{X}_1(f, k), \hat{X}_2(f, k)$ is then processed by MVDR beamformer to extract desired speech signal.

The $SM(f, k)$ is determined with the information of beampattern, which null-steering towards the direction of signal, and a function of the signal-to-noise ratio (SNR).

The differential microphone array (DIF) allows forming the null-beampattern toward the direction of speaker, θ_s.

$$Y_{DIF}(f, k) = \frac{X_1(f, k) - X_2(f, k)e^{-j\omega\tau}}{2} \tag{10}$$

$$= jS(f, k)\sin\left(\frac{\omega\tau_0}{2}\left(\cos\theta + \frac{\tau}{\tau_0}\right)\right) \tag{11}$$

where $\tau = \tau_0\cos\theta_s$.

The obtained null-beampattern is calculated by:

$$B(f, \theta) = \left|\frac{Y_{DIF}(f, k)}{S(f, k)}\right| = \left|\sin\left(\frac{\omega\tau_0}{2}\left(\cos\theta + \frac{\tau}{\tau_0}\right)\right)\right| \tag{12}$$

A method for calculating the covariance of speech $\sigma_s^2(f, k)$ is expressed as:

$$\sigma_s^2(f, k) = \frac{1}{D_s^H(f, \theta_s)\Phi_{XX}^{-1}(f, k)D_s(f, \theta_s)} \tag{13}$$

Therefore, noise covariance $\sigma_n^2(f, k)$ is determined by:

$$\sigma_n^2(f, k) = \frac{P_{X_1X_1}(f, k) + P_{X_2X_2}(f, k)}{2} - \sigma_s^2(f, k) \tag{14}$$

The temporal $SNR(f, k)$:

$$SNR(f, k) = \frac{\sigma_s^2(f, k)}{\sigma_n^2(f, k)} \tag{15}$$

The proposed author's spectral mask is derived as:

$$SM(f, k) = \left|\frac{Y_{DIF}(f, k)}{S(f, k)}\right| \times \frac{1}{1 + SNR(f, k)} \tag{16}$$

The next will perform an experiment in a real acoustic environment with a speaker, a DMA2 in presence of background noise to verify the effectiveness of suggested spectral mask.

4 Experiments

The scheme of the experiment is illustrated in Fig. 5. A target speaker stands at the direction $\theta_s = 90^0 (deg)$ relative to the axis DMA2, which located in a room conference in presence of background noise, interference and third-party speaker. For further signal processing, the author use $NFFT = 512$, overlap 50% , the smoothing parameter $\alpha = 0.1$, the sampling frequency $Fs = 16kHz$, Hamming window, the distance between two microphones $d = 5(cm)$. The experiment is aiming to compare performance of the suggested method (spmMVDR) and the conventional MVDR (conMVDR). An objective measurement SNR [11] is used for estimating the speech quality.

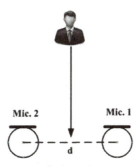

Fig. 5. The scheme of illustrated experiment

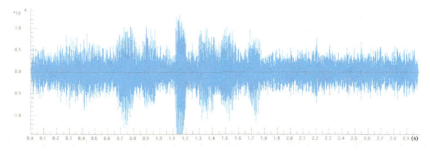

Fig. 6. The waveform of the received microphone array signal

The original microphone signal is presented in Fig. 6. Figure 7 shows the received output signal by using the conventional MVDR beamformer.

With the proposed spectral mask, the obtained processed signal as illustrated in Fig. 8.

From Figs. 7 and 8, the suggested method has the ability of preserving the original speech component of target speaker and reducing the speech distortion to 3.5 (dB) as in Fig. 9. In Table 1, the author compared the speech quality in terms of the signal-to-noise ratio (SNR). SpmMVDR increases the SNR from 5.5 to 8.0 (dB).

From the experimental results, the proposed technique was verified in a real acoustic experiment to reduce speech distortion and enhance speech quality. The proposed spectral mask has removed the desired target speech component in microphone array signals

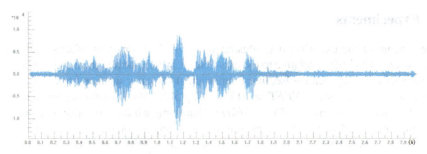

Fig. 7. The output signal by conMVDR

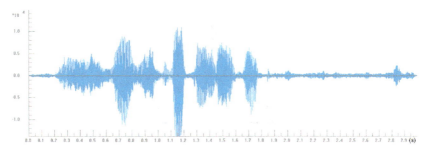

Fig. 8. The obtained signal by the proposed method (spmMVDR)

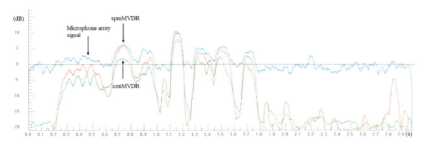

Fig. 9. The energy of microphone array signal, processed signal by conMVDR and spmMVDR

Table 1. The signal - to - noise ratio (dB)

Method Estimation	Microphone array signal	conMVDR	spmMVDR
NIST STNR	5.2	21.0	29.0
WADA SNR	3.7	18.4	24.9

and improved the performance of MVDR beamformer. Consequently, this technique can be applied into other MA digital signal processing, such as: Generalized Sidelobe Canceller, Differential Microphone Array, Linear Constrained Minimum Variance.

5 Conclusion

The evaluation of original MVDR beamformer seriously degraded in the presence of microphone mismatches, phase error, the different microphone sensitivities, the error between the assumed and actual DOA. So, in the speech enhancement applications, robustness of the beamformer is the most necessary task. In this article, the author proposed a spectral mask as a pre-processing step for signal processing to enhance the quality of output signal. The experiments have confirmed the improvement of this approach in real scenarios in the presence of noise and interference. The performance showed better reduce speech distortion and the proposed method is suitable for acoustic equipment, mobile device, teleconference, hearing aids or other compact acquisition distant in adverse noise environments. In the future, the author plans to combine the properties of environment to integrate into spectral mask for achieving more noise reduction and speech quality.

References

1. Benesty, J., Chen, J., Pan, C.: Fundaments of Differential Beamforming. p. 122. Springer (2016). https://doi.org/10.1007/978-981-10-1046-0
2. Microphone Arrays. In: Brandstein, M., Ward, D. (eds.) Springer-Verlag, Heidelberg, Germany. XVIII, p. 398 (2001). https://doi.org/10.1007/978-3-66204619-7
3. Hoshuyama, O., Sugiyama, A., Hirano, A.: A robust adaptive beamformer for microphone arrays with a blocking matrix using constrained adaptive filters. IEEE Trans. Signal Process. **47**, 2677–2684 (1999)
4. Benesty, J., Cohen, I., Chen, J.: Fundamentals of Signal Enhancement and Array Signal Processing, p. 440. Wiley, IEEE Press (2017)
5. Benesty, J., Chen, J., Huang, Y.: Microphone Array Signal Processing, p. 240. Springer-Verlag, Berlin, Germany (2008). https://doi.org/10.1007/978-3-54078612-2
6. Lockwood, M.E., et al.: Performance of time- and frequency domain binaural beamformers based on recorded signals from real rooms. The Journal of the Acoustical Society of America **115**, 379 (2004). https://doi.org/10.1121/1.1624064
7. Zhu, Y., Fu, J., Xu, X., Ye, Z.: Modified complementary joint sparse representations: a novel post-filtering to MVDR beamforming. In: Proc. 2019 IEEE International Workshop on Signal Processing Systems (SiPS), pp. 1-6. Nanjing, China (2019). https://doi.org/10.1109/SiPS47522.2019.9020522
8. Fischer, D., Doclo, S., Habets, E.A.P., Gerkmann, T.: Combined Single-Microphone Wiener and MVDR Filtering based on Speech Interframe Correlations and Speech Presence Probability. Speech Communication; 12. ITG Symposium, pp. 1–5. Paderborn, Germany (2016)
9. Sun, Z., Li, Y., Jiang, H., Chen, F., Wang, Z.: A MVDR-MWF combined algorithm for binaural hearing aid system. Proc. 2018 IEEE Biomedical Circuits and Systems Conference (BioCAS), pp. 1-4. Cleveland, OH (2018). https://doi.org/10.1109/BIOCAS.2018.8584798
10. Fischer, D., Doclo, S.: Subspace-based speech correlation vector estimation for single-microphone multi-frame MVDR filtering. In: Proc. ICASSP 2020 - 2020 IEEE International Conference on Acoustics, Speech and Signal Processing (ICASSP), pp. 856–860. Barcelona, Spain (2020). https://doi.org/10.1109/ICASSP40776.2020.9052934
11. https://labrosa.ee.columbia.edu/projects/snreval/

12. Bhatia, V., Whig, P.: Performance analysis of multi-functional bot system design using microcontroller. I.J. Intelligent Systems and Applications **02**, 69–75 (2014). https://doi.org/10.5815/ijisa.2014.02.09
13. Shcherbyna, O., Zaliskyi, M., Kozhokhina, O., Yanovsky, F.: Prospect for using low-element adaptive antenna systems for radio monitoring stations. I. J. Computer Network and Information Security **5**, 1–17 (2021). https://doi.org/10.5815/ijcnis.2021.05.01
14. Mandal, S.: Linear antenna array pattern synthesis using elephant swarm water search algorithm. I.J. Information Engineering and Electronic Business **2**, 10–20 (2019). https://doi.org/10.5815/ijieeb.2019.02.02

Modelling Smart Grid Instability Against Cyber Attacks in SCADA System Networks

John E. Efiong[1]([✉]), Bodunde O. Akinyemi[1], Emmanuel A. Olajubu[1], Isa A. Ibrahim[2], Ganiyu A. Aderounmu[1], and Jules Degila[3]

[1] Department of Computer Science and Engineering, Obafemi Awolowo University, Ile-Ife, Nigeria
{bakinyemi,emmolajubu,gaderoun}@oauife.edu.ng

[2] Department of Cybersecurity, Federal University of Technology, Owerri, Nigeria

[3] Department of Computer Science, Institute of Mathematics and Physics, University of Abomey, Calavi, Benin
jules.degila@imsp.uac.org

Abstract. The SCADA systems in the Smart Grid Network (SGN) are increasingly facing cyber threats and divers attacks due to their known proprietary vulnerabilities, most often leading to power instability and cascading failures in the Grid. This paper associates the SGN's instability with cyberattacks and models intrusions into SCADA systems using the Decentral Smart Grid Control (DSGC) model on the Stability dataset. The Time-Synchronization Attack was modelled with its effects on Transmission Line Fault Detection, Voltage Stability Monitoring and Event Locating. Classification and prediction were done on WEKA using a number of Machine Learning Algorithms and performances evaluated. Random Forest (RF) classifier was found to be more promising in all the benefit metrics yielding 92.0%, 91.9%, 92.0%, and 91.9% for Accuracy, Precision, Recall and F-measure, respectively. An ensemble classifier – Bagging, performed significantly well after the RF with 90.0%, 89.9%, 90.0% and 89.9%. The ROC Area under Curve also showed RF having 97% and Bagging 96.7%. 100 trees were constructed for the RF while considering 4 random features, and the out-of-bag error was 0.0084. RF had a TPR of 92.0%, which is the highest and an FPR of 0.11. These results provide insights into identifying appropriate ML techniques and models for the Grid's instability problem vis-à-vis modelling attacks. This is expected to help in effective security planning, threats detection, attack mitigation and risk management.

Keywords: SCADA Systems · Cyberattacks · Smart Grid Stability · Decentral Smart Grid Control · Cyber-Physical Power Systems

1 Introduction

Stability is one of the essential characteristics of the Cyber-Physical Power System (CPPS), also identified as the Smart Grid, as a heterogeneous, multi-faceted system [1]. This is in addition to the robustness, efficiency, reliability, resource management

© The Author(s), under exclusive license to Springer Nature Switzerland AG 2023
Z. Hu et al. (Eds.): ICAILE 2023, LNDECT 180, pp. 239–250, 2023.
https://doi.org/10.1007/978-3-031-36115-9_23

and load-balancing that the CPPS is expected to provide. Smart Grid stability entails maintaining a balance between the energy generated and the energy consumed [2]. It is a state of equilibrium in the business chain. The Grid ensures this by responding to volatility in voltage and frequency disturbances.

In the Smart Grid, the Phasor Measurement Unit (PMU) is a popular tool for monitoring voltage stability and analysing oscillation stability. The PMU also allows power state estimation, monitoring and control of the Wide Area Network, Volt-Amps Reactive (VAR) optimisation, blackout analysis, real-time electricity costing, and fault detection in the transmission lines. Targeted cyberattacks may result in putting the PMU off its primary functions, thereby resulting in power instability in the electric Grid.

Additionally, one of the main goals of a cyberattack on the Supervisory Control and Data Acquisition (SCADA) systems managing the Smart Grids is to interfere with the control process of the circuit breakers, thereby forcing the circuit breakers to respond to external commands other than what the control engineer would be issuing. This could be achieved by the attacker through a command injection attack, data injection attack and time-synchronisation [29] attack using primarily the interception and injection attack strategies. When circuit breakers are forced open, it results in a denial of service (instability), which may have cascading effects on several processes depending on which circuit is broken. Power stability is equivalent to the availability of resources in the grid; thus, an attack tends to cause instability or a total blackout. For instance, an attack can cause a delay in the release of the status information of the power generator to the control centre for a prompt response and necessary actions [3, 4]. Furthermore, it can trip circuit breakers at will and cripple the grid's load-balancing capabilities.

Modelling Smart Grid instabilities against cyberattacks help to examine the risks associated with cyberattacks and their propensity to affect not only power stability but the safety of personnel and security of the components of the SCADA network. According to [1], such modelling benefits power system engineers in four ways: First, they will respond to attacks based on their ability to point out the problem with specific components. Second, mitigation strategies can be developed to prevent future occurrences. Third, it will help in regular evaluation of the security posture of the organisation. Fourth, it will help the engineer to engage in a more formidable design and development of resilient Cyber-physical Systems for the Power Sector. It is important to note that power control engineers do not necessarily provide security for power assets. This role is expected to be delivered by a trained industrial cybersecurity professional in the Smart Grid. Thus, such individuals and researchers would appreciate the threat-modelling results in the grid.

Existing body of knowledge that model Smart Grid instabilities focus rather on the mechanical prediction of instability using a plethora of techniques. There is an acute shortage of works that associates instability with the possibility of cyberattacks. This paper understudies that subtle operations of attacks and their propensity to distort the stability of the Grid. Typically, in this paper, we model the Time-synchronization Attack (TSA) and implement Machine Learning algorithms for attack identification as a precursor for instability in the Smart Grid Network. The primary objective of this paper is to associate power instability with cyberattacks and provide an ML-based model for intrusion detection in the Smart Grid Network. The rest of the paper is organised as follows: Section II discusses related works. The Threat/Attack modeling technique, dataset and methods used are covered in Section III. The experimental setup is presented in Section IV, and the presentation and discussion of results are done in Section V. The last Section concludes the paper.

2 Related Works

Efforts have been made to model attacks on the Smart Grid network. Studies by [5] modelled Grid Stability with Decentral Smart Grid Control (DSGC) based on frequency by measuring the cycles per second of the alternate current. However, this was only applicable at the power generation phase of the Grid process. Reference [5] adopted the Decision Tree (CART) algorithm and space-filling designs to implement the DSGC model. Reference [6] investigated how the PMU could be used to model a Time-Synchronised attack in the Smart Grid network with the aim of revealing tampered data, possible effects on voltage stability, Automatic Generation Control (AGC) [7] and power system frequency control [8]. Instability was modelled using the delay-dependent stability analysis technique by [9] and [10] for the eigenvalue computation of the Grid Network. A directed-graph method with a dynamic system equation was utilised by [11]. The technique was used to compute the status of an individual node in the Grid network.

Reference [12] experimented with a sort of distributed denial of service as a strategy to cause transient instability in the grid using the Variable Structure system model. The theory supported a coordinated multi-switch attack capable of causing cascading failures in the Smart Grid through intrusion into the network from a single circuit breaker and extending to other nodes. Recently, an optimal partial feedback attack modelling method that relies on convex relaxation and pontryagin's maximum principle for instability was proposed by [13] to model data injection and command injection attacks. Here, the adversary modifies control signals to tweak the performance of the master unit and changes its location to avoid being traced. The proposed method revealed the attack locations [13].

Some of the works that model instability against threats and attacks in the Cyber-Physical systems of the Power sector have indicated that a multi-agent-based technique may be appropriate in detecting and spotting threats with high affinity for causing instability in the Smart Grid. For instance, in applying the multi-agent method, [14] leveraged the unique characteristics of the power system's separate physical and cyber components, which helped classify cyberattacks appropriately and separated them from the material faults of the system. Similarly, [15] proposed a control framework based on the multi-agent perspective to reduce transient instability in the grid. Also, [16] proposed a distributed averaging-based integral (DAI) controller to solve the grid's instability attack-facing problem based on time-delay and dynamic communication topology.

To leverage control of the DER and reduce transient instability in the Grid, [17] proposed a differential game-theoretic technique designed to model the attacker's behaviour in the Grid network. Other methods used in literature include linear matrix inequalities and the Lyapunov stability technique [18], Lyapunov-based time-varying multiple delayed systems method [6], time-delayed power system stability analysis by the integral quadratic constraints method [19], realistic delay modelling method [20], etc.

242   J. E. Efiong et al.

The traditional methods of modeling instability on the Smart Grid discussed above do not present formidable tools for identifying, detecting, and mitigating sophisticated threats and attacks. The availability of advanced attack-launching tools and techniques has made the adversary's job relatively more accessible and penetrative. To address these flaws, [21] proposed an optimized Deep Learning model for predicting Smart Grid instability. Although the DL model yielded an accuracy of 99.6%, the primary focus was on solving the fixed inputs (variables of the equations) of the Decentral Smart Grid Control (DSGC) system. Reference [22] proposed an optimized data-matching learning network-based model for predicting stability of a decentralized Power Grid linking electricity price formulation to Grid frequency. The model was designed for forecasting other causes of instability, excluding cyber threats.

A recent study by [23] leveraged Genetic Algorithm to predict instability in the Grid. Using a 14-feature-based dataset extracted from UCI-ML repository, the model achieved an accuracy of 98%. This dataset did not depict the DSGC system and the proposed model was not associated with cyberattack mechanisms in the Grid. Our study employs machine learning techniques to model instability, associates same with cyberattacks and provides insights that can help develop intrusion detection systems and possible prevention in the Smart Grid Network.

3 Datasets and Methods

This Section presents the attack-instability modeling, dataset, experimental setup and methods.

3.1 Attack-Instability Modelling

Attack-instability modelling can follow a dynamic system-based technique where the physical components of the CPS follow differential equations utilising energy flow, and the cyber components follow differential equations using information flow. It is expected that the cyber components would disturb the physical systems; as such, the stimulant of the states of the generator, typically frequency and angle in the rotor swing equation of the generator, is used to model the perturbations [1]. This study adopted the technique leveraged by [7].

3.1.1 Time-Synchronization Attack

In this study the Time-Synchronization Attack (TSA) is modelled. The TSA is an attack that exploits the wide area monitoring systems (WAMSs). The TSA targets vulnerable networks with dynamic data exchanges at the nodes [24]. The technique is to modify the time synchronization at the endpoints, resulting in cascading failures. In the electric Grid, a TSA attacker can build a vector to control the synchronisation generator to remotely control the Distrito control Distributed Energy Resources (DERs) remotely [30]. This enables the attacker to have lateral access to the system and enforce the instability of the Smart Grid.

We modelled the effect of TSA in Transmission Line Fault Detection was using Eq. (1):

$$\Delta D = \left(\frac{1}{2\gamma L}\right) \ln\left(\frac{(A+B)(C+D.\exp(j\Delta\theta)}{(C+D)(A+B.\exp(j\Delta\theta)}\right) \tag{1}$$

where: ΔD is the fault location error concerning TSA; $\Delta\theta$ is asynchronism of phase angle measurement between the transmitting and receiving endpoints concerning TSA; L is the length of the transmission line, γ is the attenuation constant, while A, B, C, and D are the formulas of the transmitting and receiving voltage and current respectively.

We further modelled the effects of the TSA in Voltage Stability Monitoring using Eqs. (2), (3) and (4):

$$\overline{Z'_T} = 2\frac{V_S \exp(j\Delta\theta_S) - V_R \exp(j\Delta\theta_R)}{I_S \exp(j\Delta\theta_S) - I_R \exp(j\Delta\theta_R)} \tag{2}$$

$$\overline{Z'_{sh}} = -\frac{V_S I_R + V_R I_S}{I_R^2 \exp(j2\Delta\theta_R) - I_S^2 \exp(j2\Delta\theta_S)} * (\exp j(\Delta\theta_S + \Delta\theta_R)) \tag{3}$$

$$\overline{Z'_L} = \frac{V_R \exp(j\Delta\theta_R)}{I_R \exp(j\Delta\theta_R)} \tag{4}$$

where $\overline{Z'_T}$, $\overline{Z'_{sh}}$ and $\overline{Z'_L}$ are the T-equivalent parameters for computing Voltage Stability Monitoring as affected by the TSA [31].

We then modelled the effect of the TSA on Event Locating using Eq. (5):

$$(x_i - x_e)^2 + (y_i - y_e)^2 - V_e^2(t_i - t_e)^2 = 0 \tag{5}$$

where: $t_i, i = 1, 2, 3, 4$ is the disturbing event arrival time to the ith PMU; (x_i, y_i) and (x_e, y_e) are the coordinates of the ith PMU and the disturbance event locations. V_e represents the event propagation velocity in the Smart Grid network.

3.2 Decentral Smart Grid Control (DSGC) Model

The Decentral Smart Grid Control (DSGC) model is a system that monitors frequency property of the Grid. This implies that an essential component of the Grid that the DSGC model takes due cognizance of is the frequency. This property ensures that there is a strict observance of the state of equilibrium between the power generated and consumed. The model expresses the physical dynamics of electric power generation and how it relates to the consumption of loads [5, 25, 26]. The DSGC model can be expressed mathematically as shown in Eq. 6:

$$\frac{d^2\theta_j}{dt^2} = P_j - \alpha_j \frac{d\theta_j}{dt} + \sum_k K_{jk} sin(\theta_k - \theta_j) \tag{6}$$

where:

$j = $ index representing the number of grid participant (producer(s) generating power to transmit through the grid and consumers take load from the Grid);

244 J. E. Efiong et al.

$\frac{d^2\theta_j}{dt^2}$ = grid stability indicator (negative indicates grid is unstable; positive indicates grid is stable);

P_j = mechanical power produced (e.g. by generator P_1) or consumed (e.g. by one of several consumers P_2 to P_4) (s^{-2});

α_j = damping constant related to the power dynamics of the grid;

$\frac{d\theta_j}{dt}$ = change in rotor angle for participant j relative to grid frequency ω;

K_{jk} = Coupling strength between grid participant j and k, which is proportional to line capacity (s^{-2}); and.

θ_k and θ_j are rotor angles for grid participants j and k at a specific point in time t.

The DSGC model takes into consideration the power Grid synchronization to ensure power stability [27, 28]. This becomes a critical factor which a Time Synchronization attacker leverages to exploit the Grid.

3.3 Dataset

The dataset used for experiment was the Electrical Grid Stability Simulated Dataset created by [5] to support the Decentral Smart Grid Control (DSGC) project. The dataset was eventually donated to the University of California (UCI) Machine Learning Repository. The smart grid stability (DSGC) project aimed to examine and make arrangements for both energy generation and potential utilisation by preparing for unsettling influences and vacillations that may occur in the electric business chain. The dataset was generated using a Grid stability reference 4-node star network simulation shown in Fig. 1. The dataset contains 10,000 samples.

The original dataset contained 12 features and 2 target classes for classification and prediction. The features were classified into reaction times of the participants (tau1 – tau4), nominal power generated or consumed (p1 – p4) and price elasticity coefficient for each participant on the network (g1 - g4). The reaction time takes real values between 0.5 and 10, where tau1 was assigned the supplier node, and the rest were assigned the consumer nodes. The nominal power was a real value between -2.0 to -0.5 (for p2 – p4); when generated, it was positive, and when consumed, it became negative. For the total power consumed to be equal to the total generated, Eq. (7) must hold:

$$p1(suppliernode) = -(p2 + p3 + p4) \tag{7}$$

The elastic price coefficient must also assume real values between 0.05 to 1.00, where g1 was assigned the supplier node and g2 – g4 the consumer nodes, with g representing 'gamma'.

For the dependent variables, 2 were identified as '*stab*' and '*stabf*'. The former is the maximum real part of the root of the differential equation. If it is positive, the system is said to be linearly unstable; if it is negative, it is linearly stable. The *stabf* is a binary categorical target class with labels (stable or unstable). Our models were implemented based on the *stabf* variable. We applied the Principal Component Analysis (PCA) to extract the most important features from the ranked attributes based on the computed eigenvectors. We also used the Ranker Search Method with Chi-Squared Attribute Evaluator (Chi2) and Info Gain Attribute Evaluator (InfoGain). The PCA selected tau1, g1,

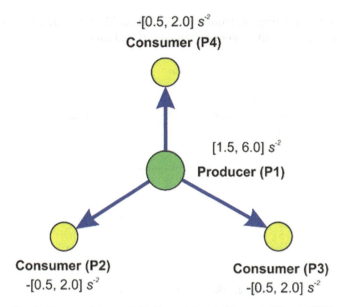

Fig. 1. Decentral Smart Grid Control 4-node Network (Wood, 2020)

tau3, tau4, g2, g4 and tau2 while both Information Gain and Ch2 selected tau1, tau2, tau4 and tau3 as the best averagely ranked attributes.

Although the dataset does not explicitly denote attack patterns, there is a strong association between the instability experienced in Grids during power generation, transmission and distribution leading to a denial of service attacks. The chance of this occurring increases as the Smart Grid is controlled and supervised by SCADA systems which lack security mechanisms to deter intrusions. Hence, the need to look at the scenarios from an intrusion point of view.

4 Experimental Setup

The experiment was carried out on the Waikato Environment for Knowledge Analysis (WEKA), a machine learning data mining tool for knowledge synthesis. The tool allows attributes selection, data preprocessing, classification, clustering, association, linking, visualisation and cross-validation. We split the dataset into 75% for training and 25% for testing and validation. We projected the "unstable" class into "Anomaly" and the "stable" into "normal" states of the system.

For training and evaluating the dataset for prediction, the following classifiers were used; Five (5) Trees-based classifiers – Random Forest (RF), Random Tree (RT), LADTree, ADTree and J48; Two (2) Rules-based – PART and JRip; One (1) Functions-based – Logistic; One (1) Bayes-based – Naïve Bayes (NB) and One (1) Meta-based – Bagging. Figure 2 shows the implemented ML classifiers. The classifiers were evaluated using Accuracy, Precision, Recall and F1-Measure. We also examined their respective performances regarding False Positive Rate (FPR) and True Positive Rate (TPR), which are helpful for computing Detection Rate. The Root Mean Square Error (RMSE) and Mean Absolute Error (MAE) were computed and evaluated.

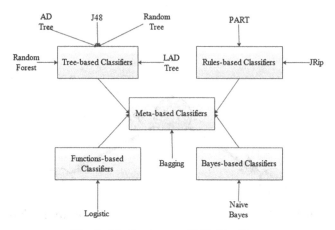

Fig. 2. The Implemented ML Classifiers

5 Results and Discussion

A total of 10 Machine Learning algorithms were implemented, and results are shown in Tables 1 and 2. All the models performed reasonably well, outputting more than 80% in all metrics. However, the Random Forest (RF) classifier demonstrated the best results in all the benefit metrics yielding 92.0%, 91.9%, 92.0%, and 91.9% for Accuracy, Precision, Recall and F-measure, respectively. The ensemble classifier closely followed this – Bagging with 90.0%, 89.9%, 90.0% and 89.9%.

The last column of Table 1 shows the ROC Area under Curve, with RF having 97% and Bagging 96.7%. The PART and Naïve Bayes had an equal result of 92%, while LADTree and ADTree had 91.0% and 91.9%, respectively. A Random Forest of 100 trees was constructed while considering 4 random features and the out-of-bag error was 0.0084. This shows the efficiency of the RF classifier in modelling instability against Smart Grid attacks. PART generated 65 rules, and JRip generated 51 rules. The J48 tree classifiers had a total of 813 trees with 407 leaves, while RandomTree had a size of 1873 trees. The bagging classifier of the Meta class had 393 trees. The LADTree classifier generated a total of 31 trees, 21 predictor nodes for the Tree size, 16 predictor nodes for Leaves, and expanded nodes of 100 with 517157 instances. ADTree had 31 nodes, and 21 Leaves used predictor nodes.

The Confusion Matrices of the classifiers, whose results are represented in Table 2, showed that RF had the True Positive Rate (TPR) of 92.0%, which is the highest and the False Positive Rate (FPR) of 0.11, which is the lowest. The MAE and RMSE of the RF were 0.18 and 0.25, respectively. The Bagging technique produced a TPR of 90.0% and an FPR of 0.13. The MAE and RMSE of Bagging were 0.17 and 0.27, respectively. It should be noted that J48 returned the lowest MAE, 0.14 but with a significantly high RMSE, 0.35.

Table 1. Model Classification and Prediction Results

Classifiers	Metrics				
	Accuracy	Precision	Recall	F-Measure	ROC Area
PART	85.9%	86.1%	85.9%	85.5%	92.0%
JRip	87.4%	87.3%	87.4%	87.3%	87.5%
Logistic	82.3%	82.1%	82.3%	82.1%	89.9%
NB	83.5%	83.5%	83.5%	83.1%	92.3%
J48	86.7%	86.7%	86.7%	86.7%	86.5%
RF	92.0%	91.9%	92.0%	91.9%	97.9%
RT	83.2%	83.3%	83.2%	83.3%	81.9%
Bagging	90.0%	89.9%	90.0	89.9%	96.7%
LADTree	84.1%	84.0%	84.1%	84.0%	91.0%
ADTree	83.8%	83.8%	83.8%	83.4%	91.9%

Table 2. Model Evaluation Results

Classifiers	Metrics			
	TPR	FPR	MAE	RMSE
PART	85.9%	0.21	0.16	0.32
JRip	87.4%	0.15	0.18	0.32
Logistic	82.3%	0.22	0.24	0.35
NB	83.5%	0.24	0.29	0.34
J48	86.7%	0.16	0.14	0.35
RF	92.0%	0.11	0.18	0.25
RT	83.2%	0.19	0.17	0.40
Bagging	90.0%	0.13	0.17	0.27
LADTree	84.1%	0.20	0.25	0.34
ADTree	83.8%	0.23	0.31	0.36

6 Conclusion

The utility service-providing domain is delicate, and requires adequate care and attention to all eventualities. This study proposes that not all power failure or instability results from mechanical faults or natural occurrences. Targeted attacks at specific components of the Grid, especially the time synchronization system of the PMU can result in serious dangers. Cyberattacks on the Smart Grid can prompt communication delays or service denials and puncture data exchanges between the PMU and the Control Centre or even take control of the Control Systems. This ultimately results in Power instability, technically referred to as 'denial of service'. Several kinds of attacks exist that target the Smart Grid.

We have modelled the Time-Synchronization Attack (TSA), which is a stealthy attack that establishes an attack vector or increases an attack surface such that the attacker takes control of the synchronisation generator, thus remotely controlling the Distributed Energy Resources (DERs) [32]. This attack technique can be catastrophic due to the several other components and processes the DER usually manages. In addition, the TSA has the propensity to open the network for distributed denial of service (DDoS). Modelling attacks in the Smart Grid network helps identify threats and develop mitigation mechanisms.

In this paper, we have demonstrated 10 ML algorithms that can help model Smart Grid Instability against Cyberattacks on Cyber-Physical Power Systems and other critical infrastructure. Random Forest classifier of the Trees Methods and Bagging classifier of the Ensemble techniques were found to be more promising and can hence be tuned to provide higher accuracy, precision, recall and F-measure. These results have implications for effective threats-modelling in the Smart Grid network. The idea portrayed in this paper would help in attacks identification, threats classification, prediction, intrusion detection and developing mitigation process. Future studies may consider modelling other kinds of attacks in the network by leveraging deep learning algorithms.

Acknowledgment. The Research was supported by funding from the Digital Science and Technology Network (DSTN), France, through the Research Partnership programme with the OAU-Knowledge Park (OAK-park), African Centre of Excellence, Obafemi Awolowo University, Ile-Ife, Nigeria.

References

1. Yohanandhan, R.V., Elavarasan, R.M., Manoharan, P., Mihet-Popa, L.: Cyber-Physical Power System (CPPS): A Review on Modeling, Simulation, and Analysis With Cyber Security Applications. IEEE Acesss **8**, 151019–151064 (2020)
2. Gopakumar, P., Jaya bharata Reddy, M., Mohanta, D.K.: Letter to the Editor: Stability Concerns in Smart Grid with Emerging Renewable Energy Technologies. Electric Power Components and Systems **42**(3–4), 418–425 (2014). https://doi.org/10.1080/15325008.2013.866182
3. Dong, C., et al.: Effective method to determine time-delay stability margin and its application to power systems. IET Gener. Transmiss. Distrib. **11**(7), 1661–1670 (2017)
4. Bevrani, H., Watanabe, M., Mitani, Y.: Power System Monitoring and Control, p. 2014. Wiley, Hoboken, NJ, USA (2014)
5. Arzamasov, V., Böhm, K., Jochem, P.: Towards Concise Models of Grid Stability. IEEE International Conference on Communications, Control, and Computing Technologies for Smart Grids (SmartGridComm), Aalborg, 1–6 (2018)
6. Zhang, F., Cheng, L., Gao, W.: Prediction based hierarchical compensation for delays in wide-area control systems. IEEE Trans. Smart Grid **9**(4), 3897–3899 (2018)
7. Sridhar, S., Govindarasu, M.: Model-based attack detection and mitigation for automatic generation control. IEEE Transactions. Smart Grid **5**(2), 580–591 (2014)

8. Anwar, A., Mahmood, A.N., Tari, Z.: Identification of vulnerable node clusters against false data injection attack in an AMI based Smart Grid. Inf. Syst. **53**(201–212), 2015 (2015)
9. Mary, T.J., Rangarajan, P.: Delay-dependent stability analysis of Microgrid with constant and time-varying communication delays. Electrical Power Components Systems **44**(13), 1441–1452 (Aug., 2016)
10. Ye, H., Gao, W., Mou, Q., Liu, Y.: Iterative infinitesimal generator discretization-based method for eigen-analysis of large delayed cyber-physical power system. Electr. Power Syst. Res. **143**, 389–399 (Feb. 2017)
11. Kundur, D., Feng, X., Mashayekh, S., Liu, S., Zourntos, T., Purry, K.L.B.: Towards modelling the impact of cyberattacks on a Smart Grid. Int. J. Secur. Netw. **6**(1), 2 (2011)
12. Liu, S., Chen, B., Zourntos, T., Kundur, D., Butler-Purry, K.: A coordinated multi-switch attack for cascading failures in smart grid. IEEE Trans. Smart Grid **5**(3), 1183–1195 (May 2014)
13. Wu, G., Wang, G., Sun, J., Chen, J.: Optimal partial feedback attacks in cyber-physical power systems. IEEE Trans. Autom. Control, early access (Mar. 19, 2020). https://doi.org/10.1109/TAC.2020.2981915
14. Rahman, M.S., Mahmud, M.A., Oo, A.M.T., Pota, H.R.: Multi-agent approach for enhancing security of protection schemes in cyber-physical energy systems. IEEE Trans. Ind. Informat. **13**(2), 436–447 (2017)
15. Farraj, A., Hammad, E., Kundur, D.: A cyber-physical control framework for transient stability in smart grids. IEEE Trans. Smart Grid **9**(2), 1205–1215 (Mar, 2018)
16. Schiffer, J., Dörfler, F., Fridman, E.: Robustness of distributed averaging control in power systems: Time delays & dynamic communication topology. Automatica **80**(1), 261–271 (Jun. 2017)
17. Srikantha, P., Kundur, D.: A DER attack-mitigation differential game for smart grid security analysis. IEEE Transactions. Smart Grid **7**(3), 1476–1485 (2016)
18. Li, J., Chen, Z., Cai, D., Zhen, W., Huang, Q.: Delay-dependent stability control for power system with multiple time-delays. IEEE Trans. Power Syst. **31**(3), 2316–2326 (2016)
19. Cai, D., Huang, Q., Li, J., Zhang, Z., Teng, Y., Hu, W.: Stabilization of time-delayed power system with combined frequency-domain IQC and time-domain dissipation inequality. IEEE Trans. Power Syst. **33**(5), 5531–5541 (2018)
20. Liu, M., Dassios, I., Tzounas, G., Milano, F.: Stability analysis of power systems with inclusion of realistic-modeling WAMS delays. IEEE Trans. Power Systems **34**(1), 627–636 (Jan. 2019)
21. Breviglieri, P., Erdem, T., Eken, S.: Predicting smart grid stability with optimized deep models. SN Computer Science **2**(2), 1–12 (2021). https://doi.org/10.1007/s42979-021-00463-5
22. Wood, D.A.: Predicting stability of a decentralized power grid linking electricity price formulation to grid frequency applying an optimized data-matching learning network to simulated data. Technology and Economics of Smart Grids and Sustainable Energy **5**(1), 1–21 (2020). https://doi.org/10.1007/s40866-019-0074-0
23. Dewangan, F., Biswal, M., Patnaik, B., Hasan, S., Mishra, M.: Chapter Five - Smart grid stability prediction using genetic algorithm-based extreme learning machine. In: Bansal, R.C., Mishra, M., Sood, Y.R. (eds.) Electric Power Systems Resiliency, pp. 149–163. Academic Press (2022). https://doi.org/10.1016/B978-0-323-85536-5.00011-4
24. Delcourt, M., Shereen, E., Dan, G., Le Boudec, J.Y., Paolone, M.: Time-synchronization attack detection in unbalanced three-phase systems. IEEE Transactions on Smart Grid **12**(5), 4460–4470 (2021). https://doi.org/10.1109/TSG.2021.3078104
25. Schäfer, B., Matthiae, M., Timme, M., Witthaut, D.: Decentral smart grid control. New J Phys **17**(1), 15002 (2015)
26. Schäfer, B., Grabow, C., Auer, S., Kurths, J., Witthaut, D., Timme, M.: Taming instabilities in power grid networks by decentralized control. The European Physical Journal Special Topics **225**(3), 569–582 (2016). https://doi.org/10.1140/epjst/e2015-50136-y

27. Nishikawa, T., Motter, A.E.: Comparative analysis of existing models for power-grid synchronization. New J Phys **17**(1), 15012 (2015)
28. Schmietendorf, K., Peinke, J., Friedrich, R., Kamps, O.: Self-organized synchronization and voltage stability in networks of synchronous machines. The European Physical Journal Special Topics **223**(12), 2577–2592 (2014). https://doi.org/10.1140/epjst/e2014-02209-8
29. Delcourt, M.M.N.: Time-Synchronization Attacks against Critical Infrastructures and their Mitigation. Lausanne, EPFL (2021). https://doi.org/10.5075/epfl-thesis-8577. https://infoscience.epfl.ch/record/288441
30. Marie, M., Delcourt, N.: Time-Synchronization Attacks against Critical Infrastructures and their Mitigation (2021)
31. Moussa, B.: Detection and Mitigation of Cyber Attacks on Time Synchronization Protocols for the Smart Grid. September (2018)
32. Zhang, Z., Gong, S., Dimitrovski, A.D., Li, H.: Time synchronization attack in smart grid: Impact and analysis. IEEE Transactions. Smart Grid **4**(1), 87–98 (Mar. 2013)

Program Implementation of Educational Electronic Resource for Inclusive Education of People with Visual Impairment

Yurii Tulashvili[1]([✉]), Iurii Lukianchuk[1], Valerii Lishchyna[1], and Nataliia Lishchyna[2]

[1] Department of Computer Science, Lutsk National Technical University, Lutsk 550000, Ukraine
y.tulashvili@lutsk-ntu.com.ua
[2] Department of Software Engineering, Lutsk National Technical University, Lutsk 550000, Ukraine

Abstract. It is substantiated that the educational software used in the educational process during the computer training of visually impaired people contributes to the formation of their correctional and compensatory devices for the use of computer technology. The peculiarities of the inclusive educational process of professional training of persons with visual impairments with the use of computer technology are pointed out. The architecture and features of creating educational software for the visually impaired are revealed. A mathematical model and algorithm for its implementation of educational software for support of inclusive educational process of professional training of persons with visual impairments have been developed. The developed training software presents the theoretical content of training in an adapted form. The developed practical tasks contain an exhaustive list for the formation of correctional and compensatory devices and the acquisition of skills and abilities to use computer technology for people with visual impairments.

Keywords: People with visual impairments · Software · Vocational rehabilitation

1 Introduction

Process of integrating people with disabilities into public relations involves a set of actions aimed at helping subjects with mental and physical disabilities to master the values of modern civilization, to be involved in active social activities through rehabilitation measures. Vocational rehabilitation of people with disabilities is defined as the process of providing rehabilitation measures for people with disabilities, taking in to account the limited scope of maintaining the ability to work, contraindications to certain types of work due to health for vocational education and rational employment.

Training of visually impaired people with the use of information technology in professional activities is an extremely important issue of socialization, is one of the main tasks of their professional rehabilitation, the solution of which will include them

---

© The Author(s), under exclusive license to Springer Nature Switzerland AG 2023
Z. Hu et al. (Eds.): ICAILE 2023, LNDECT 180, pp. 251–260, 2023.
https://doi.org/10.1007/978-3-031-36115-9_24

in society not only as socially full, but also as creatively active members [1]. This is especially with the use of ICT [2].

The development of new and the development of existing methods of using information technology for the training of the blind is especially important. Today, inclusive education is faced with the urgent task of introducing special software and hardware in the process of training future professionals.

2 Literature Analysis

The urgency of the problem of vocational rehabilitation, as the main means of social and labor adaptation of people with visual impairments, is determined primarily by the place of blind people in the social environment, the fact that all visually impaired can be included in society as socially full and creatively active its members [3].

In his work "Information technologies for the visually impaired in the system of additional education of Kabardino-Balkari" L. Shautsukova noted the possibility and necessity of using new information technologies in the educational process of people with visual impairments. She believes that computer technology, in this case, should be considered not so much as a subject of study, but as a new means of creating corrective and compensatory devices, and computer technology as a tool to empower people with visual impairments to their successful rehabilitation. Modern society [4].

Research by Ermakov V. and Yakunin G. possible areas of employment of people with visual impairments was investigated in [5]. It is determined that the ideal computer workplace for a blind programmer should have the following hardware and software: an individual computer with sufficient resources for the operation of tools and adaptive software; several "screen access" programs that have different capabilities and allow you to work in different operating systems; braille display; speech synthesizer; scanner; Internet connection.

The use of information technology in the educational process for blind people is not only wide access to new educational technologies, but also access to the global information space [6]. With the help of information technology, people with visual impairments have the opportunity to exercise their professional self-determination, as well as the opportunity to communicate freely with the world.

Research proves that success of mastering information technologies, telecommunication networks for the visually impaired creates conditions for their professional rehabilitation [7]. Digital content related to engineering, adapted using adaptation methods, allows blind students to effectively use technical means, communication channels in combination with information technology [12].

The formation of the structure of the education system on the basis of expert evaluation plays a significant role in the analysis and selection of learning content. Methods of structuring the content of learning during the mastery of information technology and the development of compensatory devices in people with visual impairments have been disclosed in previous studies [8].

3 Object, and Research Methods

In the process of forming and presenting the content of education should take into account the psycho-physiological characteristics of subjects of education with visual deprivation, such as fragmentation, verbalism and low rate of learning. Therefore, one of the conditions for the effective solution of the problem of forming the content of vocational computer training is to take into account the natural characteristics of people with visual impairments.

Research on modeling the content of learning technologies determines that the implementation of operational management of cognitive activity of educational entities should be carried out both at the didactic macro level and at the didactic micro level [10].

Therefore, among the main tasks to be solved at each level of the structural and organizational model of forming the content of professional computer training of visually impaired people, first of all is the task of adapting the content at the didactic macro level and concretizing and shaping the content of blind and partially sighted students work with the use of computers at the didactic micro level [10].

In conducting research, we emphasize the special importance of the principle of formation of compensatory devices in the teaching of information technology and its combination with traditional principles in the process of forming the educational content of professional computer training for visually impaired people.

The matrix method of selection and structuring of educational material proposed in researches is widely used for formation of the maintenance of training [13]. However, this method is not fully suitable for small learning tasks for computer training of people with visual impairments.

In conducting research, we emphasize the special importance of the principle of formation of compensatory devices in the teaching of information technology and its combination with traditional principles in the process of forming the educational content of professional computer training for visually impaired people.

The implementation of these two key principles in combination with traditional principles is to conduct a special analysis of the constituent elements of the content of education, in accordance with the defining criterion, which is characterized by the following indicators, those we have studied before [8]:

1) Perception of techniques and methods of activity through tactile and auditory analyzers, which allows you to fully determine the properties of the object or object to which the action is directed or which can be studied.
2) The formation of a complete picture of the object or subject to which the action is directed or which can be studied by repeated performance of perceptual actions and, as a consequence, the emergence of a stable information base of activity in a person with visual impairments.
3) Possibility of clear presentation of concepts, thoughts, theories and disclosure of properties of objects and objects at the verbal and tactile level of perception.

The combination of these indicators, in turn, determines the level of typical information and communication actions using modern information technology adaptation, which can be performed by subjects with visual deprivation, according to their psychophysiological indicators (blind or partially sighted).

Based on this, the main feature of the process of forming the educational content of professional computer training for visually impaired people is to ensure the effectiveness of the implementation of traditional and specific principles by adapting national requirements at the didactic macro level and specifying the content of didactic microlevel in a certain specialty.

A feature of the educational activities of people with visual impairments is the specification of the content of education at the didactic micro level to the possibilities of professional training of people with visual impairments. Therefore, in the process of determining the content of professional computer training for people with visual impairments, we focus on the specifics of building the content of the learning process.

The formation of the content of the educational task in this case is provided by the method of multiple repetition of educational objects, which is to use in the process of compiling the content of practical tasks repeatedly used selection of educational material.

Thus, in the process of forming the content of professional computer training of persons with visual impairments in a particular specialty provides structured adaptation of all disciplines and specification of their content in accordance with the invariant and variable parts of the training curriculum.

4 Results

The study of the scientific problem involves the disclosure of the features of the process of vocational training of persons with visual impairments using software training, the formation of quality knowledge of future professionals who are visually impaired, the use of computer technology in an inclusive educational environment. The compensatory devices as the main factor in the process of forming the competence of the future specialist.

In developing the training software, we relied on the aspect that information technology for people with visual impairments is not only a set of acquired levels of knowledge, skills and abilities, but, above all, the ability to access the global information space. With the help of information technology, people with visual impairments get the opportunity for their professional self-determination, the opportunity to communicate freely with the world.

In such centers, practical experience of professional training of the blind with the use of information technology has been gained [11]. Based on many years of experience, it was concluded that the most common problems faced by teachers in the training of visually impaired people in teaching them to work with computer technology are the following:

– overcoming the psychological barrier associated with uncertainty about computer technology, with great doubts about the ability to master it;
– the question of mastering the techniques of working on the keyboard. Getting to know the computer keyboard and learning the location of the keys on it and setting hands for people who have experience with a typewriter comes down to finding out the differences between typing text and entering text from the keyboard into a computer. For others, such skills are formed in the process of hard work, which is quite sufficient for mastering the keyboard;

- the need for orientation in the electronic text document, both in the self-created and in that typed by someone;
- work with operating systems is to learn the techniques of moving between computer disks, directories, files;
- awareness that the same keyboard shortcuts in different software can perform different functions;
- mastering the concept of file name and extension, their purpose and ways to create and change them, etc.

The need to solve these and other problems that arise in the process of teaching the blind information computer technology, necessitates the development of appropriate teaching methods, manuals, recommendations for working with relevant software products and more.

The professional development of people with profound visual impairments in the field of engineering and pedagogical training is associated with the study of special disciplines in the field of computer technology. First of all, in the process of professional rehabilitation, special attention was paid to the study of computer technology, information and communication technologies in classes in subjects closely related to computer science.

Studies of the problem of inclusion in higher education on the example of teaching visually impaired students through the use of special programs, educational Internet environments using so-called assistive technologies (Assistive Technologies) show the effectiveness of the use of computer tools in the educational process. Assistive technologies help the blind to obtain information for learning and work by receiving voice support when using modern processor technology [9, 14]. The basis of the success of its application is to learn how to work with a regular keyboard.

The study of new educational material should be based on the principle of consistency and gradual complication. The material should be presented interconnected, in a logical sequence of its study. For example, when studying the Word text editor, you should first get acquainted with the combinations of those "hot keys" that will open the editor, save the document to your computer disk, and then move on to the combinations associated with document formatting.

Armed with the "ten-finger method" of working on the keyboard, blind students begin to feel more confident. They form a positive motivation for further study, they understand that they have ample opportunities to access educational material, which today in many educational institutions is available on electronic media.

The next stage of rehabilitation work is extensive cooperation with teachers. The material presented in lectures, practical and laboratory classes should be transferred as much as possible to electronic media, such as Word documents. The most painstaking task is to "voice" mathematical formulas, drawings, diagrams and graphs. This requires additional time for their comprehension and correct reproduction of the text.

The study materials prepared in this way are provided to blind students before the classes for preliminary acquaintance with them on their own. In seminars, practical and laboratory classes, classrooms where students with visual impairments will study must be equipped with at least one computer workstation. This will ensure the employment of the blind student during the lesson and will not allow him to sit passively when the whole group is working on a task, which consists in calculating the results, compiling an analysis, generating a report.

Analysis of the experience of professional training of persons with visual impairments by the Laboratory of Assistive Learning Technologies and other rehabilitation centers shows that today professional and higher education is facing a humanistic problem determined by global trends in informatization of all spheres of human activity and full involvement of human potential development of the information society.

Qualitative changes in public life have led to the restructuring of technologies for collecting, storing, transmitting and presenting information. Without understanding these processes, no modern person can lead an active life today.

The intensive development of computer hardware and software has led to the emergence and further spread of the concept of using information technology as a means of developing compensatory devices for the visually impaired. Adaptive assistive computer technology provides greater access to blind and partially sighted people to information provided in electronic form, opens new opportunities for its socially useful activities.

In terms of the use of information technology as an adaptation technology for people with visual impairments, the concept of "information technology" acquires a new specific meaning that reflects the impact of computer technology on modern living conditions of people with visual impairments through the adaptive direction of modern information technology.

The software provides basic knowledge of computer science and techniques of working with computer technology. This creates the conditions for the formation of skills and abilities that people with visual impairments need to work on a personal computer, to successfully master the techniques of working with information systems.

The structure of the educational material is built in such a way that each educational module is divided into a content module of theoretical training and a content module of practical training. Theoretical training will provide visually impaired people with knowledge of the basics of computer science and the principles of computer technology, and practical training, through exercises, will allow you to acquire skills and abilities to work on a personal computer. Practical exercises should be performed with the monitor turned off using the JAWS for Windows screen reader, performing actions only on the keyboard console. This will contribute to the successful development of compensatory devices in visually impaired and blind education, which will allow them to use computer technology even when the defect has led to complete loss of vision.

The Fig. 1 shows software is designed for people with visual impairments and will be useful when studying in educational institutions in disciplines designed to develop in subjects with visual impairments knowledge, skills, and abilities to use computer technology to process information.

As a result of this approach in the system of content of professional computer training of people with visual impairments we have identified: set A, consisting of all professional industry competencies; many professional competencies of a specialist in the application of computer technology – B; set C, which corresponds to a set of professional competencies in the use of computer technology, which are available for mastery of persons with visual impairments with the use of computer adaptation tools.

Program Implementation of Educational Electronic Resource 257

Fig. 1. The interface of the developed software product (Informatyka_for_blind).

The ratio of the relationship between the many professional competencies, the content of which can be revealed through educational facilities (denote s), is available for people with visual impairments through computer means of adaptation:

$$C \in B \in A \equiv \forall s(s \in C \rightarrow \forall s(s \in B \rightarrow s \in A)). \tag{1}$$

This means that any learning object x is accessible to people with visual impairments through computer means of adaptation and that is part of the professional competence, which is an element of the set C, is an element of the set B of professional competencies, which, in its in turn, determines its inclusion in the set of professional industry competencies A.

The formation of the content of the educational task in this case is carried out using the proposed method. The method consists in the selection in the process of compiling the content of practical tasks of the selection of educational material by increasing the level of their complexity (level).

Any educational task consists of a set of elements in the form of their combinations of Si. The combination of educational objects takes the form of a tuple – a mathematically ordered and finite set of elements S.

In the process of defining the set S = {S1, S2, ..., Sn}, as a set of various objects of intellectual learning, we consider such a set of educational objects, which forms a certain amount of knowledge, skills and abilities that correspond to the competence to be mastered. a person with visual impairments, studying the thematic section of a particular content module.

The main condition for the formation of the content of practical work is that the number of training units should not exceed the dimensions of the tuple of data. To form the structure of connections in each educational task we use a polynomial generating function:

$$\sum_{k=1}^{n} \binom{n}{k} x^k = (1+x)^n. \tag{2}$$

According to the results of calculations, binomial coefficients will give possible connection structures. We will reveal the principles of this approach on the example of determining possible structures of educational tasks formed from three educational objects S1, S2, S3 provided they are combined to the number n = 5, when the following characteristics are identified: S1 occurs no more than once, S2 – no more than two, and S3 – once or twice.

Then the generating function takes the following form:

$$F(si, x) = (1 + S_1 x)(1 + S_2 x + S_2^2 x^2)(S_3 x + S_3^2 x^2) \qquad (3)$$

According to the results of calculations we get:

$$S_3 x + (S_1 S_3 + S_2 S_3 + S_3 S_3)x^2 + (S_1 S_2 S_3 + S_2 S_2 S_3 + S_1 S_3 S_3 + S_2 S_3 S_3)x^3$$
$$+ (S_1 S_2 S_3 S_3 + S_2 S_2 S_3 S_3 + S_1 S_2 S_2 S_3)x^4 + (S_1 S_2 S_2 S_3 S_3)x^5. \qquad (4)$$

From the solution we have the following possible values of the combinations educational objects. Learning tasks can be formed in the process of compiling learning tasks under such a condition that is determined by the presence of all initial actions in different variations. Therefore, learning tasks need to be formed from a set of tuples:

$$n = 3 : [S_1 S_2 S_3], [S_2 S_2 S_3], [S_2 S_3 S_3], [S_1 S_3 S_3];$$
$$n = 4 : [S_1 S_2 S_3 S_3], [S_2 S_2 S_3 S_3], [S_1 S_2 S_2 S_3];$$

This approach allows you to automate the process of content formation using the algorithm below (Fig. 2).

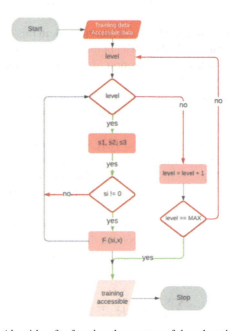

Fig. 2. Algorithm for forming the content of the educational task

Figure 3 reveals the formation and interaction of data streams that form the content of the educational task using the proposed algorithm of the software training tool designed for people with visual impairments.

The stage of software implementation of content management software provided by the software tutorial is designed using web on platform Node.js as web-site on React.js. We are currently updating the software product, as well as developing an application for a smartphone. The software is designed for people with visual impairments and will be useful when studying in educational institutions in disciplines designed to develop in subjects with visual impairments knowledge, skills and abilities to use computer technology to process information.

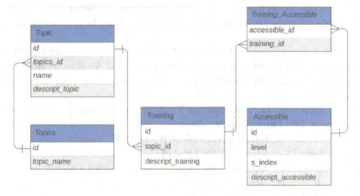

Fig. 3. Simplified data model

5 Conclusion

In the process of professional training of visually impaired people it is necessary to apply such pedagogical methods that use new achievements in the field of technical means and assistive technologies, which are constantly updated and improved, adapting the needs of the blind to the needs of modern society.

The proposed formalized approach to the formation of the content of education in the system of teaching visually impaired people to use computer tools as a means of adaptation is more effective because it does not require complex mathematical apparatus to implement simple algorithms for structuring the content of educational activities of people with visual impairments.

Mathematical model and algorithm for its implementation to determine the content of educational tasks of vocational training of people with visual impairments can be easily implemented through software. The software implementation of content management software provided as the software tutorial and is realized web on platform Node.js as web-site on React.js.

This creates an opportunity to implement in teaching technology the principles of individual and differentiated approaches, which are the basis of special pedagogical management of educational and developmental process of professional computer training of people with visual impairments, which is the subject of further research. Currently, educational materials are being prepared to improve learning outcomes by applying neural network algorithms.

References

1. Jennifer, K., Delia, F.: Enabling people with disabilities through effective accessible technology policies. In: Rune, H., Bjørn, H., Jerome, B., Delia, F., Ana, M., Guillén, R. (eds.) The Changing Disability Policy System, pp. 127–143. Routledge (2017)
2. Lewthwaite, S.: Web accessibility standards and disability: developing critical perspectives on accessibility. Disabil. Rehabil. **36**(16), 1375–1383 (2014)
3. Ferri, D., Favalli, S.: Web accessibility for people with disabilities in the european union: paving the road to social inclusion. Societies **8**(2), 40 (2018). https://doi.org/10.3390/soc8020040
4. Shautsukova, L.: Information technologies for the visually impaired in the system of additional education of Kabardino-Balkaria. In: Proceedings of the 2nd All-Russian meeting of the NMS on informatics. Actual problems of informatics in modern education, pp. 524–528. MSU Press, Moscow (2005)
5. Ermakov, V., Yakunin, G.: Development, Training and Education of Children with Visual Impairments, p. 240. Enlightenment, Moscow (2000)
6. Senjam, S.S., Foster, A., Bascaran, C., Vashist, P., Gupta, V.: Assistive technology for students with visual disability in schools for the blind in Delhi. Disabil. Rehabili.: Assistive Technol. **15**(6), 663–669 (2020). https://doi.org/10.1080/17483107.2019.1604829
7. Laabidi, M., Jemni, M., Ben Ayed, L.J., Brahim, H.B., Jemaa, A.B.: Learning technologies for people with disabilities. J. King Saud Univ. – Comput. Inform. Sci. **26**(1), 29–45 (2014). https://doi.org/10.1016/j.jksuci.2013.10.005
8. Tulashvili, Y.: Theoretical and methodical principles of professional computer training of persons with visual impairment: dis. of Dr. Ped. Sciences: [spec.] 13.00.04 "Theory and methods of professional education", p. 528. Vinnytsia State Pedagogical University Mykhailo Kotsyubynsky University. Vinnytsia (2012)
9. Viner, M., Singh, A., Shaughnessy, M.F.: Assistive technology to help students with disabilities. In: Singh, A., Viner, M., Yeh, C.J. (eds.) Special Education Design and Development Tools for School Rehabilitation Professionals, pp. 240–267. IGI Global (2019). https://doi.org/10.4018/978-1-7998-1431-3.ch012
10. Sineva, E.P.: Tiflopsychology, vol. 144. Textbook. Kyiv (2006)
11. Burgstahler, S.: 20 Tips for Teaching an Accessible Online Course. Disabilities, Opportunities, Internetworking and Technology, University of Washington (2022). https://www.washington.edu/doit/20-tips-teaching-accessible-online-course
12. Batanero, C., de Marcos, L., Holvikivi, J., Hilera, J.R., Oton, S.: Effects of new supportive technologies for blind and deaf engineering students in online learning. IEEE Trans. Educ. **62**(4), 270–277 (2019). https://doi.org/10.1109/TE.2019.2899545
13. Alekseeva, A., Antonenko, O., Zhadan, K., Lyfenko, M.: Experience in using e-learning tools in inclusive educational space of higher school. Phys. Math. Educ. **18**(4), 17–24 (2018)
14. Jones, V.L., Hinesmon-Matthews, L.J.: Effective assistive technology consideration and implications for diverse students. Comput. Schools **31**(3), 220–232 (2014). https://doi.org/10.1080/07380569.2014.932682

A Version of the Ternary Description Language with an Interpretation for Comparing the Systems Described in it with Categorical Systems

G. K. Tolokonnikov[✉]

VIM RAS, Moscow, Russia
admcit@mail.ru

Abstract. The article continues the development of a version of the language of ternary description (TDL), in which it is possible to compare the concepts of the system given by A.I. Uyemov on TDL and in the theory of categorical systems developed by the author. Categorical systems theory is an integral part of algebraic biology that predicts the properties of organisms based on the genome, including intellectual properties, which leads to artificial intelligence models. A wide range of systems approaches is formalized in TDL, so it is very important to find relationships between TDL systems and category systems theory. This paper proposes a version of the TDL, in which it is possible to present the definition of the system in the TDL in the required form, in which, in particular, the subjective interpretation of the symbol "arbitrary" in the LTO is excluded, which is unacceptable in classical logic, without which it is impossible to establish the relationship of this type of systems with the system block of algebraic biology.

Keywords: Categorical systems · Algebraic biology · Predicates · Properties · Relations · Artificial intelligence

1 Introduction

In the previous work "Ternary description language and categorical systems theory", reported at the conference, we began to develop a version of the ternary description language (TDL), in which a number of inaccuracies of the original TDL were corrected, which is necessary to clarify the general theory of systems proposed by A.I. Uyemov [1–8], and its relationship with the categorical systems theory that we are developing [9–12]. We note that the definitions introduced below agree with our definitions in that paper. In this work, we take the next step. A.I. Uyemov proceeds from the objectivity of systems: "The objectivity of connections between phenomena, their independence from our consciousness means the objectivity of systems. Material systems exist independently of human consciousness" [6]. For now, the recognition of objectivity for material systems is enough for us, although A.I. Uyemov also speaks about ideal systems. Let's draw an analogy with the concept of an atom. Atoms, like systems, exist objectively.

© The Author(s), under exclusive license to Springer Nature Switzerland AG 2023
Z. Hu et al. (Eds.): ICAILE 2023, LNDECT 180, pp. 261–270, 2023.
https://doi.org/10.1007/978-3-031-36115-9_25

The science of atoms consists in the search for true statements about atoms. We must not forget the fundamental role of the concept of truth for science. At first, we do not know the properties of atoms, but having discovered atoms with instruments, we first give a clear name, as G. Frege says, to these particles, then physicists study them, it may turn out that not all particles detected as atoms are atoms (for, example, they can sometimes be confused with molecules), this deficiency is corrected in the process of studying atoms. So, the first step in the study is to discover atoms and introduce the designation for these objects "atom", the further steps of building a theory are to search for true statements about atoms and their properties. The result is a more and more complete definition of the concept of an atom (which includes atoms of various elements, and so on), due to its inexhaustibility, like any real object, it is not possible to achieve a final definition covering all its properties. This is how a number of definitions of the atom arise: the Thomson atom, the Bohr atom, the atom of non-relativistic quantum mechanics, and the atom in quantum field theory... The search for true statements about the atom is not limited by direct experiments. As G. Frege writes: "The grounds that justify the recognition of some truth are often contained in other, already recognized truths.... Logic deals only with those grounds of the process of judgment which are truths. The process of judging, in which other truths are recognized as the grounds supporting it... is called the process of inference. There are laws related to this kind of confirmation, and the establishment of the laws of correct inference is the goal of logic" [14]. For reasoning, a language is needed, for the role of which natural language is not suitable, due to the fact that its purpose is much broader than the task of finding and substantiating the truth in the process of reasoning. Let's move on similarly (scientifically) to systems. Many researchers (in accordance with A.I. Uyemov's requirement for the objectivity of systems) discovered what they called systems. They have been given a clear name "system". Further, in the process of studying this object of reality, researchers, to the best of their abilities, put forward options for defining the concept of a system, at the moment when A.I. Uyemov gave his definition, several dozen definitions of the concept of a system accumulated, in particular, the concept of a functional system according to P.K. Anokhin, the formalization of which is carried out within the framework of categorical systems.

In his approach, however, A.I. Uyemov breaks with the natural scientific process of studying a phenomenon (system), in order to improve or replace the existing definitions with a more adequate and accurate one. Instead of studying the phenomenon, he took up the study of a set of definitions obtained by other researchers, proclaimed that it was necessary to look for something in common in their definitions, found in his opinion this something in common, and presented it as his definition of the system. This approach for the case of the atom would consist in revealing something in common (which undoubtedly exists) in the definitions of the atom by Thomson, Bohr, and others, and declaring this common definition to be more adequate to the real atom. The obvious absurdity of such an approach in the case of an atom, in our opinion, means the absurdity of such an approach in the case of any other existing object, including the system recognized by A.I. Uyemov as an objective phenomenon. It can only be argued that the definition of the system given by A.I. Uyemov, if it makes sense to consider it, is only in isolation from the method of its derivation.

For the study, inference, as G. Frege points out, and presentation of the results of study, in particular, systems, a formal language based on suitable logic is required. A.I. Uyemov does not believe that: "...Formalization based on the currently available logical apparatus can serve as the basis for a general systems theory." Moreover, in his opinion: "For the general theory of systems, a "mathematical suit" has not yet been prepared... There are no mathematical means adequate to the structural representations of systems of objects." The language of ternary description (TDL) proposed by him is such a "logical and mathematical apparatus adequate to the problems of general systems theory" [6]. Unfortunately, in many respects, the TDL is unfinished and, as indicated in [17], has incomplete and contradictory moments. We also emphasize the opinion expressed by P. Materna: "The tendency to build "non-classical" and "non-standard" logics is justified only when it is shown that "classical or "standard" logic cannot solve some logical problems [14]. I don't think anyone showed it. One of my main arguments is that the "standard" logic can easily cope with the solution to those problems, because of the solution on which the TDL was built". Unfortunately, P. Materna did not provide solutions using standard logic for those logical problems that A.I. Uyemov put forward, to justify the need to develop TDL.

A similar problem of standard logic A.I. Uyomov saw in the apparent impossibility of an unambiguous understanding of predicates from several variables. For this case, we further find (according to P. Materna) a solution to the problem of A.I. Uyemov by means of ordinary logic.

One of the main theses of A.I. Uyemov, by which he substantiates the need to construct a TDL, consists in the difficulties of ordinary logic indicated by him, in particular, which, in his opinion, exist in the definition of many-place predicates. Let us consider these problems of "different understanding of the simultaneous prediction of many things" [6]. He writes: "One or another property is attributed to a set of things $S_1,\ldots, S_n$ – the predicate P.... In what sense can the predicate P be attributed (produced) to a set of things? Various understandings of this connection are possible" [5]. It seems to us that the understanding of the formula $(S_1,\ldots, S_n)P$ (A.I. Uyemov writes the name of the predicate in the left entry) is unambiguously given by the definition, and in the text of A.I. Uyemov there are, in our opinion, various misunderstandings of the indicated formula. Since A.I. Uyemov operates with the term set, we will define of an n-place predicate from set theory. "Let A be an arbitrary set. An n-ary function f defined on A with values in the set $\{I, L\}$ is called an n-ary predicate on A. The set of those sequences $(a_1,\ldots, a_n)$ from A^n (the Cartesian product of n factors) for which is called an n-ary relation on A corresponding to the predicate refer to [16, 18]. Within the framework of axiomatic set theory (for example, NGB from [16]), this definition is unambiguous, and, we emphasize, has no alternative understandings, including those proposed by A.I. Uemov and, by the way, in which he sees logic problems. Consider the understanding offered by A.I. Uyemov. Obviously, it would not make sense to consider the same understanding of the predicate P, which is given in the above definition from mathematical logic, so one should expect some other understanding given by A.I. Uyemov.

"With one understanding, the inherentness of a predicate in a set means its inherent in each element of this set" [4]. A.I. Uyemov does not define the terms "assignment of a predicate", "intrinsicity of a predicate", unknown in mathematical logic. Let us assume that they are synonymous with "the presence of a relation (predicate)", but, here, the meaning of the "intrinsicity of its < predicate >... Element" (that is, a separate element), which is also not defined by A.I. Uyemov, is already completely unclear. Nevertheless, A.I. Uyemov gives a name to something unknown (according to G. Frege, this is already a way out of the logic in which each name has a single meaning in the form of a corresponding object). Apparently, A.I. Uyemov has an intuitive understanding of "inherence", but he did not find the opportunity to take the trouble to convey it to the reader. A.I. Uyemov proposes to guess what "inherence" is by considering an example: "For example, "all the large planets of the solar system move approximately in the plane of the ecliptic" [6]. Probably, the elements in this example correspond to the "major planets of the solar system" from the list: Mercury, Venus, Earth, Mars,..., Neptune, and the predicate P is = "Mercury, Venus, Earth, Mars,..., Neptune moving approximately in the plane of the ecliptic. In this example, if you enter a unary predicate (property) = "___ moves approximately in the plane of the ecliptic", the original predicate is the same as the predicate.

This formula clearly expresses the intuitive feeling of A.I. Uyemov. Thus, instead of the expected other understanding of the predicate P promised by A.I. Uyemov, we see a clear misunderstanding of the original predicate P, for some reason reduced to a very special case, expressed by the last formula. The predicate P does not satisfy the understanding given by A.I. Uyemov, the simplest example is the case for which there is no decomposition into unary predicates.

Let's move on to the following understanding of the predicate P, proposed by A.I. Uyemov: "In a different understanding, the predicate denotes a property inherent in the aggregates of most elements of the system, in a particular case – all elements taken together." A.I. Uyemov again does not define what a "property inherent in the aggregates of most elements" is, he thinks, for example, that the reader understands the word "majority" in his text... The work of searching for the meaning of A.I. Uyemov's text is again shifted by him on the reader who is given the example "the forest burned down" as a clue. An example that is unsuccessful, since it is easy to carry out a construction similar to the first (mis)understanding of the formula for the predicate: we introduce the predicate = "___ burned down", here, obviously, trees (bushes, etc.), the original predicate P is now represented by conjunction and negations (for unburned trees). From these conjunctions it is possible to assemble predicates on the Boolean of the forest, most likely the representation of the original predicate P by such predicates corresponds to the intuition of A.I. Uyemov in this second understanding of the predicate P.

Further, we read: "In a certain respect, the reverse <to the previous one> will be such an understanding of predication, when the predicate P is inherent in at least some of the elements of the system, for example, in the sentence "the ancient Greeks were outstanding philosophers" [6].... <previous two understandings> are not differentiated, although the difference between them is significant.

Of course, there is an ambiguity in the intuitive part of "some of the elements of the system", one should rely on the boolean.

We cannot find the meaning of the phrase "reverse < to the previous >", considered in the book [6] and [14], inverse relations make sense for binary relations, but not for arbitrary...

A direct contradiction in the phrase "they are not differentiated, although the difference between them is significant" directs A.I. Uyemov to the already mentioned term "majority" to save the text, but we will not waste time on correcting the text.

The non-differentiability of the third understanding from the second gives the same representation of the initial predicate P as in the previous understanding, so here we do not have another understanding of the predicate promised by A.I. Uyemov. The fourth understanding is described by the words: "Finally, such an understanding of predication is possible, in which the predicate characterizes the entire set by itself (as one whole)... In this case, the predicate of the system is not a predicate of a simple collection of all its elements... In some cases, such a predicate (let's call it PIV) cannot be decomposed into a set of predicates $<P_i, i = 1,2,...,n>$ related to individual elements of the system..." [6]. The fourth understanding, apparently, despite the mysterious words "the whole set in itself" (A.I. Uyemov does not notice that the essence is one element of the Cartesian product, for example, one point of three-dimensional space, and not three points x, y, z...) coincides with the definition of a predicate from mathematical logic. We also note the direct error of A.I. Uyemov "In some cases, all Pk are equal... to the predicate of the entire system" [6] (a function of several variables that explicitly depends on them cannot be equal to the function one of these variables).

A.I. Uyemov did not offer any other understanding of the predicate P (except for the one given in the definition from mathematical logic), that is, the "problems of logic" declared here by A.I. Uyemov (in particular, in the form of "differences in predication types") are absent.

There are several dozen definitions of the concept of a system, the deepest of which is the functional system.

These approaches largely fall under the language of ternary description (TDL) developed for systems [1–8], where the definition of functional systems was also considered. However, this consideration turns out to be incomplete, since the role of the system-forming factor, which also underlies categorical systems, remains unclear. This article continues the development of the TDL version, in which it is possible to present the definition of the system in TDL in a form in which it is possible to establish the relationship of this type of system with the system block of algebraic biology.

2 Ternary Description Language Version with Explicit Interpretation

The A.I. Uyemov introduces two symbols t and a into TDL, in fact, without an exact indication of their meanings, in violation of the key principle according to G. Frege "there should be no meaningless meaningless expressions in science" [18]. Indeed, he writes: "... The elementary cell of the formal apparatus of general systems theory... Consists of two "objects" – definite and indefinite (t, a)..., "t" and "a" are just things." [6]. In TDL words in which the symbols t participate, and their interpretation as a thing, relation, or property is introduced depending on the place occupied by the symbols in the word, so $(t)a$ is interpreted in natural language as the phrase "a certain thing has an indefinite property". A.I. Uyemov has nothing else in mind, both in the quoted book and in other works, using these symbols. In mathematical logic, such statements are not

266   G. K. Tolokonnikov

considered, since the phrase "certain thing" does not have any meaning, and the phrase "indefinite property" does not define any predicate. We propose the following correction of the contradictory situation in the TDL regarding "t" and "a".

We will consider the roots of the words of a certain fragment of a natural language and denote their totality by R. In the fragment under consideration, we are primarily interested in the parts of speech noun, adjective (denoting a property of an object and answering the questions "what?", "what?", "what?", "what?", "whose?" and so on) and a verb. In a language fragment, we usually consider only declarative sentences, simple declarative sentences consisting of a subject (which is a noun), a predicate (which is a verb), and an object (which is an adjective), and complex sentences (consisting of several simple sentences, connected by unions, complex and other sentences). The main members of the sentence are the subject and the predicate; the secondary members of the sentence include the addition, usually in the form of an adjective. The root is associated with the meaning (lexical meaning) of the word, just like the thought, we do not define the concept of meaning, therefore we will restrict ourselves to a typical description: the root is a morpheme that carries the lexical meaning of the word, in Russian the root is present in all independent parts of speech and is absent in many service parts of speech (for example, it is not in the union "and", the interjection "ah"). Compound words have multiple roots.

Words can be formed from the root (point to words with the same root), which can be the subject in the sentence, the resulting word will define a certain concept, for definiteness we will understand it according to G. Frege within the framework of his universe. This concept has a set or class of things that fall under it. From the root, you can also form a verb and an adjective (more precisely, in the fragment of the language under consideration, we take into account only such word roots), which will serve as a predicate and an object in the sentence. The limitations of the fragment of the language under consideration include the requirement that the matching of the root is unambiguous: the root corresponds to only one noun $noun(r)$, one $adj(r)$ adjective and one $verb(r)$. Thus, there is a set of things $V = noun(R)$ and a set of properties $S = adj(R)$. We will call a statement or judgment such declarative sentences for which it is possible to raise the question of their truth or falsity. For example, the phrase "a certain thing has an indefinite property" considered in the indicated fragment of the language is a sentence for which the truth value does not directly exist. However, for such sentences, we will offer an interpretation in which truth-values exist.

By the symbol t we will mean a variable running through the set of roots of words R. Other possible variables will be denoted using primes: $t, t', t''...$ – different variables on the set of roots.

Denote by $a, a', a'', \ldots$ The variables running through the set of variables $\{t, t'\, t''\ldots\}$, if a is one indefinite root, then a' is another indefinite root. This is our understanding of the meaning of the signs "t" and "a". So we have two kinds of variables. At the same time, there is a connection between the varieties in the form of interpretation of elements of the second kind ($a, a', a'', \ldots$) by elements of the first kind, which themselves are interpreted by the roots of words.

A Version of the Ternary Description Language   267

A.I. Uyemov, in addition to "*t*" and "*a*", introduces the symbols of brackets), (,], [, an asterisk *, the symbol *i*. Some words in the alphabet from the letters t, a,), (,], [, * it compares declarative sentences, which we call T

DL-statements, and phrases that can play the role of a subject in a sentence (the latter are obtained if the TDL-statement is enclosed in square brackets), natural language:

(*t*)*a* – "a certain thing has an indefinite property";
(*t*)*t* – "a certain thing has a certain property";
(*a*)*t* – "an indefinite thing has a certain property".
(*t**)*a* – "an indefinite property is attributed to a certain thing";
a(**t*) – "an indefinite relation is attributed to a certain thing";
[(*a*)*t*] – "an indefinite thing with a certain property";

and so on. When comparing "*t*" and "*a*", the words thing, property, and relation are replaced depending on the place in the formula (if the symbol is to the right of the symbol enclosed in parentheses, then the word property is selected, if on the left, then the relation, so that inside the parentheses, the word thing is selected).

In our version of TDL, which we will call TDLI (TDL with interpretation), two-sorted predicate letters are used to interpret LTO statements.

Let us compare in our version of TDL, that is, in TDLI, the two-sorted predicate P for (*t*)*a*:

$P(t, a) = $ "the definite thing $noun(t)$ has the indefinite property $adj(a)$" $ = $ "$noun(t)$ is $adj(a)$".

To evaluate the predicate, we select a specific pair (v, s), v is a fixed thing, for example, an explicitly specified "hand", s is a fixed property, for example, "red of a given redness level". We have fixed hand $v \leftrightarrow$ class "hands" $\leftrightarrow$ root $t = $ "hands", fixed level of red $\leftrightarrow$ property class "red" $\leftrightarrow$ root $t' = $ "red" $\leftrightarrow$ variable $a = t'$:

$P(\text{"hand"}, t') = $ "the selected particular hand is red of the selected particular level of redness".

If the selected hand is red of the selected redness level, then the predicate is 1, if not, then it is 0.

Predicates for $(\xi^*)\eta$ and $\eta(*\xi)$, where ξ, η are t or a, are introduced and evaluated similarly.

Let us consider an analogue in TDLI of the operation of enclosing TDL-statements in square brackets using the case [(*a*)*t*] as an example.

An analogue would be the introduction of the operation μ of the translation of the judgment $P(t, a) = $ "$noun(t)$ is $adj(a)$" into the root $\mu P(t, a) = \mu Pta = $ "$noun(t)$, which is $adj(a)$", corresponding to given thing $noun(t)$ having the property $adj(a)$. Here, linguistically, at each evaluation, we have two roots with the second subordinate as an adjective to the first root. For now, we will proceed formally, similarly to how a free algebra is built on a fixed set of generators and under some operation (the application of an operation to generators is declared a new element). For new roots, we leave the requirement of unambiguous mappings $noun()$, $adj()$, and $verb()$.

Let's move on to the use of the concept of "arbitrary" in TDL, in addition to the concepts of "definite" (t) and "indefinite" (a). A.I. Uyemov in TDL for the concept of "arbitrary" introduces the symbol A into the alphabet.

268 G. K. Tolokonnikov

The definition of the system given by him has the form ($\equiv$ means equal by definition, the letter i indicates that the right and left of the identity under the letter A means the same arbitrary object).

$$(iA)S \equiv ([a(*iA)])t, \text{ or } t([iA^*]a). \tag{1}$$

"Object A can be understood as "any thing". But this any thing is essentially different from a. Object a is an indefinite thing, object A is an arbitrary thing. As A, we can take any thing we want. Otherwise, we take any thing that comes across. This difference is well illustrated... by the choice of brides in a Russian fairy tale. If the brides lined up in front of the prince and he chose any one, then such a situation would be symbolized with the help of A. It is a different matter when he shoots an arrow and must marry any one that comes across.... The difference between A and a is not expressed in the syntax of the predicate calculus" [6]. Here A.I. Uyemov is mistaken, this difference can be easily expressed using limited quantifiers. Let's bring this expression and we will use it in TDL, our corrected version of TDL.

Recall what bounded quantifiers are (see, for example, [15]) and use them for the notion of "arbitrary".

Let the predicate $R(x)$ be given on the set that the variables x run through; it selects a certain subset. If we want to consider variables only on this subset, then bounded quantifiers are used, and properties are satisfied for any predicate $P(x)$.

A.I. Uyemov in TDL does not speak about the usual choice from a set of elements, but about the choice of elements that he (or the prince, or someone else) wants to choose. He may like several elements (the prince may like several brides) and from them he will already make the usual choice. Thus, for him in the set of elements, there is a subset of elements he likes, which is described by the predicate $R() = $ "I liked this ___ element". Thus reasoning, variables, quantifiers are narrowed to the specified subset, but narrowed variables, formulas, and restricted quantifiers can be applied on the narrowed subset, this is a well-known section of the predicate calculus.

So, let the predicate $R()$ be given, then in our TDL, for example, $(t)A$ has the following analogue:

$$P_R(t, a_R) = \ll noun(t) \text{is } adj(a_R) \gg, A = a_R \tag{2}$$

Here, however, we are faced with a very serious question: do the predicates R corresponding in the choice of brides, for example, to A.I. Uyemov or to some prince differ? The answer is obviously yes. As you know, "like" is an extremely "subjective" verb. In TDL, subjective judgments and concepts are clearly introduced and developed in this way, which leads, in matters of truth, to the dependence of conclusions on one or another person and his feelings. As a result, introducing the symbol A, which is clearly associated with subjective sensations, A.I. Uyemov violates the most important principle of science in TDL, justified by G. Frege and, in particular, on the basis of which he built the predicate calculus, which presented classical logic in its modern form. The choice from what is desired (likes and the like) reflects the subjective idea of the chooser about the chosen object (thing).

"The goal toward which science is striving is truth.... What is true is true regardless of our recognition (p. 287).... The states of their inner world experienced by different people

A Version of the Ternary Description Language   269

cannot be combined in one consciousness and, thanks to this, compared (p. 288)....
From the meaning and meaning of a sign, the representation associated with it should
be distinguished. If the meaning of a sign is a sensually perceived object, then my idea
of it is an internal image... The idea is subjective: the idea of one person is not the
idea of another... The idea differs significantly from the meaning of the sign in that the
meaning of the sign can be the common property of many people... (p. 232).... In order
to more clearly present all the originality of our predicate true, let us compare it with the
predicate beautiful.... The beautiful has gradations, but the truth does not... The truth
does not depend on our recognition of its truth, and the beautiful is only beautiful for the
one who perceives it as such. What's great for one isn't necessarily great for another....
How can one make the Negro in Central Africa give up the view that the narrow noses
of Europeans are ugly, while the wide Negro noses, on the contrary, are beautiful?...
Even if it were possible to give a definition of a normal person and, thereby, objectively
define the beautiful, it would still happen on the basis of subjective beauty..."(p. 311 in
[13]).

A.I. Uyemov, introducing the symbol A, associated with the concept of "arbitrary"
he defines, tries to introduce elements of subjective representations into the logic of the
DTL created by him, which, being different for different people, do not allow to attach
meaning to the symbol A and, thereby, to determine the above $R()$ predicate for bounded
variables and language quantifiers.

Some way out, which we accept for the development of the DTL, can be the construction of the predicate $R()$, based on the version indicated by G. Frege with the definition
of a normal person. It seems possible to develop a set of properties for such a definition,
while still changing the idea of a normal person over time, if one pays attention, for
example, to the solution in intuitionistic logic that is available in Kripke's semantics.

3 Conclusion

There are numerous systems theories, in particular, the general theory of systems by A.I.
Uyemov, based on TDL and covering several dozens of system approaches. Categorical
systems theory is an integral part of algebraic biology that predicts the properties of
organisms based on the genome, including intellectual properties, which leads to models
of artificial intelligence (AI) related to strong AI, in contrast to approaches based on
neural networks [16–19]. In this paper, a version of the TDL fragment is proposed, in
which the definition of the system given by A.I. Uyemov was successfully presented
in the form necessary for comparison with the system block of algebraic biology. In an
important special case of interpretation of many-place predicates, Materna's idea that
the problems of logic put forward by A.I. Uyemov to justify the need to construct a TDL
can be solved within the framework of ordinary logic is confirmed.

References

1. Uyemov, A.: The Language of ternary description as a deviant logic: Part 1. Boletim da
   Sociedade Paranaense de Matematica **15**(1), 25–35 (1995)

2. Uyemov, A.: The Language of ternary description as a deviant logic: Part 2. Boletim da Sociedade Paranaense de Matematica. **17**(1), 71–81 (1997)
3. Uyemov, A.: The Language of ternary description as a deviant logic: Part 3. Boletim da Sociedade Paranaense de Matematica. **18**(1), 173–190 (1998)
4. Uyemov, A.I.: Some issues of the development of modern logic. Sci. Notes Tauride Natl. Univ. **21**(1), 89–96 (2008)
5. Uyemov, A.I.: Things, properties and relations, M., Publishing house of the Academy of Sciences of the USSR (1963). (in Russian)
6. Uyemov, A.I.: System approach and general theory of systems, M., Thought (1978). (in Russian)
7. Uyemov, A.I., Saraeva, I.N., Tsofnas, A.Yu.: General systems theory for the humanities, W., Wydawnictwo "Uniwersitas Rediviva" (2001). (in Russian)
8. Uyemov, A.I.: Some questions of the development of modern logic. Sci. Notes Taurida Natl. Univ. **21**(1), 45–58 (2008)
9. Tolokonnikov, G.K.: Informal category systems theory. Biomachsystems **2**(4), 7–58 (2018)
10. Tolokonnikov, G.K.: Categorical gluings, categorical systems and their applications in algebraic biology. Biomachsystems **5**(1), 148–235 (2021)
11. Tolokonnikov, G.K., Petoukhov, S.V.: From algebraic biology to artificial intelligence. Adv. Intell. Syst. Comput. **1126**, 86–95 (2020)
12. Tolokonnikov, G.K.: Convolution polycategories and categorical splices for modeling neural networks. Adv. Intell. Syst. Comput. **938**, 259–267 (2020)
13. Frege, G.: Logic and logical semantics, M., Aspkt Press (2000). (in Russian)
14. Leonenko, L.L.: Definitions in the language of ternary description. Sci. Notes Taurida Natl. Univ. **24**(3), 397–404 (2011)
15. Leonenko, L.L., Tsofnas, A.Y.: On the adequacy of logical analysis to philosophical reasoning. Questions Philos. **5**, 85–98 (2004)
16. Karande, A.M., Kalbande, D.R.: Weight assignment algorithms for designing fully connected neural network. IJISA **10**(6), 68–76 (2018)
17. Dharmajee Rao, D.T.V., Ramana, K.V.: Winograd's inequality: effectiveness for efficient training of deep neural networks. Int. J. Intell. Syst. Appl. **10**(6), 49–58 (2018)
18. Hu, Z., Tereykovskiy, I.A., Tereykovska, L.O., Pogorelov, V.V.: Determination of structural parameters of multilayer perceptron designed to estimate parameters of technical systems. IJISA **9**(10), 57–62 (2017)
19. Awadalla, H.A.: Spiking neural network and bull genetic algorithm for active vibration control. IJISA **10**(2), 17–26 (2018)

A Hybrid Centralized-Peer Authentication System Inspired by Block-Chain

Wasim Anabtawi, Ahmad Maqboul, and M. M. Othman Othman[✉]

Computer Science Department, An-Najah National University, Nablus, Palestine
othman.omm@najah.edu

Abstract. The process of identifying and authenticating people is considered to be a crucial task nowadays. However, this ease of use comes with its own problems; like identity theft and identity fraud. Which, might be exploited by means like; the collusion of a corrupt authenticating organization, or internal fraud within the authenticating organization, or by a manipulative majority in case of multiple-authenticators. The proposed system will be an attempt to deflect such cases by establishing an unbiased organizational authentication system.

This work illustrates a design of a hybrid authenticating system that does not have a single actor that carries out the whole authentication process. Instead, it will borrow the concept of authenticating peers from the block-chain technology and utilize it with the architecture of centralized authenticating system. Security analysis of the proposed system shows its effectiveness against various types of attacks and scenarios.

Keywords: Authentication · Peer-Authentication · Centralized-Authentication · Block-Chain · Trust · Digital certificates

1 Introduction

The use of the Internet and computers in the past years had eased the way transactions occur. A vast majority of these transactions include exchanging high valued assets, which are done over the Internet; according to [1]. And thus, the security of this process is considered to be a vital enabler for modern transactions.

It is well-known that the Internet from security prospective has two main features; obscurity & openness [2, 3]. Which, provides an attractive circumstances for malicious users to commit frauds; by spoofing data or identity (Identity Theft). Recently there has been a continuous increase in the number of such frauds and identity thefts, as can be seen by the "Consumer Sentinel Network Data Book 2020″ issued by the United States' Federal Trade Commission (FTC) [4]. Which, tracks consumer fraud and identity theft complaints. It showed 45% increase over 2019 in fraud cases, which is correlated to the 113% increase in the identity theft cases. It is known that improper measures or improper protection of privacy by governments or organizations can be used directly or indirectly by malicious users in order to carry out identity thefts' attacks. This can be more serious issue if we consider corruption in addition to the improper measures. According to

© The Author(s), under exclusive license to Springer Nature Switzerland AG 2023
Z. Hu et al. (Eds.): ICAILE 2023, LNDECT 180, pp. 271–287, 2023.
https://doi.org/10.1007/978-3-031-36115-9_26

the"Transparency International" in their"Corruption Perceptions Index 2020" [5]; more than two-thirds of the 180 scored countries scored 50/100 on the Corruption Perceptions Index (CPI), and the average score is 43/100.

Given the previously mentioned reasons; the users/systems of the Internet will be left without enough guarantee on the honesty of the identity of the users. That's why user identification and authentication have become an essential technology.

User identification and authentication systems originally began as centralized systems; with digital certificates and keys being provided by centralized organization systems. In such systems of identification and authentication; digital certificates are used for users and systems in order to prove their identity. For example, [3] had declared that the whole purpose of an identity authentication system is to generate a digital certificate, which will be a proof for the identity of the subject. This has provided a type of protection to the user's identity to some degree. However, systems faces various shortcomings. One of them is not guaranteeing the integrity or fairness of the centralized organizations.

The users in the mentioned centralized systems rely solely on authority of these systems in order to carry out transactions. This means that the central authority has the ability to alter any data; because no distribution of authority is present. In addition to that, the centralized systems are bound to be the single point of failure.

For the previously mentioned issues, it became more evident that there is a need for a decentralized system. One of the efforts to was the introduction of the block-chain technologies [6, 7]. Blok-chain is a decentralized system that distributes authority between the users of the system. Making it suitable for applications such as"Distributed Ledger Management for an Organization using Blockchains" [8]. Block-chain also implements the concept of"cooperative maintenance", which makes all the nodes in the system share the same information regarding the identities of users and their authentication as mentioned by [1]. In addition, block-chain have an immutable database infrastructure; i.e. if any record in the block-chain is tampered with, the rest of the peer nodes would cancel the transaction related to that block-chain.

Even though block-chain based systems provide a secure, transparent, un-breakable records of data according to [9], but it still has its disadvantages. The main disadvantage being the need of high energy consumption and big computational load in order to keep track of real-time ledger as mentioned by [10].

Another disadvantage of block-chain according to [12], that it's vulnerable to the 51% attack; in which a group of miners exist where they control more than 50% of the network's mining computing and hashing power. Which gives them the ability to prevent new transactions from being confirmed, and the ability to manipulate the block-chain records, and reverse transactions.

Many efforts have been put to make use of the decentralized Block-Chain based authentication systems. Lin, He, and Huang et al. [13] made a secure mutual authentication system based on block-chain, in order to execute an access control policy. Cui, XUE, and Zhang et al. [14] Proposed a hybrid Block-Chain based identity authentication system regarding multi-wireless sensor networks, in which local block-chain was used to authenticate the identity of the nodes in the system. In 2017 [15] proposed an identity authentication system based on Bitcoin in block-chain, which suggested the identity management system to be built on the Bitcoin block-chain. Sen, Mukhopadhyay, et al.

[16] in 2021 proposed a Blockchain based Framework for Property Registration System. In which, the system relies on multiple endorsing peers that are different governmental and organizational authorities.

As for the previously mentioned decentralized block-chain based systems, even though there is no single point of failure. However, those systems are vulnerable to the 51% attack, in addition to the burden of high computation requirements.

The rest of this paper is organized as follows. Section 2 explains the motivation and goals sought in this work. Section 3 shows the detailed design of the proposed system. After that, Sect. 4 sheds more light about some special cases that might accrue during the use of the proposed system, and shows how the proposes system remedies those cases. Section 5 shows the evaluation of the proposed system in terms of applicability; by evaluating generated data, and the needed cryptographic operation. Furthermore, evaluation in terms of security is done in Sect. 6. Finally, Sect. 7 concludes the work done in this paper.

2 Motivation and Goals

This paper proposes a user authentication system that borrows the concept of block-chain, while keeping the central governmental organization updated on what's going on for legal purposes. The identification system is used to prove a Subject's identity, without needing a certain central side that controls everything. The proposed system consists of three main actors, these actors all depend on each other. The first actor being the Subject asking for a certificate to prove their identity. Second actor will be a central government which will verify the Subject from its own side. Third actor will be a certain number of peers assigned to Subject which will verify the Subject's identity for the Subject. In addition to the use of digital certificates, the distribution of authority mentioned in the paper will be an assurance that there won't be any modifications from any side during a transaction. Moreover, the proposed system provides immunity to identity theft and identity fraud. And, preventing any exploits by means like; the collusion of a corrupt authenticating organization, or internal fraud within the authenticating organization, or by a manipulative majority in case of multiple-authenticators.

3 System Architecture and Design

The proposed system design is based on a mobile application, which when the user asks for a certificate to be verified the application will produce an authenticated certificate having the user's information. Which will be used for transactions and identity proof.

3.1 User Sign up

For users to be added in the system, they need to sign up via the device, or they can go to a governmental organization which uses the proposed system. Subjects which are going to be identified will give their national ID, Full name, Date of birth, Address, Mobile number, Username, and password which will be used to log in the mobile application.

Adding users from a governmental side allows the government to do background checks and check for the preciseness of the given information, the user's information will be added into the database of the central government and the proposed system as shown in Fig. 1, even though the users are added in the database, they won't have the ability to request certificates until they're validated from the governmental organization.

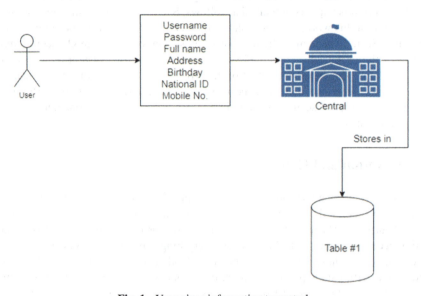

Fig. 1. User gives information to central

3.2 Initial User Validation

When a new user logs in the application using their device, the application will generate a public and a private key to the Subject wanting to identified, which will both be stored in the Subject's device. For an Subject to be validated in order to ask for certificates and participate in the system, after logging in the application, they will be asked to provide a password different from the one used to login the system, in addition to a bio-metric identifier, which will either be a finger ID, or a face ID. If the users don't have a device with a touch ID, or a face ID scanner, they'll go to a governmental center to provide them. After the user provides these private information, they will be hashed, and encrypted with their private key, and sent to the central governmental side. The user's public key will be sent with these information as well, in which they're saved in a table#2 containing them. Afterwards the central side will send a file to the user's device containing the following:

1) Information of the peers that will verify the user.
2) Public keys of the peers that will verify the user.

3) Digital Signature #1 from the central side.

Digital Signature #1 will contain:

1) Plain text which has ID of peers that will verify the users & Timestamp of when the peers got sent to the user.
2) Encrypted hash of the plain text using the central side private key.
3) Public key of the central side.

This process is all illustrated in Fig. 2, and the contents of digital signature #1 are in Fig. 3.

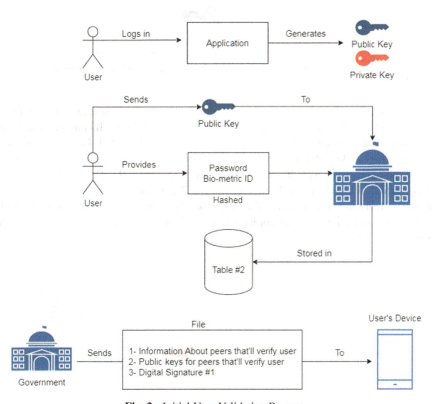

Fig. 2. Initial User Validation Process.

When this process is finished, the user will exchange public keys with the central side successfully, the user is validated, he/she will be able to participate in the system, ask for certificates, and validate other peers.

3.3 Certificate from Central Server

For users to obtain the authenticated certificate that verifies their identity, they firstly need to request a certificate from the central governmental organization, they could

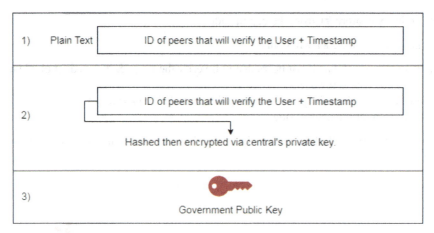

Fig. 3. 1) and 2) are the contents of digital signature #1 from the central

either go to one of the governmental organizations located in their area, after that they'll request the certificate in which they should give their bio-metric ID and the transaction password. The biometric ID and password will both get hashed then encrypted with the user's private key and sent to the central server, the central server will then use the user's public key to decrypt the password and the bio-metric ID. These information will be checked and compared with the data in table #2, if they match the user will have a certificate from the central server side and it will be stored on the user's device. And if they don't match, the user will not have the certificate. All of this process is illustrated in Fig. 4.

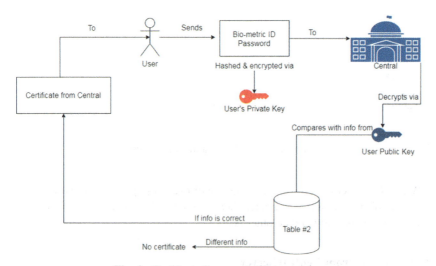

Fig. 4. Certificate from central side process.

3.4 Live Message #1

One of the essential parts in the design of the proposed system are the live messages, because it's the way the user communicates with his/her generated peers, and the central. Side. Live Message #1 is sent from the user to the central side periodically each day, in order to check who will user verify while participating in the system. The message will contain the user's ID, and after sending that message to the central side, the central side checks in table #3 for peers that are generated to the user. If there are peers that get verified by the user, then their public key, and ID are both sent to the user from the central side. This process is illustrated in Fig. 5. The purpose of this message is to generate peers that need verification from the user, which are known as the peers that will the user verify. When this process is done, the user will have both peers that verify him/her, and peers that needs verification by him/her.

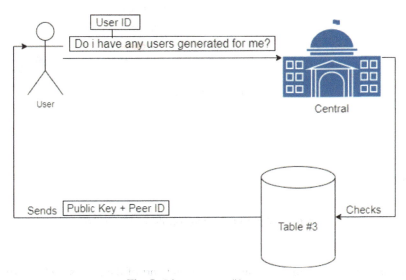

Fig. 5. Live message #1 process.

3.5 Certificate from peers & Live message #2

This subsection has three parts, in which the process of getting the certificate from peers is explained, all three parts depend on each other, and for a user to get the certificate from the peers, he/she need to go through all three parts without any complications.

1) *First Part of the process:* The first part consists of the process of obtaining the authenticated certificate to get a certificate from peers, in which the user requests the certificate from the peers through the central side. The user starts by requesting a certificate from peers through the central side, in which the user firstly sends a digital signature that got sent to the user in the initial validation process. In addition to a message asking for a certificate, which is hashed and encrypted using the private key. Afterwards

the central side decrypts the signature and checks it, after checking for its authenticity, the central adds the ID of the user that wants the certificate, his/her peers, and the message the user sent into table #4. The message will contain a string with the following text: "I'm User A, and i need certificate from peers side".

The first half is illustrated in Fig. 6. The use of digital signature #1 in this process ensures that user really exists, and the use of the encrypted message is to request users without any complications.

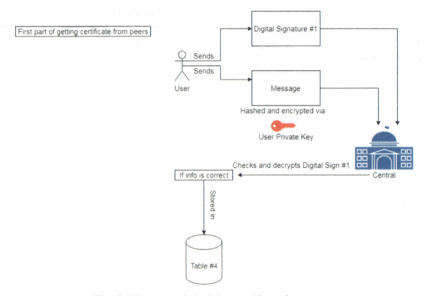

Fig. 6. First part of obtaining certificate from peers.

2) *Second part of the process:* This is the second part of the process, which will talk about live message #2 and its connection to getting the certificate from the peers. Live Message #2 is a message that's sent periodically over the day from peers, the purpose of this message is to ask the central side if there are any users that need verification. At this stage, the peers have users to verify, but these users didn't ask for certificate, which makes it different than Live Message #1. When the central side receives the messages from the peers, the central side checks in table #4, when it finds users asking for verification in order to get certificate from peers, the central side will send the hashed & encrypted message from table #4 with the ID of the user asking to be verified to the peers in order to verify him/her. The application will check the files of the peers, and compare it to the received ID, if they match then the peers will decrypt the message using the user's public key in their file. Afterwards the peers will send a message hashed & encrypted with their private keys, in addition to a timestamp to the central side. The central side will store the ID of the user who asked for a certificate, the ID of the peers that will verify him/her, and the confirmation message from the peers in table #5. This half is all illustrated in Fig. 7. The contents of the message that

the peers encrypt, hash and send to the central will consist of the following text:"The user who asked for peers certificate is verified by (Peer ID)".

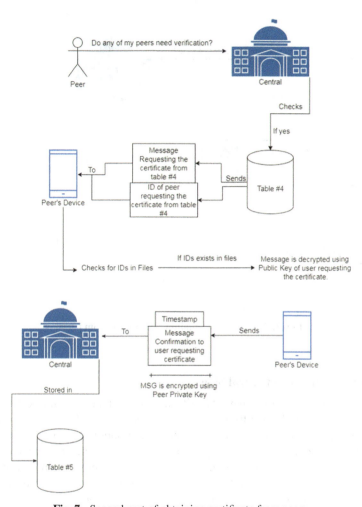

Fig. 7. Second part of obtaining certificate from peers.

3) *Third Part of the process (Live Message #3):* This sub-subsection is the third and final part of getting the certificate from the peers side. Live Message #3 is used by users in order to check if any peer had sent them any certificate. This message is sent periodically over the day to the central governmental side, in which the user asks the government to check for any new certificates. After the user asks the central to check for messages sent to him/her, the central will check table #5 mentioned before, and search it for messages. When the central takes the message from table #5, the message is firstly encrypted with the peer's private key, then the central side private

key. Afterwards the message is sent to the user who needs the certificate from peers, the user decrypts it using the central public key then the peers' public keys. This process is illustrated in Fig. 8.

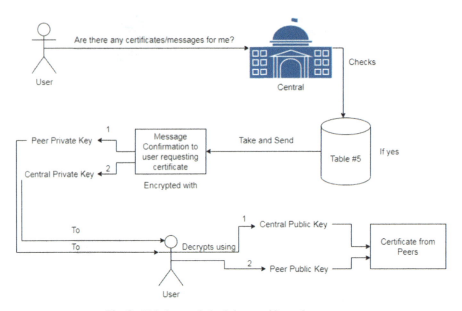

Fig. 8. Third part of obtaining certificate from peers.

3.6 Confirmation with Central Side

In this section, the user should have 2 certificates, the first one from the central side, and the second one being the one from the peers that verified him/her. After obtaining the certificates, the user will receive an automated confirmation message, in which it asks him/her to confirm the certificates. Upon hitting confirm in the message, an encrypted message will be sent to the central side which contains the user's info, his/her certificates, and the timestamp of receiving them. The central side will then decrypt it using the user's public key, afterwards the central side will save the message, the ID of the user, the beginning time of the certificate, and the end time of the certificate in table #6. Table #6 is used as a history for certificates, and can be checked in case of any possible complications. All of this process is illustrated in Fig. 9. Lastly a final message is sent to the user via the central side, that message will be later converted into an authenticated certificate which is explained in the next section, and the contents of the message are shown in Fig. 10.

3.7 Creation of the Authenticated Certificate

This is the final process of obtaining the authenticated certificate, which contains:

1) Certificate from the central side.

A Hybrid Centralized-Peer Authentication System 281

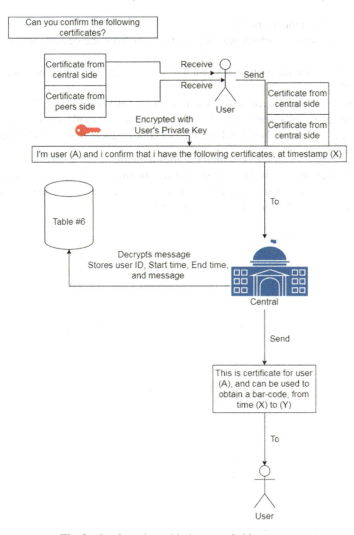

Fig. 9. Confirmation with the central side process.

Fig. 10. Contents of the final message the user gets, which will be used in the making of the authenticated certificate.

2) Certificate from the peers side.
3) Digital Signature #1 which was sent to the user from the central side, as shown in Fig. 3.
4) The confirmation message that's sent to user from table #5 via central side as shown in Fig. 8.
5) The final confirmation message from central side as shown in Fig. 9.
6) The message that's sent from the user to the central side upon confirming the certificates, which is shown in Fig. 9 in the confirmation with central process.

Those elements will all be converted into an authenticated certificate, which will have a life span, in which the user should use before it expires. The contents of the authenticated certificate are illustrated in Fig. 11, Each number represents the element corresponding to the list mentioned in this section..

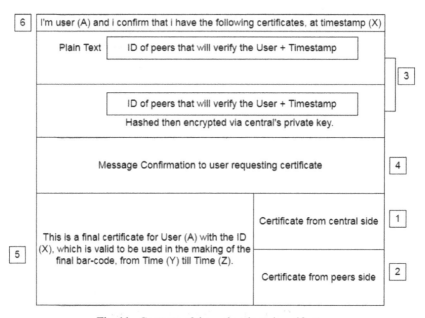

Fig. 11. Contents of the authenticated certificate.

4 Scenarios

Different scenarios occur while users are using the system, this section will mention and discuss some of them with their solution in the proposed system.

4.1 No Certificate from the Central Side

When the Subject wanting to be verified requests a certificate from the central governmental side, and the governmental side doesn't send them one, the information the

A Hybrid Centralized-Peer Authentication System   283

Subject provided could be wrong, or simply the central side doesn't want to provide them with a certificate for any reason. That's why governmental signatures exist, these signatures are generated to the user's device as soon as they install the application on his device.

These signatures are a definite proof that the user has peers and could ask for certificates from the central side and the peers' side. The user can check what's wrong with the certificate process from central side by showing them the signature on his device. The user might take legal action if the governmental side doesn't want to give him/her the certificate for no reason. If the user is not authorized to do transactions because of his suspicious background, they can simply do a background check before he's added to the system, which won't allow him/her to ask for certificates and get an authenticated certificate in the first place.

4.2 Peers Might be Offline

When users ask for a certificate from peers side and the users might be offline, that's why the peers send different types of messages during the period of the day. So when a user requests a certificate from the peers and they don't respond for a while or don't respond at all, the user will check what's going on by asking the central side. Because the user has their information, and the governmental signature that proves that the peers in fact exist and should send him the message back in order to get the certificate..

5   Evaluation and Discussion

The evaluation process for the proposed system began by testing multiple scenarios using two machines, the first machine which was a mac-book pro 2013 using the mac-OS Catalina V 10.15.7, with 16GB of RAM and 2.7 GHz Quad- Core Intel Core i7 Processor. The Mac-book pro was used as the hosting server for the scenarios. The second machine used was a Dell Latitude E7250, with a 2.3 GHz Core i5 processor, with 8GB of ram, using the Windows 10 64-bit OS. The Dell laptop was used to log in to the user and ask for certificates from peers generated to him/he.

5.1   Message Size Before and After the Execution of Some Functions

The second test was held by comparing the sizes of the messages when certain functions get executed to show the effect of the functions and services used in the proposed system. The functions tested were the following:

1) Validation function
2) Get Certificate function
3) Live MSG#2
4) Get Peer Certificate

The results with their differences are shown in Table 1. The results show the differences between the messages before and after the executions of functions, these results show the impact of using certain applications and services during the run of the proposed system. Because we used Flask application for the web server application, its effect is clear on the size of the message in Table 1.

284 W. Anabtawi et al.

Table 1. Size Comparison

Function Name	MSG SIZE BEFORE	MSG SIZE AFTER	Difference
Validate	232	367	135
Get Certificate	232	1214	982
Live MSG #2	232	28	-204
Get Peer Certificate	232	1696	1464

5.2 Number of Hashing Functions and Encryption – Decryption Functions

The final evaluation process was by counting the number of hashing functions and encryptions during the scenarios that require them. The results of this test are shown in Table 2.

The results in Table 2 show number of hashing functions and encryptions which shows how many are needed for each scenario. For example, the initial user validation process requires the following:

1) 2 Hashing functions (1 from the user side, 1 from the central side, & none from the peers' side).
2) 1 Encryption function from the central side only.
3) 1 Decryption function from the user side only.

Table 2 also shows the sum of all operations for each side, for example in order for the user to all operations, the following amount of functions is required:

1) 6 Hashing functions.
2) 5 Encryption functions.
3) 6 Decryption functions.

These results can be used to study the effect of the hashing algorithms used in the proposed system and how many are required.

6 System Security Analysis

6.1 Identity Fraud

If a user's device is stolen and the person that stole the device requests a certificate to get an authenticated certificate and do transactions based on a stolen identity, the proposed system requires him to log in using the username and password of the user, and if he knows them, he would still need the transaction password and the bio-metric ID in order to get any certificate. The proposed has multiple layers of security which provide a secure environment for the users to use.

A Hybrid Centralized-Peer Authentication System 285

Table 2. Number of Hash, Encryption & Decryption Function

Operation	D: Decryption, E: Encryption, H: Hash Function			Total Sum on All Sides		
	User-Side	Central Server	Peer-Side	# Hashings	# Encryptions	# Decryptions
Initial User Validation	1D, 1H	1E, 1H	/	2	1	1
Certificate from Central to User	1E, 1H	1D	/	1	1	1
Certificate from Peers to User	2D, 1E, 1H	1D, 2E, 1H	1D, 1E	2	4	4
User wants Final Full Certificate	3D, 3E, 3H	3D, 3E, 2H	1D, 2E, 1H	5	8	7
Sum of All Operations	6D, 5E, 6H	5D, 6E, 4H	2D, 3E, 1H			

6.2 Network Spoofing

If someone decided to spoof the network while a request for any certificate was going on, if the person spoofing the network in a way gets their hand on any information. The information is all hashed and encrypted using the user's private key. The encryption of messages and hashing them prevents any kind of tampering, and editing on the messages, and the use of timestamps in certificates and messages prevents any manipulation from happening. Which guarantees the inability for anyone to decrypt and unhash these information to use them against users and hack their accounts.

6.3 Replays in Certificates

The proposed system uses timestamps with digital signatures from the governmental organizations side in order to prevent users from getting multiple certificates. For example, if a user finds a way to manipulate the proposed system in order to get more than one certificate from the central side for example, the proposed system will prevent him from doing that by checking through the timestamps and certificates the users get, in addition to the certificate history table mentioned in the design section.

7 Conclusion

The whole purpose of founding an unbiased citizen authentication system is to find solutions for problems like identity theft fraud, and the central government knowing too much about its citizens. This paper started by explaining different types of authentication systems, their types, and which type of methods they have used. Then it talked about the technology of block-chain, when not to use it, and why. Afterwards, the paper talked about the proposed system, which borrows the concept of block-chain, and utilized it with the use of centralized governmental systems. The paper then talked about the architecture of the system, which is based on three main actors, all helping each other in order to achieve an unbiased citizen authentication system. Which aims to assure that no one uses the system for their own advantage,

The paper then talked about different scenarios which may occur when using the system. Following that, the paper then explained the security analysis for the proposed system by explaining different features, which allow the relationship between the main three actors to be unbiased towards anyone. In the end, the paper talks about the evaluation/discussion process, in addition to an analysis regarding the security of the proposed system. The proposed system grants the users the ability to get certificates in order to obtain an authenticated certificate to complete transactions and prove their identity using a mobile device.

References

1. Fan, P., Liu, Y., Zhu, J., Fan, X., Wen, L.: Identity management security authentication based on blockchain technologies. Int. J. Netw. Secur. **21**, 912–917 (2019)
2. Pfleeger, C., Pfleeger, S., Margulies, J.: Security in Computing. Pearson Education (2015). https://books.google.ps/books?id=VjMqBgAAQBAJ
3. Lin, Q., Yan, H., Huang, Z., Chen, W., Shen, J., Tang, Y.: An id-based linearly homomorphic signature scheme and its application in blockchain. IEEE Access **6**, 20632–20640 (2018)
4. F. T. C. (FTC): Consumer Sentinel Network Data Book 2020 (2020). https://www.ftc.gov/system/files/documents/reports/consumer-sentinel-network-data-book-2020/csn_annual_data_book_2020.pdf, Accessed 1 Sep-2021
5. T. International: Corruption Perceptions Index 2020 (2020). https://images.transparencycdn.org/images/CPI2020_Report_EN_0802-WEB-1_2021-02-08-103053.pdf. Accessed 1 Sep 2021
6. Nakamoto, S.: Bitcoin: A peer-to-peer electronic cash system. In: Decentralized business review, p. 21260 (2008)
7. Anwar, S., Anayat, S., Butt, S., Butt, S., Saad, M.: Generation analysis of blockchain technology: bitcoin and ethereum. Int. J. Inform. Eng. Electr. Bus. (IJIEEB) **12**(4), 30–39 (2020)
8. Pawade, D., Jape, S., Balasubramanian, R., Kulkarni, M., Sakhapara, A.: Distributed ledger management for an organization using blockchains. Int. J. Educ. Manag. Eng. **8**(3), 1 (2018)
9. Jacobovitz, O.: Blockchain for identity management. In: The Lynne and William Frankel Center for Computer Science Department of Computer Science. Ben-Gurion University, Beer Sheva (2016)
10. Chakraborty, R., Pandey, M., Rautaray, S.: Managing computation load on a blockchain based multi-layered internet of things network. Procedia Comput. Sci. **132**, 469–476 (2018)
11. Ali Syed, T., Alzahrani, A., Jan, S., Siddiqui, M.S., Nadeem, A., Alghamdi, T.: A comparative analysis of blockchain architecture and its applications: problems and recommendations. IEEE Access **7**, 176838–176869 (2019)
12. Ye, C., Li, G., Cai, H., Gu, Y., Fukuda, A.: Analysis of security in blockchain: Case study in 51percent-attack detecting. In: Proceedings – 2018 5th International Conference on Dependable Systems and Their Applications, DSA 2018, ser. Proceedings – 2018 5th International

Conference on Dependable Systems and Their Applications, DSA 2018. United States: Institute of Electrical and Electronics Engineers Inc., Dec. 2018, pp. 15–24, 5th International Conference on Dependable Systems and Their Applications, DSA 2018; Conference date: 22-09-2018 Through 23-09-2018

13. Lin, C., He, D., Huang, X., Choo, K.-K.R., Vasilakos, A.: Bsein: A blockchain-based secure mutual authentication with fine-grained access control system for industry 4.0. J. Netw. Comput. Appl. **116**, 42–52 (2018)

14. Cui, Z., Xue, F., Zhang, S., Cai, X., Cao, Y., Zhang, W., Chen, J.: A hybrid blockchain-based identity authentication scheme for multi-wsn. IEEE Trans. Serv. Comput. **13**(2), 241–251 (2020)

15. Augot, D., Chabanne, H., Chenevier, T., George, W., Lambert, L.: A user-centric system for verified identities on the bitcoin blockchain. In: Garcia-Alfaro, J., Navarro-Arribas, G., Hartenstein, H., Herrera-Joancomartí, J. (eds.) Data Privacy Management, Cryptocurrencies and Blockchain Technology: ESORICS 2017 International Workshops, DPM 2017 and CBT 2017, Oslo, Norway, September 14-15, 2017, Proceedings, pp. 390–407. Springer International Publishing, Cham (2017). https://doi.org/10.1007/978-3-319-67816-0_22

16. Sen, S., Mukhopadhyay, S., Karforma, S.: A blockchain based framework for property registration system in e-governance. Int. J. Inform. Eng. Electron. Bus. **13**(4), 30–46 (2021)

Heuristic Search for Nonlinear Substitutions for Cryptographic Applications

Oleksandr Kuznetsov[1,2(✉)], Emanuele Frontoni[1,3], Sergey Kandiy[1], Oleksii Smirnov[4], Yuliia Ulianovska[5], and Olena Kobylianska[2]

[1] Department of Political Sciences, Communication and International Relations, University of Macerata, Via Crescimbeni, 30/32, 62100 Macerata, Italy
kuznetsov@karazin.ua, emanuele.frontoni@unimc.it
[2] Department of Information and Communication Systems Security, Faculty of Comupter Science, V. N. Karazin Kharkiv National University, 4 Svobody Sq., Kharkiv 61022, Ukraine
[3] Department of Information Engineering, Marche Polytechnic University, Via Brecce Bianche 12, 60131 Ancona, Italy
[4] Cybersecurity & Software Academic Department, Central Ukrainian National Technical University, 8, University Avenue, Kropyvnytskyi 25006, Ukraine
[5] Department of Computer Science and Software Engineering, University of Customs and Finance, Vernadskogo Street, 2/4, Dnipro 49000, Ukraine

Abstract. Heuristic algorithms are used to solve complex computational problems quickly in various computer applications. Such algorithms use heuristic functions that rank the search alternatives instead of a full enumeration of possible variants. The algorithm selects, at each iteration, an alternative with the best value for the heuristics. In this paper, we investigate the complex computational problem of finding highly nonlinear substitutions (S-boxes) in the space of 8-bit permutations. The generation of S-boxes is an important field of research, since nonlinear substitutions are widely used in various cryptographic applications. For instance, S-boxes in symmetric ciphers are responsible for cryptographic strength to linear, differential, algebraic, and other types of cryptanalysis. We propose new heuristics in the form of a cost function, calculated with the Walsh-Hadamard transform. We use the Hill Climbing Algorithm to find highly nonlinear substitutions. Our experiments demonstrate that the new heuristics give good results – it takes about 80,000 iterations of the algorithm to generate 8-bit S-boxes with a nonlinearity of 104. For the optimized parameters, the number of iterations of the algorithm is comparable to the best known results, which confirms the significance and value of the research.

Keywords: Heuristic Search · Nonlinear Substitutions · Hill Climbing Algorithm · Cost Function · S-boxes Generation

1 Introduction

Heuristic algorithms are used to rapidly solve complex computational problems in various computer science applications [1, 2]. The algorithms are based on the use of heuristic functions (or simply heuristics), which rank possible search alternatives. Heuristics are

© The Author(s), under exclusive license to Springer Nature Switzerland AG 2023
Z. Hu et al. (Eds,): ICAILE 2023, LNDECT 180, pp. 288–298, 2023.
https://doi.org/10.1007/978-3-031-36115-9_27

commonly represented as a cost function of the search [3–6]. In this case, the possible alternatives are ranked in ascending order of cost, i.e. the best alternative has the lowest value of the cost function. The search algorithm selects the best alternatives and iteratively repeats this process. Thus, instead of going through all the alternatives completely, the heuristic algorithm selects only some alternatives – the best ones according to the cost function value. This significantly reduces search complexity. However, it also does not guarantee the best, optimal solution. As a rule, heuristic algorithms give an approximate solution, but this solution would be found quickly enough [1, 2, 7].

In this paper, we discuss the complex computational problem of generating highly nonlinear substitutions (S-boxes) for cryptographic applications [8–10]. This is an important field of research, since S-boxes are used in many cryptographic algorithms to enhance strength against analytical attacks [11–13]. For example, the nonlinearity of S-boxes provides resistance to linear cryptanalysis [14]. Substitutions are required to be random, in order to be resistant to algebraic cryptanalysis. Consequently, substitutions should not contain hidden mathematical constructs, which describe the cipher with simple algebraic equations [15–17]. In this sense, the generation of random highly nonlinear S-boxes is a relevant computational problem [13, 18, 19].

We consider the generation of bijective permutations in the space of all 256! possible permutations of 8-bit numbers. This is a huge space of possible states, and it is computationally impossible to find a solution by trying all $256! \approx 10^{507}$ alternatives. We use a heuristic search to find highly non-linear substitutions and propose new heuristics, based on calculating the Walsh-Hadamard spectrum.

The goal of our research is to optimize the generation of random 8-bit S-boxes with non-linearity 104. Existing techniques require a large number of iterations of heuristic search algorithms. We hope that the new heuristics will improve the generation speed.

2 Related Works

The problem of rapid generation of random S-boxes was considered in many related papers. For example, [8, 20–23] and many others investigated various heuristics for cryptographic applications, in particular iterative techniques for generating Boolean and vector functions, as well as cost functions for ranking different search alternatives. In further works these techniques were generalized and extended to various computational algorithms: simulated annealing [10, 24], hill climbing [8, 25, 26], genetic algorithms [27–30], artificial immunity techniques [31], etc.

The search for various heuristic functions was also investigated in [30, 32, 33]. The WHS cost function optimisation was explored in [34, 35], and various algorithms with WHS and other heuristics were studied in [30, 36]. Reference [32] proposed a new cost function, which proved to be a more effective heuristic for finding S-boxes. Moreover, the authors used genetic algorithms and local search methods. Another heuristics, proposed in [33], are one of the most effective to date. As shown in [30], the use of this cost function gives the lowest computational complexity of the heuristic search. For instance, a local search algorithm requires about 150,000 iterations; a combination of a genetic algorithm and a search through a decision tree requires about 116,000 iterations. Meanwhile, the Hill Climbing (HC) algorithm has the lowest computational complexity. Combined with

290   O. Kuznetsov et al.

the cost function from [33], the HC algorithm requires about 70,000 iterations and this appears to be the best known result.

Examination of the related works reveals that, in most cases, the authors solve the problem of rapid generation of highly nonlinear bijective S-boxes. Consequently, the authors search for a substitution with nonlinearity of 104 in the space of $256! \approx 10^{507}$ possible permutations of 8-bit numbers. The nonlinearity of the substitution is calculated by the Walsh-Hadamard transform. In order to increase the nonlinearity of the S-box, the absolute values of the Walsh-Hadamard coefficients should be reduced [37]. Therefore, it is important to estimate the spectrum of coefficients, and this is implemented in all known heuristics.

3 Bijective 8-bit S-boxes

In combinatorics, a permutation is an ordered set of numbers $0, 1, ..., N - 1$. Such a set of $0, 1, ..., N - 1$ corresponds to each number i with an i-th element from the se [38, 39]. Thus, a permutation is a bijection on the set $\{0, 1, ..., N - 1\}$.

The number N is called the order (degree) of the permutation. We consider permutations of 8-bit numbers, i.e. for $N = 2^8 = 256$.

Denote each element in the input of the permutation as $X_i, i = 0, 1, ..., 255$, and the elements in the output as $Y_j, j = 0, 1, ..., 255$. Each X_i and Y_j requires 8 bits to encode, as indicated in Fig. 1. The input binary vector $(x_0, x_1, ..., x_7)$ is fed to the select lines of the Multiplexer. Consequently, this vector specifies the number of one of the 256 outputs of the $X_i, i = 0, 1, ..., 255$ Multiplexer. Each $X_i, i = 0, 1, ..., 255$ is connected to one of the 256 permutations $Y_j, j = 0, 1, ..., 255$ outputs. The demultiplexer performs the reverse conversion, encoding the output number $Y_j, j = 0, 1, ..., 255$ of the permutation as a binary vector $(y_0, y_1, ..., y_7)$.

Thus, a permutation of the set $\{0, 1, ..., N - 1\}, N = 2^8 = 256$ can be represented as a substitution of

$$S = \begin{pmatrix} x_0 \ x_1 \ ... \ x_7 \\ y_0 \ y_1 \ ... \ y_7 \end{pmatrix},$$

which implements a vector mapping $S : (x_0, x_1, ..., x_7) \rightarrow (y_0, y_1, ..., y_7)$.

The mathematical apparatus of Boolean functions is used to describe the S-box structure analytically [37]. Each binary substitution output is written as the value of a Boolean function:

$$y_0 = f_0(x_0, x_1, ..., x_7),$$
$$y_1 = f_1(x_0, x_1, ..., x_7),$$

$$\cdots$$

$$y_7 = f_7(x_0, x_1, ..., x_7).$$

The apparatus of Boolean functions is very useful for evaluating the cryptographic properties of S-boxes. For example, the nonlinearity of the substitution $N(S)$ is calculated using the Walsh-Hadamard transform.

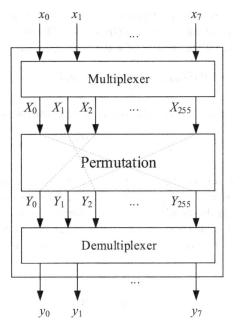

Fig. 1. Data conversion scheme in an 8-bit bijective S-box

4 The Walsh-Hadamard Transform and Nonlinearity of Substitution

Denote the scalar product of the binary vectors $w = (w_0, w_1, ..., w_7)$, and $x = (x_0, x_1, ..., x_7)$ as

$$\langle w, x \rangle = w_0 x_0 + w_1 x_1 + ... + w_7 x_7$$

Therefore, the Walsh-Hadamard transform of the Boolean function $f(x)$ is given by the next expression:

$$WHT(f(x), w) = \sum_{x \in \{0,1\}^8} (-1)^{f(x) \oplus \langle w, x \rangle} \qquad (1)$$

Expression (1) allows the calculation of Walsh-Hadamard coefficients $WHT(f(x), w)$ for all 256 possible values of $w = (w_0, w_1, ..., w_7)$.

In order to calculate the nonlinearity $N(S)$, the Walsh-Hadamard transform should be performed for all linear combinations of coordinate Boolean functions $f_0, f_1, ..., f_7$.

Denote the set of coordinate functions as a vector

$$F = (f_0, f_1, ..., f_7).$$

Then the set of all linear combinations of Boolean functions $f_0, f_1, ..., f_7$ is given by the scalar product of all binary vectors $u = (u_0, u_1, ..., u_7)$ and the vector F:

$$\langle u, F \rangle = u_0 f_0 + u_1 f_1 + ... + u_7 f_7.$$

For each of the 255 values $\langle u, F \rangle$ (except $u = (0, 0, ..., 0)$), perform the Walsh-Hadamard transform and find all 256 values of $WHT(\langle u, F \rangle, w)$. The maximum absolute value of the Walsh-Hadamard coefficients over all $u \neq (0, 0, ..., 0)$ and w, denote as

$$WHT_{\max} = \max_{u \neq 0, w} |WHT(\langle u, F \rangle, w)|.$$

Hence, the nonlinearity $N(S)$ is calculated according to the formula:

$$N(S) = 2^7 - \frac{WHT_{\max}}{2} \tag{2}$$

Table 1 gives an example of the correspondence between some $WHT_{\max}$ and $N(S)$ values. We are interested in highly nonlinear S-boxes. In particular, the related works most frequently solved the problem of finding S-boxes with $N(S) \geq 104$, i.e. with $WHT_{\max} < 48$. In this sense, the used heuristics should consider the values of the Walsh-Hadamard coefficients (1), in order to minimise the cost of finding highly nonlinear substitutions.

Table 1. Example of matching some $WHT_{\max}$ and $N(S)$ values

$WHT_{\max}$	$N(S)$
80	88
76	90
72	92
68	94
64	96
60	98
56	100
52	102
48	104
44	106

5 New Heuristics for Finding Nonlinear Substitutions

One of the first and most studied heuristic functions for finding highly nonlinear substitutions is the CF_{WHS} cost function [24]:

$$CF_{WHS} = \sum_{u \neq 0} \sum_{w} ||WHT(\langle u, F \rangle, w)| - X|^R, \tag{3}$$

where X and R – selectable heuristics parameters.

Research of function eq. (3), with optimisation of X и R parameters, was carried out in [27, 34, 35] and many other studies. However, to date, the CF_{WHS} function is considered as an inefficient one [32, 33].

Another example of a cost function is the heuristics, proposed in [32]:

$$CF_{PCF} = \sum_{i=1}^{N} 2^{-i} W(S)_{k-i}, \tag{4}$$

where $W(S)$ is a value vector $|WHT(\langle u, F \rangle, w)|$, in which the i-th position contains the number of coefficients with values $|4i|$; k is a maximum position number with a non-zero value.

The cost function (4) performed very well in [32]. At the same time, the authors used genetic algorithms and local search methods.

The most effective heuristics for finding S-boxes, with $N(S) \geq 104$, is the cost function from [33]:

$$CF_{WCF} = \sum_{u \neq 0} \sum_{w} \prod_{z \in C} ||WHT(\langle u, F \rangle, w)| - z|, \tag{5}$$

where $C = \{0, 4, ..., 32\}$.

The use of function (5) in the local search algorithm requires about 150 thousand iterations, according to [33]. A genetic algorithm with a decision tree search requires about 116,000 iterations [33]. The HC algorithm showed the lowest computational complexity, requiring about 70,000 iterations [33].

In this paper, we offer a new cost function for the heuristic search for highly nonlinear substitutions. The proposed heuristics are as follows:

$$CF_{new} = \sum_{\substack{u \neq 0 \\ |WHT(\langle u,F \rangle, w) > X}} \sum_{w} 2^{\frac{|WHT(\langle u,F \rangle, w)| - Y}{4}}, \tag{6}$$

where X and Y – selectable parameters.

The logic behind using parameter X in function (6) is to consider only those coefficients, which could potentially reduce the nonlinearity of the substitution. For instance, in order to achieve $N(S) \geq 104$, it is necessary to reach $WHT_{max} < 48$, according to Table 1. Meanwhile, the values with $|WHT(\langle u, F \rangle, w)| << 48$ cannot change the nonlinearity of $N(S)$ and can be disregarded in the search algorithm.

It makes sense to use the Y parameter to adjust the $|WHT(\langle u, F \rangle, w)|$ value as part of the calculation of the cost function. We believe, this will allow the flexibility of heuristic parameters and optimise heuristics for rapid search of nonlinear substitutions.

6 Experimental Results

We implemented a HC algorithm to test the new cost function, as described in the papers [30, 33].

294 O. Kuznetsov et al.

By the definition of the Walsh-Hadamard transform, the coefficient values of $|WHT(\langle u, F \rangle, w)|$ are always multiples of 4. We used only integer values of heuristic (6) to enable a rapid calculation of the cost function. In this regard, we used the following parameters:

- $X \in \{0, 4, 8, ..., 32\}$;
- $Y \in \{-48, -44, -40, ..., X\}$.

We ran the heuristic search algorithm 100 times, for each combination of X and Y, during the experiments. Table 2 shows the average results for the number of iterations of the algorithm. This number of iterations was required to find substitutions with $N(S) = 104$. The last row shows the averages for each column. The best findings are highlighted in bold.

Table 2. Experimental results

Y	X								
	0	4	8	12	16	20	24	28	32
-48	85422.0	80838.1	79138.5	73825.4	83562.6	86979.6	94338.1	110166.0	195612.3
-44	83774.9	85538.7	77852.6	78690.4	84786.1	77609.5	97099.8	118036.8	200007.5
-40	82643.5	78212.0	75316.2	73847.4	78611.7	84321.7	94367.2	118641.3	194665.3
-36	82072.2	79546.9	73394.3	74295.6	77930.3	88315.8	92394.9	116221.9	191935.5
-32	83186.5	85567.5	82312.2	85569.5	83112.2	85196.5	97609.9	123604.3	199450.9
-28	86873.0	81251.6	81279.5	81928.4	75135.9	80347.4	101076.5	116314.3	193298.3
-24	84868.9	79824.1	81598.6	83278.6	83655.9	86218.9	96098.9	111999.5	202497.7
-20	82638.5	80950.7	86224.3	**72664.1**	78963.5	87201.8	98080.6	119439.6	199461.0
-16	80310.3	86568.9	84965.1	78817.2	85600.6	78384.7	96479.4	119236.5	188040.4
-12	80191.2	79117.8	80638.7	77344.3	78307.5	83894.2	91427.1	120797.1	201318.4
-8	87627.3	86069.4	82611.0	73585.6	73606.7	81490.3	95740.1	113123.2	203048.6
-4	73051.8	82376.5	78928.2	74525.6	84984.3	86219.5	94873.3	116409.0	192092.6
0	86420.7	82060.4	80937.1	**71907.7**	83205.1	79284.4	96674.7	116712.3	193454.0
4		82517.3	82075.0	77207.3	80112.7	86983.8	96276.9	118212.3	195199.7
8			78264.8	79414.9	88172.7	80514.4	93364.8	114364.7	203120.6
12				77904.6	76233.8	86039.6	93817.1	120395.0	201043.3
16					82576.6	72349.2	95295.9	116427.4	196346.7
20						85482.2	94475.8	119483.8	196285.3
24							92488.9	118142.0	195413.9
28								116663.5	198115.4
32									195733.5
	83006.2	82174.3	80369.1	77175.4	81091.7	83157.4	95367.4	117219.5	196959.1

7 Discussion of Results and Conclusions

Based on our analysis of the results, we conclude that the new cost function (6) shows stable heuristic search results over a wide range of X and Y parameters:

- the search algorithm needs on average about 80,000 iterations for the $X \in \{0, 4, 8, ..., 20\}$, and $Y \in \{-48, -44, -40, ..., X\}$ values;
- for $X > 20$ the search complexity increases, and for $X > 32$ we were unsuccessful in finding any substitutions with $N(S) = 104$.

The observations reveal that:

- it is necessary to consider almost all values of the coefficients of $|WHT(\langle u, F \rangle, w)|$, in order to find highly nonlinear substitutions, i.e. the parameter X cannot be large;
- the search complexity is almost independent of the Y parameter;
- obtained results, for the number of iterations of the algorithm, look like a random process realisation.

According to Table 2, the smallest number of iterations (on average) is observed for $X = 12$. In some cases (in bold) we managed to find substitutions with $N(S) = 104$ in about 70,000 iterations of the algorithm. This compares with the best known results from [30, 33].

Thus, this work advances the field of knowledge in the direction of improving cost functions for heuristic search of random S-boxes. This can be useful in various fields of computer science and cybersecurity [40–43], including the improvement of symmetric key encryption algorithms and hash functions [44].

Notably, in this paper, we used only the HC algorithm. Possibly, the cost function (6) will give an alternative result for other heuristic search algorithms. This is the subject of our further research.

Acknowledgment. 1. This project has received funding from the European Union's Horizon 2020 research and innovation programme under the Marie Skłodowska-Curie grant agreement No. 101007820.

2. This publication reflects only the author's view and the REA is not responsible for any use that may be made of the information it contains.

References

1. Banzhaf, W., Hu, T.: Evolutionary computation. In: Banzhaf, W., Hu, T. (eds.) Evolutionary Biology. Oxford University Press (2019). https://doi.org/10.1093/obo/9780199941728-0122
2. Gilli, M., Maringer, D., Schumann, E.: Chapter 13 – Heuristics: a tutorial. In: Gilli, M., Maringer, D., Schumann, E. (eds.) Numerical Methods and Optimization in Finance (Second Edition), pp. 319–353. Academic Press (2019). https://doi.org/10.1016/B978-0-12-815065-8.00025-X
3. Gandomi, A.H., Yang, X.-S., Talatahari, S., Alavi, A.H.: Metaheuristic algorithms in modeling and optimization. In: Metaheuristic Applications in Structures and Infrastructures, pp. 1–24. Elsevier (2013). https://doi.org/10.1016/B978-0-12-398364-0.00001-2

4. Çataloluk, H., Çelebı, F.V.: A heuristic algorithm for Chan-Vese model. In: 2018 26th Signal Processing and Communications Applications Conference (SIU). pp. 1–4 (2018). https://doi.org/10.1109/SIU.2018.8404820
5. van der Stockt, S.A.G., Engelbrecht, A.P., Cleghorn, C.W.: Heuristic space diversity measures for population-based hyper-heuristics. In: 2020 IEEE Congress on Evolutionary Computation (CEC), pp. 1–9 (2020). https://doi.org/10.1109/CEC48606.2020.9185719
6. Tunç, A., Taşdemir, Ş., Sağ, T.: Comparison of heuristic and metaheuristic algorithms. In: 2022 7th International Conference on Computer Science and Engineering (UBMK). pp. 76–81 (2022). https://doi.org/10.1109/UBMK55850.2022.9919459
7. Salhi, S.: Hybridisation search. In: Heuristic Search, pp. 129–156. Springer, Cham (2017). https://doi.org/10.1007/978-3-319-49355-8_5
8. Burnett, L.D.: Heuristic Optimization of Boolean Functions and Substitution Boxes for Cryptography (2005). https://eprints.qut.edu.au/16023/
9. Álvarez-Cubero, J.: Vector Boolean Functions: applications in symmetric cryptography (2015). https://doi.org/10.13140/RG.2.2.12540.23685
10. McLaughlin, J.: Applications of search techniques to cryptanalysis and the construction of cipher components. https://etheses.whiterose.ac.uk/3674/ (2012)
11. Rodinko, M., Oliynykov, R., Gorbenko, Y.: Optimization of the high nonlinear s-boxes generation method. Tatra Mt. Math. Publ. **70**, 93–105 (2017). https://doi.org/10.1515/tmmp-2017-0020
12. Biham, E., Perle, S.: Conditional linear cryptanalysis – cryptanalysis of DES with less than 242 complexity. In: IACR Transactions on Symmetric Cryptology, pp. 215–264 (2018). https://doi.org/10.13154/tosc.v2018.i3.215-264
13. Freyre Echevarría, A.: Evolución híbrida de s-cajas no lineales resistentes a ataques de potencia (2020). https://doi.org/10.13140/RG.2.2.17037.77284/1
14. Mihailescu, M.I., Nita, S.L.: Linear and differential cryptanalysis. In: Mihailescu, M.I., Nita, S.L. (eds.) Pro Cryptography and Cryptanalysis with C++20: Creating and Programming Advanced Algorithms, pp. 387–409. Apress, Berkeley, CA (2021). https://doi.org/10.1007/978-1-4842-6586-4_19
15. Ars, G., Faugère, J.-C.: Algebraic Immunities of functions over finite fields. In: INRIA (2005)
16. Bard, G.V.: Algebraic Cryptanalysis. Springer US, Boston, MA (2009). https://doi.org/10.1007/978-0-387-88757-9
17. Courtois, N.T., Bard, G.V.: Algebraic cryptanalysis of the data encryption standard. In: Galbraith, S.D. (ed.) Cryptography and Coding 2007. LNCS, vol. 4887, pp. 152–169. Springer, Heidelberg (2007). https://doi.org/10.1007/978-3-540-77272-9_10
18. Lisitskiy, K., Lisitska, I., Kuznetsov, A.: Cryptographically properties of random s-boxes. In: Proceedings of the 16th International Conference on ICT in Education, Research and Industrial Applications. Integration, Harmonization and Knowledge Transfer. Volume II: Workshops, Kharkiv, Ukraine, 06–10 Oct 2020, pp. 228–241 (2020)
19. Gorbenko, I., Kuznetsov, A., Gorbenko, Y., Pushkar'ov, A., Kotukh, Y., Kuznetsova, K.: Random s-boxes generation methods for symmetric cryptography. In: 2019 IEEE 2nd Ukraine Conference on Electrical and Computer Engineering (UKRCON), pp. 947–950 (2019). https://doi.org/10.1109/UKRCON.2019.8879962
20. Clark, A.J.: Optimisation heuristics for cryptology. https://eprints.qut.edu.au/15777/ (1998)
21. Millan, W., Clark, A., Dawson, E.: Heuristic design of cryptographically strong balanced Boolean functions. In: Nyberg, K. (ed.) EUROCRYPT 1998. LNCS, vol. 1403, pp. 489–499. Springer, Heidelberg (1998). https://doi.org/10.1007/BFb0054148
22. Millan, W., Burnett, L., Carter, G., Clark, A., Dawson, E.: Evolutionary heuristics for finding cryptographically strong s-boxes. In: Varadharajan, V., Mu, Y. (eds.) ICICS 1999. LNCS, vol. 1726, pp. 263–274. Springer, Heidelberg (1999). https://doi.org/10.1007/978-3-540-47942-0_22

23. Millan, W., Clark, A., Dawson, E.: Boolean function design using hill climbing methods. In: Pieprzyk, J., Safavi-Naini, R., Seberry, J. (eds.) ACISP 1999. LNCS, vol. 1587, pp. 1–11. Springer, Heidelberg (1999). https://doi.org/10.1007/3-540-48970-3_1
24. Clark, J.A., Jacob, J.L., Stepney, S.: The design of S-boxes by simulated annealing. New Gener. Comput. **23**, 219–231 (2005). https://doi.org/10.1007/BF03037656
25. Freyre-Echevarría, A., Martínez-Díaz, I., Pérez, C.M.L., Sosa-Gómez, G., Rojas, O.: Evolving nonlinear s-boxes with improved theoretical resilience to power attacks. IEEE Access **8**, 202728–202737 (2020). https://doi.org/10.1109/ACCESS.2020.3035163
26. Kavut, S., Yücel, M.D.: Improved cost function in the design of boolean functions satisfying multiple criteria. In: Johansson, T., Maitra, S. (eds.) INDOCRYPT 2003. LNCS, vol. 2904, pp. 121–134. Springer, Heidelberg (2003). https://doi.org/10.1007/978-3-540-24582-7_9
27. Tesar, P.: A new method for generating high non-linearity s-boxes. Radioengineering **19**, 23–26 (2010)
28. Ivanov, G., Nikolov, N., Nikova, S.: Reversed genetic algorithms for generation of bijective s-boxes with good cryptographic properties. Cryptogr. Commun. **8**(2), 247–276 (2016). https://doi.org/10.1007/s12095-015-0170-5
29. Kapuściński, T., Nowicki, R.K., Napoli, C.: Application of genetic algorithms in the construction of invertible substitution boxes. In: Rutkowski, L., Korytkowski, M., Scherer, R., Tadeusiewicz, R., Zadeh, L.A., Zurada, J.M. (eds.) ICAISC 2016. LNCS (LNAI), vol. 9692, pp. 380–391. Springer, Cham (2016). https://doi.org/10.1007/978-3-319-39378-0_33
30. Freyre-Echevarría, A., et al.: An external parameter independent novel cost function for evolving bijective substitution-boxes. Symmetry **12**, 1896 (2020). https://doi.org/10.3390/sym12111896
31. Ivanov, G., Nikolov, N., Nikova, S.: Cryptographically strong s-boxes generated by modified immune algorithm. In: Pasalic, E., Knudsen, L.R. (eds.) BalkanCryptSec 2015. LNCS, vol. 9540, pp. 31–42. Springer, Cham (2016). https://doi.org/10.1007/978-3-319-29172-7_3
32. Picek, S., Cupic, M., Rotim, L.: A new cost function for evolution of s-boxes. Evol. Comput. **24**, 695–718 (2016). https://doi.org/10.1162/EVCO_a_00191
33. Freyre Echevarría, A., Martínez Díaz, I.: A new cost function to improve nonlinearity of bijective S-boxes (2020)
34. Kuznetsov, A., et al.: WHS cost function for generating S-boxes. In: 2021 IEEE 8th International Conference on Problems of Infocommunications, Science and Technology (PIC S T), pp. 434–438 (2021). https://doi.org/10.1109/PICST54195.2021.9772133
35. Kuznetsov, A., et al.: Optimizing the local search algorithm for generating s-boxes. In: 2021 IEEE 8th International Conference on Problems of Infocommunications, Science and Technology (PIC S T), pp. 458–464 (2021). https://doi.org/10.1109/PICST54195.2021.9772163
36. Kuznetsov, A., Wieclaw, L., Poluyanenko, N., Hamera, L., Kandiy, S., Lohachova, Y.: Optimization of a simulated annealing algorithm for s-boxes generating. Sensors **22**, 6073 (2022). https://doi.org/10.3390/s22166073
37. Carlet, C.: Vectorial Boolean functions for cryptography. Boolean Models and Methods in Mathematics, Computer Science, and Engineering (2006)
38. Sachkov, V.N., Vatutin, V.A.: Probabilistic Methods in Combinatorial Analysis. Cambridge University Press (1997). https://doi.org/10.1017/CBO9780511666193
39. Sachkov, V.N., Kolchin, V.: Combinatorial Methods in Discrete Mathematics. Cambridge University Press (1996). https://doi.org/10.1017/CBO9780511666186
40. Beletsky, A.: Generalized galois-fibonacci matrix generators pseudo-random sequences. IJCNIS **13**, 57–69 (2021). https://doi.org/10.5815/ijcnis.2021.06.05
41. Krasnobayev, V., Kuznetsov, A., Kuznetsova, K.: Synthesis of the structure of a computer system functioning in residual classes. Int. J. Comput. Netw. Inform. Secur. **15**(1), 1–13 (2023). https://doi.org/10.5815/ijcnis.2023.01.01

42. Iavich, M., Kuchukhidze, T., Gnatyuk, S., Fesenko, A.: Novel certification method for quantum random number generators. IJCNIS **13**, 28–38 (2021). https://doi.org/10.5815/ijcnis.2021.03.03
43. Shekhanin, K., Kuznetsov, A., Krasnobayev, V., Smirnov, O.: Detecting hidden information in fat. Int. J. Comput. Netw. Inf. Security. **12**, 33–43 (2020). https://doi.org/10.5815/ijcnis.2020.03.04
44. Kuznetsov, A., et al.: Performance analysis of cryptographic hash functions suitable for use in Blockchain. IJCNIS **13**, 1–15 (2021). https://doi.org/10.5815/ijcnis.2021.02.01

Dangerous Landslide Suspectable Region Forecasting in Bangladesh – A Machine Learning Fusion Approach

Khandaker Mohammad Mohi Uddin[1(✉)], Rownak Borhan[1], Elias Ur Rahman[1], Fateha Sharmin[2], and Saikat Islam Khan[1]

[1] Department of Computer Science and Engineering, Dhaka International University, Dhaka 1205, Bangladesh
`jilanicsejnu@gmail.com`
[2] Department of Chemistry, University of Chittagong, Chittagong, Bangladesh

Abstract. Destruction caused by landslides in Bangladesh's south-east highlights the importance of landslide-hazard mapping and a better understanding of the geomorphic development of landslide hazardous landscapes. Suitable relief, steep terrain gradients, Permiancyclothems and Pennsylvanian that weather into fine-grained soils which have extensive clay, and appropriate precipitation are present in Chittagong, Bandarban, and other hillside regions of Bangladesh. In order to map landslide susceptibility for some of the areas in Rangamati District, Bangladesh, this study compares and evaluates a variety of machine learning models, including Random Forest (RF), Support Vector Machine (SVM), Naive Byes (NB), Random Forest (RF), Logistic Regression (LR), Decision Tree (DT), and the hybrid approaches. This study's combination of logistic regression and random forest demonstrated 91.1% accuracy in landslide prediction. For risk management and disaster planning, this paper provides a useful analysis for selecting the best model for determining landslide susceptibility.

Keywords: Landslide · Machine Learning fusion · geohazards · logistic regression · random forest

1 Introduction

Landslides cause loss of life and property worldwide since they are one of the most dangerous geohazards. As a result, government and non-government organizations must move quickly to assess landslide vulnerability and mitigate its negative effects [1]. Bangladesh is particularly vulnerable to a wide variety of natural hazards and calamities. Particularly vulnerable to landslides in Bangladesh is the Chittagong Hill Tracts (CHT) [2]. Heavy rains during the monsoon season are the most likely cause of landslides in the hillside area. People play a unique part in the instability of hillslopes, which was previously unknown. As a result of this, there has been a rise in population, which has led to deforestation, unsustainable farming methods, urban expansion in Bangladesh, and a lack of effective government [2]. Landslide Risk in Bangladesh is shown in Fig. 1.

© The Author(s), under exclusive license to Springer Nature Switzerland AG 2023
Z. Hu et al. (Eds.): ICAILE 2023, LNDECT 180, pp. 299–309, 2023.
https://doi.org/10.1007/978-3-031-36115-9_28

A landslide susceptibility map is a crucial tool for geohazard management through better decision in prone areas and planning for land use. These geographical traits are often referred to as "cause factors." The landslide hazard assessment considers the determination of these cause elements to be the keystone [3]. Future landslides are thought to be possible in regions that have already experienced landslides and where the variables contributing to such landslides have recently produced favorable conditions for the triggering of landslides [4]. Several natural and man-made things, such as volcanic activity, groundwater extraction, rapid snow melting, long periods of rain, deforestation, hill cutting, changes in land usability, etc., can cause landslides, which is why the term "triggering factors" is used to describe them [5, 6]. Landslide risk can be measured in two ways: qualitatively and quantitatively [7].

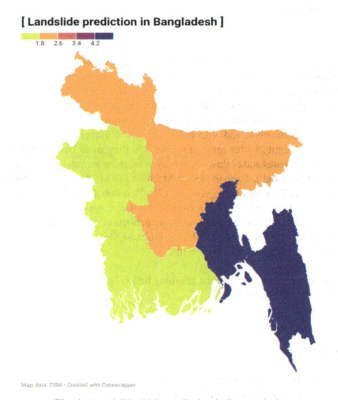

Fig. 1. Landslide Risk prediction in Bangladesh

Expert analyses of the causes of the factors provide the foundation for qualitative assessments of landslide susceptibility. Quantitative methods, however, take use of the mathematical connections between the sites of landslides and their underlying causes. Mixed and semi-quantitative methodologies are frequently used in qualitative assessments to process the viewpoints of experts. Systems that employ machine learning are thought to be superior to methods based on expert judgment and analytical procedures for the spatial prediction of landslides [1]. These methodologies evaluate landslide susceptibility by using algorithms for machine learning to look at the spatial correlation between previous landslide occurrences and a set of conditioning variables, from which the probable likelihood of landslide occurrence is determined. In many parts of the world, various machine learning techniques have been developed and put to use for creating landslide susceptibility maps, and artificial neural networks based on biological neural networks have been used to predict the geographic distributions of landslides [8]. Fuzzy inference techniques have been used to analyze the geographical distribution of landslides. Additionally, new strategies have been created, including Logistic Regression (LR) and Support Vector Machines (SVM). This work makes an effort to identify landslide hotspots in Bangladesh's tropical region using machine learning approaches.

2 Literature Review

The majority of natural hazards in mountainous regions are landslides. Landslides represent 9% of all natural disasters and are ranked as the third-worst natural disaster in the world. Researchers from all across the world have been attempting to forecast landslides. On the other hand, multivariate statistical tools determine how landslides are associated with a variety of potential causes. Multivariate approaches [9] include general additive models, logical regression, adaptive regression spline, and simple decision trees. Normality and collinearity assumptions make it difficult for bivariate and multivariate models to perform their intended functions. Machine learning-based structures are less restricted by these presumptions than traditional models are [10], which enables them to take into account the fact that landslides are not linear [11]. Bivariate and multivariate statistical models, according to some, frequently perform worse than machine learning-based models like gradient boosting, random forest, and support vector machines [12, 13].

LR has been extensively used to investigate landslides and has proven to be an effective method for predicting where landslides will occur. LR is a better method for mapping the susceptibility of landslides than probability value and multi-criteria decision – making process.

Among the most widely used techniques for categorizing complex data is FLDA (Fisher's linear discriminant analysis), which is straightforward enough to be used to a formal analysis of each data point in the projected region. Only a few researches have used FLDA to address landslide issues.

Haojie Wang et al. [14] showed that the BN is seen as a promising way to figure out how dangerous something is. But it is still rarely used to figure out how dangerous a landslide is. Some of the studies on assessing landslides have also used the NB method successfully. It was claimed that this technique, which was used to forecast landslides in space, is an efficient machine learning method for determining landslide susceptibility.

Random Forest (RF), K-Nearest Neighbor (KNN), and Extreme Gradient Boosting were three different methods of e-machine learning (ML) that were utilized by Yasin Wahid Rabbyet al [2]. The XGBoost model is the most recent of these three, and it is a brand-new method for modeling landslides that delivers highly precise results. However, both the RF and KNN models are often used and have adequate accuracy. The performance of the cutting-edge machine learning algorithm Extreme Gradient Boosting (XGBoost) has not been compared to the published research nearly as frequently as it should be, despite the fact that RF and KNN have been used extremely frequently.

3 Methodology

The techniques used in this study include data collection, data analysis, forecasting, and the creation and validation of landslide susceptibility maps. In data preparation, some data preprocessing technique, prediction machine learning approach. **Algorithm 1** depicts the proposed method of this experiment to forecast the landslide in the Rangamati area. Step by step algorithm shows that data collection, data preprocessing, feature selection, and machine learning algorithm are the main functions and these functions will be described briefly later. The block diagram of this study is given Fig. 2.

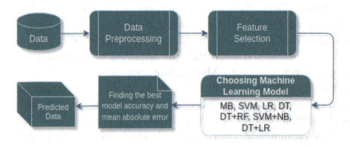

Fig. 2. Block diagram of Landslide susceptibility's proposed method.

Algorithm 1: Algorithm for the forecast landslide
1. Begin
2. Predict_Landslide( ):
3.   dataset(13 metadata)
4.   data_describe( )
5.   data_precprocess( )
6.   x= (x - mean(x))/std(x)
7.   feature_selection( ):
8.   StandardScaler( )
9.   PCA analysis ( )
10.   Machine learning classifier ( ):
11.   SVM( )
12.   LR( )
13.   NB ( )
14.   DT ( )
15.   RF ( )
16.   Ensemble ( ):
17.   DT,RF ( )

18.	SVM, NB ( )
19.	DT, LR ( )
20.	if accuracy is max:
21.	Show_accuracy()
22.	Show_predicted_data()
23.	end if
24.	End

3.1 Data Collection

The Rangamati district is in the southeast of the Chittagong Hill Tracts. It is 6116.19 km^2 in size (CHT). It is between the latitudes of $22°27'$ and $23°44'$ north and the longitudes of $91°56'$ and $92°33'$ east. There are 243,999 persons living in the research region, of whom 48% are Bengalis and 52% are tribes. There are 97 individuals per km^2 and a male to female ratio of 1.1. (BBS 2011). The research region is 70.39 m above sea level on average. The study area has a tropical monsoon environment with moderate temperatures, significant rainfall, frequently high humidity, and obvious seasonal fluctuations because the country is in a tropical monsoon region.

Rabby et al. [2] used the Rangamati landslide dataset and this dataset is actually used in this experiment also. Datasets metadata description is given in Table 1.

3.2 Data Preprocessing

The likelihood of a landslide Data from Rangamati is used to forecast the location of the next landslide. Preprocessing has been done using the core Python modules NumPy, Pandas, and Matplotlib. The computation approach of mean, average, and max was utilized to locate and handle the missing data. To scale the characteristics in this research, a standardizing procedure was applied. With this technique, a dataset's independent variables are kept inside a predetermined range. The standardized equation for scaling features is shown in (Eq. 1), where mean() returns the average value of feature x and std() returns the feature's standard deviation.

$$x = (x - mean(x))/std(x) \tag{1}$$

3.3 Machine Learning Models

Predicting the class of a set of data points is the process of classification. There are numerous categorization algorithms available right now, but it is impossible to say which one is best. It relies on the application and type of data set that is readily available. Several machine classification algorithms were utilized in this investigation to find the intended outcome.

Table 1. Metadata description of the Dataset

Metadata	Description
Aspect	An aspect map shows both the direction and grade of terrain at the same time [15]
Curvature	Plan curvature refers to the outlines on a topographic map or the curve of a slope in a horizontal surface [16]
Earthquake	In the event of an earthquake, the ground's surface may shake and vibrate because of the propagation of seismic waves
Elevation	For the assessment of landslide vulnerability, elevations are frequently used. Elevation differences may be attributed to different environmental conditions as rainfall and plant species [9]
Flow	Landslides known as flows occur when fluid-like material slides down a slope [10]
Lithology	According to the results of several research, Lithology significantly affects the regional variance of landslide prevalence, kind, and depth [11]
NDVI	The RS readings are analyzed and the normalized difference vegetation index (NDVI), a straightforward graphical indication, is frequently used to determine whether or not the target being examined includes green, healthy vegetation [12]
NDWI	Surface water features can be distinguished using the normalized difference water index (NDWI) [13]
Plan Curvature	The curvature of a slope in a horizontal plane, or the curvature of contours on a topographic map, is referred to as plan curvature [14]
Profile Curvature	The maximum slope's direction is parallel to the profile's curvature. Negative value depicts an upwardly convex surface at that particular cell. A positive profile shows that the cell's surface is concave upward [17]
Slope	Hillsides can have straight contours known as planar zones, convex outward plan curvatures known as noses, and concave outward plan curvatures known as hollows. Landslide debris congregates in hollows near the slope's base in a small area [18]
Landslide	This metadata indicates the landslide risk [19]

3.3.1 Random Forest (RF)

Random Forest is a widely used machine learning algorithm for supervised methods. It may be used in ML to solve classification and regression-related issues. The model's performance is combined and improved using an ensemble of classifiers, which is based on the idea of ensemble learning. A group of decision trees are trained using the "bagging" approach to create the "forest." The fundamental idea of the bagging technique is that by mixing many instructional modalities, pupils may perform better [19].

$$\sum_{i=1}^{c} fi(1 - fi) \tag{2}$$

3.3.2 Decision Tree (DT)

The decision Tree method belongs to the class of algorithms known as "supervised learning algorithms". The ability of the decision tree approach to address both regression and classification issues set it apart from other supervised learning techniques. By learning straightforward decision rules from prior data, a Decision Tree is used to build a training model that may be used to predict the class or value of the target variable (training data). In Decision Trees, we begin at the root of the tree to forecast the class label of a record. The root attribute and the record's attribute values are contrasted. We go to the following node along the branch that corresponds to that value based on the comparison [20].

$$E(S) = \sum_{i=1}^{c} -p_i log_2 p_i \tag{3}$$

3.3.3 Ensemble Technique

An ensemble model is a method of using multiple machine learning models in a single model. It is used when we are not confident with our single model for prediction. It is a collection of weak models to produce more reasoning than a single model. It is a quite popular model for exploiting the strength of different models. Ensemble models can be of two types such as Homogeneous ensemble model and Heterogeneous ensemble model. The homogeneous ensemble model uses the same type of classifiers whereas the heterogeneous ensemble model uses different types of classifiers. Bagging and boosting are examples of a homogeneous ensemble model and stacking is of a heterogeneous ensemble model. Stacking is also known as blending. It is a stacked generalization that considers heterogeneous weak learners and learns them in parallel. The stacking model can be divided into two parts. One is called level 1 and another is level 2. The models that are used in level 1 are also known as the lower-level models and the model used in level 2 is also called the meta-model. The number of the model in lower-level depends on the model creator but the main fact that needs to be in mind is that the bigger the model the slower it will become. The output of the lower-level model is then passed down to the meta-model as an input. The meta-model can be any machine learning model of the creators' choice. It can even be the same as one of the models that are used in the lower level. The input will be used to train the meta-model and the output of the meta-model will be considered as the final output.

3.3.4 Accuracy Measurement

The confusion matrix has four outcomes that evaluate each classifier's performance on positive and negative classes separately. These outcomes are true positive (TP), false-negative (FN), true negative (TN), and false positive (FP). Equation 4 helps to calculate accuracy and Eq. 5 and 6 are used to calculate Precision, and Recall, respectively. F-1 score and Sensitivity are calculated by Eqs. 7 and 8, respectively [21, 22].

$$Accuracy = \frac{(TP + TN)}{(TP + FP + FN + TN)} \tag{4}$$

$$Precision = \frac{TP}{(TP + TN)} \tag{5}$$

$$Recall = \frac{TP}{(TP+FN)} \tag{6}$$

$$F1\_score = \frac{2 \times Recall \times Precision}{Recall + Precision} \tag{7}$$

$$Error\ rate = \frac{FP+FN}{TP+TN+FP+FN} \tag{8}$$

4 Experiment and Results

The experiment is carried out on a local computer using Jupyter Notebook, NumPy (1.0.1), Matplotlib (1.0.1), Scikit-Learn (1.0.1), and Pandas (1.11.0). Based on the experiment, two different sets of results have been analyzed. This experiment used different machine learning algorithms, including a decision tree (DT), a random forest (RF), and a Gaussian Naive Bayes (NB). After preprocessing the data and selecting the features to apply, SVM surpassed other machine learning classifiers with an accuracy of 75.99 percent. Then, to improve the accuracy, this study used a fusion method of machine learning algorithms. This experiment shows that this study uses NB, DT and RF, and DT and LR, DT and GB together to find the best solution.

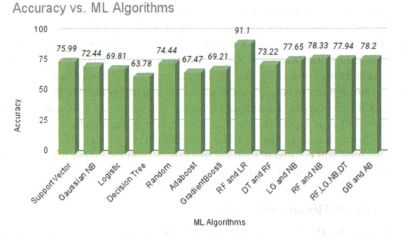

Fig. 3. Comparison among machine learning classifiers based on accuracy for landslide forecasting in the tropical region in Bangladesh

For the purpose of forecasting landslides in Bangladesh's tropical area, comparison of machine learning classifiers based on accuracy is shown in Fig. 3 and **Table 2** depicts the performs of the different ML algorithms. By using the voting classifier, this classifier ensemble approach was made possible. With a 91.1 percent accuracy rate, the fusion of RF and LR did better than other fusions of machine learning classifiers.

Dangerous Landslide Suspectable Region Forecasting   307

Table 2. Accuracy, f-1 score, recall, precision and sensitivity of different ML algorithms.

ML Algorithms	Accuracy	F-1 score	Recall	Precision	Sensitivity
Support Vector Machine	75.99	0.77	0.77	0.78	76%
Gaussian NB	72.44	0.75	0.76	0.76	62%
Logistic Regression	69.81	0.73	0.73	0.71	64%
Decision Tree	63.78	0.69	0.69	0.7	39%
Random Forrest	74.44	0.69	0.69	0.7	46%
Adaboost	67.47	0.73	0.81	0.81	58%
GradientBoostingClassifier	69.21	0.69	0.69	0.71	65%
RF and LR Fusion	91.1	0.86	0.84	0.85	74%
DT and RF	73.22	0.75	0.76	0.78	72%
LG and NB	77.65	0.76	0.77	0.78	60%
RF and NB	78.33	0.76	0.77	0.79	61%
RF,LG,NB,DT	77.94	0.78	0.78	0.8	73%
GB and AB	78.2	0.77	0.78	0.8	59%

Table 3 shows that the proposed fusion method performed better than various neural network approaches and other machine learning model approaches. The profile curvature and slope have the highest correlation of prediction in landslides, according to the proposed fusion technique approach. In every indicator, the suggested model outperformed the six benchmarks. The proposed model achieved the highest percentage of accuracy, scoring 91.1% overall.

Table 3. Comparison table between various approach of finding landslide prediction

References	Landslide Prediction Technique	Area	Accuracy
Mohammad Azarafza et al. [21]	CNN–DNN model	Iran	90.9%
Haojie Wang et al. [14]	CNN model	Hong Kong	92.5%
BinhThaiPham et al. [1]	NB model	India	90.1%
Proposed Method	Fusion approach of RF and LR	Bangladesh	91.1%

5 Discussion

In this work, landslide susceptibility mapping in certain locations of Bangladesh was modeled using different machine learning techniques. Data preparation is essential for employing these techniques to get the better performance. The main objective of these models is to combine the output of different algorithms to obtain maximum accuracy. The results demonstrate a considerable increase in landslide susceptibility prediction accuracy when compared to the benchmark models. Modeling the system was challenging because there weren't many reference landslides in the data, the triggering factors' data depended heavily on the spatial resolutions of satellite sensor imagery, and the quality of the DEM data directly impacted the quality of the input database, and the predictive model needed powerful processors to handle the inputs during landslide susceptibility assessments.

SVM has 75.99 percent accuracy after preprocessing the data and selecting the features to use, outperforming other commonly used machine learning classifiers. The study then used a fusion method of machine learning algorithms to improve accuracy. This experiment demonstrates how this study combines SVM and NB, DT and RF, RF and LR, DT and GB to find the best solution. This study suggests that the areas which are in great risk of landslides are the South-East areas of Bangladesh.

This study could pave the way for future machine learning research in geological data. This study employed a geological dataset from Bangladesh's southeast that included information on landslide features. As a consequence, this study illustrates how machine learning algorithms work when a geological dataset is employed. Then again, this study reveals how altering some characteristics might change when different natural disasters occur.

6 Conclusions

One of the most difficult jobs in geohazard assessments is mapping the vulnerability of slides. In this study, a novel fusion strategy of a prediction model is used to analyze the susceptibility to landslides in Rangamati, Bangladesh. Various triggering conditions and historical landslide data were used to fit the model. The suggested machine learning fusion model outperformed a wide range of benchmark approaches and provided accuracy of 91.1%. The statistical analysis and presentation of the predicted data show that Bangladesh's hillside faces a serious risk of landslide. The study's findings can be utilized to prepare in advance for any potential landslides in the south-east of the country. So, the study has the potential to save innumerable lives as well as the general public's wealth.

References

1. Pham, B.T., Pradhan, B., Bui, D.T., Prakash, I., Dholakia, M.B.: A comparative study of different machine learning methods for landslide susceptibility assessment: a case study of Uttarakhand area (India). Environ. Model. Softw. **84**, 240–250 (2016)
2. Rabby, Y.W., Hossain, M.B., Abedin, J.: Landslide susceptibility mapping in three Upazilas of Rangamati hill district Bangladesh: application and comparison of GIS-based machine learning methods. Geocarto Int. **37**(12), 3371–3396 (2022)
3. Rabby, Y.W., Ishtiaque, A., Rahman, M.S.: Evaluating the effects of digital elevation models in landslide susceptibility mapping in Rangamati district, Bangladesh. Remote Sens. **12**(17), 2718 (2020)

4. Guzzetti, F., Mondini, A.C., Cardinali, M., Fiorucci, F., Santangelo, M., Chang, K.T.: Landslide inventory maps: New tools for an old problem. Earth Sci. Rev. **112**(1–2), 42–66 (2012)
5. Arora, M.K., Das Gupta, A.S., Gupta, R.P.: An artificial neural network approach for landslide hazard zonation in the Bhagirathi (Ganga) Valley, Himalayas. Int. J. Remote Sens. **25**(3), 559–572 (2004)
6. Althuwaynee, O.F., Pradhan, B., Park, H.J., Lee, J.H.: A novel ensemble bivariate statistical evidential belief function with knowledge-based analytical hierarchy process and multivariate statistical logistic regression for landslide susceptibility mapping. CATENA **114**, 21–36 (2014)
7. Aleotti, P., Chowdhury, R.: Landslide hazard assessment: summary review and new perspectives. Bull. Eng. Geol. Env. **58**(1), 21–44 (1999)
8. Zare, M., Pourghasemi, H.R., Vafakhah, M., Pradhan, B.: Landslide susceptibility mapping at Vaz Watershed (Iran) using an artificial neural network model: a comparison between multilayer perceptron (MLP) and radial basic function (RBF) algorithms. Arab. J. Geosci. **6**, 2873–2888 (2013)
9. Dou, J., et al.: Optimization of causative factors for landslide susceptibility evaluation using remote sensing and GIS data in parts of Niigata, Japan. PLOS ONE **10**(7), e0133262 (2015)
10. Achour, Y., Pourghasemi, H.R.: How do machine learning techniques help in increasing accuracy of landslide susceptibility maps? Geosci. Front. **11**(3), 871–883 (2020)
11. Henriques, C., Zêzere, J.L., Marques, F.: The role of the lithological setting on the landslide pattern and distribution. Eng. Geol. **189**, 17–31 (2015)
12. Pettorelli, N.: The normalized Difference Vegetation Index. Oxford University Press (2013)
13. Ji, L., Zhang, L., Wylie, B.: Analysis of dynamic thresholds for the normalized difference water index. Photogramm. Eng. Remote. Sens. **75**(11), 1307–1317 (2009)
14. Wang, H., Zhang, L., Yin, K., Luo, H., Li, J.: Landslide identification using machine learning. Geosci. Front. **12**(1), 351–364 (2021)
15. Cellek, S.: The Effect of Aspect on Landslide and Its Relationship with Other Parameters. In Landslides. IntechOpen (2021)
16. Ohlmacher, G.C.: Plan curvature and landslide probability in regions dominated by earth flows and earth slides. Eng. Geol. **91**(2–4), 117–134 (2007)
17. Dey, S.K., et al.: Prediction of dengue incidents using hospitalized patients, metrological and socio-economic data in Bangladesh: a machine learning approach. PLoS ONE **17**(7), e0270933 (2022)
18. Schuldt, C., Laptev, I., Caputo, B.: Recognizing human actions: a local SVM approach. In: Proceedings of the 17th International Conference on Pattern Recognition, 2004. ICPR 2004, vol. 3, pp. 32–36. IEEE (2004)
19. Akyol, K., Karacı, A.: Comparing the performances of ensemble-classifiers to detect eye state. I.J. Inform. Technol. Comput. Sci. **14**, 33–38 (2022)
20. Maharjan, M.: Comparative analysis of data mining methods to analyze personal loans using decision tree and naïve bayes classifier. In. J. Educ. Manage. Eng. **12**(4), 33–42 (2022). https://doi.org/10.5815/ijeme.2022.04.04
21. Latif, S., Dola, F.F., Afsar, M.D.M., Esha, I.J., Nandi, D.: Investigation of machine learning algorithms for network intrusion detection. Int. J. Inform. Eng. Electr. Bus. **14**(2), 1–22 (2022)
22. Rahman, M.M., Rana, M.R., Alam, M.N.A., Khan, M.S.I., Uddin, K.M.M.: A web-based heart disease prediction system using machine learning algorithms. Netw. Biol. **12**(2), 64–80 (2022)

New Cost Function for S-boxes Generation by Simulated Annealing Algorithm

Oleksandr Kuznetsov[1,2(✉)], Emanuele Frontoni[1,3], Sergey Kandiy[2], Tetiana Smirnova[4], Serhii Prokopov[5], and Alisa Bilanovych[2]

[1] Department of Political Sciences, Communication and International Relations, University of Macerata, Via Crescimbeni, 30/32, 62100 Macerata, Italy
kuznetsov@karazin.ua, emanuele.frontoni@unimc.it
[2] Department of Information and Communication Systems Security, Faculty of Computer Science, V. N. Karazin Kharkiv National University, 4 Svobody Sq., Kharkiv 61022, Ukraine
[3] Department of Information Engineering, Marche Polytechnic University, Via Brecce Bianche 12, 60131 Ancona, Italy
[4] Cybersecurity & Software Academic Department, Central Ukrainian National Technical University, 8, University Ave, Kropyvnytskyi 25006, Ukraine
[5] Department of Economic and Information Security, Dnipropetrovsk State University of Internal Affairs. Ave. Gagarina 26, Dnipro 49005, Ukraine
prokopovsergejua@gmail.com

Abstract. The simulated annealing algorithm relates to heuristic techniques for approximating optimization problems. It is well suited to finding a solution in a discrete state space. This algorithm simulates the physical processes that occur during metal annealing. While the temperature of the metal is high, the atoms of the substance can pass between the cells of the crystal lattice. As the substance cools, the probability of transitions decreases, and the process freezes. This is simulated in the computational algorithm for finding the global optimum. At each iteration, the simulated annealing algorithm forms several alternatives and chooses one of them. While the temperature is high, the adoption of worsening alternatives is allowed. As the temperature drops, the probability of worsening steps decreases. Cost functions are used to evaluate and compare possible alternatives. We are looking at the computational problem of generating S-boxes in the space of all 8-bit permutations. This is a difficult task, because the search space is very large – there are 256! permutations. We offer a new cost function that reduces the difficulty of finding S-boxes by an algorithm simulated annealing. Generated S-boxes can be used in cryptographic applications, such as symmetric key encryption algorithms and hash functions. Our experiments show that the new cost function allows you to quickly generate highly nonlinear S-boxes for cryptographic applications. For the simulated annealing algorithm, we obtained the best result, and this is the main contribution of this work.

Keywords: Global Optimization · Nonlinear substitution · Simulated Annealing Algorithm · Cost Function · S-boxes Generation

© The Author(s), under exclusive license to Springer Nature Switzerland AG 2023
Z. Hu et al. (Eds.): ICAILE 2023, LNDECT 180, pp. 310–320, 2023.
https://doi.org/10.1007/978-3-031-36115-9_29

1 Instruction

Simulated annealing algorithm refers to the heuristic class of global optimum search [1, 2]. This algorithm is inspired by the physical processes that occur during the cooling of metals. After the metal is heated, slow cooling gives the atoms enough time to organize into stable structures of the crystal lattice. While the temperature of matter is high individual atoms can pass from one cell of the crystal lattice to another. However, as the metal cools, such events occur less frequently and gradually the process is frozen.

For the computational algorithm for global optimum, the metal cooling process is simulated as simple formulas [3, 4]. Possible alternative in the search space is evaluated and ranked using the cost function. We set the initial temperature and calculate the probability of transition between the cells of the crystal lattice. In the search algorithm, this probability will define the adoption of the worst alternative. We are gradually reducing the temperature and the likelihood of adopting deteriorating alternatives is also diminishing. As a result, only improving alternatives will be accepted at low temperature and the process is frozen.

Therefore, to implement heuristic search it is necessary to set several parameters of imitation annealing:

- Initial temperature;
- Cooling coefficient;
- Cost function for comparison and ranking of alternatives.

In this work we offer a new cost function to solve the problem of generating nonlinear substitutions (S-boxes) in the space of all 8-bit permutations. This is a difficult task, because the search space is very large – there are 256! permutations. We want to find high nonlinear substitutions that are used in cryptographic applications. For example, they are useful in symmetric key encryption algorithms, hashing functions, etc.

Thus, the goal of this work is to develop a new cost function that is expected to reduce the computational complexity of the simulation annealing algorithm when searching for S-boxes.

2 Related Works

The problem of generating highly nonlinear S-boxes for cryptographic applications has been solved in many related works.

The works [5–7] used various heuristic techniques to search for cryptographic boolean functions and nonlinear substitutions.

The [8] work used the simulated annealing algorithm to solve this problem. The use of the Walsh-Adamar transform value function was suggested as a heuristic.

[9–11] uses genetic algorithms to generate S-boxes. [12, 13] uses local optimization methods. In [14, 15] the hill climbing algorithm was investigated. The [14] work examined techniques of artificial immunity. Last works summarize and develop this direction. In particular, the work [12] proposes a new cost function that has proved very effective for genetic algorithms and local optimization techniques. The works [13, 16] offer another cost function, which also showed high efficiency with the local search algorithm and

312 O. Kuznetsov et al.

with the hill climbing algorithm. [17] investigated several heuristic algorithms using different heuristics.

In our last work [18] the simulated annealing algorithm was optimized. A simple cost function was used, based on the calculation of the maximum absolute values of the Walsh-Adamar coefficients.

It should be noted that the difficulty of heuristic search of highly nonlinear S-boxes is still very high. For example, for the simulated annealing algorithm, the best known result from [18] is about 450,000 iterations to generate S-boxes with a nonlinearity of 104.

In this paper we continue to investigate the algorithm of simulated annealing. We offer a new cost function and optimize its parameters for fast generation of 8-bit S-boxes with non-linearity 104.

3 Methods

The simulated annealing algorithm is motivated by physical considerations.

The most advantageous energy state of the thermodynamic system is determined by the Boltzmann distribution [19, 20]:

$$p(E_i) = \frac{1}{Z} e^{-\frac{E_i}{k_B T}},\qquad(1)$$

where:

- E_i – the energy of the i-th state;
- k_B – is Boltzmann constant;
- Z – normalization constant defined in such a way that the sum of probabilities for all i is 1.

The probability ratio of two states is called the Boltzmann factor, it usually depends only on the difference of energies of states [19, 20]:

$$\frac{p(E_i)}{p(E_j)} = e^{\frac{E_j - E_i}{k_B T}}.\qquad(2)$$

The Boltzmann distribution shows that states with lower energy will always be more likely. Therefore, the Boltzmann distribution is usually used to solve a variety of tasks [19, 20].

Simulated annealing algorithm simulates physical processes of thermodynamic system [1–3]. We consider the task of finding the optimal solution in the discrete space of possible states. Let's label each state S_i. For example, for the non-linear substitution problem, S_i will denote a specific 8-bit S-box from the space of all 256! permutations.

Consider the cost function F_{CF}, whose values rank states S_i. In this way, $F_{CF}(S_i)$ is the energy analogue of the i-th state of E_i in formula (1). The probability of finding the system in the i- th state by analogy with formula (1) write in the form:

$$p(S_i) = \frac{1}{Z} e^{-\frac{F_{CF}(S_i)}{T}}.\qquad(3)$$

The probability ratio (Boltzmann factor analogue) should be written as.

$$\frac{p(S_i)}{p(S_j)} = e^{\frac{F_{CF}(S_j) - F_{CF}(S_i)}{T}}.$$ (4)

Formula (4) calculates the probability of a system moving from one state to another.

Suppose that the current state of the system is S_j and we have the value of the cost function $F_{CF}(S_j)$. If for an alternative state S_i the value of the cost function satisfies the condition

$$F_{CF}(S_j) - F_{CF}(S_i) > 0,$$

then S_i is accepted as the current state:

$$S_j \leftarrow S_i.$$

If the alternative state S_i is degrading, i.e. if

$$F_{CF}(S_j) - F_{CF}(S_i) \leq 0,$$

then S_i is accepted as the current state with probability (4).

The probability (4) decreases when the temperature T drops (see Fig. 1).

Figure 1 shows the cases:

- $\delta = -(F_{CF}(S_j) - F_{CF}(S_i)) = 1;$
- $\delta = -(F_{CF}(S_j) - F_{CF}(S_i)) = 10;$
- $\delta = -(F_{CF}(S_j) - F_{CF}(S_i)) = 100;$
- $\delta = -(F_{CF}(S_j) - F_{CF}(S_i)) = 1000.$

Thus, to initiate the simulated annealing algorithm, it is necessary to select the cost function F_{CF}, whose values rank the states S_i, as well as to optimize the algorithm by the temperature value T.

High T temperatures allow the algorithm to accept deteriorating $S_j \leftarrow S_i$ states, for which $F_{CF}(S_j) - F_{CF}(S_i) \leq 0$. This is useful for leaving the local optimum. At the same time, the adoption of all worsening states leads to an accidental transition from one state to another state, without optimizing the cost function. For this reason, the T temperature is continuously reduced in proportion to the cooling coefficient α:

$$T_i = \alpha \cdot T_{i-1},$$

where T_i is the temperature used to calculate the probability eq. (4) on the i-th iteration of the algorithm.

In the work [18] we optimized the algorithm of simulated annealing on various parameters. In particular, the most optimal parameters were:

- Initial temperature $T_0 = 20000;$
- Cooling coefficient $\alpha = 0.95.$

For comparison and ranking of alternatives in the work [18] we used the cost function

$$F_{\max WHS}(S) = \sum_{v \in \{0,1\}^8, v \neq 0} \left| \max_{u \in \{0,1\}^8} |WHT(v \cdot S(x), u)| - X \right|^R,$$ (5)

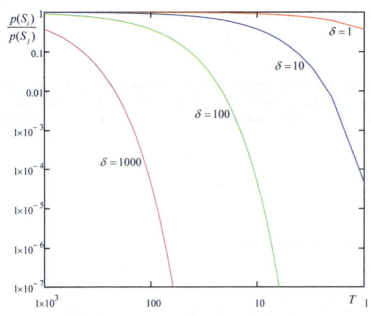

Fig. 1. Probability of taking deterioration $S_j \leftarrow S_i$ for different values of $\delta = -(F_{CF}(S_j) - F_{CF}(S_i))$

where:

- X and R - configurable parameters;
- $WHT(f(x), u)$ - Walsh-Adamar transform coefficients computed with [21]:

$$WHT(f(x), u) = \sum_{x \in \{0,1\}^8} (-1)^{f(x) \oplus u \cdot x}. \quad (6)$$

We consider the generation of 8-bit S-boxes, which are given as a vector function

$$S(x) = \{f_0(x), f_1(x), ..., f_7(x)\}, x = (x_0, x_1, ..., x_7).$$

Cost function (5) inspired by heuristic from work [8]:

$$F_{WHS}(S) = \sum_{v \in \{0,1\}^8, v \neq 0} \sum_{u \in \{0,1\}^8} ||WHT(v \cdot S(x), u)| - X|^R, \quad (7)$$

where X and R - configurable parameters.

In this paper we offer a new cost function and show that its use allows to significantly reduce the number of iterations of the simulated annealing algorithm for generation of 8-bit S-boxes with nonlinearity 104.

4 New Cost Function and Experimental Results

To solve the problem of finding non-linear 8-bit S-boxes, we offer to use the cost function:

$$F_{new}(S) = \sum_{\substack{v \in \{0,1\}^8, v \neq 0 \\ |WHT(v \cdot S(x),u)| > X}} \sum_{u \in \{0,1\}^8} 2^{\frac{|WHT(v \cdot S(x),u)| - Y}{2}}, \tag{8}$$

where X and R - configurable parameters.

We have conducted numerous experiments with the cost function (8). To find substitutions with nonlinearity 104 we used the algorithm of simulated annealing with parameters of [18]. The aim of our experiments was to optimize the parameters X and R in (8), i.e. to find the combinations of parameters that require the least number of iterations of the search algorithm.

The results of the experiments are shown in Tables 1, 2 and 3. In Fig. 2 the results of the experiments are visualized as a diagram.

The Walsh-Adamar transform coefficients calculated by formula (6) are always multiples of 4. For this reason we only consider values of X and R in (8), that are multiples of 4. Moreover, values of $|WHT(v \cdot S(x), u)|$ can only be positive, therefore we only consider $X \geq 0$.

Table 1. The results of experiments

X	Y						
	-48	-44	-40	-36	-32	-28	-24
0	81510.5	82007.1	80159.0	84573.8	82512.3	83204.2	84098.6
4	76745.7	80152.2	77775.5	87041.4	78836.1	82196.8	72215.4
8	77896.1	77147.5	77016.4	78604.4	81195.6	84373.4	85636.3
12	81306.0	80311.2	76357.3	77183.3	85566.4	79498.4	86121.4
16	69922.7	80535.2	72299.1	72995.8	80595.1	75198.7	78454.0
20	78454.8	82545.1	76416.0	81484.9	77798.0	80743.3	81526.4
24	72819.7	74015.6	80954.0	**67748.8**	71932.4	73653.7	77709.8
28	78504.1	75821.3	88405.7	81132.8	83522.8	79287.4	77723.8
32	83205.8	80829.9	75084.4	82512.9	72930.8	80382.0	80026.1
36	82917.9	77044.3	81486.0	82037.7	78891.2	76541.9	78188.4
40	85203.4	74981.2	78713.3	80840.9	78588.0	83632.2	86047.3
44	145775.0	153850.1	140859.2	140564.6	143037.3	149241.1	142471.1
48	–	–	–	–	–	–	–

In this case, we consider only those terms in formula (8) for which.

$$|WHT(v \cdot S(x), u)| - Y \geq 0,$$

Table 2. The results of experiments (continuation)

X	Y						
	−20	−16	−12	−8	−4	0	4
0	80791.1	78309.7	80615.1	87944.2	77346.1	80537.1	0.0
4	74735.9	81232.5	72797.6	76619.7	73580.5	78049.7	70135.7
8	68934.1	80125.6	82580.5	82322.4	73958.8	77840.1	83358.6
12	81978.5	80578.8	84741.8	86711.7	76006.8	84126.6	75120.2
16	74791.3	83375.3	76082.4	75094.1	87580.5	79944.8	80719.5
20	82538.9	82066.4	82771.2	81455.4	86835.6	80526.6	75460.4
24	76279.5	71695.3	81258.0	83892.5	73082.3	73204.3	73171.7
28	78874.1	80055.2	73827.4	74075.7	85799.7	83360.9	76422.2
32	77249.3	76160.7	81596.4	76832.6	79585.0	88636.2	76864.1
36	83097.9	87752.7	84910.5	79248.4	76908.2	78458.9	75015.3
40	84667.1	90001.9	77207.2	81686.5	84297.8	84692.8	79382.6
44	145443.0	143764.6	146710.3	147690.3	155602.7	143380.4	147893.6
48	–	–	–	–	–	–	–

Table 3. The results of experiments (continuation)

X	Y										
	8	12	16	20	24	28	32	36	40	44	48
0											
4											
8	79440.0										
12	82124.2	82615.9									
16	87901.7	75372.5	68098.1								
20	78759.4	83094.7	84156.3	119882.9							
24	68431.8	73465.9	79901.7	111417.2	–						
28	78445.8	77738.1	80959.3	112657.6	–	–					
32	81088.8	84978.2	78308.8	121169.3	–	–	–				
36	80594.5	84572.1	84167.9	117877.9	–	–	–	–			
40	81973.8	79641.8	82003.1	119002.4	–	–	–	–	–		
44	145833.9	141513.1	146733.6	207419.9	–	–	–	–	–	–	
48	–	–	–	–	–	–	–	–	–	–	–

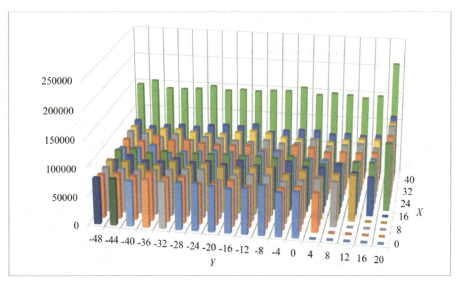

Fig. 2. Average number of iterations of the simulated annealing algorithm with new cost function

i.e. only cases when

$$Y \leq X \geq 0.$$

Tables 1, 2 and 3 provide estimates of the number of iterations of the simulated annealing algorithm for generating 8-bit S-boxes with a nonlinearity of 104. Each value in the table is the average of 100 algorithm runs. Cells marked with the «-» symbol correspond to cases when in all 100 runs of the algorithm no S-boxes with non-linearity 104 were found.

5 Comparison of Results and Discussion

The obtained results show that the new cost function gives high efficiency of search in a wide range of parameters X and R. On average, the simulated annealing algorithm requires about 80,000 iterations to find 8-bit S-boxes with a nonlinearity of 104. For $X > 40$ and $Y > 16$ the difficulty is increasing. For example, with $X > 44$ and $Y > 20$ we could not find any S-boxes with non-linearity 104.

Table 4 shows the results of the comparison with known results. We compare the efficiency of the simulated annealing algorithm by the average number of iterations, as well as the probability of finding S-boxes.

As can be seen from Table 4, the use of the new cost function allowed us to significantly reduce the complexity of S-boxes generation. Compared to the best known result for the simulated annealing algorithm of [18] we managed to reduce the average number of iterations by 5.6 times. The probability of finding 8-bit S-boxes with non-linearity 104 was 100%.

318   O. Kuznetsov et al.

Table 4. Comparison of 8-bit S-boxes search results using simulated annealing algorithm

	Nonlinearity of S-boxes	Generation probability	Average number of iterations
Works [14, 22]	102	0.5%	–
Work [23]	92	–	–
Work [24]	104	–	3 000 000
Work [18]	104	56.4%	450 000
Our work	104	100%	80 000

The best obtained result (highlighted in Table 1 in bold) was about 67,750 iterations, which roughly corresponds to the best known result for the hill climbing algorithm from [13, 16]. So we showed that the simulated annealing algorithm can produce very good results in generating nonlinear substitutions for cryptographic applications.

6 Conclusion

Nonlinear substitution (S-boxes) play an important role in cryptography. They provide nonlinear data transformation and protect against some cryptanalytic attacks. For example, differential, linear algebraic, and many other types of cryptanalysis. Thus, high-nonlinear substitution generation is a topical field of research in modern computer science [25–27].

In this paper heuristic methods of search in space of discrete states were considered. We were looking at the simulated annealing algorithm, inspired by the physics of thermodynamic systems.

Our research consisted in developing a new cost function for heuristic search of 8-bit S-boxes with nonlinearity of 104. We have conducted numerous experiments and have shown that the computational complexity of the simulated annealing algorithm can be significantly reduced. By the number of iterations of the search, our cost function shows better values (for the simulated annealing algorithm) and some better than other heuristic algorithms. A promising area for further research is the use of a new value function in other heuristic search algorithms, for example, in local optimization and hill climbing algorithms [28, 29].

Acknowledgment. 1. This project has received funding from the European Union's Horizon 2020 research and innovation programme under the Marie Skłodowska-Curie grant agreement No. 101007820,

2. This publication reflects only the author's view, and the REA is not responsible for any use that may be made of the information it contains.

References

1. Delahaye, D., Chaimatanan, S., Mongeau, M.: Simulated annealing: from basics to applications. In: Gendreau, M., Potvin, J.-Y. (eds.) Handbook of Metaheuristics. ISORMS, vol. 272, pp. 1–35. Springer, Cham (2019). https://doi.org/10.1007/978-3-319-91086-4_1
2. Eremia, M., Liu, C.-C., Edris, A.-A.: Heuristic optimization techniques. In: Advanced Solutions in Power Systems: HVDC, FACTS, and Artificial Intelligence, pp. 931–984. IEEE (2016). https://doi.org/10.1002/9781119175391.ch21
3. Kirkpatrick, S.: Optimization by simulated annealing: quantitative studies. J Stat Phys. **34**, 975–986 (1984). https://doi.org/10.1007/BF01009452
4. Aarts, E.H.L., van Laarhoven, P.J.M.: Statistical cooling: a general approach to combinatorial optimization problems. Philips J. Res. **40**, 193–226 (1985)
5. Millan, W., Clark, A., Dawson, E.: Heuristic design of cryptographically strong balanced Boolean functions. In: Nyberg, K. (ed.) EUROCRYPT 1998. LNCS, vol. 1403, pp. 489–499. Springer, Heidelberg (1998). https://doi.org/10.1007/BFb0054148
6. Millan, W., Burnett, L., Carter, G., Clark, A., Dawson, E.: Evolutionary heuristics for finding cryptographically strong s-boxes. In: Varadharajan, V., Mu, Y. (eds.) ICICS 1999. LNCS, vol. 1726, pp. 263–274. Springer, Heidelberg (1999). https://doi.org/10.1007/978-3-540-47942-0_22
7. Millan, W., Clark, A., Dawson, E.: Boolean function design using hill climbing methods. In: Pieprzyk, J., Safavi-Naini, R., Seberry, J. (eds.) ACISP 1999. LNCS, vol. 1587, pp. 1–11. Springer, Heidelberg (1999). https://doi.org/10.1007/3-540-48970-3_1
8. Clark, J.A., Jacob, J.L., Stepney, S.: The design of S-boxes by simulated annealing. New Gener Comput. **23**, 219–231 (2005). https://doi.org/10.1007/BF03037656
9. Tesar, P.: A new method for generating high non-linearity s-boxes. Radioengineering **19**, 23–26 (2010)
10. Ivanov, G., Nikolov, N., Nikova, S.: Reversed genetic algorithms for generation of bijective s-boxes with good cryptographic properties. Cryptogr. Commun. **8**(2), 247–276 (2016). https://doi.org/10.1007/s12095-015-0170-5
11. Kapuściński, T., Nowicki, R.K., Napoli, C.: Application of genetic algorithms in the construction of invertible substitution boxes. In: Rutkowski, L., Korytkowski, M., Scherer, R., Tadeusiewicz, R., Zadeh, L.A., Zurada, J.M. (eds.) ICAISC 2016. LNCS (LNAI), vol. 9692, pp. 380–391. Springer, Cham (2016). https://doi.org/10.1007/978-3-319-39378-0_33
12. Picek, S., Cupic, M., Rotim, L.: A new cost function for evolution of s-boxes. Evol. Comput. **24**, 695–718 (2016). https://doi.org/10.1162/EVCO_a_00191
13. Freyre-Echevarría, A., et al.: An External parameter independent novel cost function for evolving bijective substitution-boxes. Symmetry **12**, 1896 (2020). https://doi.org/10.3390/sym12111896
14. Ivanov, G., Nikolov, N., Nikova, S.: Cryptographically strong s-boxes generated by modified immune algorithm. In: Pasalic, E., Knudsen, L.R. (eds.) BalkanCryptSec 2015. LNCS, vol. 9540, pp. 31–42. Springer, Cham (2016). https://doi.org/10.1007/978-3-319-29172-7_3
15. Freyre-Echevarría, A., Martínez-Díaz, I., Pérez, C.M.L., Sosa-Gómez, G., Rojas, O.: Evolving nonlinear s-boxes with improved theoretical resilience to power attacks. IEEE Access **8**, 202728–202737 (2020). https://doi.org/10.1109/ACCESS.2020.3035163
16. Freyre Echevarría, A., Martínez Díaz, I.: A new cost function to improve nonlinearity of bijective S-boxes (2020)
17. McLaughlin, J.: Applications of search techniques to cryptanalysis and the construction of cipher components https://etheses.whiterose.ac.uk/3674/ (2012)
18. Kuznetsov, A., Wieclaw, L., Poluyanenko, N., Hamera, L., Kandiy, S., Lohachova, Y.: Optimization of a simulated annealing algorithm for s-boxes generating. Sensors **22**, 6073 (2022). https://doi.org/10.3390/s22166073

19. Klenke, A.: Wahrscheinlichkeitstheorie. Springer Berlin Heidelberg, Berlin, Heidelberg (2020). https://doi.org/10.1007/978-3-662-62089-2
20. Landau, L.D., Lifshitz, E.M.: Statistical Physics, vol. 5. Elsevier (2013)
21. Carlet, C.: Vectorial Boolean functions for cryptography. Boolean Models and Methods in Mathematics, Computer Science, and Engineering (2006)
22. Clark, J.A., Jacob, J.L., Stepney, S.: The design of s-boxes by simulated annealing. In: Proceedings of the 2004 Congress on Evolutionary Computation (IEEE Cat. No.04TH8753), vol. 2, pp. 1533–1537 (2004). https://doi.org/10.1109/CEC.2004.1331078
23. Wang, J., Zhu, Y., Zhou, C., Qi, Z.: Construction method and performance analysis of chaotic s-box based on a memorable simulated annealing algorithm. Symmetry **12**, 2115 (2020). https://doi.org/10.3390/sym12122115
24. McLaughlin, J., Clark, J.A.: Using evolutionary computation to create vectorial Boolean functions with low differential uniformity and high nonlinearity. arXiv (2013). https://doi. org/10.48550/arXiv.1301.6972
25. Beletsky, A.: Generalized galois-fibonacci matrix generators pseudo-random sequences. IJCNIS **13**, 57–69 (2021). https://doi.org/10.5815/ijcnis.2021.06.05
26. Kuznetsov, A., et al.: Performance analysis of cryptographic hash functions suitable for use in Blockchain. IJCNIS **13**, 1–15 (2021). https://doi.org/10.5815/ijcnis.2021.02.01
27. Iavich, M., Kuchukhidze, T., Gnatyuk, S., Fesenko, A.: Novel certification method for quantum random number generators. IJCNIS **13**, 28–38 (2021). https://doi.org/10.5815/ijcnis. 2021.03.03
28. Kuznetsov, A., et al.: Optimizing hill climbing algorithm for S-boxes generation. Electronics **12**, 2338 (2023). https://doi.org/10.3390/electronics12102338
29. Kuznetsov, A., et al.: Optimizing the local search algorithm for generating s-boxes. In: 2021 IEEE 8th International Conference on Problems of Infocommunications, Science and Technology (PIC S T), pp. 458–464 (2021). https://doi.org/10.1109/PICST54195.2021.977 2163

Enriched Image Embeddings as a Combined Outputs from Different Layers of CNN for Various Image Similarity Problems More Precise Solution

Volodymyr Kubytskyi and Taras Panchenko$^{(\boxtimes)}$

Taras Shevchenko National University of Kyiv, Kyiv, Ukraine
taras.panchenko@knu.ua

Abstract. The number of images is growing daily. Thus, the task of image similarity detection (also known as the de-duplication problem) becomes more and more strong and important for a wide range of applications. The complexity of the two images comparison is too high because of the structure of the data and the general complication of similarity understanding. There are a lot of approaches mentioned in this work have been developed to solve this class of tasks, including SIFT, SURF, ORB, key points, ResNet50-based and many others. But the quality of these methods measured by F_1 score is still non-satisfactory. In this work, we propose a solution for specified tasks, which exceeds the known state-of-the-art models in quality for image pairs similarity detection. The new image embedding, namely the descriptive image vector built as a combination of outputs at different layers of the proposed pre-trained and fine-tuned Convolutional Neural Network, based on ResNet50, is presented here. The model based on a Neural Network was trained over the introduced image embedding vectors over the pairs of similar images from prepared datasets to detect input images similarity. The measurements show that the proposed approach outperforms the state-of-the-art results by F_1 score for image similarity or images near-duplicate detection tasks, described in this work. The real-world applications for three specified sub-tasks from the real estate domain are considered in this work. The proposed model has significantly improved the existing solution for the considered tasks. It is also promising to help solve more general and connected tasks with higher result quality.

Keywords: Image embedding · Image descriptor · Image near duplicates · Image near similarity · Convolutional Neural Network · Intermediate layer

1 Introduction

Image processing technologies have been developing rapidly in recent years. Among the vivid examples are DALL-E [1] and DALL-E 2 [2] by Open AI – two versions of the AI system that can create realistic images and art, expand existing images beyond the original canvas, and make realistic edits to images. AI-based methods and models for image identification [3] exceeded the accuracy of human classification ability on average.

© The Author(s), under exclusive license to Springer Nature Switzerland AG 2023
Z. Hu et al. (Eds.): ICAILE 2023, LNDECT 180, pp. 321–333, 2023.
https://doi.org/10.1007/978-3-031-36115-9_30

By having this tremendous progress in AI-based approaches (primarily artificial neural networks of modern structures) for solving different kinds of tasks with images (classification, clustering, object detection, and other kinds of processing), there are lots of other tasks, targeted at more specific aspects. For example, the detection of sketches, schematical layouts, near-duplicates (similar images) [4–7], images of the same scene or object from different perspectives or angles of view (a kind of similarity again), and many more. In this work, we concentrate on some of these specific tasks, highlight their importance, and propose a solution.

Image representation is a crucial component in computer vision tasks, such as image retrieval and classification. The representation of an image should capture its essential features while being compact enough to be processed efficiently. One of the most commonly used image representations is hand-crafted features, such as SIFT [8, 9], which are not always perfect for tasks that require complex patterns in image recognition.

In recent years, Convolutional Neural Networks (CNNs) have been shown to be effective in extracting powerful features from images, making them a popular choice for image representation. However, using the output of the final layer of a pre-trained CNN as image embeddings can result in large feature vectors with reduced interpretability. To overcome this issue, this paper focuses on using intermediate layers of pre-trained CNNs for image embeddings, which have been shown to provide compact representations that retain the essential features of the image while being computationally efficient. (Image embeddings are compact representations of an image that capture its essential features.)

We explore using intermediate layers of pre-trained CNNs for image embeddings. This approach can provide powerful features compared to traditional image representations, as the intermediate layers have learned to recognize complex patterns in images during training on large datasets. The paper investigates the effectiveness of using intermediate layers as image embeddings in various tasks, such as image retrieval and classification. We also support the proposed heuristics with the calculated F_1 measure score.

A wide range of methods for image similarity analysis were proposed in recent years [10–38]. An overview of these existing solutions is presented in the next section. They are ranged from key point-based, geometric invariant, CNN-based, embeddings-based, local and global features oriented. But these methods do not always show acceptable results for real-world specific tasks.

In this paper, we present the structure of the solution proposed for the image de-duplication problem (also known as near-similar images search) [4–7], introduce new image embedding, describe the developed model in detail, and analyze the results obtained, and its applicative benefits for the LUN.ua, a well-known Ukrainian company in real estate domain.

The main contribution of this work is a newly developed and proposed approach for the de-duplication problem (or image similarity detection task). The main idea of the model is a descriptive image vector ("embedding") introduced as a combination of outputs at different layers of the proposed pre-trained and fine-tuned Neural Network, based on ResNet50 [3].

The obtained result is promising because it exceeds the state of the art for the image de-duplication task [4, 5, 27–33] by using a new combination of outputs at different layers to construct an image embedding.

2 Tasks to Solve and State-of-the-Art Methods Overview

In the real estate realm, there are many tasks with objects and their images. One of the important issues is the de-duplication problem, which means merging separate advertisements of the same object (building, flat, and other types). Different kinds of similarity should be applied here, including image similarity. As practice shows, this could help to fix this issue in good measure.

We will consider the following three types of tasks.

2.1 Near Duplicate Images

Identification of near duplicate images (in general) of various graphical contexts (see Fig. 1 for example). This is an important task to better understand that the real-estate object (flat or some other type) is the same in two different advertisements.

Fig. 1. Near duplicate images

2.2 Multi-Angle Views

The second task is the identification of the same object viewed from different angles (different perspectives) like in Fig. 2. This also helps to identify the same real estate object from different advertisements and photos.

2.3 Schematical Layouts

It is important to classify the images of a real estate object – is it a façade, a schematical plan, or a living room photo? Identification of floorplans like in Fig. 3 is an important task, but out of the scope of this paper. Here we concentrate on floorplan images similarity, which is our third task.

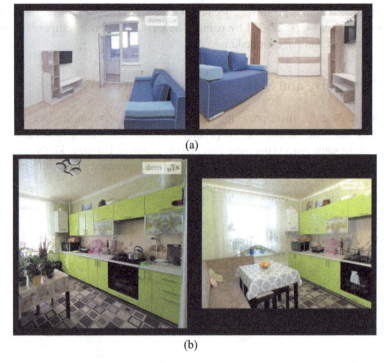

Fig. 2. Multi-angle views

2.4 Existing Approaches and the State-of-the-Art Overview

To solve the issues mentioned above, a wide range of approaches has been developed in recent years. Let us overview the known methods and ideas behind them:

- sub-image retrieval [10, 11];
- local-based binary representation [12];
- keyframe identification with interest point matching and pattern learning [13];
- keypoint-based with scale-rotation invariant pattern entropy analysis [14];
- geometric invariant features [15];
- color histogram, local complexity based on entropy [16];
- min-hash and TF-IDF weighting [17], and other signatures [18];
- affinity propagation [19];
- CNN-based methods and ideas [20–22]: global and local features matching, and intermediate layers aggregation;
- color histograms and locality-sensitive hashing [9], SIFT [8, 9], approximate set intersections between documents computing [9, 17].

Several datasets and benchmarks for progress tracking also appeared [23–26].

Some methods proposed recently are based on image embeddings [27–33]. It implies that a numeric vector, a kind of description, should be calculated (or built) for every image. This embedding (vector) should be distinctive to "catch" the specifics of each

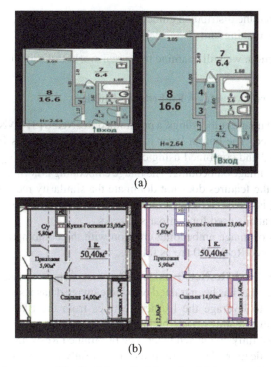

Fig. 3. Schematical layouts of flats

image. We can find similar models (namely, "embeddings") for texts, audio, and other kinds of media to be used in data science widely in our days for different analyses.

The best-known results [38] by the commonly used measure F_1 (which is the harmonic mean of precision and recall for image pairs true near-duplicates detected by the evaluated method) could be obtained by the following approaches [38]:

- by image embedding formed by taking previous before the last layer of pre-trained ResNet50 (the image feature vector),
- SIFT / SURF / ORB descriptors,
- perceptual DCT hash,
- image embedding formed by the combination of intermediate layers of ResNet50 (the proposed method).

Now we are going to describe the proposed approach in detail.

3 Methodology. Image Embedding Constructing and Using to Solve the Tasks

To construct the new proposed image embeddings, and thus to solve the tasks stated in the previous section, we have done the following:

1) built an image feature extractor (embedding builder),

326   V. Kubytskyi and T. Panchenko

2) adopted and used the distributed embeddings storage,
3) built a decision-making unit.

In the next section, we will examine these real-world tasks application results.

3.1 Image Feature Extractor – Embedding Builder

To create proposed image embeddings, a pre-trained ResNet50 [3] CNN model for image classification is used. The output layer is removed, and activations from intermediate layers are extracted and combined using concatenation to form a single feature vector, which serves as the image embedding. The image embedding is normalized to ensure that the magnitude of the features does not dominate the similarity measure. The resulting image embedding is robust to image transformations, brightness/contrast changes, and noise/watermarks and is used for comparison tasks.

Detailed steps of image embedding creation with the combination of intermediate layers of a pre-trained ResNet50 are as follows.

1. Load the pre-trained ResNet50 model. The first step is to load the pre-trained ResNet50 model that has been trained on a large image dataset such as ImageNet [3]. This allows us to leverage the learned features from the training data to extract high/mid/low-level features from the input images.
2. Remove the final fully connected layer. The pre-trained ResNet50 model is designed for image classification. Therefore, it has a final fully connected layer that outputs the class probabilities. This layer is removed as we do not need it for our image embedding task.
3. Extract intermediate layer activations. We extract the activations of the intermediate layers of the ResNet50 model, as these layers contain different types of features at different levels of abstraction. These activations can be seen as feature maps representing the input image's feature activations at each layer. To extract the intermediate layer activations, we pass the input image through the ResNet50 model and record the activations at each intermediate layer.

   Choosing which intermediate layers to extract. The choice of which intermediate layers to extract is based on the desired level of abstraction and the complexity of the features needed for a specific task. In general, the activations from early layers of the CNN contain lower-level features such as edges, textures, and basic shapes, while the activations from later layers contain higher-level features such as objects and parts.

   The number of intermediate layers to extract depends on the desired size of the image embedding and the computational resources available. A larger image embedding vector may contain more information but may also require more computational resources. On the other hand, a smaller image embedding vector may contain less information but may be computationally less expensive. Generally, a trade-off is made between the size of the image embedding and the computational resources available. Experimentation with different numbers of intermediate layers can help determine the optimal size of the image embedding.

   For image comparison tasks, extracting activations from both early and late layers is beneficial, as this ensures that the image embedding captures both low-level and high-level features. We do also use mid-level features to enrich image embedding with more details.

4. Combining the intermediate layer activations. The intermediate layer activations are concatenated into a single vector, which forms the image embedding. The concatenation of activations from multiple intermediate layers is important as it captures the rich and diverse set of features from different levels of abstraction. The activations from each layer are first flattened into a one-dimensional vector and then concatenated along the feature dimension to form the final image embedding.

   There are several ways to combine intermediate layer features to form an image embedding:

   - Concatenation. Concatenation is the process of combining the intermediate layer features by appending them end-to-end to form a single long feature vector. The resulting feature vector is then used as the image embedding. Concatenation is a simple and straightforward method and has been used successfully in many image comparison tasks. The main advantage of concatenation is that it captures the rich and diverse set of features from different levels of abstraction in CNN.
   - Summation. Summation combines the intermediate layer features by adding them elementwise to form a single feature vector. Summation is computationally efficient as it requires fewer operations compared to concatenation. However, summation may lead to information loss as it combines features from different levels of abstraction into a single feature vector.
   - Maximum activation. Maximum activation combines the intermediate layer features by selecting the maximum activation value from each layer to form a single feature vector. Maximum activation is computationally efficient and can capture the most discriminative features from the intermediate layers. However, maximum activation may not capture the rich and diverse set of features from different levels of abstraction in the CNN.
   - Average activation. Average activation combines the intermediate layer features by averaging the activations from each layer to form a single feature vector. Average activation is computationally efficient and can capture the overall features from the intermediate layers. However, average activation may not capture the rich and diverse set of features from different levels of abstraction in the CNN.

   So, the choice of combining methods depends on the task and the desired level of abstraction in the image embedding. Concatenation is a widely used method for image comparison tasks as it captures the rich and diverse set of features from different levels of abstraction in the CNN, so we also use it.

   The resulting image embedding vector serves as a compact representation of the input image and can be used for comparison tasks. The image embedding vector is resistant to linear image transformations, brightness and contrast changes, and is robust to image noise and watermarks. By combining the intermediate layer activations, we can obtain a rich and diverse set of features that represent the input image.

5. Normalize the image embedding. The image embedding vector is normalized to ensure that the magnitudes of the elements do not dominate the similarity measure

used for comparison. This is important as it ensures that the embeddings are not affected by the scale of the features.

As the result, the process of creating image embeddings involves inheriting a pre-trained CNN for image classification, removing the output layer responsible for image classification, and combining intermediate layer features using a method such as concatenation to form a single feature vector that serves as an image embedding, which can then be used for image comparison tasks.

Image embeddings can be used for various image comparison tasks such as image retrieval, image similarity, and image classification. For example, in an image retrieval task, given a query image, the image embeddings can be used to retrieve similar images from a database. In an image similarity task, image embeddings can be used to measure the similarity between two images. In an image classification task, the image embeddings can be used as input features for a classifier to classify images into different categories.

3.2 Embeddings Comparison, Search Optimization, and an Appropriate Storage

For efficient retrieval of similar images, it is important to index the image embeddings in the database. Indexing is a technique that allows for fast searching of data in a database, and it can be implemented using various algorithms such as KD-Tree, Ball-Tree, or Annoy or DCT hash comparison [39].

DCT hash comparison can be used as a pre-filter for multi-million image comparison tasks to eliminate images that are significantly different from the query image. The DCT hash is a compact and robust representation of an image, obtained by applying the Discrete Cosine Transform (DCT) to a small, downscaled version of the image. The DCT coefficients are then quantized, and the resulting values are concatenated to form the DCT hash [40].

The DCT hash comparison can be performed by calculating the hamming distance between the query image's DCT hash and the DCT hashes of the images stored in the database. The hamming distance is a measure of the difference between two binary strings and can be calculated as the number of positions at which the corresponding bits are different.

Images with a high hamming distance can be considered significantly different and can be eliminated from the database before the image embedding comparison. This can greatly reduce the size of the database that needs to be searched, leading to improved efficiency.

As for storing the image embeddings, a document-oriented database, such as MongoDB, is a suitable solution for storing the data. MongoDB is a NoSQL database that stores data in JSON-like documents and can handle large amounts of unstructured data. The image embeddings can be stored as values in the database, with unique identifiers, such as file names or image hashes, serving as the keys.

By using MongoDB, the image embeddings can be easily queried and retrieved for comparison without the need for complex relational database management. Indexing can also be applied to the image embeddings to further improve the efficiency of the comparison process.

The comparison results can be further processed and filtered based on the required similarity threshold or can be sent to the next block – the decision-making unit, which helps to apply the same image embeddings to different comparison tasks.

3.3 Decision-Making Unit

To apply the previously built and stored image embeddings to a specific task of supervised image comparison, we propose to follow the next steps.

1. Collect a dataset of image pairs. First, we need to collect a dataset of image pairs that we want to compare. For example, if we want to compare near-duplicate images, we need to collect pairs of images that are considered to be near-duplicates and which are not considered as such. If we want to compare photos of rooms taken from different angles, we need to collect pairs of photos of the same room taken from different angles and photos of various rooms taken from random view angles.
2. Generate image embeddings. For each image in the sample dataset, we run the image embedding builder and form a vector that describes the image features. These image embeddings will be used as input for the decision-making unit.
3. Form a new vector for each image pair. Each pair from the sample forms a new vector, which is obtained by combining the image embeddings of the two images. The combination can be constructed using Euclidean distance, L_1 distance, cosine similarity, or a combination of these metrics.
4. Train a neural network. The new image pair vector thus formed should be used to train a new fully-connected N-layer neural network with one output neuron. The output of the neural network will be used as a decision on the similarity between the two images in the pair.
5. Fine-tune the network. The trained neural network can be further fine-tuned based on specific task requirements, such as the choice of a distance metric, the number of hidden layers, and the number of neurons in each layer (a set of hyper-parameters).

By following these steps, we can apply the previously built image embeddings to a specific task of supervised image comparison, where the goal is to compare pairs of images and decide on their similarity.

4 Real-World Tasks Application Results and Comparison

Let's examine the results on three practical applications mentioned in the Introduction section, namely: finding near duplicate images, clustering rooms based on photos from different shooting angles, and identifying similar floorplans (or schematical layouts).

To solve the proposed problem of near-duplicate identification, collecting a dataset of pairs of sample images was necessary.

In the first case, we need to collect pairs that are considered to be near-duplicates and which are not considered as such. In the second, there are pairs where the photo of the same room is taken from different angles, and various rooms are taken from random view angles. For the third, we need pairs of floorplans of the same flat (real estate object).

LUN.ua have prepared 3 datasets for these three sub-types of the image comparison tasks:

- near duplicate images of various graphical contexts (dataset size: 80 000 image pairs),
- multi-angle photos of the rooms (dataset size: 12 500 image pairs),
- schematical layouts / floorplans (dataset size: 8 800 image pairs).

Then we run the embeddings builder and form a vector for each image in the sample dataset. Each pair from the sample forms its new vector, which is obtained by combining the Euclidean metrics, L_1, and cosine distance. The new image pair vector thus formed has been used then to train a new fully-connected N-layer neural network with one output neuron. As a result of training, the decision block has been trained to compare pairs of images to get an expected result of similarity.

Building in a proposed way the image embedding and applying it to the task of the image near duplicate detection showed incredible results on the private dataset (LUN real estate images) – 8–10 times fewer mistakes in near-duplicate determination compared to SIFT [8, 9], SURF [41], ORB [42] key points algorithms, and 3–6 times fewer mistakes in comparison with ResNet50 [3], the state of the art result in the field. The obtained comparative experimental results are presented in Table 1, where one can ensure that the proposed model *outperforms the best-known results* for these 3 tasks.

Table 1. Results comparison (F_1 measure score) for the 3 tasks

Task \ Method	$(1)^{**}$	(2)	(3)	(4)
Near duplicate images of various graphical contexts (dataset size: 80 000 image pairs)	$\mathbf{0.94}^*$	0.84	0.82	0.76
Multi-angle photos of the rooms (dataset size: 12 500 image pairs)	$\mathbf{0.87}^*$	0.79	0.77	0.66
Schematical layouts / floorplans (dataset size: 8 800 image pairs)	$\mathbf{0.79}^*$	0.70	0.67	0.66

* the best results are in **bold** for each task (the higher F_1 measure score – the better)
** (1) the proposed method's F_1 score, (2) ResNet50 features vector F_1 score, (3) SIFT/SURF/ORB descriptors F_1 score, (4) Perceptual DCT hash F_1 score

Hence, the proposed approach demonstrates acceptable results and outperforms the best-known models by F_1 score, and now is used in production at LUN.ua. Detailed measurements of the F_1 score and other metrics can be found in [38]. The dataset description as well as more details on calculations could also be viewed there. We will not concentrate on these questions here.

5 Summary and Conclusion

The three specific tasks of the image similarity detection general problem (or image de-duplication problem) were the research topic in this paper.

The new image embedding was introduced in this work. The method of data preparation as well as the step-by-step algorithm for the proposed model construction was presented here. The usage of a combination of different layers outputs from a pre-trained ResNet50 model to obtain new image embeddings (with calibration and normalization) helped to train a decision-making unit (a neural network), which gave an acceptable solution for images similarity task, and specifically three applicative sub-tasks for the real-estate domain.

The main difference between the proposed method from the related research is the new combination of outputs at different layers of the proposed pre-trained and fine-tuned Convolutional Neural Network, based on ResNet50, for the construction of a descriptive image vector called "image embedding", which helped to exceed the state of the art results for the tasks specified by the commonly used F_1 measure. The proposed and described in the paper new enriched image embedding is the main contribution to the research field.

Measurement of the solution quality, namely the F_1 score, assures that the suggested approach is appropriate and applicable to the considered tasks. The optimized model proved its high efficiency and has been used in LUN.ua Ukrainian company for real-world tasks solutions from the real estate domain.

The obtained results demonstrated a new way to solve an important image similarity problem with better quality. This way the presented model advances the existing methods toward results with higher precision, which means more accurate image near-duplicate detection.

References

1. Ramesh, A., et al.: Zero-shot text-to-image generation. In: International Conference on Machine Learning, pp. 8821–8831. PMLR, (2021)
2. Ramesh, A., Dhariwal, P., Nichol, A., Chu, C., Chen, M.: Hierarchical text-conditional image generation with clip latents. arXiv preprint arXiv:2204.06125 (2022)
3. Deng, J., Dong, W., Socher, R., Li, L.-J., Li, K., Fei-Fei, L.: ImageNet: a large-scale hierarchical image database. In: IEEE Computer Vision and Pattern Recognition, CVPR (2009)
4. Thyagharajan, K.K., Kalaiarasi, G.: A review on near-duplicate detection of images using computer vision techniques. Arch Computat. Methods Eng. **28**(3), 897–916 (2021)
5. Kaur, G., Devgan, M.S.: Data deduplication methods: a review. Int. J. Inform. Technol. Comput. Sci. **9**(10), 29–36 (2017)
6. Mohidul Islam, S.M., Debnath, R.: A comparative evaluation of feature extraction and similarity measurement methods for content-based image retrieval. Int. J. Image, Graph. Sig. Process. **12**(6), 19–32 (2020)
7. Er. Numa Bajaj, Er. Jagbir Singh Gill, Rakesh Kumar. An Approach for Similarity Matching and Comparison in Content based Image Retrieval System[J]. IJIEEB, 2015, 7(5): 48–54
8. Lowe, D.G.: Distinctive image features from scale-invariant keypoints. Int. J. Comput. Vision **60**(2), 91–110 (2004)
9. Chum, O., Philbin, J., Isard, M., Zisserman, A.: Scalable near identical image and shot detection. In: Proceedings of the 6th ACM International Conference on Image and Video Retrieval, pp. 549–556 (2007)
10. Ke, Y., Sukthankar, R., Huston, L.: Efficient near-duplicate detection and sub-image retrieval. ACM Multimedia **4**(1), 1–5 (2004)

11. Ke, Y., Sukthankar, R., Huston, L.: An efficient parts-based near-duplicate and sub-image retrieval system. In: Proceedings of the 12th Annual ACM International Conference on Multimedia, pp. 869–876 (2004)
12. Nian, F., Li, T., Wu, X., Gao, Q., Li, F.: Efficient near-duplicate image detection with a local-based binary representation. Multimed. Tools Appl. **75**(5), 2435–2452 (2016)
13. Zhao, W.L., Ngo, C.W., Tan, H.K., Wu, X.: Near-duplicate keyframe identification with interest point matching and pattern learning. IEEE Trans. Multimed. **9**(5), 1037–1048 (2007)
14. Zhao, W.L., Ngo, C.W.: Scale-rotation invariant pattern entropy for keypoint-based near-duplicate detection. IEEE Trans. Image Process. **18**(2), 412–423 (2009)
15. Lei, Y., Zheng, L., Huang, J.: Geometric invariant features in the Radon transform domain for near-duplicate image detection. Pattern Recogn. **47**(11), 3630–3640 (2014)
16. Li, Y.: A fast algorithm for near-duplicate image detection. In: 2021 IEEE International Conference on Artificial Intelligence and Industrial Design, AIID: IEEE, pp. 360–363 (2021)
17. Chum, Q., Philbin, J., Zisserman, A.: Near duplicate image detection: min-hash and TF-IDF weighting. BMVC **810**, 812–815 (2008)
18. Liu, L., Lu, Y., Suen, C.Y.: Variable-length signature for near-duplicate image matching. IEEE Trans. Image Process. **24**(4), 1282–1296 (2015)
19. Xie, L., Tian, Q., Zhou, W., Zhang, B.: Fast and accurate near-duplicate image search with affinity propagation on the ImageWeb. Comput. Vis. Image Underst. **124**, 31–41 (2014)
20. Zhou, Z., Lin, K., Cao, Y., Yang, C.N., Liu, Y.: Near-duplicate image detection system using coarse-to-fine matching scheme based on global and local CNN features. Mathematics **8**(4), 644 (2020)
21. Kordopatis-Zilos, G., Papadopoulos, S., Patras, I., Kompatsiaris, Y.: Near-duplicate video retrieval by aggregating intermediate CNN layers. In: Amsaleg, L., Þór Guðmundsson, G., Gurrin, C., Þór Jónsson, B., Satoh, S. (eds.) Multimedia Modeling, pp. 251–263. Springer International Publishing, Cham (2017). https://doi.org/10.1007/978-3-319-51811-4_21
22. Zhang, Y., Zhang, S., Li, Y., Zhang, Y.: Single- and cross-modality near duplicate image pairs detection via spatial transformer comparing CNN. Sensors **21**(1), 255 (2021)
23. Barz, B., Denzler, J.: Do we train on test data? Purging Cifar of near-duplicates. J. Imaging **6**(6), 41 (2020)
24. Matatov, H., Naaman, M., Amir, O.: Dataset and case studies for visual near-duplicates detection in the context of social media. arXiv preprint arXiv:2203.07167 (2022)
25. Tralic, D., Zupancic, I., Grgic, S., Grgic, M.: CoMoFoD: New database for copy-move forgery detection. In: Proceedings of the 55th International Symposium ELMAR, pp. 49–54 (2013)
26. Morra, L., Lamberti, F.: Benchmarking unsupervised near-duplicate image detection. Expert Syst. Appl. **135**, 313–326 (2019)
27. Barz, B., Denzler, J.: Hierarchy-based Image Embeddings for Semantic Image Retrieval. (2019). https://arxiv.org/pdf/1809.09924v4.pdf
28. Berman, M., J´egou, H., Vedaldi, A., Kokkinos, I., Douze, M.: MultiGrain: a unified image embedding for classes and instances. arXiv preprint (2019). https://arxiv.org/pdf/1902.055 09v2.pdf
29. Yu, Z., Zheng, J., Lian, D., Zhou, Z., Gao, S.: Single-Image Piece-wise Planar 3D Reconstruction via Associative Embedding. arXiv preprint (2019). https://arxiv.org/pdf/1902.097 77v3.pdf
30. Rau, A., Garcia-Hernando, G., Stoyanov, D., Brostow, G.J., Turmukhambetov, D.: Predicting Visual Overlap of Images Through Interpretable Non-Metric Box Embeddings. arXiv preprint (2020). https://arxiv.org/pdf/2008.05785v1.pdf
31. Feng, G., Hu, Z., Zhang, L., Lu, H.: Encoder Fusion Network with Co-Attention Embedding for Referring Image Segmentation. arXiv preprint (2021). https://arxiv.org/pdf/2105.01839v1. pdf

32. Asadi-Aghbolaghi, M., Azad, R., Fathy, M., Escalera, S.: Multi-level Context Gating of Embedded Collective Knowledge for Medical Image Segmentation. arXiv preprint (2020). https://arxiv.org/pdf/2003.05056v1.pdf
33. He, K., Zhang, X., Ren, S., Sun, J.: Deep Residual Learning for Image Recognition. arXiv preprint (2015). https://doi.org/10.48550/arXiv.1512.03385
34. Lytvynenko, T.I., Panchenko, T.V., Redko, V.D.: Sales forecasting using data mining methods. Bull. Taras Shevchenko Natl. Univ. Kyiv, Ser.: Phys.-Math. Sci. **4**, 148–155 (2015)
35. Bieda, I., Panchenko, T.: A systematic mapping study on artificial intelligence tools used in video editing. Int. J. Comput. Sci. Netw. Secur. **22**(3), 312–318 (2022)
36. Bieda, I., Kisil, A., Panchenko, T.. An approach to scene change detection. In: Proceedings of the 11th IEEE International Conference on Intelligent Data Acquisition and Advanced Computing Systems: Technology and Applications (IDAACS'2021), vol. 1, pp. 489–493 (2021)
37. Bieda, I., Panchenko, T.: A Comparison of scene change localization methods over the open video scene detection dataset. Int. J. Comput. Sci. Netw. Secur. **22**(6), 1–6 (2022)
38. Kubytskyi, V., Panchenko, T.: An effective approach to image embeddings for e-commerce. In: Information Technologies and Implementations (2022)
39. Tang, Z., Yang, F., Huang, L., Zhang, X.: Robust image hashing with dominant DCT coefficients. Optik **125**(18), 5102–5107 (2014)
40. Zeng, J.: A novel block-DCT and PCA based image perceptual hashing algorithm. arXiv preprint (2013). arXiv:1306.4079
41. Bay, H., Tuytelaars, T., Van Gool, L.: SURF: speeded up robust features. In: Leonardis, A., Bischof, H., Pinz, A. (eds.) ECCV 2006. LNCS, vol. 3951, pp. 404–417. Springer, Heidelberg (2006). https://doi.org/10.1007/11744023_32
42. Rublee, E., Rabaud, V., Konolige, K., Bradski, G.: ORB: An efficient alternative to SIFT or SURF. In: 2011 International Conference on Computer Vision, pp. 2564–2571. ICCV, Barcelona, Spain (2011)

A Novel Approach to Network Intrusion Detection with LR Stacking Model

Mahnaz Jarin[1(✉)] and A. S. M. Mostafizur Rahaman[2]

[1] Department of ICT, Bangladesh University of Professionals, Dhaka, Bangladesh
mahnaz.jarin@gmail.com
[2] Department of CSE, Jahangirnagar University, Dhaka, Bangladesh

Abstract. IDS is one of the most researched subjects of network security. Single-classifier IDS tends to fail in many scenarios where ensemble or hybrid approaches have been adopted. In the ensemble approach, many techniques have been adopted that give high efficiency yet complicate the system even when to make the system less complicated widely used classifiers like SVM, KNN, RF etc. is vastly used while others lag behind. Keeping a view to that, we propose a stacking ensemble approach with Logistic regression classifier for network-based anomaly detection. Also, using it to detect both binary and multiclass classification, shows its multiclass usability. The model was evaluated on the NSL-KDD dataset, for both binary and multiclass classification. Achieving an accuracy of 98.80%, a precision of 98.80%, and a recall of 98.80% for binary classification. 98.66%, 98.63%, And 98.66% were respectively the accuracy, precision, and recall for the multiclass classification.

Keywords: Logistic regression · intrusion detection · multiclass classification

1 Introduction

Intrusion detection systems come in various forms. Some are built for host-based detection, and some are for network intrusion detection. These categories can be further divided into branches like signature-based detection, anomaly detection, etc. different machine learning, deep learning, and unsupervised learning techniques have been widely used in developing these systems.

Ensemble models have been one of them. Researchers from [1–6] adopted many ensemble techniques that can produce a high detection rate, but with that it can make the system computationally complex, and heavy for many devices. Even they vastly used machine learning classifiers like SVM, RF, KNN, and gradient boost (XGB).but a few of them has explored another promising yet simple classifier like logistic regression (LR). The logistic regression classifier has lagged behind in use of intrusion detection especially as the Meta classifiers. Some approaches have adopted it in building the ensemble model like in [7] but it is used just as a weak learner. Again. [8] Used it for Host based detection specifically for cloud environment. These shows a gap of utility of LR in network based intrusion detection that can unfold improved detection rates.

© The Author(s), under exclusive license to Springer Nature Switzerland AG 2023
Z. Hu et al. (Eds.): ICAILE 2023, LNDECT 180, pp. 334–343, 2023.
https://doi.org/10.1007/978-3-031-36115-9_31

Also fill the gap that is sustaining. Leveraging the advantages of LR like fast training, efficient discrete output in network-ids is the primary objective of this contribution.

With a view to that, this paper focused on achieving three goals, first building a network-based ensemble model with LR as the final estimator. Secondly, securing a decent detection rate both for binary and multiclass classification as LR works best for binary detections. Finally to build a system that would be lightweight and won't jam daily use devices.

[1] uses DT-ensemble SVM for binary classification on NSL-KDD dataset achieving an accuracy of 99.36% and false alarm rate(FAR) of 0.38%. [2] Proposes an approach using Particle Swarm optimizer (PSO) with SVM and KNN for intrusion detection having an accuracy of 92.82% on KDD Cup99 dataset. Kunal and Mohit [3] demonstrated a rank-based attribute selection method to reduce attributes and then used an ensemble approach with RF, KNN, j48graft, and REP Tree.99.68% and 99.72% accuracy was achieved for multiclass and binary classification accordingly. [4] proposes CFS-BA, using the ensemble of C4.5, Forest PA, and RF algorithms. This model was evaluated on the AWID, NSL-KDD, and CICIDS2017 datasets. Getting the highest accuracy of 99.89% for their approach in the CICIDS2017 dataset. [5] built an GBM-RF model for multiclass classification with highest 99.77% accuracy for DoS attack. [6] presents a KNN, extreme machine learning (ELM) based method for intrusion detection with 84.29% accuracy.

Abbas et al. [8] presented an ensemble model for IoT that used LR, NB, and DT. They achieved 88.96% accuracy for binary classification and 88.92% accuracy for multiclass classification on the CICIDS2017 dataset was used. [8] Uses LR for host based detection applicable for cloud environment which has 97.51% accuracy. [9] Introduced a stacking model with SVM, KNN, RF, and LR. Achieving accuracy of 94% for emulated and 97% for the real-time dataset. Kumar [3] proposed a hybrid model using NN and MOGA. The model works in two phases and uses the major voting technique for the final output. It has an accuracy of 88% and 97% respectively for NSL_KDD and ISCX-2012. Bhati et al. [10] debated that XGBoost can smoothen the bias-variance tradeoff, by using this they proposed an ensemble approach evaluating on KDD CUP99 dataset. That achieved an accuracy of 99.95% for detecting normal data.

Gao et al. [11] took the NSL-KDD dataset and proposed a MultiTree algorithm by coping up the size of the training set and building several decision trees. They used several base classifiers like RF, and KNN and introduced an adaptive voting algorithm. The MultiTree had an accuracy of 84.2%, final accuracy for the voting algorithm was 85.2%. [12] Presents a RF feature selection with XGB hybrid. From [13–17] shows different ensemble approaches where [16] had the highest accuracy of 99.81% for binary classification using adaboost reptree.

[18] Experiments with RF- DT ensemble and feature selection for IoT on NF-ToN-IoT-v2 NF-BoT-IoT-v2 datasets. [19] Uses the algorithms and Fusion of Multiple Classifier (FMC) and Fuzzy Ensemble Feature selection (FEFS). Kumar et al. [20] conducted a broad review on ensemble IDS, where they concluded that modern-day IDS needs the ability to handle a wide range of traffic on speedy networks. Also, they mentioned the challenges such as the lack of an ideal distributed test bed to examine the ids. They stressed the use of ensemble models in the field of IDS as they have the potential to

deliver more effective results. [20] Proposed white box attacks against NIDS and their defense mechanisms. [21] On the NSL KDD dataset, a cost-effective feature selection IDS was proposed. Hybrid, wrapper, and filter techniques are applied in [22] for feature selection and SVM to classify for NIDS.

1.1 Stacking Model

Among the three types of ensemble learning, we choose stacking approach for our model. Since the objective was to build an ensemble model giving utmost importance to Logistic Regression, the stacking model would help us to approach our work in two phases.

Phase one consists of using base classifiers, the goal was to choose classifiers that would complement each other's weaknesses, and won't make the training set difficult to train. Since our final estimator was fixed, the base classifiers have to be efficient enough to make the final outcome better on average to avoid the poor choice of any classifier. As Logistic regression can't work better with variables that aren't linearly separable, the multiclass classification could raise an issue.

Our model uses eight machine-learning classifiers as base models, these classifiers are Linear SVM since it's computationally easy gives accuracy while there is a clear margin, and polynomial SVM helps to expand the classifier to a higher degree. KNN is used as its ideal for non-linear data, handles multiclass problems efficiently, easily trainable. RF for its class balance tendency and high performance. DT has the ability to handle multi output data, the using cost of the tree is logarithmic making it lightweight. XGB for its flexibility and accuracy performance. Also it can boost any loss function. LightGBM for faster training speed, giving better results for XGB. Finally, LR is used both in as a base classifier also a final estimator.

2 Methodology

Our stacked approach uses eight classifiers to build new set of training data that would be used to identify unknown data. The base classifiers for say $A_1, A_2, \ldots, A_8$ are trained using the training dataset and each of the classifiers creates their own individual predictions for say $P_1, P_2, \ldots, P_n$, these predictions form a new training dataset for the second classifier od also known as the meta classifier. The meta-classifier is trained on the newly made dataset and optimized it for errors. After training, the output of the meta classifier performs as a generalized model.

In our model the base classifiers, for example, $A_1, A_2, \ldots, A_8$ are SVM (linear and polynomial), KNN, RF, LR, DT, XGB, and LightGBM. As the final estimator, we used LR. That works in,

$$p = e^{(c^0 + c^1 * x)}/(1 + e^{(c^0 + c^1 * x)}) \tag{1}$$

The p is the predicted outcome and c^0 is the intercept term or generally known as bias and c^1 is the coefficient term for each single value input that is generated by the base learners. Figure 1 shows the flow of our system.

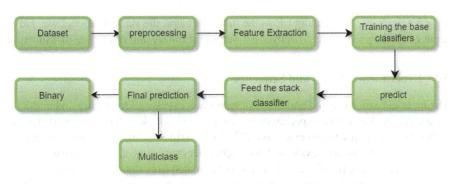

Fig. 1. Stacking ensemble model

2.1 Dataset and Preprocessing

To conduct the experiment the KDDtrain[+] was chosen. This is from the NSL-KDD dataset group. This dataset consists of a total of 42 attributes, were 4 of which were non-numeric. It has 23 different attacks under the label attribute.

For preprocessing, the dataset was checked for the number of rows and columns. Looked for if any null values were present. The appropriate column names were given to it. The 23 attacks were classified under their corresponding attack type. The four types were: DoS, probe, U2R, and R2L. Then, the numeric attributes were normalized. Among the four non-numeric attributes, the difficulty column was dropped. The remaining three, namely flag, protocol_type, and service went through one-hot encoding and a categorical dataframe was built. Two datasets were built, one was for binary classification, and the other was for multiclass. For the first one, the labels were changed to "normal" and "abnormal". A binary dataframe was created with them and labels were encoded after that. A dataset with labels and encoded columns was created.

For multiclass, the same process was repeated. Only here the attack labels are kept the same as the four categories assigned to them at first. Another label was "normal". A dataset with multi-labels and encoded columns was created.

2.2 Feature Extraction and Model Training

In this stage, from the binary dataset, the numeric attributes and the encoded label attributes were taken to create a dataframe. Using Pearson correlation, the features having greater than 0.5 correlations with encoded attack label features were selected. 9 attributes were found here. These attributes were taken and combined with the categorical dataframe of one-hot-encoded labels. Finally, for the final binary dataset real attack labels, one-hot-encoded and encoded features were combined.

Again, the same steps were repeated for the multi-class dataset. The final dataset was saved. These two sets were read to be machine fed. For this experiment, these datasets were fed to SVM (Linear and Polynomial), KNN, RF, LR, DT, GradiantBoost, and LightGBM serially in two rounds with a ratio of 0.75:0.25 for training testing split respectively. First for binary classification, and second for multi-class. These models were cross-validated using tenfold cross-validation. RF had 500 estimators. KNN had

$K = 5$ neighbors. Then, using LR as the final estimator, the stacking model was fed the input of all the predictions to deliver the output for binary and multiclass classification.

2.3 Experimental Setup and Sequence

The meta-classifier was fed and the final result was calculated for binary and multiclass classification. The whole experiment was executed in google colab on a corei5 machine with 8 GB ram and with no external GPU. It was written using the python language.

The experiment was evaluated for higher accuracy, low false negative rate. The accuracy, precision, recall and $F1score$ was calculated first, then the confusion matrices were generated to show the exact number of correctly identified intrusions, missed intrusion etc. and were expected that would run smoothly on the machine. Also, first the accuracy.

Evaluation Metrics: Accuracy and recall are the most crucial evaluation indicators for our IDS evaluation. Greater accuracy is achieved by correctly identifying intrusion across all samples; the lower the false intrusion rate, the higher the recall rate. However, confusion metrics were generated for each classifier and both classes. In addition to the accuracy, the F1 score and precision were calculated too.

$$Accuracy = (TP + TN)/(TP + TN + FP + FN) \tag{2}$$

$$Recall(R) = TP/(TP + FN) \tag{3}$$

$$Precision(P) = TP/(TP + FP) \tag{4}$$

$$F1score = (2.R.P)/(R + P) \tag{5}$$

$$FNR = 1 - recall \tag{6}$$

Confusion Matrix: In this matrix, the $TN = Intrusion\ Correctly\ Detected$, $FP = Incorrectly\ Detected$, $FN = Intrusion\ Missed$, $TP = Intrusion\ Detected$ (Table 1).

Table 1. Confusion matrix

predict real	Negative	Positive
Negative	True negative (TN)	False positive (FP)
positive	False negative (FN)	True positive (TP)

3 Result Analysis

Accuracy is the most important parameter for our result analysis. Secondly, we considered the recall rate from there the False Negative Rate (FNR) is calculated and finally from the confusion matrix the number of classified intrusions, missed intrusions is observed. Figure 2 shows the overall accuracy, recall, precision and F1 score for the stacking model for binary and multiclass classification.

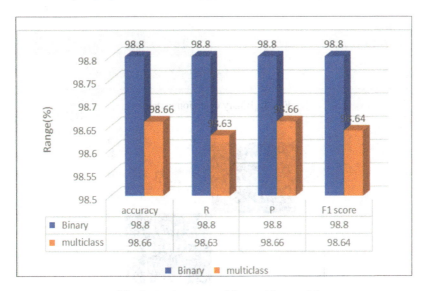

Fig. 2. Performance of the stacking model

All the metrics showed the same results for both the classification that was 98.80%, from the recall the FNR for binary classification was 1.2%, and for multiclass the accuracy and precision was 98.66%, recall was 98.63% and F1 was 98.64%.was it was 1.37%. The training dataset had an accuracy of 99.08% for binary classification. For multiclass it was 98.99%.

In Fig. 3, the precision, recall, and f1 scores for detecting normal and abnormal data detection are given for binary classification. This showed the model performed well in terms of accuracy and had the highest recall rate for the abnormal classification.

In Fig. 4, from the confusion matrix for binary classification, it has shown that, among 16,674 samples it detected 16,546 intrusions correctly, also detected 219 false intrusions and missed 228 intrusions. The model performs well in detecting attacks, but the number of missed intrusions can be lower.

For multiclass classification, based five categorical types the performance of the model is shown in Fig. 5. Also Fig. 6 shows the confusion matrix for multiclass classification.

As per Fig. 5, the precision for detecting DoS attacks was 99.33%, for probe it was 96.99%, for R2L was 90.97%. Normal samples had precision of 98.64% and for U2R attack it was 20%. The recall rate was highest for the detection of normal data with

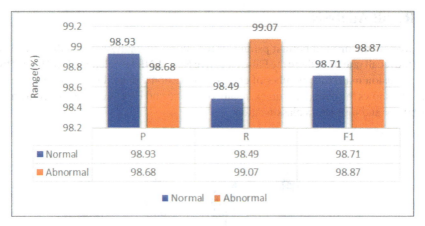

Fig. 3. Binary detection metrics

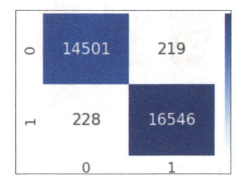

Fig. 4. Confusion matrix of stacking Model (Binary)

99.14%.after that consequently there were Dos with 98.65%, probe with 97.35%, R2L with 88.92% and U2R with 6%. From the recall rate, the FNR for normal sample were 0.86%. Dos had FNR of 1.35%. For probe the FNR was 11.68%. It was highest for U2R attacks with 94%.

In Fig. 6 the confusion matrix shows among the 15 U2R samples it detected only two correctly. 11,322 DoS attacks were identified correctly. 237 R2L attacks, 2862 probe attacks and 16,537 normal data correctly.

The model was compared with some pre-existing models, in which it showed better performance in terms of accuracy for multi-class classification. The comparison is shown in Table 2. In Table 3, the comparison with some pre-existing models for binary classification in shown. From there it can be seen that, [16] had higher accuracy then our model with adaboost reptree approach.

Analysis: Our stacking approach showed excellent results in terms of accuracy for binary and multiclass detection. It ran smoothly on the machine. It had high precision, recall and F1 score for both normal and abnormal data samples in binary classification. In multiclass classification, the model performed best in detecting the normal data then

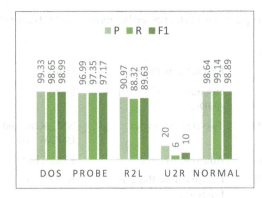

Fig. 5. Attack detection metrics

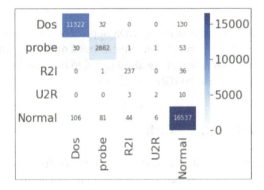

Fig. 6. Confusion matrix of stacking Model (Multiclass)

Table 2. Comparison with other ensemble approaches for Multiclass classification

Multiclass classification			
Author	Ensemble methods	Dataset	Accuracy
Adeel Abbas [8]	NB (M), DT, LR	CICIDS2017	88.96%
Xianwei Gao [11]	Ensemble voting	KDDtest+	85.2%
Majd latah [6]	Elm and KNN	KDDtest+	84.29%
Simone A. Ludwig [13]	Neural network ensemble	NSl-KDD	85.93% and 98.28%
Our proposed model	Stacking ensemble with LR	KDDtrain+	98.66%

DoS attacks, probe and R2L was moderately good but it showed terrible performance in detecting U2R attacks since the recall was very low and FNR rate was so high that interprets that the model would miss most of the U2R categorical attacks.

While compared with four pre-existing ensemble approaches, LR- stcaking outperformed neural network ensemble, KNN and ELM, NB (M), DT, LR ensemble. It shows

Table 3. Comparison with other ensemble approaches for Binary classification

Binary Classification

Author	Ensemble methods	Dataset	Accuracy
M A Jabbar [14]	ensemble classifier (RF + AODE)	kyoto	90.51%
Bayu Adhi Tama [15]	Two-step ensemble	NSL_KDD and UNSW-NB15	85.8%
Celestine Iwendi [16]	Adaboost Reptree	KDD CUP99	99.80%
Poulmanogo Illy [17]	Decision Tree Bagging Ensemble	KDDTest+	85.81%
Our proposed model	Stacking ensemble with LR	KDDtrain+	98.80%

the strong hold of LR to be used in other NIDS approaches. In binary classification, it got close to adaboost reptree approach, but outperformed others. Again, there are a lot of pre-existing approaches available that are also needed to be explored for having a better understanding of the quality of the IDS.

4 Conclusion and Future Work

This paper proposed an ensemble-based stacking approach for intrusion detection with LR. It showed excellent performance not only in binary classification but also in the multiclass approach. Together Host-based approaches LR is fully capable of using in other Network-IDS. With other vastly used classifiers, this model showed the capability of LR to be of use too. It was fast, ran smoothly, and provides a high detection rate with low FNR.

With the low detection rate of U2R attacks, we envision increasing the U2R detection rate. Also, the use of different datasets and introducing other ensemble techniques with LR is our goal eventually. We want the system to be tested in real-life network traffic.

References

1. Gu, J., et al.: A novel approach to intrusion detection using SVM ensemble with feature augmentation. Comput. Secur. **86**, 53–62 (2019)
2. Aburomman, A.A., Ibne Reaz, M.B.: A novel SVM-KNN-PSO ensemble method for intrusion detection system. Appl. Soft Comput. **38**, 360–372 (2016)
3. Kumar, G.: An improved ensemble approach for effective intrusion detection. The J. Supercomput. **75**, 275–291 (2020)
4. Zhou, Y., Cheng, G., Jiang, S., et al.: Building an efficient intrusion detection system based on feature selection and ensemble classifier. Comput. Netw. **174**, 107247 (2020)
5. Rajadurai, H., Gandhi, U.: A stacked ensemble learning model for intrusion detection in wireless network. Neural Comput. Appl. **34**, 15387–15395 (2022)

A Novel Approach to Network Intrusion Detection   343

6. Latah, M., Toker, L.: Towards an efficient anomaly-based intrusion detection for software-defined networks. IET Netw. **7**(6), 453–459 (2018)
7. Besharati, E., Naderan, M., Namjoo, E.: LR-HIDS: logistic regression host-based intrusion detection system for cloud environments. J. Ambient Intell. Human. Comput. **10**(9), 3669–3692 (2019)
8. Abbas, A., Khan, M.A., Latif, S., et al.: A new ensemble-based intrusion detection system for internet of things. Arab. J. Sci. Eng. **47**, 1805–1819 (2022)
9. Rajagopal, S., Kundapur, P.P., Hareesha, K.S.: A Stacking ensemble for network intrusion detection using heterogeneous datasets. Secur. Commun. Netw. **2020**, 1–9 (2020)
10. Bhati, S., Chugh, G., Al-Turjman, F., Bhati, N.S.: An improved ensemble based intrusion detection technique using XGBoost. Trans. Emerg. Telecommun. Technol. **32**, e4076 (2021)
11. Gao, X., Shan, C., Hu, C., Niu, Z., Liu, Z.: An adaptive ensemble machine learning model for intrusion detection. IEEE Access **7**, 82512–82521 (2019)
12. Faysal, J.A., et al.: XGB-RF: a hybrid machine learning approach for IoT intrusion detection. Telecom **3**(1), 52–69 (2022)
13. Ludwig, S.A.: Applying a neural network ensemble to intrusion detection. J. Artif. Intell. Soft Comput. Res. **9**(3), 177–188 (2018)
14. Jabbar, M.A., Aluvalu, R., et al.: RFAODE: a novel ensemble intrusion detection system. Procedia Comput. Sci. **115**, 226–234 (2017)
15. Tama, B.A., Comuzzi, M., Rhee, K.-H.: TSE-IDS: a two-stage classifier ensemble for intelligent anomaly-based intrusion detection system. IEEE Access **7**, 94497–94507 (2019)
16. Iwendi, C., Khan, S., Anajemba, J.H., Mittal, M., Alenezi, M., Alazab, M.: The use of ensemble models for multiple class and binary class classification for improving intrusion detection systems. Sensors **20**(9), 2559 (2020)
17. Illy, P., Kaddoum, G., Miranda Moreira, C., Kaur, K., Garg, S.: Securing fog-to-things environment using intrusion detection system based on ensemble learning. In: IEEE Wireless Communications and Networking Conference (WCNC), pp. 1–7. Marrakesh, Morocco (2019)
18. Le, T.-T.-H., Kim, H., et al.: Classification and explanation for intrusion detection system based on ensemble trees and SHAP method. Sensors **22**(3), 1154 (2022). https://doi.org/10.3390/s22031154
19. Priyadarsini, P.I., Anuradha, G.: A novel ensemble modeling for intrusion detection system. Int. J. Electr. Comput. Eng. (IJECE) **10**(2), 1963 (2020)
20. Kumar, G., Thakur, K., Ayyagari, M.R.: MLEsIDSs: machine learning-based ensembles for intrusion detection systems—a review. J. Supercomput. **76**(11), 8938–8971 (2020)
21. Mukeri, A.F., Gaikwad, D.P.: Adversarial machine learning attacks and defenses in network intrusion detection systems. Int. J. Wireless Microwave Technol. **12**(1), 12–21 (2022)
22. Khine, P.T.T., Win, H.P.P., et al.: New intrusion detection framework using cost sensitive classifier and features. Int. J. Wireless Microwave Technol. **12**(1), 22–29 (2022). https://doi.org/10.5815/ijwmt.2022.01.03
23. Sakr, M.M., Tawfeeq, M.A., El-Sisi, A.B.: An efficiency optimization for network intrusion detection system. Int. J. Comput. Netw. Inform. Secur. **11**(10), 1–11 (2019)

Boundary Refinement via Zoom-In Algorithm for Keyshot Video Summarization of Long Sequences

Alexander Zarichkovyi[✉] and Inna V. Stetsenko

Igor Sikorsky Kyiv Polytechnic Institute, 37 Prospect Peremogy, Kyiv 03056, Ukraine
alexander.zarichkovyi@gmail.com

Abstract. This study presents a novel algorithm for keyshot video summarization in the context of real-world videos, such as those found on platforms like YouTube or Facebook. Our proposed boundary refinement algorithm via zoom-in addresses the problem of finding keyshots in long sequences, which is a major challenge for real-world video summarization. Existing state-of-the-art techniques tend to sub-sample such videos with small frame rate, leading to a coarse prediction of the summarization boundaries, or make several predictions on overlapping frames, which results in significant computation overhead and degraded performance due to loss of the context. In contrast, our approach refines these boundaries by zooming in on selected fragments, resulting in a more accurate and efficient summary. Additionally, we propose a coarse-to-fine score alignment procedure that minimizes the score discrepancy between the different stages of algorithm resulting in improved quality. The developed algorithm was evaluated on two standard datasets, SumMe and TVSum, and achieved state-of-the-art performance with scores of 56.2% (+0.6%) and 63.1% (+0.5%), respectively. Our proposed method demonstrates significant improvements in keyshot video summarization and has the potential to enhance the effectiveness of video summarization in real-world applications.

Keywords: Computer Vision · Video Processing · Keyshot Frame Selection · Long Video Summarization · Algorithms

1 Introduction

Video summarization is a crucial task in computer vision, as it involves creating a condensed version of a video that effectively captures the most important and relevant information. The process is illustrated in Fig. 1.

There are several reasons why video summarization is important in the field of computer vision. Firstly, with the exponential growth of video content available on the internet [1, 2], it has become increasingly difficult for people to keep track of all the videos they watch or want to watch. Video summarization provides a solution to this problem by enabling users to get a quick and concise overview of a video, which can help them determine whether or not they want to watch the full video.

© The Author(s), under exclusive license to Springer Nature Switzerland AG 2023
Z. Hu et al. (Eds.): ICAILE 2023, LNDECT 180, pp. 344–359, 2023.
https://doi.org/10.1007/978-3-031-36115-9_32

Secondly, video summarization can improve the efficiency and effectiveness of video search engines. By creating a summary of a video, search engines can better match search queries with relevant video content, leading to a more user-friendly experience [3, 4].

Thirdly, video summarization can enhance the accessibility of video content for individuals with disabilities. For instance, individuals who are deaf or hard of hearing can use video summarization to obtain a written summary of the audio content in a video.

Finally, video summarization has numerous practical applications, such as video surveillance. In this context, it can assist security personnel in quickly identifying and analyzing suspicious activities.

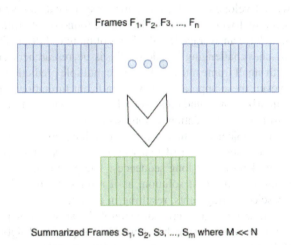

Fig. 1. The process of video summarization. N number of frames in the video is summarized to M number of frames where M is far smaller than N

Overall, video summarization is an important task in the field of computer vision because it can help people access, understand, and make use of video content more in a convenient and efficient way. In addition to the benefits mentioned above, video summarization has the potential to revolutionize the way we consume and interact with video content [1]. For example, video summarization can be used to create personalized video highlights or summaries based on an individual interests or preferences. It could allow people to watch only the most relevant and interesting parts of a video, rather than having to sit through the entire thing [5–7]. Video summarization can also be used to automatically create subtitles or closed captions for videos, which can make them more accessible to a wider audience. This is especially important for people who are deaf or hard of hearing, as well as for people who are learning a new language and want to improve their listening and comprehension skills. In addition, video summarization can be used to create video summaries for educational purposes. For example, teachers could use video summarization to create shorter, more focused videos that cover specific topics or concepts, which could be more engaging and effective for students. The potential applications of video summarization are vast, and as technology continues to advance, it is likely that we will see even more innovative and creative uses for this important tool in the field of computer vision.

According to a recent study of the literature [1], most video summarization methods use RNNs [8], LSTMs [9], GRUs [10], or Wavelet transformation [11] to model temporal dependencies and predict the importance of frames in a video. However, these methods have shown weaknesses related to modeling long sequences and backpropagating signals. To address these limitations, some works [12–14] have proposed using attention mechanisms to model temporal dependencies without gradient vanishing or explosion. Additionally, future works have improved these methods by modeling temporal dependencies on different levels of granularity [15–18]. Existing solutions for video summarization were developed to solve problems for specific datasets that contain short sequences. However, real-world videos, such as those found on platforms like YouTube or Facebook, tend to have an average length of tens of minutes [19], making the application of these approaches challenging. To address this issue, researchers tend to subsample long sequences with small frame rates [20] or make multiple predictions on overlapping frames [21]. However, these approaches have significant limitations, including coarse prediction of summarization boundaries, loss of context, and significant computation overhead and performance degradation on long sequences.

The major research objective of this study is to develop a novel algorithm for keyshot video summarization in the context of real-world videos, specifically addressing the challenges of finding keyshots in long sequences. The proposed boundary refinement algorithm via zoom-in is the best solution, as it refines the boundaries of keyshots by zooming in on selected fragments, resulting in a more accurate and efficient summary with smaller computational overhead. State-of-the-art performance was achieved in keyshot video summarization using this approach. The proposed algorithm was tested on the TVSum [22] and SumMe [23] datasets, resulting in state-of-the-art performance of 56.2% (+0.6%) and 63.1% (+0.5%), respectively. The author hopes that this approach will enhance the effectiveness of video summarization in real-world applications, such as YouTube or Facebook, where long video sequences are common.

2 Related Work

Most of the early work on video summarization uses hand-made heuristics for unsupervised training. When choosing representative frames for the video summary, they use the importance score to determine which frames are important or representative [1]. For video summarization, supervised learning techniques have recently been investigated. These methods make use of training data that consists of human-produced ground truth for clips. Since they can implicitly learn high-level semantic knowledge that is used by humans to produce summaries, these supervised learning approaches frequently outperform early work on unsupervised methods [1].

Deep learning techniques [12–18, 24–36], have recently become more popular for video summarization by modeling the variable-range temporal dependence among frames and learning how to assign priority to them based on ground truth annotations. For this, various techniques make use of Fully Convolutional Sequence Networks [18] or RNNs [24–26]. The idea behind employing them is to efficiently capture long-range dependencies between video frames, which are essential for the development of meaningful summaries. As shown in [25] variable-range dependency can be modeled using two LSTMs treating the video summarizing task as a structured prediction problem on sequential data.

Other studies try to address concerns with the constrained memory of RNNs and make use of either external storage [27] or the hierarchical stacking of numerous LSTM and memory layers [28]. Some algorithms incorporate customized attention mechanisms into classic [29] or sequence-to-sequence [30] RNN-based architectures to represent the evolution of the users' interest. Particularly, in [29] was proposed to embed an LSTM-based attention layer to model temporal dependencies and form new representations that allow producing diverse video summary. In [30] problem of video summarization viewed as a seq-to-seq learning task and was proposed to integrate an attention layer into an LSTM-based encoder-decoder network. This layer gets the encoder's output and the previous hidden state of the decoder and computes a vector with attention values, which subsequently affects the video decoding process.

Furthermore, several methods aimed to model the dependencies between frames by utilizing variations of the self-attention mechanism were found in Transformer Networks [31]. The first method [12] combines a soft self-attention mechanism with a two-layer fully connected network to regress the frame importance scores. Liu et al. [13] employ a hierarchical approach, first defining shot-level candidate key-frames and then utilizing a multi-head attention model to evaluate their importance and select the key-frames for the summary. Li et al. [32] enhance the training pipeline of the standard self-attention mechanism by incorporating a step that increases the diversity of the visual content in the summary using the computed attention values and human annotations. Ghauri et al. [33] present a variation of the architecture from [12] that incorporates additional representations of the video content, including CNN-based features from the pool5 layer of GoogleNet [34] trained on ImageNet and motion-related features from the Inflated 3D ConvNet [35] trained on Kinetics. These features are fed into self-attention mechanisms, and the outputs are combined to form a common embedding space for representing the video frames. Finally, the representation is used to learn frame importance. The PGL-SUM model [36], which is closely related to existing self-attention-based approaches, captures frame dependence at different levels of granularity using global and local multi-head attention mechanisms. Furthermore, unlike other methods that ignore the sequential nature of video, the PGL-SUM model [36] incorporates temporal information about the frame position, a crucial factor in estimating frame importance.

Existing solutions for video summarization were designed to handle specific datasets with short sequences, making them challenging to apply to real-world videos found on platforms such as YouTube or Facebook, which tend to be tens of minutes long on average [19]. The most practical ways to address this issue are subsampling with small frame rates [20] or making multiple predictions on overlapping frames [21], but these approaches have significant limitations, including loss of context, coarse prediction of summarization boundaries, and high computation overhead.

To overcome these challenges, some researchers have developed architectural level solutions for video summarization, such as the clustering-based approach presented in [37]. In this work, a graph-based hierarchical clustering method is used to compute a video summary by selecting salient frames to represent the video content. A weight map is generated from the frame similarity graph, and clusters or connected components of the graph can easily be inferred on a minimum spanning tree of frames, with a weight map based on hierarchical observation scales computed over that tree.

The closest prior work to our research is the Distilled Kernel Temporal Segmentation (D-KTS) method presented in [38]. D-KTS detects shot boundaries in state-of-the-art video solutions and exploits the temporal locality of long videos. Existing state-of-the-art solutions for video summarization [15, 36–41] typically follow the common paradigm of using two modules for processing long sequences: a preprocessing module and a model inference module. However, their preprocessing modules rely on the Kernel Temporal Segmentation [42, 43] (KTS) method. It performs an automatic Kernel-based Temporal Segmentation based on state-of-the-art video features automatically selecting the number of segments. Then, equipped with an SVM classifier for importance scoring that was trained on videos for the category at hand, we score each segment in terms of importance. Finally, the KTS approach outputs a video summary composed of the segments with the highest predicted importance scores, see Fig. 2 for more clarity.

The major drawback of KTS is time-consuming under long videos (accounting for about 68% of the total processing time). The KTS computes the variances of every interval and applies the dynamic programming technique to get the shot boundaries. To consider every possible interval and obtain precise boundaries, the complexity of KTS is $O(N^2)$, and it becomes the bottleneck for processing long videos. To handle a 10-min video, it even needs more than 8 h while using the baseline KTS method.

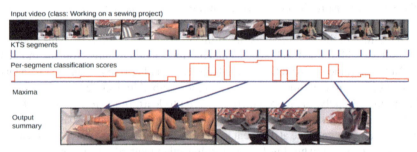

Fig. 2. Overall scheme of Kernel-based Temporal Segmentation (KTS) for Video Summarization presented in [42]

So, [38] in their research, the authors presented the Distribution-based Kernel Temporal Segmentation (D-KTS) method, which leverages a new dynamic programming algorithm to improve upon the Kernel Temporal Segmentation (KTS) method used in existing state-of-the-art solutions for video summarization. Additionally, they introduced the Hash-based Adaptive Frame Selection (HAFS) to pre-process frames and construct an adaptive frame selection mechanism, improving the feature extraction performance. Specifically, they observed that the visual similarity of two intra-shot frames is higher than that of two inter-shot frames. To reduce the cost of feature extraction, they proposed for the first frame to record its hash vector through the d-hash [44]. Then, compute the hash vector for each frame and make a comparison with the record vector. If the Hamming distance is larger than our threshold, they replaced the record with the current hash vector and extract the feature vector through the neural network. Otherwise, they repeat the feature vector of the record frame without making feature extraction on the current frame. See Fig. 3 that visualize D-KTS pipeline with HASF mechanism. With D-KTS and HAFS they achieved state-of-the-art results on the TvSum and SumMe datasets when using a high frame-rate setup.

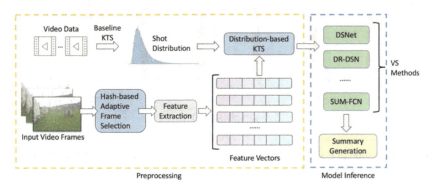

Fig. 3. Overview of Distilled Kernel Temporal Segmentation (D-KTS) pipeline with Hash-based Adaptive Frame Selection (HAFS) mechanism presented in [38]

The authors' work is particularly noteworthy for its findings on the relationship between frame rate and F-score (see Fig. 4 and Fig. 5). Their research demonstrated that there is a positive correlation between frame rate and F-score, with down-sampling to 1 fps (as used in most related works) resulting in negative impacts on the result. Specifically, compared to no sub-sampling (30 fps), the F-score decreases by an average of 20.92% for the TVSum dataset and 20.72% for the SumMe dataset. This significant finding motivated us to research question: "Why does increasing the frame rate have such a positive impact on the F-score?" We hypothesized that the increased frame rate results in more precise identification of borders, which yields improvements compared to the heuristics used in other works [8–18, 24–27, 36, 37, 39–41].

It is worth noting that our approach differs from the D-KTS [38] method in that we do not require additional neural networks for frame pre-selection. Instead, we use a coarse-to-fine score alignment procedure that minimizes the score discrepancy between different stages. Furthermore, our algorithm recommends using the original 1 fps frame rate, which is more applicable for real-world problems where end-users have GPUs with limited memory compared to researchers' GPUs that often have 40GB or more. As a result, our approach is simpler to train and requires fewer computational resources during both training and inference compared to DKTS.

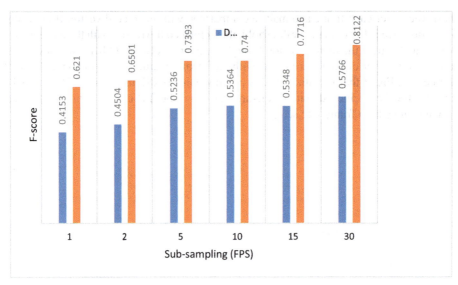

Fig. 4. Correlation of F-score and FPS for D-KTS [38] approach using DR-DSN [39] and DSNet [15] backones on TVSum [23] dataset

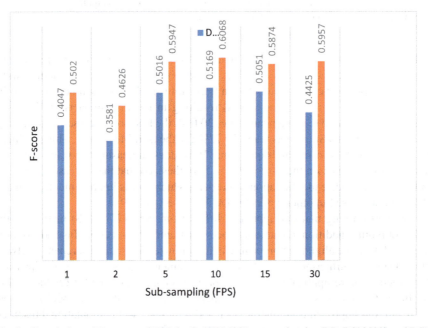

Fig. 5. Correlation of F-score and FPS for D-KTS [38] approach using DR-DSN [39] and DSNet [15] backones on SumMe [22] dataset

3 Algorithm Description

The proposed algorithm for video summarization is comprised of two distinct phases: key-frame pre-selection and frame border refinement. The key-frame pre-selection stage aims to identify the most relevant frames, or S', by maximizing the likelihood of their selection, as is common in traditional methods for key-shot video summarization. Traditional methods then incorporate additional context in the form of short clips, referred to as padding P, to provide additional information about the video content. However, this padding may also include irrelevant information. The frame border refinement phase of the proposed method addresses this issue by cutting the irrelevant information from the tails of the padding to produce more accurate and shorter clips. To achieve this, the model is run on top of the selected frames S' with padding P, but at a higher frame rate to enable the processing of more accurate borders. This process is formalized as Algorithm 1. Example is depicted in Fig. 6.

Algorithm 1. Proposed algorithm for fine-grained video summarization

---

Inputs: keyframe selection model M, pre-defined processing sequence length T, the set of video frames $F = \{F_1, F_2, ..., F_n\}$, the set of corresponding labels for each frame $L = \{L_1, L_2, ..., L_n\}$, refinement padding length P, loss function F_{loss}

    1. If $n < T$, then pad F with $T - n$ empty frames

    2. From F subsample equal frequency T frames $F' = \{F'_1, F'_2, ..., F'_T\}$

        // *Pre-selection*

    3. Get importance scores $S' = M(F')$

        // *Border Refinement*

    4. From F subsample equally T frames $F'' = \{F''_1, F''_2, ..., F''_T\}$:

        4.1. $F'' = \{\}$

        4.2. While $len(F'') < T$:

            4.2.1. Select frame f with largest score in S'

            4.2.2. Add frame f to F'' and remove it from S'

                // *Frame iteration order* $f, f - 1, f + 1, f - 2, f + 2, ..., f - n, f + n$

            4.2.3. For frames p in range from $f - P$ to $f + P$:

                4.2.3.1. If p not in F'' then add to F'' and remove it from S'

    5. Get importance scores $S'' = M(F'')$

        // *Training*

    6. $loss = F_{loss}(S', L) + F_{loss}(S'', L) - \alpha * (S'' - S')^2$

In the algorithm, we acknowledge that the varying distribution of data between the pre-selection and border refinement stages could lead to inconsistent scores for the same frames. To address this issue, we introduce a coarse-to-fine score alignment procedure that aims to minimize the discrepancy between the scores obtained from the two stages. This, in turn, leads to a more robust network that can effectively handle both distributions. It can be done efficiently by minimizing Euclidian distance between scores from different stages.

4 Experiments and Results

4.1 Datasets

To compare proposed algorithm with previous work, experiments were conducted on four datasets: TvSum [28], SumMe [29], OVP [45], and YouTube [45]. The datasets OVP and YouTube were used solely for the purpose of augmenting the training dataset. TvSum and SumMe are the only currently available datasets with suitable labeling for keyshot video summarization, although they remain relatively small for the training of deep models. A summary of the main properties of these datasets can be found in Table 1.

The TvSum dataset is annotated by frame-level importance scores, while the SumMe with binary keyshot summaries. OVP and YouTube are annotated with keyframes and need to be converted to the frame-level scores and binary keyshot summaries.

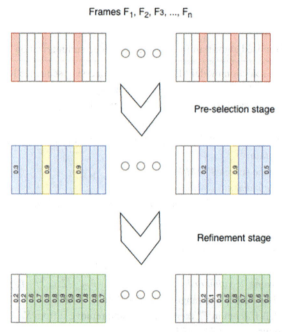

Fig. 6. Processing workflow. In grey boxes not selected frames are represented, red boxes are frames selected for processing in pre-selection stage, yellow boxes are selected high confidence frames from pre-selection stage, blue boxes are frames added for border refinement stage as padding frames, in green boxes are frames selected after refinement stage and final result. Numbers inside the boxes are the corresponding model's score to this frame after processing at each stage.

Table 1. Overview of the datasets properties

Dataset	# videos	Annotation	Annotation type	Video length (sec)		
				Min	Avg	Max
TvSum [22]	50	20	Frame-level importance scores	83	235	647
SumMe [23]	25	15–18	keyshots	32	146	324
OVP [45]	50	5	keyframes	46	98	209
YouTube [45]	39	5	keyframes	9	196	572

4.2 Evaluation Protocol

For fair comparison with the previous state-of-the-arts, we follow evaluation protocol from [16]. To assess the similarity between the machine and user summaries we use the harmonic mean of precision and recall expressed as the F-score in percentages:

$$F = 2(p \cdot r)/(p + r) \cdot 100, \tag{1}$$

p p is precision value, r r is recall value.

True and false positives and false negatives for the F-score are calculated per-frame as the overlap between the ground truth and machine summaries, as shown in Fig. 7.

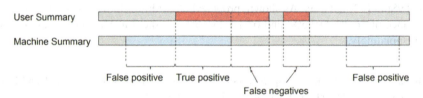

Fig. 7. True positives, False positives and False negatives are calculated per-frame between the ground truth and machine binary keyshot summaries

The machine summary is limited to 15% of the original video length and then evaluated against multiple user summaries. Precisely, on the TvSum [22] benchmark, for each video, the F-score is calculated as an average between the machine summary and each of the user summaries Average F-score over videos in the dataset is then reported. On the SumMe [23] benchmark, for each video, a user summary most similar to the machine summary is selected.

4.3 Implementation Details and Setup

The experiments are carried out under Python 3.9.13, PyTorch 1.13.1, CUDA 11.7, OpenCV 4.5.2. The operating system is Ubuntu 20.04. Our device has 8x Nvidia A100 80GB GPUs, 2x Intel(R) Xeon(R) Platinum 8275CL CPU @ 3.00GHz, and 1152GB memory.

In our study, we adhered to the established protocol in the field and utilized a 1 FPS framerate subsampling of videos for the preselection stage. The deep representations of frames were obtained through extraction of the output from the Video Swin Transformer [46], Swin-T specifically. The input frames were resized to a resolution of 224x224, resulting in an input token size of 16x56x56.

The model was initialized with pre-trained weights for the backbone and randomly initialized weights for the head. Optimization was carried out using the AdamW [47] algorithm with an initial learning rate of 3e-4, and the model was trained for a total of 200 epochs, including a 2.5 epoch warm-up phase. The effective batch size during training was set to 64. Consistent with prior research, the gradients of the backbone were multiplied by a factor of 0.1. An increasing degree of stochastic depth rate and weight decay was employed for larger models, as recommended by [46]. The stochastic depth rate was set to 0.1, and the weight decay was set to 0.02. All experiments used $\alpha = 0.5\alpha = 0.5$ if not specified otherwise.

The best-trained model was selected using a model selection criterion based on the loss curves, as outlined in [36]. If rapid changes in the loss value were not observed, the model was selected based on minimization of the loss value.

All experiments were conducted using the publicly available PGL-SUM implementation available at https://github.com/e-apostolidis/PGL-SUM.

4.4 Results

Table 2 presents the comparison of the state-of-the-art approaches for SumMe [23] and TvSum [22] datasets. As shown, proposed algorithm achieved state-of-the-art performance of 56.2% (+0.6%) and 63.1% (+0.5%) on the SumMe and TVSum datasets correspondingly.

Table 2. Comparison different supervised summarization approaches on SumMe [23] and TvSum [22] datasets. All methods are tests with fixed 1FPS framerate.

Approach	SumMe, F1-score	TvSum, F1-score
VasNet [12]	49.7	61.4
RR-STG [14]	53.4	63.0
DSNet [15]	50.2	62.1
PGL-SUM [36]	55.6	62.7
D-KTS w/ DSNet backbone [38]	50.2	62.3
Ours	**56.2**	**63.2**

4.5 Ablation Study

In the ablation study section, our objective is twofold. Firstly, we aim to evaluate the impact of each modification introduced in our study on the final performance. Secondly, we aim to compare the performance of our algorithms with that of the D-KTS [38] approach at higher framerates. This comparative analysis will provide insights into the efficacy of our approach and its potential to outperform the state-of-the-art technique. Additionally, by isolating the contribution of each change, we can better understand the relative importance of each modification and its impact on the overall performance of the systems.

4.6 Study on Importance of Introduced Components

Within this study, we proposed a novel algorithm that incorporates three major modifications to the PGL-SUM [36] framework. First, we replaced the backbone with the Video Swin Transformer, a more effective feature extractor for video summarization tasks. Second, we introduced boundary refinement to improve the accuracy of the extracted keyframes. Third, we implemented a coarse-to-fine score alignment procedure to better align the score predictions across different stages of the algorithm. The effectiveness of each modification is evaluated and reported in Table 3. Our results indicate that the introduction of the Video Swin Transformer backbone leads to approximately 50% of the overall improvement in the performance, while the remaining 50% can be attributed to the proposed boundary refinement and coarse-to-fine score alignment techniques.

Table 3. Impact of different component of proposed algorithm

Method	SumMe, F1-score	TvSum, F1-score
PGL-SUM [36] (baseline)	55.6	62.7
+ Video Swin Transformer as backbone	55.9 (+0.3)	62.9 (+0.2)
+ Boundary refinement	56.1 (+0.2)	63.1 (+0.2)
+ Coarse-to-fine alignment	56.2 (+0.1)	63.2 (+0.1)

4.7 Comparison with D-KTS at Higher Framerates

In order to compare our algorithm with the D-KTS approach [38], we increased the frame rate for pre-selection phase of proposed algorithm and linearly increased the temporal shape of the input dimension for the Video Swin Transformer. As shown in Fig. 8 and Fig. 9, our proposed algorithm outperformed the D-KTS approach at lower frame rates. However, this difference became negligible at higher frame rates of 15FPS and 30FPS, resulting in a difference of only $+0.1\%$ and $0.yyy\%$ on the SumMe and TVSum datasets, respectively. We attribute this to the fact that at higher frame rates, both approaches have access to more frames, leading to a rapid decrease in the difference in boundary detection accuracy. However, as we stated earlier, the primary goal of this study is to provide a practical solution for handling long video sequences without adding complexity or overhead during inference. In our opinion, the proposed algorithm performs better in this regard than D-KTS [38]. It can be used at lower frame rates and can be deployed on smaller GPU accelerators with 12GB of VRAM for processing sequences that are tens of minutes long, which is common on platforms such as YouTube or Facebook [19]. Therefore, we believe that our algorithm is a more practical and feasible approach for real-world applications.

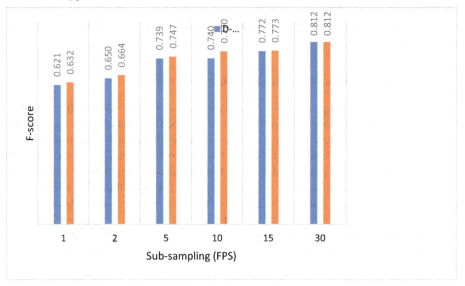

Fig. 8. Comparison of proposed algorithm with D-KTS [38] at higher framerates on TVSum dataset [22]

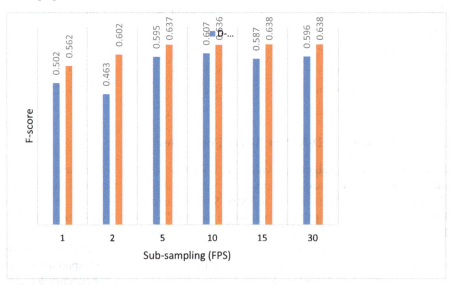

Fig. 9. Comparison of proposed algorithm with D-KTS [38] at higher framerates on SumMe dataset [23]

5 Conclusions

In this study, a novel algorithm is proposed for video summarization that improves upon the state-of-the-art PGL-SUM approach. The algorithm introduces a boundary refinement stage that further reduces the number of non-relevant frames identified by the pre-selection stage, through the use of coarse prediction. Also coarse-to-fine score alignment technique is implemented to ensure consistency between the scores obtained from each stage. Compared to D-KTS, closes competitor, the proposed algorithm is more practical for real world problems as it achieves superior performance at lower frame rates, making it applicable with smaller GPU accelerators with only 12GB of VRAM. Our approach incorporates a Video Swin Transformer backbone and proposed algorithm into PGL-SUM pipeline. The significant improvements on the SumMe and TVSum datasets, with a 56.2% (+0.6%) and 63.1% (+0.5%) improvement, respectively, are shown. These results demonstrate the effectiveness of our algorithm in addressing the challenges of video summarization for long sequences, making it a valuable tool for practical applications in the field.

References

1. Evlampios, A., Eleni, A., Alexandros, I.M., Vasileios, M., Ioannis, P.: Video summarization using deep neural networks. CoRR, vol. abs/2101.06072, http://arxiv.org/abs/2101.06072 (2021)
2. Fu, H., et al.: Self-attention binary neural tree for video summarization. Pattern Recognit. Lett. (2021)
3. Sahu, A., et al.: First person video summarization using different graph representations. Pattern Recognit. Lett. (2021)
4. Fei, M., et al.: Memorable and rich video summarization. J. Vis. Commun. Image Represent (2017)
5. Nagendraswamy, H.S., Chethana kumara, B.M., Guru, D.S., Naresh, Y.G.: Symbolic Representation of Sign Language at Sentence Level. IJIGSP 7(9), 49–60 (2015). https://doi.org/10.5815/ijigsp.2015.09.07
6. Shivanand, S., Gornale, A.K.B., Yannawar, P.L.: Analysis and detection of content based video retrieval. Int. J. Image Graph. Sig. Proc. (IJIGSP) 11(3), 43–57 (2019). https://doi.org/10.5815/ijigsp.2019.03.06
7. Rahman, A., Hasan, S., Rafizul Haque, S.M.: Creation of video summary with the extracted salient frames using color moment, color histogram and speeded up robust features. Int. J. Info. Technol. Comp. Sci. (IJITCS) 10(7), 22–30 (2018). https://doi.org/10.5815/ijitcs.2018.07.03
8. Fu, T.-J., et al.: Attentive and adversarial learning for video summarization. In: IEEE Winter Conf. on Applic. of Comp. Vision HI, USA: IEEE, pp. 1579–1587 (2019)
9. Hochreiter, S., et al.: Long short-term memory. Neural Comput. 9(8), 1735–1780 (1997)
10. Cho, K., et al.: Learning Phrase Representations using RNN Encoder–Decoder for Statistical Machine Translation. In: 2014 Conf. on Empirical Methods in Natural Language Processing. Doha, Qatar: ACL, pp. 1724–1734 (2014)
11. Kavitha, J., Arockia Jansi Rani, P.: Design of a video summarization scheme in the wavelet domain using statistical feature extraction. IJIGSP 7(4), 60–67 (2015). https://doi.org/10.5815/ijigsp.2015.04.07

12. Fajtl, J., et al.: Summarizing Videos with Attention. In: Asian Conf. on Comp. Vision 2018 Workshops, pp. 39–54. Springer Int. Publishing, Cham (2018)
13. Liu, Y.-T., et al.: Learning hierarchical self-attention for video summarization. In: 2019 IEEE Int. Conf. on Image Processing. IEEE, pp. 3377–3381 (2019)
14. Zhu, W., Han, Y., Lu, J., Zhou, J.: Relational reasoning over spatial-temporal graphs for video summarization. IEEE Trans. Image Process. **31**, 3017–3031 (2022). https://doi.org/10.1109/TIP.2022.3163855
15. Zhu, W., Lu, J., Li, J., Zhou, J.: DSNet: a flexible detect-to-summarize network for video summarization. IEEE Transactions on Image Processing **30**, 948–962 (2021). https://doi.org/10.1109/TIP.2020.3039886 (2020)
16. Zhu, W., Lu, J., Han, Y., Zhou, J.: Learning multiscale hierarchical attention for video summarization. Pattern Recognition **122**, 108312 (2022). ISSN 0031-3203, https://doi.org/10.1016/j.patcog.2021.108312 (2021)
17. Zhu, W., Han, Y., Lu, J., Zhou, J.: Relational reasoning over spatial-temporal graphs for video summarization. IEEE Transactions on Image Processing **31**, 3017–3031 (2022). https://doi.org/10.1109/TIP.2022.3163855
18. Rochan, M., et al.: Video summarization using fully convolutional sequence networks. In: Europ. Conf. on Comp. Vision 2018, pp. 358–374. Springer Int. Publishing, Cham (2018)
19. Ceci. L.: Statista. Average YouTube video length as of December 2018 (Aug 2021). www.statista.com/statistics/1026923/youtube-video-category-average-length
20. Sah, S., Kulhare, S., Gray, A., Venugopalan, S., Prud'Hommeaux, E., Ptucha, R.: Semantic text summarization of long videos. In: 2017 IEEE Winter Conference on Applications of Computer Vision (WACV), pp. 989–997. Santa Rosa, CA, USA (2017). https://doi.org/10.1109/WACV.2017.115
21. Wenyi, L., Wenjing, L.: Sliding Window Recurrent Network for Efficient Video Super-Resolution. CoRR, vol. abs/2208.11608, http://arxiv.org/abs/2208.11608 (2022)
22. Song, Y., Vallmitjana, J., Stent, A., Jaimes, A.: Tvsum: summarizing web videos using titles. In: Proceedings of the IEEE CVPR, pp. 5179–5187 (2015)
23. Gygli, M., Grabner, H., Riemenschneider, H., Van Gool, L.: Creating Summaries from User Videos. In: Fleet, D., Pajdla, T., Schiele, B., Tuytelaars, T. (eds.) ECCV 2014. LNCS, vol. 8695, pp. 505–520. Springer, Cham (2014). https://doi.org/10.1007/978-3-319-10584-0_33
24. Zhao, B., et al.: Hierarchical recurrent neural network for video summarization. In: 25th ACM Int. Conf. on Multimedia NY, pp. 863–871. ACM, USA (2017)
25. Zhang, K., et al.: Video summarization with long short-term memory. In: Europ. Conf. on Comp. Vision 2016, pp. 766–782. Springer Int. Publishing, Cham (2016)
26. Zhao, B., et al.: HSA-RNN: Hierarchical Structure-Adaptive RNN for Video Summarization. In: 2018 IEEE Conf. on Comp. Vision and Pattern Rec., pp. 7405–7414 (2018)
27. Feng, L., et al.: Extractive video summarizer with memory augmented neural networks. In: 26th ACM Int. Conf. on Multimedia, pp. 976–983. ACM, NY, USA (2018)
28. Wang, J., et al.: Stacked memory network for video summarization. In: 27th ACM Int. Conf. on Multimedia, pp. 836–844. ACM, NY, USA (2019)
29. Lebron Casas, L., et al.: Video Summarization with LSTM and Deep Attention Models. In: 25th Int. Conf. on Multimedia Modeling, pp. 67–79. Springer Int. Publishing, Cham (2019)
30. Ji, Z., et al.: Deep attentive and semantic preserving video summarization. Neurocomputing **405**, 200–207 (2020)
31. Vaswani, A., et al.: Attention is all you need. In: 31st Int. Conf. on Neural Information Processing Systems, pp. 6000–6010. Red Hook, NY, USA: Curran Associates Inc. (2017)
32. Li, P., et al.: Exploring global diverse attention via pairwise temporal relation for video summarization. Pattern Recognition **111**(107677) (2021)
33. Ghauri, J., et al.: Supervised Video Summarization Via Multiple Feature Sets with Parallel Attention. In: 2021 IEEE Int. Conf. on Multimedia and Expo, pp. 1–6. IEEE, CA, USA (2021)

34. Szegedy, C., et al.: Going deeper with convolutions. In: 2015 IEEE Conf. on Comp. Vision and Pattern Rec., pp. 1–9
35. Carreira, J., et al.: Quo vadis, action recognition? a new model and the kinetics dataset. In: 2017 IEEE Conf. on Comp. Vision and Pattern Rec., pp. 4724–4733
36. Apostolidis, E., Balaouras, G., Mezaris, V., Patras, I.: Combining global and local attention with positional encoding for video summarization. In: 2021 IEEE International Symposium on Multimedia (ISM), pp. 226-234. Naple, Italy (2021). https://doi.org/10.1109/ISM52913.2021.00045
37. Belo, L., Caetano, C., Patrocínio Jr, Z., Guimarães, S.: Summarizing video sequence using a graph-based hierarchical approach. Neurocomputing **173**, 1001-1016 (2016). https://doi.org/10.1016/j.neucom.2015.08.057
38. Ke, X., Chang, B., Wu, H., Xu, F., Zhong, S.: Towards practical and efficient long video summary. In: ICASSP 2022 - 2022 IEEE International Conference on Acoustics, Speech and Signal Processing (ICASSP), pp. 1770–1774. Singapore, Singapore (2022). https://doi.org/10.1109/ICASSP43922.2022.9746911
39. Zhou, K., Qiao, Y., Xiang, T.: Deep reinforcement learning for unsupervised video summarization with diversity-representativeness reward. In: Proceedings of the AAAI Conference on Artificial Intelligence, vol. 32 (2018)
40. Zhang, Y., Kampffmeyer, M., Zhao, X., Tan, M.: Dtr-gan: Dilated temporal relational adversarial network for video summarization. In: Proceedings of the ACM Turing Celebration Conference-China, pp. 1–6 (2019)
41. Ji, Z., Zhao, Y., Pang, Y., Li, X., Han, J.: Deep attentive video summarization with distribution consistency learning. IEEE transactions on neural networks and learning systems **32**(4), 1765–1775 (2020)
42. Potapov, D., Douze, M., Harchaoui, Z., Schmid, C.: Category-Specific Video Summarization. In: Fleet, D., Pajdla, T., Schiele, B., Tuytelaars, T. (eds.) ECCV 2014. LNCS, vol. 8694, pp. 540–555. Springer, Cham (2014). https://doi.org/10.1007/978-3-319-10599-4_35
43. Lei, Z., Sun, K., Zhang, Q., Qiu, G.: User video summarization based on joint visual and semantic affinity graph. In: Proceedings of the 2016 ACM Workshop on Vision and Language Integration Meets Multimedia Fusion, pp. 45–52. New York, NY, USA (2016). Association for Computing Machinery
44. Hacker Factor: The hacker factor blog (January 2013). http://www.hackerfactor.com/blog/?/archives/529-Kind-of-Like-That.html
45. De Avila, S.E.F., Lopes, A.P.B., da Luz Jr, A., de Albuquerque Ara'ujo, A.: Vsumm: a mechanism designed to produce static video summaries and a novel evaluation method. Pattern Recognition Letters **32**(1), 56–68 (2011)
46. Liu, Z., Ning, J., Cao, Y., Wei, Y., et al.: Video Swin Transformer. CoRR, vol. abs/2106.13230, http://arxiv.org/abs/2106.13230 (2021)
47. Kingma, D.P., Ba, J.: Adam: A method for stochastic optimization. CoRR, vol. abs/1412.6980, http://arxiv.org/abs/1412.6980 (2014)

Solving Blockchain Scalability Problem Using ZK-SNARK

Kateryna Kuznetsova[1,2(✉)], Anton Yezhov[1], Oleksandr Kuznetsov[1,2,3], and Andrii Tikhonov[1]

[1] Zpoken, OU, Harju Maakond, Kesklinna Linnaosa, Sakala tn 7-2, 10141 Tallinn, Estonia
kate7smith12@gmail.com

[2] Department of Information and Communication Systems Security, School of Computer Sciences, V.N., Karazin Kharkiv National University, 4 Svobody Sq., Kharkiv 61022, Ukraine

[3] Department of Political Sciences, Communication and International Relations, University of Macerata, Via Crescimbeni, 30/32, 62100 Macerata, Italy

Abstract. Modern users demands for digital information processing and storage systems, which provide reliable, safe and consistent reproduction of recorded facts and issues. We are well acquainted with distributed ledger technology (DLT) that provides the information storage on independent and unaccountable nodes which interact in complete distrust with each other. It yet provides reliable, secure and historically consistent storage. Due to the increase of blockchain network users it is necessary to provide a resource availability service for billions of users in different parts of the world. Moreover, everyone who interacts with the blockchain network in the mode of complete distrust forces to repeatedly check long chain of blocks containing cryptographic integrity and involvement marks of the party that generates them. Thus, the main recent problem of DLTs is scalability. In this paper, we present the main idea of our solution using new cryptographic technology, known as zero-knowledge proof system. We substantiate the general scheme of fast, reliable and secure DLT verification systems and provide time and measurement result of the classic DLT verification and the new one to demonstrate the perspective of the chosen direction.

Keywords: Blockchain · Scalability Problem · Distributed Ledger · ZK-SNARK Technology

1 Introduction

Blockchain is a distributed decentralized data storage, implemented by linking information into a continuous chain of blocks [1–3]. Each next block is formed using a cryptographic connection using hash function as it is shown in the figure below [4].

In blockchain, there is the well-known trilemma, which formulates three main requirements: security, decentralization and scalability [5].

The security property avoids information theft and misused. It also provides the right to privacy and anonymity. Security is the basic requirement, which attracts investors or help to win approval of partners.

© The Author(s), under exclusive license to Springer Nature Switzerland AG 2023
Z. Hu et al. (Eds.): ICAILE 2023, LNDECT 180, pp. 360–371, 2023.
https://doi.org/10.1007/978-3-031-36115-9_33

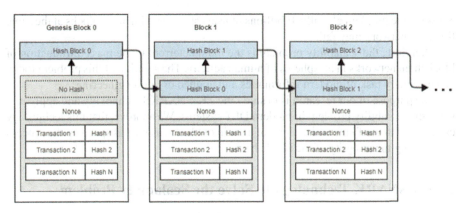

Fig. 1. Chain of linked blocks built using hashing

In addition to security, we require decentralized system, i.e. which avoids control by a single decision-making center, server, node or subject. This is a problem of reliability, because a single point of failure always attracts swindlers and attackers. Decentralization also protects from censorship. It is a manifestation of freedom and independence and democracy of decisions.

Scalability is the property of a system to continuously evolve as the amount of work increases. For example, if the number of users in a system increases, then their service time does not increase significantly. This requirement is a difficult to implement because it is not always possible to provide extensive scalability (only by increasing the number of computers).

During the development every large DLT project faces the limitations in the number of users, resources and time that can be processed by a system [6]. Everyone needs to check long chain of cryptographic primitives, and the longer the project exists, the more entities are involved in it, the more difficult and costly these verifications are.

The scheme shown in the Fig. 1 solves the problems of security and decentralization. Complex cryptographic transformations are used to link blocks together and confirm their authenticity. All blocks are accepted by the consensus of the participants, that is there is no main node (single decision-making center) in the system. But there is a lot of work to validate the records in such DLT.

The mostly used method for verifying the correctness of blocks in a blockchain is native registry check, which consists in recalculation of a chain of linked hashes and all signatures of all blocks. This method requires a great amount of time. For instance, if we take a millionth block, it is necessary to calculate a million hashes and a million signatures. It is really computationally hard. In addition, native check includes viewing all records, which violates the privacy.

Modern blockchain projects solve this problem in their own way [7–10]. Some of them try new technologies to create a network like directed acyclic graph (DAG) [11–14] or residual number system (RNS) [15]. The NEAR project, for instance, partially solves the scalability problem by segmenting (sharding) the ledger [16, 17]. However, this does not solve the problem globally. If billions of users start checking their blocks at the same time, the network may fail.

Therefore, the scalability problem is the main deterrent in the mass introduction of blockchain networks in all spheres of human activity. The solution of this problem today is the primary task of modern informatics and applied crypto-engineering.

The purpose of our research is to study the ZK-SNARKS technology and apply it to solve the scalability problem in modern DLT systems. We approve, that this technology can be used to verify the authenticity of a chain of blocks, as well as speed up the process of verifying the integrity of the blockchain.

2 ZK-SNARK Technology to Solve the Scalability Problem

ZK-SNARK is an acronym that stands for «Zero-Knowledge Succinct Non-Interactive Argument of Knowledge» [18].

ZK-SNARK is a cryptographic proof that allows one party to prove it possesses certain information without revealing that information.

This technology was originally designed to provide privacy [19]. However, such properties as the short proofs and the logarithmic dependence of their verification, instead of native check, which provides the linear one, make it possible to use ZK-SNARK to solve the global problem of scalability of blockchain systems. Suppose we are dealing with some computational algorithm that requires t steps. For example, this is an algorithm for generating a chain of blocks (or their native verification) in modern DLT. Asymptotic estimates show that ZK-SNARK technology allows generating a proof approximately in $O(t \cdot log(t))$ steps and verifying them in $O(log^2(t))$ steps.

Thus, ZK-SNARK divides the all calculations into two steps:

- A complex process of generating proof of computational integrity;
- A fast verification of computational integrity by everyone using cheap and low-resource computers.

Computational Integrity (CI) guarantees (with some, very small error) the correctness of the performed calculations. For example, there is a proof that the block in the DLT does indeed contain all cryptographic integrity and involvement marks, and they are formed according to the specified protocol.

Consider an example of binary logarithm, where $t = 2^{20} = 10^6$. We need to perform $20 \cdot 10^6$ steps to make a proof and approximately $20^2 \approx 400$ steps to check it. The verification is significantly faster than the native registry check, which requires $2^{20} \approx 1$ million steps. So this solves the scalability problem.

On Fig. 2 we present an asymptotic estimate for native check, proof generation and verification due to the above reasoning.

Native check requires constant recalculation of all blocks. This is computationally intensive task for long chain of blocks. Generating proofs takes time. It is even slightly longer than the native check. The good point is that you can generate proofs once, using some powerful server with a multi-core architecture. Verification significantly reduces the required processing time. It is achieved due to the logarithmic dependency. In this case for all users, even with low-powered gadgets, computations will be available.

As a result, you have a service that stores proofs for each block and allows billions of users to verify their blocks at the same time very quickly.

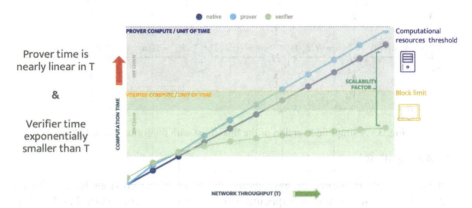

Fig. 2. Asymptotic estimates of computational complexity of native verification (dark graph), proof generation (sky-blue graph) and proof verification (green graph) [20]

3 Our Solution

We implement a recursive proof of the computational integrity of a chain of linked hashes with digital signature verification. This protects against possible data changes and the imposition of a false chain of related hashes.

Initial data. For a test case, consider a simplified version, where each block contains three fields (Fig. 3):

- Unique block number: $Nonce_i$;
- Hash of the previous block: h_{i-1};
- Digital signature: EDS_i.

To simplify, we accept $Nonce_i = i$. The result of the n-th hashing is:

$$h_n = H(h_{n-1}||n) = H(H(h_{n-2}||n-1)||n) = \ldots =$$
$$= H(H(H(\ldots H(H(0)||1)\ldots||)n-2)||n-1)||n),$$

Additionally, each hash h_i is encrypted with a secret key sk, i.e. we form a signature $EDS_i = EDS(h_i, sk)$. The public key pk is used to verify the signature, i.e. we decrypt EDS_i and check for equality $h_i = D(EDS_i, pk)$. We use only the signature verification algorithm $h_i = D(EDS_i, pk)$ to generate proofs and CI verification.

While implementing a proving scheme, our team decided to divide the research into two directions: aggregation and recursion.

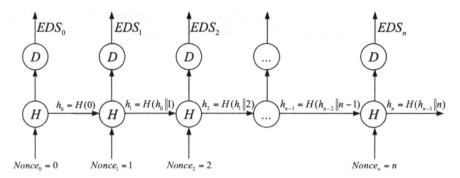

Fig. 3. A simplified version of the linked list with the formation of EDS

3.1 Aggregation

A chain of proofs is created by aggregating the proof of the previous block with the current one. As a result, for each specific block, we can check the proof that the block hash and signatures were calculated correctly and the proof chain for previous blocks was verified correctly. This method requires digital signatures to confirm computational integrity of the block, otherwise a possible attacker can simply substitute some blocks.

To implement a recursive proof of CI of a chain, it is necessary to consistently implement the following tasks:

1. Implement a hash chain $h_0 = H(0)$, $h_i = H(h_{i-1}||i)$, $i = 1,...n$;
2. For each hash $h_0,... h_n$:
   a. Create a circuit $CH_i(xH_i, wH_i)$ of the hash algorithm H, where the public input $xH_i = h_i$ is the result of hashing, the witness wH_i is the hash preimage: $wH_0 = 0$, $wH_i = h_{i-1}||i, i = 1,...n$;
   b. Form public settings $(SH_{pi}, SH_{vi}) = S(CH_i(xH_i, wH_i))$, where SH_{pi} are public prover settings, SH_{vi} are public verifier settings;
   c. Form a proof CI for hashing $\pi H_i = P(SH_{pi}, xH_i, wH_i)$;
   d. Implement verification algorithm $V(SH_{vi}, xH_i, \pi H_i)$ takes values $\{0, 1\}$ (accept or reject);
   e. Proof verification, i.e. to make sure that $V(SH_{vi}, xH_i, \pi H_i) = accept$.
3. For each signature $EDS_i = E(h_i, sk)$, $i = 0,...n$:
   a. Create a circuit $CD_i(xD_i, wD_i)$ of proof verification $h_i = D(EDS_i, pk)$, where the public input $xD_i = h_i$ is the result of hashing, the witness $wD_i = (EDS_i, pk)$ are the signature and the public key;
   b. Form public settings $(SD_{pi}, SD_{vi}) = S(CD_i(xD_i, wD_i))$, where SD_{pi} are public prover settings, SD_{vi} are public verifier settings;
   c. Form a proof CI for signature verification $\pi D_i = P(SD_{pi}, xD_i, wD_i)$;
   d. Implement proof verification algorithm $V(SD_{vi}, xD_i, \pi D_i)$ takes values $\{0, 1\}$ (accept or reject);
   e. Proof verification, i.e. to make sure that $V(SD_{vi}, xD_i, \pi D_i) = accept$.
4. For every triple of proofs $\prod_{i-1} = P(S_{Pi-1}, X_{i-1}, W_{i-1})$, $\pi H_i = P(SH_{pi}, xH_i, wH_i)$ and $\pi D_i = P(SD_{pi}, xD_i, wD_i)$, $i = 1,...n$:
   a. Create a circuit $C_i(X_i, W_i)$ verification algorithm V, where:

- $X_i = (V(S_{Vi-1}, X_{i-1}, \prod_{i-1})), V(SH_{vi}, xH_i, \pi H_i), V(SD_{vi}, xD_i, \pi D_i)$, for all $i = 1,...n$.
- $W_1 = (\pi_0, h_0, \pi_1, h_1, EDS_1, pk), W_1 = (\prod_{i-1}, X_{i-1}, \pi_i, h_i, EDS_i, pk)$, for all $i = 2,...n$.

b. Form public settings $(S_{Pi}, S_{Vi}) = S(C(X_i, W_i))$, where S_{Pi} are public prover settings, S_{Vi} are public verifier settings;
c. Form a proof of CI $\prod_i = P(S_{Pi}, X_i, W_i)$;
d. Implement proof verification algorithm $V(S_{Vi}, X_i, \prod_i)$ takes values {0, 1} (accept or reject);
e. Proof verification, i.e. to make sure that $V(S_{Vi}, X_i, \prod_i) = accept$.

Thus, each proof $\prod_i = P(S_{Pi}, X_i, W_i)$, $i = 1,...n$ is the aggregation of three other proofs:

1 Proof CI of previous chain of linked hashes $\prod_{i-1} = P(S_{Pi-1}, X_{i-1}, W_{i-1})$;
2 Proof CI of current hash $\pi H_i = P(SH_{pi}, xH_i, wH_i)$;
3 Proof CI of current signature verification $\pi D_i = P(SD_{pi}, xD_i, wD_i)$.

Condition fulfillment $V(S_{Vi}, X_i, \prod_i) = accept$ for all $i = 1,...n$ means that the proof verification $\prod_{i-1} = P(S_{Pi-1}, X_{i-1}, W_{i-1})$, $\pi H_i = P(SH_{pi}, xH_i, wH_i)$ and $\pi D_i = P(SD_{pi}, xD_i, wD_i)$ were calculated correctly. If $V(S_{Vi-1}, X_{i-1}, \prod_{i-1}) = accept$, $V(SH_{vi}, xH_i, \pi H_i) = accept$ and $V(SD_{vi}, xD_i, \pi D_i) = accept$, it means that:

1. There is a proof of CI of previous chain, i.e. the verification $V(S_{Vi-2}, X_{i-2}, \prod_{i-2}) = accept$ is computed correctly;
2. There is a proof of CI of current hash, i.e. the value $h_i = H(h_{i-1}||i)$ is computed correctly;
3. There is a proof of CI of current signature, i.e. verification $h_i = D(EDS_i, pk)$ is computed correctly.

The scheme of forming a chain of recursive proofs of computational integrity with verification of the correctness of electronic digital signatures is shown in Fig. 4.

3.2 Recursion

This time the next proof in a chain is created based on the proof of the previous block. We generate a single proof for hash and signatures of the current block, not aggregating them, just making an enlarged circuit.

A proof of the next block is formed by verifying the previous block (i.e., by verifying the correctness of hash and signatures of the previous block) and calculating a proof for the current block.

This scheme (Fig. 5) allows checking the previous hash so that we solve the block substitution problem. The scheme does not require a digital signature to provide the computational integrity of a chain.

To implement a recursive proof of CI of a chain, it is necessary to consistently implement the following tasks:

1. Implement a block chain based on linked hashes $h_0 = H(0)$, $h_i = H(h_{i-1}||i)$, $i = 1,...n$;

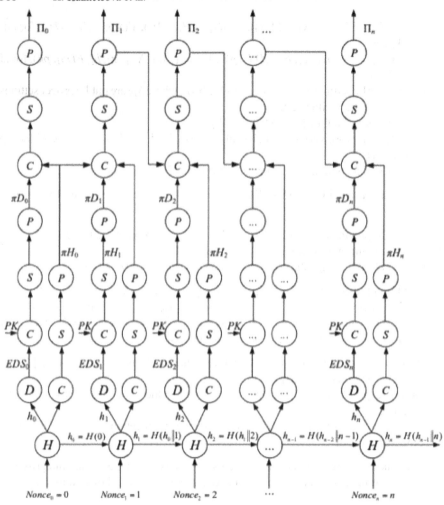

Fig. 4. Scheme of forming a chain of proofs of CI with digital signature verification (first variant)

2. For each block create a circuit $C_i(X_i, W_i)$ of the hash algorithm H, signature verification and the verification algorithm V for the previous block $\pi_{i-1} = P(S_{pi-1}, x_{i-1}, w_{i-1})$. $X_i = (V(S_{vi-1}, x_{i-1}, \pi_{i-1}), x_i)$: $V(S_{vi-1}, x_{i-1}, \pi_{i-1})$ is the result of verification (accept or reject) of the previous proof π_{i-1}, $x_i = h_i$ is the result of hashing. $W_i = (\pi_{i-1}, h_{i-1}, wH_i, wD_i)$: π_{i-1}, h_{i-1} are the input data for verification of proof and w_i are the input data (witness, hash-preimage) to calculate hashes: $w_0 = 0$, $w_i = (h_{i-1} || i)$, $i = 1,...n$, the witness $wD_i = (EDS_i, pk)$ are the signature and the public key;
3. Form public settings $(S_{pi}, S_{vi}) = S(CH_i(xH_i, wH_i), CD_i(xD_i, wD_i))$, where S_{pi} are public prover settings, S_{vi} are public verifier settings;
4. Implement proof verification algorithm $V(S_{vi}, x_i, \pi_i)$ takes values {0, 1} (accept or reject);
5. Proof verification, i.e. to make sure that $V(S_{vi}, x_i, \pi_i) = accept$.

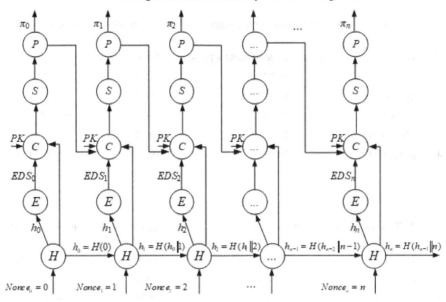

Fig. 5. Scheme of forming a chain of proofs of CI with digital signature verification (second variant)

Condition fulfillment $V(S_{Vi}, X_i, \prod_i) = accept$ for all $i = 1,...n$ means that there is a proof of CI of previous chain, i.e. the verification $V(S_{vi-1}, x_{i-1}, \pi_{i-1}) = accept$ is computed correctly, CI of current hash, i.e. the value $h_i = H(h_{i-1}\|i)$ is computed correctly and CI of current signature, i.e. verification $h_i = D(EDS_i, pk)$ is computed correctly.

4 Time and Measurement Results

All computations were made on 2,40GHz Intel(R) Xeon(R) CPU E5–2680 v4.

4.1 Aggregation Method

Table 1. Time and measurement results for a chain of proofs for hashes

№	Hash	Time to build a circuit, s	Time to make a proof, s	Proof size, bytes	Verification, s
0	5FECEB66FFC86F38D952786C6D696C79 C2DBC239DD4E91B46729D73A27FB57E9	1.111688	0.9969	132640	0.0224
1	F3C1D27925BF4262AE19BFCBCCE1B7612 4F54A4DBF62DC4B09E6B353F34D277D	2.2134259	1.6811	150268	0.0282
2	9C808663A3D1A47F3FCFE26BA11AFD 90D0F00A41984F8058DE3CE8552C6BEF77	2.2696147	1.7133	150268	0.1521
3	B3CC82D30374FC4FA584032CF35E86 A54E70E08A9E3E9982BFE06A6022A4937A	2.3223338	1.7160	150268	0.1571
4	4D23AF03289C3EB09E4C888999102EE53 E6A49D49FD06326A09ED924D22D34A9	2.8574667	2.6339	150268	0.0621

K. Kuznetsova et al.

Table 2. Time and measurement results for a chain of proofs for signatures

№	Time to build a circuit, s	Time to make a proof, s	Proof size, bytes	Verification, s
0	63.335045	75.2214	208376	0.0170
1	58.583714	61.1627	208376	0.0173
2	64.45872	57.5522	208376	0.0225
3	71.951866	72.8148	208376	0.0182
4	102.80682	59.8327	208376	0.0168

Table 3. Time and measurement results for a chain of aggregated proofs for signatures & hashes

№	Time to build a circuit, s	Time to make a proof, s	Proof size, bytes	Verification, s
0	2.6056433	1.4931	146348	0.0122
1	2.6340964	1.7561	146348	0.0120
2	2.5258436	1.6699	146348	0.0171
3	3.8990123	2.5616	146348	0.0178
4	2.6605875	1.7653	146348	0.0138

4.2 Recursion Method

4.3 Discussion

The obtained results fully confirm the asymptotic estimates presented in Sect. 2. Thus, the time for creating a proof increases with the increase in the amount of data. We

Table 4. Time and measurement results for a chain of recursive proofs

№	Hash	Time to build a circuit, s	Time to prove, s	Verification, s	Proof size, bytes
0	81DDC8D248B2DCCDD3FDD5E8 4F0CAD62B08F2D10B57F9A831C13451E5C5C80A5	1.119945	1.4125	0.0213	149084
1	AC8BE15C3CC494661BC32FF34 57E273A98896338195A812702108574D9930EAD	72.49396	74.2432	0.0457	211432
2	3730496E35571584A2839C24348 1A461AA7E304203666CB6A358247CB2818914	53.94758	46.4669	0.0337	211432
3	C9C25FCF5183B139A462ECD0 6919572A88059355D3271F5CFDBBCD46F32C6723	50.751015	48.2629	0.0339	211432
4	81DDC8D248B2DCCDD3FDD5E84 F0CAD62B08F2D10B57F9A831C13451E5C5C80A5	52.462822	44.4447	0.0379	211432

see from the aggregation example that creating a proof for a digital signature is much more difficult than for a hash. We also note that we use the SHA256 algorithm for the cryptographic connection of blocks, i.e. hashing, and EDDSA25519 curve and the SHA512 hashing algorithm for signature.

The verification time is very low, which is explained by the logarithmic dependence. There is practically no difference between verifying a proof for a signature and a hash.

For aggregation we first implement proofs for hashes (Table 1) and signatures (Table 2) for each block and aggregate them with the proof from the previous block (Table 3).

As you can see the proof size for signature is a little bigger than for hash, but we use aggregation, which compress the final proof.

The final recursive proof is much bigger that in the case of aggregation. This is explained by the fact that in the aggregation scheme we create a separate circuit for each entity: hash, signature and aggregation. And the recursion scheme requires only one circuit for all these three entities. Creating such an extended scheme takes more memory and, as a result, the final proof is also large.

As you can see from the tables for both implementations verification is really fast. Proof generation takes time, but its speed depends on the number of processor cores, i.e. the task is well parallelized.

It is also worth considering that computations were made on the system, which also performs other processes. Because of this there are some measurement errors.

5 Conclusion

On the graph below there are the results of computation times for native verification (or recalculation of all hashes), verification of proofs generated for each block and recursive proof. We see that even the proof for a separate block significantly reduces the cost of verification. Recursion combines all these proofs into one. Verification is very fast, which, in fact, solves the main problem of mass introduction of distributed systems, which is the scalability.

Thus, zero-knowledge proofs are a great solution to solve the problems of scalability and confidentiality. This is especially important in large distributed computing projects. We can introduce a proof system and replace native block verification with it. This significantly speeds up verification and facilitates the work of a system.

In this article, we have shown how ZK-SNARK technology can be used to prove the integrity of the blockchain. As an example, we took a simplified version of the block, that is, without taking into account transactions and other auxiliary data structures necessary for the safe and reliable operation of the blockchain system. Our task was only to demonstrate the prospect of this direction. In the future, this technology can be applied to a specific blockchain ledger, using real network data as blocks. Moreover, the obtained results can be also useful in various computer science applications, including the development and implementation of computationally efficient systems for information protection, reliability and security (Fig. 6).

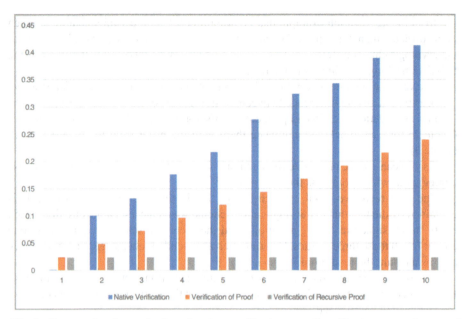

Fig. 6. Computational complexity of native verification, proof verification, recursive proof verification

Acknowledgment. This project is supported by Zpoken, OU, Harju maakond, Tallinn, Kesklinna linnaosa, Sakala tn 7–2, 10141, Estonia.

References

1. Nakamoto, S.: Bitcoin: A Peer-to-Peer Electronic Cash System
2. Lee, S.-W., Singh, I., Mohammadian, M. (eds.): Blockchain Technology for IoT Applications. BT, Springer, Singapore (2021). https://doi.org/10.1007/978-981-33-4122-7
3. Xu, K., Zhu, J., Song, X., Lu, Z. (eds.): CBCC 2020. CCIS, vol. 1305. Springer, Singapore (2021). https://doi.org/10.1007/978-981-33-6478-3
4. Distributed Ledger Technology: beyond block chain. UK: Government Office for Science (2016)
5. Buterin, V.: Why sharding is great: demystifying the technical properties. Vitalik Buterin's website 7 (Apr 2021)
6. Chowdhury, M.J.M., et al.: A comparative analysis of distributed ledger technology platforms. IEEE Access **7**, 167 930–167 943 (2019)
7. Krishnan, S., Balas, V.E., Golden, J., Robinson, Y.H., Balaji, S., Kumar, R. (eds.) Handbook of Research on Blockchain Technology. Academic Press (2020)
8. Oliynykov, R., Kuznetsov, O., Lemeshko, O., Radivilova, T. (eds.): Information Security Technologies in the Decentralized Distributed Networks. LNDECT, vol. 115. Springer, Cham (2022). https://doi.org/10.1007/978-3-030-95161-0
9. Shapoval, O., Kuznetsov, A., Poluyanenko, N., Yakovenko, V., Prokopovych-Tkachenko, D., Kavun, S.: The decentralized voting model using the hyperledger platform paper. In: 2019 International Conference on Information and Telecommunication Technologies and Radio

Electronics (UkrMiCo), pp. 1–5 (2019). https://doi.org/10.1109/UkrMiCo47782.2019.916 5368

10. Poluyanenko, N., Kuznetsov, A., Lisickiy, K., Datsenko, S., Nakisko, O., Rudenko, S.: The problem of double costs in blockchain systems. In: Hu, Z., Petoukhov, S., Dychka, I., He, M. (eds.) Advances in Computer Science for Engineering and Education III, pp. 640–652. Springer International Publishing, Cham (2021). https://doi.org/10.1007/978-3-030-55506-1_57

11. Ferraro, P., King, C., Shorten, R.: On the stability of unverified transactions in a DAG-based distributed ledger. IEEE Trans. Autom. Control. **65**, 3772–3783 (2020). https://doi.org/10.1109/TAC.2019.2950873

12. Devarajan, A., Karabulut, E.: Directed Acyclic Graph Based Blockchain Systems (2020). https://doi.org/10.13140/RG.2.2.18968.67849

13. Kotilevets, I.D., Ivanova, I.A., Romanov, I.O., Magomedov, S.G., Nikonov, V.V., Pavelev, S.A.: Implementation of directed acyclic graph in blockchain network to improve security and speed of transactions. IFAC-Pap. **51**, 693–696 (2018). https://doi.org/10.1016/j.ifacol.2018.11.213

14. Benčić, F.M., Žarko, I.P.: Distributed Ledger Technology: Blockchain Compared to Directed Acyclic Graph, http://arxiv.org/abs/1804.10013 (2018). https://doi.org/10.48550/arXiv.1804.10013

15. Guo, Z., Gao, Z., et el.: Design and optimization for storage mechanism of the public blockchain based on redundant residual number system. IEEE Access **7**, (22 Jul 2019)

16. Yu, G., Wang, X., et al.: Survey: Sharding in Blockchains. IEEE Access **8** (9 Jan 2020)

17. Near protocol. Sharding Design: Nightshade

18. Jo, T.: An Exploration of Zero-Knowledge Proofs and zk-SNARKs (2019)

19. Nitulescu, A.: zk-SNARKs: A Gentle Introduction

20. StarkWare: STARK Math: The Journey Begins, https://medium.com/starkware/stark-math-the-journey-begins-51bd2b063c71, last accessed 15 February 2023

Interactive Information System for Automated Identification of Operator Personnel by Schulte Tables Based on Individual Time Series

Myroslav Havryliuk[✉], Roman Kaminskyy, Kyrylo Yemets, and Taras Lisovych

Lviv Polytechnic National University, S. Bandera Street, 12, Lviv 79013, Ukraine
myroslav.a.havryliuk@lpnu.ua

Abstract. The selection of operator personnel to manage technological processes is an important problem in various industries. A large percentage of accidents at enterprises occur due to the fault of workers serving the production. One of the techniques for determining a person's suitability for such work is reading Schulte's tables. Today, there is no freely available specialized software to implement this technique. In this work, the authors developed an information system for the automated identification of operator personnel. It implements the method of multidimensional average on the indicators of descriptive statistics of an individual time series. The created system allows users to compare candidates and form their ratings directly. The software product improves the quality of professional selection by automating the process and using an advanced analysis method.

Keywords: Time-series Data; Automated System · Professional Selection · Schulte Tables · Personnel Identification · Software Product

1 Introduction

The selection of highly qualified specialists, in particular operators, to manage technological processes is an actual problem in many industries. However, information technology has a much faster rate of development than the selection processes of employees of the corresponding level.

Special literature reports that a large percentage of accidents at enterprises occur due to the fault of workers serving the production. One of the main reasons for this is the insufficient level of professional selection and training of personnel. Improving the efficiency of personnel selection and certification by automating it with the help of the latest information technologies is an urgent problem in the field of personnel policy of enterprises and institutions.

Special techniques are used to select applicants for the position of the operator. They were developed by specialists in the field of psychology and adapted to the specifics of the operators' activities. One of these techniques is reading Schulte tables. It is designed to determine the stability of attention and working capacity in dynamics. Its main advantage is that the reading process is an example of finding similar images of objects of a given

© The Author(s), under exclusive license to Springer Nature Switzerland AG 2023
Z. Hu et al. (Eds.): ICAILE 2023, LNDECT 180, pp. 372–381, 2023.
https://doi.org/10.1007/978-3-031-36115-9_34

class on the monitor, which is a common task in operator work. The technique is also quite simple and has several varieties.

In [1], the authors implemented the Schulte table methodology using web development tools. However, this program only has a classic version of the reading table with small modification possibilities. Methods of analyzing the results in this development are also traditional.

In [2], the results of the experiment of reading Schulte tables by a group of people are given. Based on the received data, a rating was formed where the participants were placed according to the level of stability of attention. However, only classical methods were used for the analysis.

[3] describes the role of the Schulte method in the procedure of professional selection for the operators' positions. Paper Schulte tables were used for the experiments, and the analysis was also traditional.

An attempt to build a model of operator personnel identification is given in [4]. For this, statistical indicators of time series obtained based on experimental data in image processing systems were used. The authors propose using the multivariate average method to assess the working capacity and build a rating of the participants.

This study aims to create an interactive information system for the automated identification of operator personnel according to Schulte tables using individual time series.

The main contribution of this paper is the practical implementation of the multidimensional average method for analysing results obtained by the Schulte tables.

2 Materials and Methods

2.1 Schulte Tables Methodology

The selection of qualified specialists for the position of the operator is an important part of the company's work. Currently, there are many different psychological methods and criteria for evaluating job applicants. The key indicators for this kind of work are stability of attention and dynamics of working capacity. Using Schulte's tables technique is convenient for studying these parameters. The simplest Schulte table is a square table divided into 25 identical square cells (5 rows and 5 columns), in which numbers from 1 to 25 are entered in random order (Fig. 1) [5].

18	7	22	25	1
17	5	16	8	14
6	3	12	15	13
19	9	4	24	23
21	20	11	2	10

Fig. 1. An example of Schulte table

The essence of conducting the test according to this method is as follows: the subject is offered the Schulte table. The controller uses a stopwatch to record the time the subject spends searching for and pointing to a number of numbers in the table. The obtained data are used for further analysis. The way to search and identify the required objects is illustrated in Fig. 2.

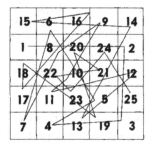

Fig. 2. The way to search for objects

However, this is only the most common version of the technique. There are many modifications to it: the table can have a different size; instead of a sequence of numbers, any other ordered sequence can be used, such as alphabetic characters; the background color of the cells may vary, most often red and black tables are used [6]; if automation is possible, the table may change after each element found. The use of different types of tables is another reason for the relevance of process automation.

2.2 Operator Activity Modeling

Passing tests according to the Schulte method is one of the types of operator activity. In general, the operator transforms the received information into a decision regarding the control of the system's operation, process, etc. This activity consists in finding the right image on the screen. After finding the image, the operator makes a decision - he presses a button to confirm it. Let G^* be an object of a certain defined class that should be detected (in this case, a number or a symbol), which can be randomly placed on the provided image of the training situation (Schulte table) and which can be characterized by a vector of recognition features $g_i^* \in G^*$. We denote the set of objects of the same class that are similar in shape to the searched object as G.

The operator's task is to find, detect and recognize the necessary object (number or symbol) among similar ones and make a decision about its placement on the image. During the process of the operator's work, at the moments t_i, specified by the training situation $Z(t)$ monitoring parameters (in this case – at the moments of finding the necessary element of the table), the operator receives images for processing, which are located on the monitor of the automated system. Images $x_i \in X$ where

$$X = \{x_i : x_i = x(t_i), t_i \in T, i = \overline{1, N}\} \tag{1}$$

are training models of the environment (Schulte tables) in which the situation $Z(t)$ arose, and illustrate the state of the situation at moments t_i i.e. $Z(t) \to Z(t_i)$. Each object belonging to the set G has its vector of features $g_i = \langle g_i, \bullet, g_r \rangle, g_i \in G$, besides, among them, there may be such as those of the desired object, but there must also be signs of difference from the desired object.

At the moment of finding the desired object, the operator compares its features with normative features $g_i^* \in G^*$, stored in the human memory as a result of educational experience in the form of "standard images" $x_i^* \in X^*$. As a result, the operator makes an appropriate decision from several alternatives. Therefore, comparing images x_i and x_i^*, operator identifies the situation given to him, chooses and confirms the correct, in his opinion, decision $y_j \in Y$, where

$$Y = \{y_j : y_j = y(t_j), j = \overline{1, N}, t_j = t_i + t_i; t_i, t_j \in [0, T]\} \tag{2}$$

and t_j is the moment of implementation of the decision made by the operator, which the operator selects with a set of commands that are the control vector $u_j = \langle u_1, \bullet, u_h \rangle, u_j \in U$, where U is the set of vectors of control commands, and $h = 1, 2, \bullet$ is the number of commands in the j-th situation.

Continuous mental tension creates a sense of increased responsibility. When the operator works for a long time, obviously, a person is exposed to a large psychological load. Visual sensitivity decreases, reactions and speed of actions deteriorate, the number of errors increases, and as a result, nervous tension arises. Of course, this condition worsens the efficiency of the operator's activity. Therefore, the quality of operator work must be linked to certain states c_k. A set of such states

$$C = \{c_k : c_k = c(\delta t_k), \delta t_k = t_{k+1} - t_k; t_k, t_{k+1} \in [0, T]\}, \tag{3}$$

can be considered discrete.

So, having described the general model of individual intellectual work in the system for recognizing and detecting objects among a certain class on training images with a situation $Z(t)$ using the example of Schulte's methodology, it is possible to formulate the task of operator activity: to find the desired object on the image $x_i \in X$ among the objects of a certain class, for which the expression $\left| g_i^* - g_i \right| \equiv 0$ is true and then make a decision and implement it using the appropriate command from the vector u_j. The goal of operator activity is maximum reliability $p \to 1$ with minimum time expenditure $t \to 0$, i.e.

$$X(g_i - g_i^*) \underset{\substack{p \to 1 \\ t \to 0}}{\longrightarrow} optU(x_t, c_t, y_t) \to \min \tag{4}$$

This expression illustrates the operator activity's main task in optimizing the situation's solution $Z(t)$[7].

2.3 Multidimensional Average Method

Since, at the moment, there is no general model for determining the assessment of operator activity, statistical methods can be used to identify applicants [8]. One of the most common methods that allow getting a generalized score based on several features is the search for a multidimensional average [9]. Let there be N objects for identification by

M features. Then, $x_{ij}, i = \overline{1,N}, j = \overline{1,M}$ are the values of the i-th object by the j-th feature. Accordingly, $\overline{x_j}, j = \overline{1,M}$ are arithmetic average values. Then the multidimensional average $a_i, i = \overline{1,N}$ for each object can be calculated as:

$$a_i = \frac{\sum_{j=1}^{M} \frac{x_{ij}}{\overline{x_j}}}{N} \qquad (5)$$

The features that are used to search for a multidimensional average in our system are the measures of the descriptive statistics of the time of finding each object (for the Schulte method - a number/symbol):

- arithmetic average;
- median;
- standard deviation;
- average deviation;
- interquartile range;
- coefficient of variation;

The resulting multidimensional averages can be used to rank objects. The essence of Schulte method involves the minimization of features, so the object with the lowest value of the multidimensional average will be the highest in the rating. This method is advanced for evaluating the operator's performance, as it allows better use of time series data compared to the classical one.

3 Results

The main menu is implemented on the main page of the application. The user has four options for further action (Fig. 3):

- create new Schulte table;
- read existing Schulte table;
- see results;
- exit.

Fig. 3. The main page of the application

The user can define parameters for the desired Schulte table on the page for creating a new table. Among the parameters (Fig. 4):

- name of the table;
- type of elements (numbers or symbols of the Ukrainian/English alphabet);
- number of sequence iterations;
- table size;
- static/dynamic location of objects after each found one;
- background color of cells (white/red-black).

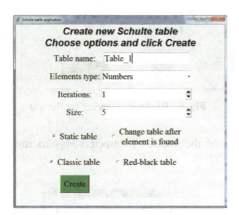

Fig. 4. Page for creating a new table.

After creating a table, all its parameters (including the generated locations of objects on the table) are saved. This is necessary so that all test participants receive absolutely identical Schulte tables. Thus, the comparison of results is direct.

All previously created tables are shown on the Schulte table selection page for reading (Fig. 5).

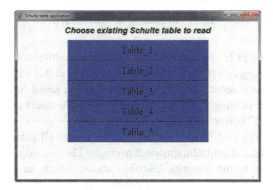

Fig. 5. Page for selecting a table from the available ones

After selecting the Schulte table, the user gets to the page of preparation for reading the table. Table parameters are displayed here so that the participant clearly understands the task. Before reading the Schulte table, the user must enter his name (Fig. 6).

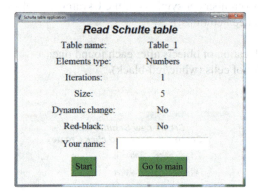

Fig. 6. Page for preparing for the test

The Schulte table with the relevant parameters appears immediately after pressing the "Start" button (Fig. 7).

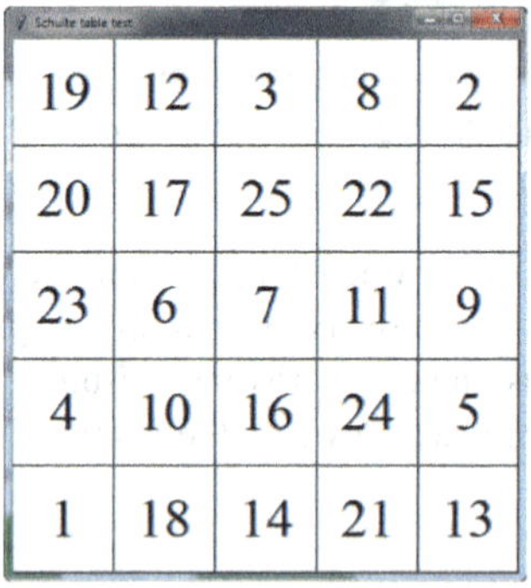

19	12	3	8	2
20	17	25	22	15
23	6	7	11	9
4	10	16	24	5
1	18	14	21	13

Fig. 7. An example of Schulte table

The required object is considered found when the participant clicks on it. In this case, the corresponding cell is highlighted in green. Also, at this moment, the time spent on finding this object is recorded. All obtained results are saved for further analysis.

On the page for viewing the results, the user needs to select the Schulte table and press the "See rating" button (Fig. 8).

The rating is displayed in the form of a table, where all participants are ordered according to the defined multidimensional average. The multidimensional average is calculated according to the features described above, which, in turn, are determined based on stored data on reading the Schulte table (Fig. 9).

With the help of the formed rating, it is possible to compare applicants for the operator position, which is the main purpose of creating the system.

Fig. 8. Page for selecting a table to view the result

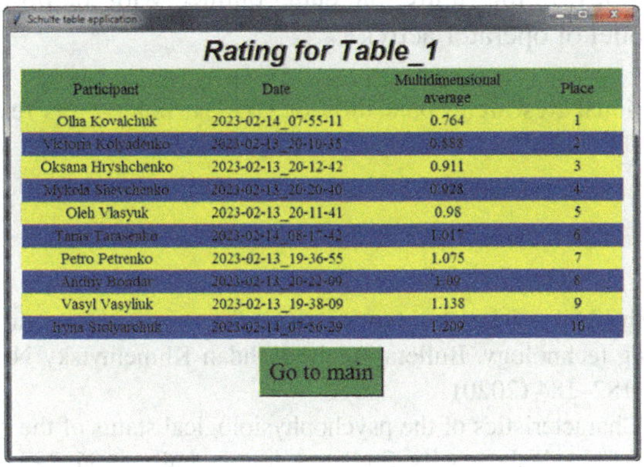

Fig. 9. Page for viewing results

4 Discussion

Exact analogs of the system for determining psychological characteristics, according to Schulte tables, were not found in free access. In general, many applications with Schulte tables are used as a training game for developing attention[10, 11]. However, they cannot fully fulfill the role of a system for the identification of operator personnel [12]. Their main advantage is a large number of implemented modifications of the technique [13, 14]. Among their main shortcomings: a lack of detailed statistics of passing the test [15, 16]; an inability to save results [17, 18].

When comparing applicants for the operator position, the purity of the experiment is important [19]. Each person must deal with the same Schulte table [20]. The developed system provides such an opportunity for direct comparison of applicants. The program makes it possible to collect and save individual time series for further analysis. The system implements an identification method based on the calculation of a multidimensional average. The multidimensional average method provides rank-based comparisons using the relative values of identified features.

5 Summary and Conclusion

The selection of operator personnel to manage technological processes is an important problem in various industries. One of the techniques for determining a person's suitability for such work is reading Schulte's tables. Today, there is no freely available specialized software to implement this technique. In this work, the authors developed an information system for the automated identification of operator personnel. It implements the method of multidimensional average on the indicators of descriptive statistics of an individual time series. This method is advanced for evaluating the operator's performance, as it allows better use of time series data compared to the classical one. The created system allows users to compare candidates and form their ratings directly. The software product can be used for systems of professional selection of operator personnel, in training systems to improve the level of concentration and work capacity, and in diagnostic systems in psychology. In the future, the data obtained with the program's help can be used to build a model of operator activity.

Funding. The National Research Foundation of Ukraine funded this research under project number 2021.01/0103.

References

1. Fedota, M.V., Shasholko, S.I.: Use of interactive trainers as modern means of implementation of speed reading technology. Bulletin of the Bohdan Khmelnytsky National University of Cherkasy **8**(2), 182–184 (2020)
2. Kovalchuk, I.: Characteristics of the psychophysiological status of the body in persons with different levels of physical capacity. Student scientific bulletin **36**, 7–9 (2015)
3. Hohoman, T.S., Chumayeva, Y.V.: Effectiveness of psychocorrection of mental fatigue of maritime transport operators by means of multifunctional sound-color regulation. Actual issues of psychology in the modern innovative space **10**(4), 89–92 (2022)
4. Kaminsky, R.M., Nych, L.Y.: Identification of the intellectual activity of the operator personnel based on experimental data. Bulletin of the Lviv Polytechnic National University. Information systems and networks **715**, 134–149 (2011)
5. Nechyporenko, A.S.: New intelligent-based approach for the early detection of disorders: use on rhinological data. International Journal of Image, Graphics and Signal Processing (IJIGSP) **9**(8), 1–8 (2017)
6. Bernacki, J., Kołaczek, G.: Anomaly detection in network traffic using selected methods of time series analysis. IJCNIS **7**(9), 10–18 (2015)
7. Nych, L.Y., Kaminsky, R.M.: Evaluation of the effectiveness of individual intellectual activity of operators in image processing systems based on experimental data. Bulletin of the Lviv Polytechnic National University. Information systems and networks **699**, 193–204 (2011)
8. Hassan, M.M., Mirza, T.: Using time series forecasting for analysis of GDP growth in india. Int. J. Edu. Manage. Eng. (IJEME) **11**(3), 40–49 (2021)
9. Mishra, N., Soni, H.K., Sharma, S., Upadhyay, A.K.: Development and analysis of artificial neural network models for rainfall prediction by using time-series data. Int. J. Intelli. Sys. Appli. (IJISA) **10**(1), 16–23 (2018)
10. Jain, E.G., Mallick, B.: A study of time series models ARIMA and ETS. Int. J. Modern Edu. Comp. Sci. (IJMECS) **9**(4), 57–63 (2017)
11. Izonin, I., Tkachenko, R., Holoven, R., Yemets, K., Havryliuk, M., Shandilya, S.K.: SGD-based cascade scheme for higher degrees wiener polynomial approximation of large biomedical datasets. Machine Learning and Knowledge Extraction **4**(4), 1088–1106 (2022)
12. Ramakrishna, M.T., Venkatesan, V.K., Izonin, I., Havryliuk, M., Bhat, C.R.: Homogeneous adaboost ensemble machine learning algorithms with reduced entropy on balanced data. Entropy **25**(2), 245 (2023)

13. Basystiuk, O., Melnykova, N.: Multimodal approaches for natural language processing in medical data. In: Proceedings of the 5th International Conference on Informatics & Data-Driven Medicine. Lyon, France, November 18–20, CEUR-WS.org, pp. 246–252 (2022)
14. Basystiuk, O., Shakhovska, N., Bilynska, V., Syvokon, O., Shamuratov, O., Kuchkovskiy, V.: The developing of the system for autimatic audio to text conversion. Proc. of the IT&AS'2021 Symposium on Information Technologies & Applied Sciences, CEUR-WS.org, pp. 1–8 (2021)
15. Mochurad, L., Hladun, Y.: Modeling of psychomotor reactions of a person based on modification of the tapping test. Int. J. Comp. **20**(2), 190–200 (2021)
16. Mochurad, L., Yatskiv, M.: Simulation of a human operator's response to stressors under production conditions. ICEUR Workshop Proceedings **2753**, 156–169 (2020)
17. Zomchak, L, Nehrey, M.: Economic growth and capital investment: the empirical evidence. Lecture Notes on Data Engineering and Communications Technologies **135**, 645–652 (2022)
18. Auzinger, W., Obelovska, K., Stolyarchuk, R.: A revised gomory-hu algorithm taking account of physical unavailability of network channels. Communications in Computer and Information Science **1231**, 3–13 (2020)
19. Oleksiv, I.B.: Selection of important company stakeholders: theory and practice. Naukovyi Visnyk Natsionalnoho Hirnychoho Universytetu **1**, 128–134 (2013)
20. Kuzmin, O.Y., Oleksiv, I.B., Mykhailyak, G.V.: Integral evaluation of company employees' competence system. Actual Problems of Economics **155**(5), 506–513 (2014)

DIY Smart Auxiliary Power Supply for Emergency Use

Nina Zdolbitska[✉], Mykhaylo Delyavskyy, Nataliia Lishchyna, Valerii Lishchyna, Svitlana Lavrenchuk, and Viktoriia Sulim

Lutsk National Technical University, Lutsk 43000, Ukraine
ninazdolb@gmail.com

Abstract. Implementation of Internet of Things technologies in various fields is one of the priority areas of modern development. Energy efficiency is one of the urgent problems in Ukraine today. As part of this study, the use of IoT will help monitor and control energy consumption data. Auxiliary power supplies are devices that provide supplemental power to an existing power source. They are often used in automotive, electrical, and industrial applications. DIY auxiliary power supplies are those that are built by the user instead of purchasing a ready-made product. Building your own power supply can be a cost-effective option, as well as an enjoyable DIY project. The purpose of the research is to develop a device based on the existing electrical engineering components, which allows you to completely or partially replace the functionality of industrial analogs of power supplies.

Keywords: Auxiliary power supply · uninterruptible power supply · inverter · battery · DC-DC converter · communication · lighting

1 Introduction

1.1 Problem Statement

The creation of an alternative auxiliary power source for communication devices and the organization of emergency lighting, which will be available for self-production, will not contain expensive components and will have advantages in compared to industrial models, is topical task in emergency situations.

The need for lower price devices arises under unforeseen circumstances, such as war, natural disasters, limited resources, low purchasing power of consumers.

You may have mission-critical life support systems that must continue to function even during an extended blackout.

The key challenges. The auxiliary power supply must simultaneously provide safe and reliable power, high performance, low power consumption, and low bill-of-materials.

An auxiliary power supply is an additional power source used to supplement the main power supply of a system or device. It is used to provide extra power when it is needed or when the main power source fails. Auxiliary power sources can range from

© The Author(s), under exclusive license to Springer Nature Switzerland AG 2023
Z. Hu et al. (Eds.): ICAILE 2023, LNDECT 180, pp. 382–392, 2023.
https://doi.org/10.1007/978-3-031-36115-9_35

small battery backups to large fuel-burning generators. They are often used to power critical security systems and other electronics that must remain operational even if the main power source fails.

1.2 Analysis of Recent Research and Publications

Implementation of Internet of Things technologies in various fields is one of the priority areas of modern development. Energy efficiency is one of the urgent problems in Ukraine today. As part of this study, the use of IoT will help monitor and control energy consumption data [1]. The Internet of Things in the energy sector optimizes work at all levels: from the moment of data transfer from the user to the state of the power grid and generating capacities [2, 3]. Modern smart home automation technologies are becoming more and more advanced, able to make life easier and save money. Home automation technologies using low-cost components such as the Arduino microcontroller, control modules and sensors can completely or partially replace the functionality of industrial IOT devices [4–6].

An auxiliary power supply is required for many devices, so they are the object of research in many scientific and practical articles. Technologies that include auxiliary power supply systems are used on an industrial scale for suburban and regional electric transport and trains, energy-efficient street lighting systems [7] that are parts of a smart city, in wind turbines technology [8], solar panels [9], in microturbines and other elements of aircraft and automobile structures (provides electrical and pneumatic power to various on-board subsystems), in alarm systems [10] (fire alarm, air alarm, video surveillance). Numerical modeling and analysis using the Matlab/Simulink Simulation System including an auxiliary power supply (lithium battery and a supercapacitor), a DC-DC converter, a DC-AC inverter were carried out in [11].

UPS has found many applications, for example, they are used to power telecommunications and security systems, computers, and devices that require increased reliability of power supply [12, 13]. For ordinary citizens, the use of additional power sources is also very important.

In this difficult time for Ukraine, the problem of providing small and large gadgets with uninterrupted power supply is especially relevant. Smartphones, tablets, and laptops need special attention. There is a need for emergency lighting of apartments, houses, and other premises in order to maintain life comfort during scheduled and emergency power outages.

Inefficient, cheap, unreliable solutions are available on the market, while expensive ones are not available for mass use. For example, industrial designs of such companies as RIELLO, BLUETTI, and EcoFlow, which offer functionality and high reliability, but also have inflated prices, are popular among European brands on the market. Backup UPS kits from the Ukrainian manufacturer LogicPower are also popular.

Wartime has an essential impact on the excessive hype in the Ukrainian market for small auxiliary power devices such as PowerBank and UPS. They are often impossible to buy because they are not physically available from sellers, and product deliveries are delayed and irregular due to various circumstances. Therefore, the question arises of how to solve the problem of powering communications equipment and organizing any emergency lighting with the help of available improvised means and materials [14].

1.3 Purpose of the article

The purpose of the research is to develop a device based on the existing electrical engineering components, which allows you to completely or partially replace the functionality of industrial analogs of power supplies.

The object of research is the technology for the development of alternative sources of power supply.

2 Development Requirements

Auxiliary power supplies can provide additional reliability to critical systems by allowing them to continue operating even if the main power fails. They can also provide additional power during peak demand or when the primary power is operating at lower peak power. Auxiliary power can be used to provide transient power during startup or shutdown operations.

It is necessary to take into account the safety requirements for the operation of devices and systems. Comprehensive measures are needed to protect both the person and the device to prevent accidents. The following aspects are especially important: protection against malfunctions, continuity of power supply, operational features.

When creating an auxiliary household power supply, it is necessary to provide for forced air cooling, reduce the weight and lower the price of components, and ensure a minimum noise level.

3 Research Results

3.1 Possibility of Use

When the power of the main network is turned off, the Internet, which is necessary for work, education, entertainment, and other communications, usually disappears. It often happens that to access the Internet, it is enough to connect to power the router, since on the provider's side the corresponding communication equipment can have its emergency power supply, thanks to which the provider will deliver the service even in the absence of electrical energy.

This method is most likely available from providers with optic fiber lines connected to consumers. Another example is a mobile communication channel through a mobile phone or modem. They also need to be powered from an auxiliary power supply to get a long Internet connection, since the built-in battery usually does not last long, and mobile operators provide 4G/3G service for a rather long period since this is a critical infrastructure.

In particular, there is a question of emergency main or additional lighting. Ordinary incandescent light bulbs are not very suitable for such purposes, energy-efficient LED lamps are usually used. LED lamps are the best solution today for both the main and emergency lighting of premises. The only problem is the voltage level that is needed for the lighting network. Usually, this is a mains voltage of 230 V, which is not available in simple power banks or batteries that are used for auxiliary power gadgets.

It's necessary to calculate the power and find a compromise between consumption and system autonomy. Additional costs are also required for the organization of an integrated solution in the existing lighting system.

Therefore, an urgent task is to create such an auxiliary power source, available for self-manufacturing, based on inexpensive components.

Such a device can be a homemade (DIY) auxiliary power supply based on 12-V AGM batteries and a combination of computer components with automotive gadgets.

In this device, you can reuse secondary components, such as batteries from uninterruptible power supplies, which are periodically replaced at critical sites of enterprises, security alarms, and transport companies. The main emphasis in such a system is on the reuse of elements, due to which economic and environmental benefits are achieved, as well as versatility.

An auxiliary power supply with voltages of 5/12/230 V has been developed for charging mobile gadgets and organizing emergency lighting based on the existing component base of electrical engineering, which will allow replacing the functionality of industrial analogs of power supplies in full or in part.

3.2 Advantages and Disadvantages of Auxiliary Power Supply

Advantages of auxiliary power supply:

- can be used as an additional power supply along with the main power supply (reducing the load on the power grid);
- can be used to provide emergency power during a power outage;
- it is easy to install and maintain;
- reliable and cost-effective solution for providing power to remote areas or locations without grid access;
- scalable based on power requirements;
- it is an eco-friendly source of renewable energy, can help reduce the use of fossil fuels, and does not produce any emissions;
- it is a cost-effective and reliable way to provide electricity and reduce your bills;
- can be used portably for a short period;
- can be used in emergencies as a reliable backup power source;
- the device is quiet (low noise level);
- option of voltage selection (5/12/230 V);
- a safe power supply that can shield gadgets from sudden voltage changes;
- lower cost than industrial counterparts;
- ease of maintenance.

Disadvantages of auxiliary power supply:

- limited energy storage capacity;
- relatively short battery off-line life;
- the need for investment of money and time;
- difficulty finding compatible parts;
- sensitivity to weather conditions (must be used indoors);
- unpredictable performance (due to component reuse).

3.3 The Architecture of the Developed Device (MVP)

The developed auxiliary power supply (Fig. 1) contains several USB ports and can charge small gadgets with a power consumption of up to 15 watts.

The available power of the lighting system will depend on the built-in 230 V inverter and the capacity of the battery. For example, in the proposed system, a low-power automobile inverter is used, which allows healing a maximum load of 150 watts, and the capacity of the installed batteries is sufficient for emergency lighting at maximum power for 4 h.

Fig. 1. Auxiliary power supply

To display the general device structure, its main nodes, and the links between them, as well as to understand its functionality and operating modes, we present a block diagram of the device (Fig. 2).

The main nodes of the designed system shown in Fig. 1 are:

- 12 V battery pack with 42 Ah capacity of AGM structure (6 elements);
- 12–220 V inverter with an approximate sine wave at the output;
- DC-DC converter 12 V to 5 V;
- a charger that allows you to charge a 12 V battery pack from a 220 V network;
- a switching scheme that ensures interaction between blocks;
- the control panel, which is used to switch between work modes;
- a control and display unit that allows you to visually control the operating mode of the device.

Lead acid batteries are sold in 6 V and 12 V. You need to connect them in series to increase the voltage, or in parallel to increase the available amp-hours (Fig. 3). Maintenance of lead acid batteries: it's necessary to top it off periodically with distilled water to keep your lead acid battery well maintained and get at least its minimum life expectancy.

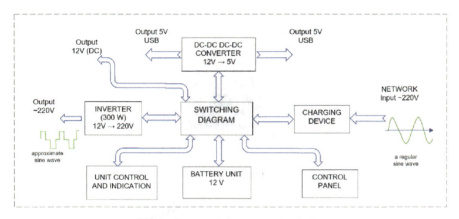

Fig. 2. Structural diagram of the device

Fig. 3. Lead acid battery CSB 12V/7.2Ah

Advantages of lead acid batteries:

- wide range of sizes and specifications;
- relatively low cost;
- high output power;
- widely available, many producers worldwide;
- being able to deliver high currents,
- being tolerant of high-rate discharges,
- and having a longer life cycle than other battery technologies.

Disadvantages of lead-acid batteries:

- big weight and bulky;
- low cycle life;
- high self-discharge rate;
- maintenance required;
- corrosive electrolyte.

A power inverter (Fig. 4) converts 12-V DC power to standard household 230-V AC power. It enables quick USB charging for different devices, AC power socket can be used as an emergency power supply. It has silent cooling, produces modified sine wave.

DC-DC buck converters have compact size, high reliability, low cost, wide input voltage range, making them ideal for applications [15, 16].

Fig. 4. Inverter

Fig. 5. Lead-acid charger

The charger (Fig. 5) can supply enough current to charge the battery and keep up with the inverter's load. It suitable for battery lead-acid maintenance battery.

3.4 Remote Control and Indication

The sensors and structural modules listed in Table 1 were used to expand the functionality of the developed device.

For more comfortable use of the auxiliary power source, it can be equipped with a Bluetooth interface for remote control and management. The easiest way is to use a ready-made Bluetooth module like HC-05/HC-06 and an Arduino Uno board (Fig. 6).

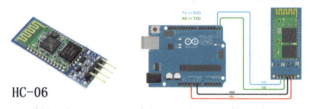

Fig. 6. HC-06 Bluetooth module and its connection diagram to Arduino UNO

To connect a Bluetooth module to the Arduino, necessary to connect the module's TX pin to the Arduino RX's pin, the module's RX pin to the Arduino's TX pin, and the

DIY Smart Auxiliary Power Supply for Emergency Use

Table 1. Sensors and structural modules

The sensors and structural modules	Functionality
	1-bit module AC 230V voltage detect board with optocoupler isolation function: testing whether AC 230V is existed
	INA219 module with I2C interfaces and zero drift bi-directional function: current power supply sensor
	INA3221 triple-channel module with I2C interfaces and zero drift bi-directional function: triple-channel current power supply sensor
	ACS712 Hall current sensor module function: single channel current sensor 5A
	LM2596 power converter step down module DC-DC 1.5V-35V
	1way 5V relay module with optocoupler
	LCD display – measure and display battery capacity and voltage function: indication
	KCD-11 push button switch 10x15mm 3A 250V / 30A 12V
	Car cigarette 12V lighter socket made of heat-resistant plastic

module's ground pin to the Arduino's ground pin. It may be necessary to connect the module's VCC pin to the Arduino's VCC pin, depending on the module's requirements.

We can send commands to the Arduino board or receive data from the HC-05 module using our own program. The base software for the Arduino platform was developed in the Arduino IDE, and the Android Studio software was used for mobile development.

The Bluetooth module allows remote control of the DIY device:

– to connect or disconnect it to the batteries using a relay;
– to receive data from sensors;

- notify about the presence/absence of 230 V network voltage;
- report the main battery voltage;
- inform power consumption of the main connected devices on one/three channels of the controller.

3.5 System Load Testing

To test the system, an 11W LightMaster LB-680 LED lamp (LightMaster LB-680 11 W/A60/E27/230 V/50 Hz) was used. The lamp was connected to the inverter with a two- wire copper cable 8 m long and with a cross-sectional area of 2.5 mm^2.

The measured inverter self-consumption (no-load current) was 0.26 A at a nominal voltage of 12.8 V. The measured inverter cut-off voltage (load off) was 10.6 V. The measured total self-consumption of display devices was 0.055A at a voltage of 12.8 V. A fully charged battery pack has a voltage of 13.7 V.

The measured values of the duration of the system's operation until complete shutdown are shown in Fig. 7. The data are correlated by the number of cells in the battery.

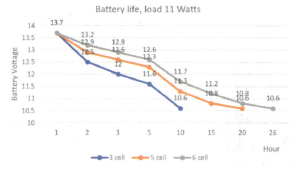

Fig. 7. System load testing

The experimental results are in good agreement with theoretical calculations.

Since the batteries actually used for our system were not new, but used ones, they usually have a smaller capacity than the nominal value, so the obtained experimental results differ downward from the theoretically calculated ones.

4 Summary and Conclusion

The developed auxiliary power supply is perfect for home use as an alternative power source. This device may be the only solution for households (especially for residents of high-rise buildings) during planned power outages, voltage fluctuations, and systematic blackouts.

The DIY system has an open architecture, so there is a possibility for improvement in consumer characteristics, for instance, an increase in battery capacity will increase the duration of the entire system.

The environmental component should also be noted. DIY device does not have harmful emissions, does not create additional noise, unlike generators with internal combustion engines, and meets improved fire safety standards, so it can be placed indoors.

Our device is cheaper than manufactured due to the reuse of secondary components (power supplies and a computer system unit case).

References

1. Tsindeliani, D., Povstyana, Y., Lishchyna, N., Yashchuk, A.: Latency reduction in real-time GPS tracking in Android and the web-based GPS Monitoring System. In: 2022 12th International Conference on Dependable Systems, Services and Technologies (DESSERT) (2022)
2. Wikiman, O., Thomas, S., Nzerem, P., Koyunlu, G.: Design and construction of a prototype wireless power transfer device. Int. J. Eng. Manuf. **2019**(9), 16–30 (2019)
3. Adamu, S., Bature, U.I., Nasir, A.Y., Hassan, A.M., Jahun, K.I., Toro, U.S.: IOT controlled home automation technologies. In: 2019 2nd International Conference of the IEEE Nigeria Computer Chapter (NigeriaComputConf), pp. 1–7. Zaria, Nigeria (2019)
4. Ali Shah, S.K., Mahmood, W.: Smart home automation using IOT and its low cost implementation. Int. J. Eng. Manuf. **10**(5), 28–36 (2020)
5. Cruz, L., Griño, M., Tungol, T., Bautista, J.: Development of a low-cost air quality data acquisition IoT-based system using arduino leonardo. Int. J. Eng. Manuf. **9**(3), 1–18 (2019)
6. Akwu, S., Bature, U.I., Jahun, K.I., Baba, M.A., Nasir, A.Y.: Automatic plant Irrigation control system using Arduino and GSM module. Int. J. Eng. Manuf. **10**(3), 12–26 (2020). https://doi.org/10.5815/ijem.2020.03.02
7. Jha, A., Maharjan, M.: Smart lighting system using LoRa WAN technology. Int. J. Eng. Manuf. (IJEM) **12**(1), 48–53 (2022)
8. Rebello, E., et al.: Developing, implementing and testing up and down regulation to provide AGC from a 10 MW wind farm during varying wind conditions. J. Phys.: Conf. Ser. **1102**, 012032 (2018). https://doi.org/10.1088/1742-6596/1102/1/012032
9. Holovan, M., Zdolbitska, N., Lishchyna, V., Hrinyuk, S.: The analysis of the productivity of the solar panels automatic positioning system. Comput.-Integr. Technol.: Educ. Sci. Prod. **41**, 23–29 (2020)
10. Terletskyi, T., Kaidyk, O., Tkachuk, A., Zabolotnyi, O., Cagáňová, D.: Ensuring the reliability of functioning of non-addressed fire alarm. EAI Endorsed Trans. Energ. Web **22**, e11 (2021)
11. Kuo, J.-K., Huang, P.-H., Wang, C.-F.: Numerical modeling and analysis of PEMFC integrated with auxiliary power source. Int. J. Energ. Res. **37**, 1635–1644 (2013)
12. Raju, E.S.N.P, Jain, T.: Distributed energy resources and control. In: Distributed Energy Resources in Microgrids (2019)
13. Ahmed, U., Ali, F., Jennions, I.: A review of aircraft auxiliary power unit faults, diagnostics and acoustic measurements. Prog. Aerosp. Sci. **124**, 100721 (2021). https://doi.org/10.1016/j.paerosci.2021.100721
14. Ostapchuk, O.V., Zdolbitska, N.V.: Auxiliary power supply of means of communication and the organization of emergency lighting. In: The role of innovations in the transformation of the image of modern science: Materials of the VI International Scientific and Practical Conference, pp. 139–143. Kyiv, 23–24 Dec 2022

15. Zapata, J.W., Kouro, S., Carrasco, G., Renaudineau, H., Meynard, T.A.: Analysis of partial power DC–DC converters for two-stage photovoltaic systems. IEEE J. Emerg. Sel. Topics Power Electron. **7**(1), 591–603 (2019)
16. Zdolbitska, N.V., Zdolbitskyy, A.P., Sopizhuk, R.V., Suprunyuk, V.V.: DC-AC converter with microcontroller-controlled frequency inverter. Comput.-Integr. Technol.: Educ. Sci. Prod. (23), 25–31

A Computerised System for Monitoring Water Activity in Food Products Using Wireless Technologies

Oksana Honsor[✉] and Roksolana Oberyshyn

Lviv Polytechnic National University, Lviv 79070, Ukraine
{oksana.y.honsor,roksoliana.r.oberyshyn}@lpnu.ua

Abstract. The food industry needs constant monitoring of both the technological process and the quality and safety of finished products. Control of the water activity value is essential to ensure the appropriate quality level of food products during their production because it affects the growth of bacteria and microorganisms, affecting the quality indicators, shelf life and proper storage of products. Modern enterprises must have an automated quality monitoring system. A specialised computer system for monitoring water activity should be an integral part of the overall quality monitoring system. This makes it possible to quickly react to deviations in the value of the water activity indicator and take preventive actions. The structural scheme of such a specialised computer system will be presented in this article. It is also important to correctly choose the type of water activity analyser based on the main parameters of the sensors, which are described in this article. The basic principles of developing the necessary software and a block diagram of the data flow algorithm in the system are also presented.

Keywords: Water activity · computer system · food · quality monitoring

1 Introduction

Water activity a_w is a parameter that determines the ratio between the water vapour pressure of the product and the saturation pressure of pure water at the same temperature. It is an essential indicator of product quality level in various industries, in particular, food, pharmaceuticals, industrial production, tobacco, and seed storage. The water activity measurement limits are from (0 to 1) a_w.

The activity of water affects the following properties of the product: microbiological stability, chemical stability, enzyme stability, colour and taste, nutritional value, protein and vitamin content, composition stability, expiration date, solubility and texture, storage and packaging.

That is why constant monitoring of this indicator within the general product quality control system is important and necessary. A specialised computer system using wireless technology is best suited for this.

The value of water activity in food products can be controlled with the help of various additives (moisturisers), using satisfactory packaging materials and maintaining

© The Author(s), under exclusive license to Springer Nature Switzerland AG 2023
Z. Hu et al. (Eds.): ICAILE 2023, LNDECT 180, pp. 393–403, 2023.
https://doi.org/10.1007/978-3-031-36115-9_36

favourable ripening and storage conditions. However, the most critical point of control is the final stage of food product manufacturing. If too much water is available, there is a risk of microbial growth and water migration. Food manufacturers today must prove to food quality and safety organisations that the product's water activity has been reduced to the point where bacteria cannot grow.

The diagram shown in Fig. 1 can help better understand the relationship between the value of product moisture and water activity and their influence on the leading chemical and biological indicators of product quality [1, 2].

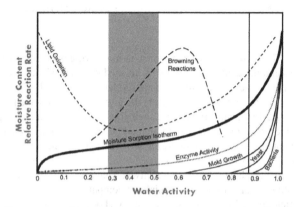

Fig. 1. The effect a_w on oxidation, browning reactions, enzymatic and microbial activity.

The grey zone in Fig. 1, which corresponds to values from (0.3 to 0.5) a_w, demonstrates the lowest chemical and biological activity, which leads to less spoilage and an increase in the shelf life of products.

There are a number of devices for measuring water activity. They have high accuracy, efficiency, compatibility and ease of calibration, which are modern and have good characteristics (for instance, the device for measuring water activity "AwTherm" (accuracy ± 0.005a_w, ± 0.1 K), laboratory instrument "Hygrolab" (accuracy ± 0.008 a_w at 23 °C), station probe HC2-AW-USB). However, these devices are primarily stationary and make it possible to record measurement results manually, without storing them in a database, processing and graphical presentation.

Modern food enterprises (in particular, those that make cookies and crackers) strive to reach a new level of quality control and apply non-destructive determination of the content of active water in products without interrupting the production process [3]. However, non-destructive methods have unfortunately not been studied at a sufficient level.

2 Related Work

The problem of measuring water activity is deeply studied nowadays, but it started many years ago [4, 5]. In particular, the general properties of this indicator and the main methods of its measurement are considered in [6, 7]. The moisture transfer rate during drying and through the packaging film or edible coating of food products during storage can be estimated using water activity. As a result, drying conditions, packaging, or coating material can be selected based on the identified properties of the food products [8].

The importance of correctly choosing methods and means for determining water activity is highlighted in [9, 10] since this indicator is sensitive to temperature and the content of volatile substances in the product under study [11, 12]. Features of the application of types of measuring devices for determining the activity of water, in the case that the tested substance contains volatile substances, are considered in [13, 14].

It is also important to correctly choose the method of measuring water activity for its introduction into the enterprise's general computerised quality monitoring system. Similar systems for monitoring quality parameters are considered in [15, 16]. Based on a thorough analysis of related information, new ways of solving the problem are proposed.

3 Methodology and Regulatory Principles for Determining Water Activity in Food Products

3.1 Regulation of the Water Activity Indicator in the Main International Regulatory Documents

The growing recognition of the principle of water activity is illustrated by its inclusion in FDA and USDA regulations, GMP and HACCP requirements, and most recently, in NSF International Standard 75. Let's consider the main ones.

1. The HACCP system describes the need to apply a_w for food control and safety measures through risk analysis and critical control points [17].
2. The system of standards Good Manufacturing Practice (GMP) regulates the use of the water activity indicator as a means of microbiological control;
3. The Food and Drug Administration (FDA) addresses the interaction of aw and pH under certain conditions of heat treatment and packaging. This synergism controls or eliminates pathogens that would otherwise be ineffective when used alone.
4. The United States Department of Agriculture (USDA) and the Food Safety and Inspection Service (FSIS) have included water activity as a microbial control in their Generic HACCP Model 10«Heat-treated, Shelf-stable Meat and Poultry Products».
5. Standard ANSI/NSF 75–2000 «Non-potentially hazardous foods». The standard regulates test methods and evaluation criteria (in particular, water activity) that allow determining that a food product meets the requirements of the FDA Food Code for "potentially hazardous food" and does not require refrigeration for safety [18, 19].
6. International standard ISO 18787:2017. Foodstuffs – Determination of Water Activity – establishes basic principles and specifies requirements for the methods of determining water activity (a_w) of food products for human consumption and animal feed within a measurement range of 0 to 1 [20]. The methods described in this standard are based on the measurement of the dew point or on the determination of the change in the electrical conductivity of the electrolyte or the dielectric constant of the polymer. However, it should be noted that the methods for measuring water activity described

396   O. Honsor and R. Oberyshyn

in the standard cannot be applied to frozen products, to products in the state of water-in-fat emulsion, and also to crystalline products such as sugar, salt or minerals. Also, when the tested product contains volatile compounds, such as alcohols, the method may require the adaptation of special equipment.

3.2 Main Methods and Instruments for Water Activity Measuring

Water activity in food products is often referred to as "free water" or "non-chemically bound water". It is important not to confuse this concept with the mass fraction of moisture in the product because these are conceptually different product properties that reflect other quality indicators. The mass fraction of moisture characterises the moisture content in the product, and its value depends on the mass of the product (1):

$$\omega = \frac{m_{moisture}}{m_{moisture} - m_{drysubstance}}.$$ (1)

The value of water activity does not depend on the mass of the product and reflects the presence of "free" water in the product:

$$a_w = p/p_0,$$ (2)

p – water vapour pressure above the product.
p_0 – water vapour pressure above pure water [2].

An important feature of the water activity indicator is its temperature sensitivity. Measurements can only be made when the product sample, sample holder and measurement sensors are at stable temperatures. Many standards require measurements to be made at a specific temperature [16].

Three main modern methods for measuring water activity should be distinguished, on which the principle of operation of modern laboratory equipment is based – the method using an adjustable diode laser, the cooled mirror method, and the capacitive (or conductivity) method. Simplified structural diagrams of devices using these methods are given below.

Figure 2 shows the structural diagram of measuring water activity by the capacitive method. The capacity of the sensor varies depending on the number of water molecules in the air. The sample of the researched product should be placed in a special cup, which in turn is placed in a steel holder (sample holder). After that, the cover with the measuring head must be tightly closed. It is essential to ensure a tight connection between the measuring head and the sample holder because only then will the system be closed, and equilibrium in the system will be achieved. Water activity can be measured in two ways: either by using a predictive model or by waiting for the water vapour pressure and temperature to reach equilibrium in the measuring chamber [8]. This measuring circuit also includes a temperature sensor and a humidity sensor. This method has a reasonably high accuracy but requires a long time for determination a_w. Commercially available instruments measure over the entire water activity range with an accuracy of $\pm 0.015 a_w$.

Figure 3 shows the structural diagram of the device for determining the activity of water by the method of a cooled mirror and using an optical sensor.

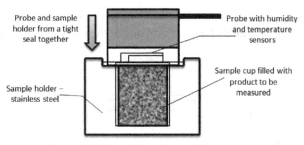

Fig. 2. Structural diagram of a laboratory device for determining the activity of water by the capacitive method

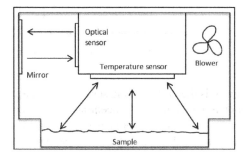

Fig. 3. Structural diagram of a laboratory device using the chilled mirror method

As air is passed over a chilled surface whose temperature is gradually reduced, the temperature at which the water vapour begins to condense on the surface is the dewpoint temperature, which is directly related to the vapour pressure of the air. The system consists of a sensor block containing a dewpoint sensor, an infrared thermometer, and a fan. A sample is placed in a sample cup which is sealed against the sensor block. The dewpoint sensor measures the dewpoint temperature of the air, and the infrared thermometer measures the sample temperature. The relative humidity of the headspace is calculated as the ratio of dewpoint temperature saturation vapour pressure to saturation vapour pressure at the sample temperature. When the water activity of the sample and the relative humidity of the air is in equilibrium, measurement of the headspace humidity gives the water activity of the sample. Chilled-mirror instruments make accurate (± 0.003 a_w) measurements in less than 5 min without the need for calibration [21].

The method using a diode laser (TDL – Tunable Diode Laser) has a high measurement speed and meets the requirements for measurement accuracy [12, 13].

The structural diagram of the device for measuring water activity using TDL is shown in Fig. 4.

The sensor emits a precisely tuned infrared laser beam through the space above the sample. A laser beam less than one nanometer wide is specific for a common isotope of water. Other vapour molecules, including alcohols, gasoline, organic solvents, and propylene glycol, do not affect readings. The attenuation of the beam is measured by the LSD laser receiver, and directly from this value, the concentration of water molecules in the air of the chamber is determined. The vapour pressure determined by the TDL is then divided by the saturated vapour pressure at the sample temperature to obtain a water activity value.

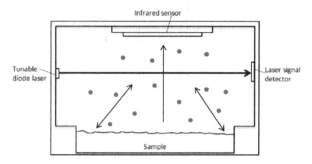

Fig. 4. Structural diagram of water activity measurement using TDL.

The diode laser sensor is the only existing sensor that can accurately measure water activity in samples containing significant concentrations of volatile substances.

The temperature of the sample is measured using an infrared sensor. The temperature value should be maintained at 25 °C, as the water activity indicator is temperature sensitive.

3.3 Analysis of the Main Criteria for Choosing a Method of Measuring Water Activity

Several criteria should be considered to correctly choose the method and type of sensor for measuring water activity for a specific product and under particular conditions. Table 1 shows several main criteria characterising the measurement methods discussed in this article above [6, 13].

Table 1. Main characteristics of sensors

Criteria/sensor type	Capacitive sensor	Mirror dew point sensor	Diode laser TDL
Measuring range	0.0 – 1.00 a_W	0.050 – 1.000 a_W	0.000 – 1.000 a_W
Measurement accuracy	± 0.02 a_W	± 0.003 a_W	± 0.005 a_W
Measuring time	≤ 5 min	≤ 5 min	≤ 5 min
Temperature control	Not necessary	15 – 50 °C ± 0.1°C	15 – 50 °C ± 0.1°C
Resolution	± 0.0001 a_W	± 0.0001 a_W	± 0.0001 a_W
Volatile substances influence	Yes	Yes	No
Calibration	Necessary	Necessary	Not necessary

First, you should determine the type of product and its consistency, packaging methods and terms, as well as conditions of storage. Also, an important parameter is the measurement's accuracy and the method's speed. The content of volatile substances in the product can also affect the measurement result.

As expected, the water activity indicators of the mirror dew point sensor were significantly affected by the presence of volatile substances. The capacitive sensor is also sensitive to some types of samples containing volatile substances. In addition, the capacity sensor, being a secondary method of determining water activity, is not as fast and accurate as the cooled mirror sensor. The performance of the TDL water activity sensor against water activity standards was equal to or better than the best water activity testing options currently available. The highest performance in water activity testing combined with ease of use and lack of sensitivity to any type of sample makes TDL the preferred method for measuring water activity [12].

4 A Specialised Computer System for Monitoring Water Activity

4.1 Hardware design

Modern enterprises must have a computerised control system for all quality indicators to ensure the appropriate quality of the products produced. It will make it possible to monitor quality indicators, analyse them and promptly respond to possible deviations. According to the HACCP system, analysis of indicator deviations will make it possible to effectively prevent their occurrence at critical control points.

Such a system has three subsystems and channels that connect them together (Fig. 5).

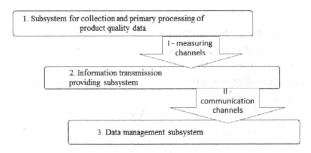

Fig. 5. A structural diagram of a computerised system for quality control at the enterprise.

At the first level, the collection and initial processing of data on quality indicators from the relevant control points is carried out. This subsystem consists of multi-parameter sensors, controllers and additional wireless communication devices for transmitting information from the sensor to the controller. Controllers collect data and process it.

The second layer is a data transmission subsystem consisting of wireless communication devices with built-in security features that transmit data from the controller to a cloud storage environment.

The third level is the data management subsystem. It contains an application that accesses the cloud storage environment and displays it to the end user.

Connecting link I – measuring channels, which include all measuring devices and communication lines from the metering point to the controller.

Connecting link II – communication channels. Physical wired and wireless communication lines are used as communication channels.

Water activity is an important comprehensive indicator that affects foods' taste, smell and texture, microbiological stability (bacterial growth) and shelf life. Permanent control, analysis and processing of data on water activity should be an integral part of the computerised quality control system at the enterprise.

Consider the structural diagram of a specialised computer system for monitoring the water activity indicator (Fig. 6).

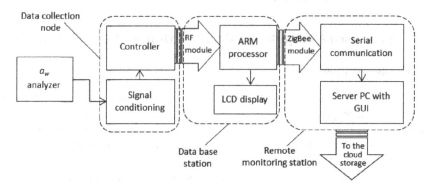

Fig. 6. Block diagram of proposed Wireless Measuring System

The data received from the analyser is passed through a measuring transducer to form an analogue signal that meets the requirements of the next station for further work with it. After that, the converted data is sent to the controller (PIC16F877A can be used, for example). The controller has a built-in ADC that converts an analogue signal into a digital one for further processing.

The received digital data is sent to the ARM processor using the RF module, as shown in Fig. 6. The ARM processor model can be used LPC2148, for example. At this stage, the received data can be displayed on the LCD for initial control of their compliance with the requirements. The results are also sent to the remote monitoring station using the ZigBee module [14, 15].

In the third stage, the converted digital data is sent via serial communication to the server equipped with specialised programs and a Graphic User Interface (GUI), as shown in Fig. 6. The received data is presented in a graphical interpretation for better visual perception by specialists. Special programming and numeric computing platforms (for ex, MATLAB) are used and will be saved for further reference. In addition, the obtained data must be compared with the standard values of the water activity. The specialist and authorised persons are notified if unacceptable deviations are detected to take the necessary preventive and corrective actions. One of the options for sending a message can be an SMS notification.

4.2 Software Design

A detailed block diagram of the algorithm of the entire system, as well as software development, is shown in Fig. 7.

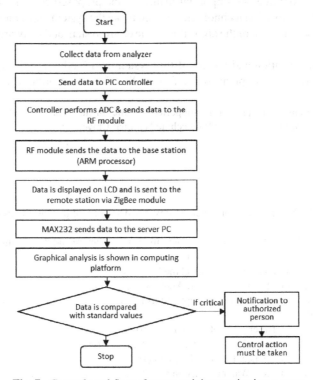

Fig. 7. General workflow of water activity monitoring system

The software for the water activity control system can be conditionally divided into three parts, the main part of which is the programming of the microcontroller. Next, the software for the ARM processor and the design of the graphical user interface for the interpretation of the received data are developed. It is crucial to ensure the correct interaction of the GUI software with the hardware at the remote monitoring station.

5 Conclusion

Regular control of water activity in food products is essential and necessary. According to the requirements of regulatory documents, the control of this indicator should be introduced into the general product quality monitoring system. Typically, this indicator is monitored using laboratory methods followed by manual registration and subjective analysis of the results.

This article proposes a specialised computer system that makes it possible to monitor water activity by the most appropriate method, to carry out the primary transformation of the received data, to transfer them to an ARM processor for further processing, analysis and storage. The results are transmitted to the remote station to the relevant specialists, where it is possible to display and graphically interpret them using a specialised user interface. Immediate notification of responsible persons in case of critical values of water activity is also foreseen. An important feature of the proposed computerised system is wireless data transmission technologies, which make the system convenient and unified. A detailed block diagram of the algorithm of the entire system and software development is proposed.

The implementation and testing of this system at a food enterprise (production of cookies and crackers) are planned as a further direction of scientific work.

Acknowledgement. This project is supported by the Institute of Computing Technology, Automation and Metrology of Lviv Polytechnic National University.

References

1. Simons, C.: Determination of water activity. Food science. https://cwsimons.com/determina tion-of-water-activity
2. Mathlouthi, M.: Water content, water activity, water structure and the stability of foodstuffs. Food Control **12**(7), 409–417 (2001)
3. Saqaeeyan, S., Rismantab, A.: A novel method in food safety management by using case base reasoning method. Int. J. Intell. Syst. Appl. **7**(10), 48–54 (2015)
4. Sandulachi, E.: Water activity concept and its role in food preservation. Meridian ingineresk 39–48 (2012)
5. Van den Berg, C.: Description of water activity of foods for engineering purposes by means of the GAB model of sorption. In: McKenna, B.M. (eds.) Engineering and Food, pp. 311–321. Elsevier Applied Science, London, England (1984)
6. Sahin, S., Sumnu, S.G.: Water activity and sorption properties of foods. In: Physical Properties of Foods, pp. 193–228. Springer New York, New York, NY (2006). https://doi.org/10.1007/0-387-30808-3_5
7. Water activity: Solutions for laboratories. https://www.hlr.ua/storage/editor/files/2486ca5be fa76fdd0c325a24820c613a.pdf
8. Mitrevski, V., Geramitcioski, T., Mijakovski, V., Andreevski, I.: Water activity VS. equilibrium moisture content. J. Process. Ener. Agric. **20**, 69–72 (2016)
9. How to carry out a water activity (Aw) measurement. https://www.processsensing.com/en-us/blog/how-to-do-water-activity-measurements
10. Devices for determining water activity. https://labtime.ua/uk/produkciya-c2/laboratornye-pribory-c6/oborudovanie-obschelaboratornoe-c31190/pribory-dlya-opredeleniya-aktivnosti-vody-c10069
11. Rahman, M.S., Shyam, S.S.: Water activity measurement methods of foods. Food Properties Handbook 9–32 (2009)
12. Ren, Y., Jiewen, G., Sicheng, S., Shyam, S.: Tang Juming Understanding water activity change in oil with temperature. Current Res. Food Sci. **3**, 158–165 (2020)
13. Campbell, G.S., Galloway, M., Campbell, Z.: Measurement of water activity in the presence of high volatile concentrations using a tunable diode. laser-single laboratory validation, first action 2021.04. J. AOAC Int. **105**(3), 649–656 (2021)

14. Carter, B.P., Brown, G.: Superior Water Activity Measurement of Oils and Lubricants Using a Tunable Diode Laser. AquaLab Univercity (2015). https://usermanual.wiki/Document/146 44SuperiorWaterActivityMeasurementofOilsandLubricantsUsingaTunableDiodeLaser
15. Barabde, M.N., Danve, S.R.: A review on water quality monitoring system. Int. J. VLSI Embedded Syst.-IJVES **06**, 1475–1479 (2015)
16. Li, Z., Wang, K., Liu, B.: Sensor-network based intelligent water quality monitoring and control. Int. J. Adv. Res. Comput. Eng. Technol. **2**(4) (2013)
17. Food quality and safety systems – a training manual on food hygiene and the hazard analysis and critical control point (HACCP) system, 167 p. Food and agriculture organization of the united nations, Rome (1998)
18. NSF International: 2000 Nov. 10. Non-potentially hazardous foods, 12 p. NSF International. Report nr ANSI/NSF 75-2000, Ann Arbor (MI)
19. Powitz, R.W.: Water Activity: A New Food Safety Tool. Food Safety Magazine (2007). https://www.food-safety.com/articles/3960-water-activity-a-new-food-safety-tool
20. ISO 18787:2017: Food Stuffs—Determination of Water Activity, 9 p. ISO, Geneva, Switzerland (2017)
21. Neil, H.: Mermelstein measuring moisture content & water activity. Food Technol. Magazine **63**(11) (2009)

Reengineering of the Ukrainian Energy System: Geospatial Analysis of Solar and Wind Potential

Iryna Doronina[1], Maryna Nehrey[2]($\boxtimes$), and Viktor Putrenko[3]

[1] Kyiv National Economic University named after Vadim Hetman, Kyiv 03057, Ukraine
[2] National University of Life and Environment Science of Ukraine, Kyiv 03041, Ukraine
marina.nehrey@gmail.com
[3] American University Kyiv, Kyiv 04070, Ukraine

Abstract. The potential for using renewable energy sources is a priority in the development of decentralized energy and ensuring the country's energy security. The aim of the study is to calculate the technical potential of using the territory for renewable energy (solar and wind). The potential industrial area was calculated and the technical potential for the introduction of solar and wind energy in Ukraine was calculated. The technical potential of the sun is determined at the maximum level of 6.84 GW and the optimal level of production of 45.43 GW, and correspondingly of the wind -28.76 GW and 190.99 GW. Geospatial analysis has identified favorable areas for renewable energy infrastructure development and estimates of technical potential can be used to inform policy and investment decisions. The issue of the combination of food security and energy security in the process of the formation of the security model of Ukraine is examined. The study is an important contribution to the promotion of sustainable economic growth and the reduction of carbon dioxide emissions in Ukraine.

Keywords: Renewable energy · Security · Energy transformation · Geospatial analysis · ArcGis · Potential · Sustainability

1 Introduction

Energy is the core of the economy, a crucial component of the efficient development of industrial production, the functioning of housing and municipal services, and the organization of human activity. The energy transition, based on the principles of sustainable development, involves the gradual replacement of fossil energy resources with renewable ones. The concept of the "Brown" economy is being replaced by new ones, such as the "Green Economy" and "Circular Economy", which are based on the assessment of the area and the availability of resources in this area.

Ukraine's energy system is centralized and unprepared for the challenges caused by Russia's military aggression. Damage to the energy infrastructure and transmission lines has led to massive power outages in all regions of the country and blackouts. Therefore, measures are being taken at the regional level to decentralize the energy sector with a focus on using local renewable resources.

© The Author(s), under exclusive license to Springer Nature Switzerland AG 2023
Z. Hu et al. (Eds.): ICAILE 2023, LNDECT 180, pp. 404–415, 2023.
https://doi.org/10.1007/978-3-031-36115-9_37

The new technologies' application and opportunities for renewable resources on the ground require a general understanding of the country's renewable energy potential. Another issue is the balance of using the territory's potential. It should be pointed out that Ukraine consolidates about 30% of the world's black soil reserves.

The focus only on technical capacity is the main limitation of this study. To get a complete picture, an assessment of local feasibility of renewable energy projects should be added. Further research will be needed to assess the economic and social impacts of farmland replacement, and the feasibility and perception of such projects at local levels.

We aim to provide a scientific basis for developing environmental policies and implementing energy decentralization in Ukraine. The paper demonstrates the significant potential of renewable energy sources in Ukraine, in particular solar and wind energy, and their potential contribution to sustainable economic growth and decarbonization of the economy. It also identifies the most favourable areas for the implementation of renewable energy infrastructure, quantifies the technical potential for solar and wind energy in Ukraine, and proposes a sustainable energy planning tool that can be replicated in other regions.

2 Materials and Methods

An analysis of government regulations and documents of international organizations for the period from 2014 to 2022 showed that there is a significant difference in the approaches and results of assessing the potential of renewable resources in Ukraine. Therefore, different methodologies, assumptions, and timeframes are used for research.

Government documents and the 2015 report of the International Renewable Energy Agency (IRENA) [1] estimate the technically achievable renewable energy potential based on scientific and economic research and legally defined targets. The indicators include targets for the replacement of brown generation, following the Paris Climate Agreement, and considering domestic demand in Ukraine.

The REmap 2030 assessment methodology [2] analyzes renewable energy development and substitution. It includes cost analysis and technical efficiency (standard plant capacity, utilization rate and conversion efficiency) of renewable energy technologies and traditional (fossil fuel, nuclear power) for each sector: industry, construction, transportation, electricity, and heat. This methodology identifies possible technically and potentially achievable targets for renewable energy facilities. Ukraine's total onshore wind potential is estimated at 24 GW. The solar energy potential is estimated at 4 GW.

The state program documents set out an economically and logically sound approach to determining the technical potential of renewable energy, which was calculated as an indicator of oil equivalent replacement per kWh.

The National Renewable Energy Action Plan for the period up to 2020 [3] defines the annual technically achievable energy potential of renewable energy sources, which reaches 68.6 million tons of oil equivalent per year. The reasonable installed capacity of solar energy in Ukraine is 4 GW. At the same time, the potential economically feasible installed capacity of wind energy in Ukraine reaches 15 GW. The Action Plan considered the direction of energy security and the independence of Ukraine from imported energy resources.

406 I. Doronina et al.

As the green business and green investment environment has developed over the past 10 years, based on the highest feed-in tariff in Europe until 2030 [4], private companies have called on Ukrainian government agencies to revise the renewable energy potential (upward). Subsequently, this indicator was used as an argument to increase the investment attractiveness of private businesses in Ukraine. Also, simplified procedures for environmental impact assessment [5] and procedures for obtaining permits at the regional and local without special approval from the public, were introduced. All of this has led to rapid growth and uneven distribution of installed renewable energy capacities in Ukraine and created an imbalance in the Integrated Power System of Ukraine. As of December 31, 2021, the installed capacity of renewable energy facilities with a green tariff (Feed-in-tariff) is 8434.12 MW. The idea that a relatively small share of renewables cannot significantly affect the operation and balancing of the energy system is incorrect [6].

Energy policy in the 2017–2018 analysis showed that the state faced the need to introduce strict regulation of renewable energy to promote sustainable development and energy security in the country. During this period, to increase financial opportunities, private investors focused on the possible export of surplus renewable energy and the formation of Ukraine's image as an "energy exporter to the EU". However, it should be emphasized that Ukraine, both then and now, lacks the technical capabilities for full-scale energy exports to the EU.

Table 1. Comparative analysis of the Ukrainian RES potential assessment

Document	Author	Year	Wind onshore, GW	Solar plants, GW
REMAP 2030	IRENA	2015	24	4
Prospects for the development of renewable energy in Ukraine until 2030	State Agency on Energy Efficiency and Energy Saving	2015	16	4
About the National Renewable Energy Action Plan for the period up to 2020	The Ministry of Energy	2015	15	4
Atlas Renewable Energy 2010	IEE of the National Academy of Sciences of Ukraine	2015	24	4
Cost-competitive renewable power generation	IRENA	2017	320,58	70,61
Atlas Renewable Energy 2020	IEE of the National Academy of Sciences of Ukraine	2020	438	82,76

The IRENA report [7] estimates Ukraine's renewable energy potential at 414.78 GW. However, the highest figures for Ukraine's renewable energy potential are presented in the Atlas of Renewable Energy Potential of Ukraine [8], developed by The Institute of Renewable Energy of the National Academy of Sciences of Ukraine (Table 1).

In January 2022, the Cabinet of Ministers of Ukraine launched a public discussion of an action plan for implementing and developing renewable energy. The National Renewable Energy Action Plan for the period up to 2030 [9] outlines the following opportunities

- increase in electricity production through the use of more powerful wind turbines and commissioning of new capacities of onshore wind power plants up to 15.3 TWh in 2030 (total capacity of 4.7 GW);
- increase in electricity production from solar energy to 10.5 TWh in 2030 (with a total capacity of about 10 GW, including 7 GW of industrial producers and 3 GW of consumers).

Ukraine is viewed internationally as an importer of electricity and technology, caused by the large-scale destruction of energy infrastructure due to Russian military aggression. Therefore, the task of decentralized energy and renewable sources primarily for the country's domestic needs is urgent today. According to the energy balance for 2020 (calculated by the rules of Directive 2009/28/EC), Ukraine's total final energy consumption amounted to 50.5 million tons of oil equivalent (of which only about 4 million tons from renewable sources). Thus, if Ukraine maintains its level of energy consumption in 2020, it will be able to fully meet its energy needs with energy from renewable sources going forward.

Scenarios for renewable energy development and the corresponding benefits of greening development and reducing CO_2 emissions are actively discussed in academic circles [10–14]. Scientists are studying both the impact of renewable energy on the economy [15–17] and the promotion of environmental sustainability [18–23]. At the same time, the development of renewable energy from a technical point of view and the possibility of introducing modern technologies in the energy sector is crucial [24–27].

Our study is based on the methodology for determining the geographic and technical potential of Renewable Energy IRENA 2014, 2017 [7, 28], which requires several processing steps and assumptions. Shifting from the theoretical capacity (e.g., the total amount of solar radiation reaching a country or the total theoretical wind energy in a given area) to the technical or implementation potential requires a series of additional data assumptions (Fig. 1).

The study aims to determine the geographical and technical potential of renewable energy in Ukraine. Advances in geographic information system programming can help achieve this goal [29]. GIS is a powerful tool with a variety of applications in analyzing and presenting spatial data (Table 2).

The use of ArcGIS software allowed us to model maps that can be used to identify ideal project sites in the future and initiate a dialogue with regional and local organizations and communities.

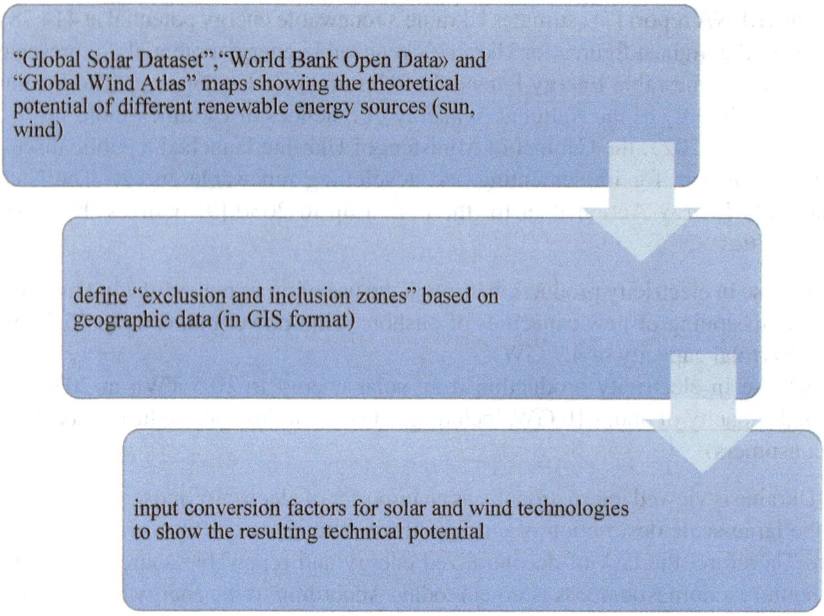

408 I. Doronina et al.

"Global Solar Dataset","World Bank Open Data» and "Global Wind Atlas" maps showing the theoretical potential of different renewable energy sources (sun, wind)

define "exclusion and inclusion zones" based on geographic data (in GIS format)

input conversion factors for solar and wind technologies to show the resulting technical potential

Fig. 1. Stages of technical potential assessment (author's illustration)

Table 2. The dimensions considered for the analysis, dataset selection for the limitation

Map km^2	GHI	Speed wind	Grids buffer zone	Slope %	Protected area	Budlings	Water	Forest	Roads, Railways	Black soil
Solar	+	-	-	>35%	-	-/+	-	-	-	?
Wind	-	+	-	>20%	-	-	-	-	-	?

• "-" exclusion zones, • "+" inclusion zones

3 Results

We found that 75% of the territory of Ukraine has a global horizontal irradiation GHI of more than 1169 kWh/m^2/year. This global solar radiation is sufficient for the widespread introduction of both photovoltaic (solar power plants) and heat and power (solar collectors) equipment. Map data (min-max range) GHI for Ukraine per year from 1095–1461 kWh/m^2/year is presented in Fig. 2.

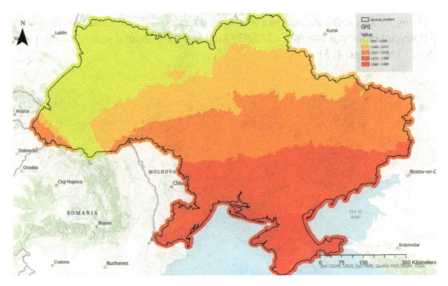

Fig. 2. GHI Ukraine

Given that more than 90% of winds have a speed of 7–8 m/s, which is sufficient for high-performance wind generation. The strongest winds are observed along the coast of the Azov and Black Seas. The average wind speed throughout the territory is from 2 to 10 m/s (Fig. 3).

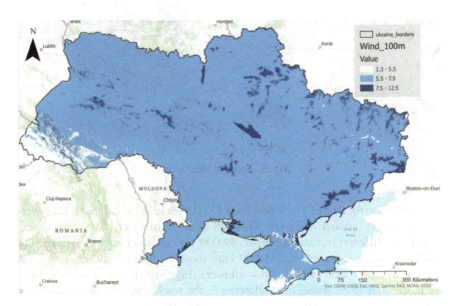

Fig. 3. Wind speed 100 m altitude

Using ArcGis software and data [30, 31] for analysis, we identified the territories of the Potential industrial sun area - 325,839 km^2 and the Potential industrial wind area – 325,023 km^2 (Figs. 4, 5).

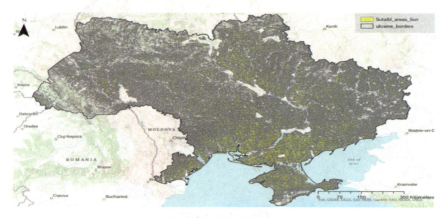

Fig. 4. Suitable sun area

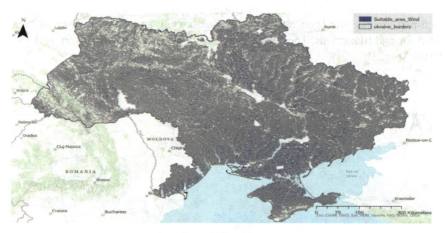

Fig. 5. Suitable wind area

According to the data of The World Bank [32] Ukraine has 1/3 of the world's organic black soil stocks, around 70% (410 000 km^2) of the territory of Ukraine, of these: 36 000 km^2 are different recreational areas; 89 000 km^2 are used for various agricultural purposes (grasses, fruits, etc.); 285 000 km^2 (outside of cities and protected areas) are used for arable land (wheat and sunflower) (Fig. 6). Ukraine is also the world's leading exporter (52% share) of sunflower oil, the world's second-largest exporter of grain (wheat, oats, …) and the world's third-largest exporter of corn.

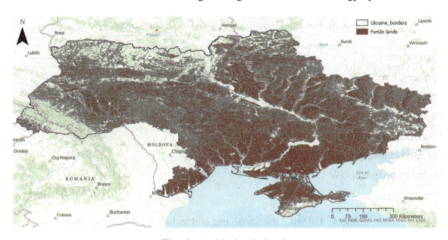

Fig. 6. Arable\fertile lands

Until 2021, there was a complete moratorium on the use of agricultural land in Ukraine [33], not for its intended purpose. However, there were changes after 2021 [34] regarding the rules for the use of agricultural land. Therefore, it is possible to consider changing the purpose of privately owned agricultural land but only after proving its economic feasibility. State-owned land can be changed only if the CMU and state collegial bodies decide to do so, proving the public benefit of the project.

As of 2022, according to Geocadastre statistics, letter No. 281/...69–22 dated 01.12.2022, privately owned agricultural land covers an area of 235,113 km^2, state-owned land - 40,940 km^2, and communal land $-8,100$ km^2.

If we take into account the change in the rules for the use of agricultural land and do not take into account the decrease in Ukraine's productive capacity, we can consider changing the intended use of privately owned agricultural land. We emphasize that only Ukrainian citizens can own agricultural land.

After subtracting the territories that cannot be used for RES industrial facilities and the lands of the agricultural land fund (state, municipal, private), we obtained the Potential industrial area maximal extraction for solar 49,786 km^2 and wind 48,970 km^2 (Tables 3, 4).

Potential industrial area optimal extraction (including privately owned land) for solar 276,799 km^2 and wind 275,983 km^2 (Tables 3, 4).

To determine the technical potential, we rely on the IRENA 2017 methodology [7] for the allocation of industrial land for renewable energy, in the amount of 1% of the geographical potential and calculate the technical potential according to their methodology (Tables 3, 4).

412   I. Doronina et al.

Table 3. Potential industrial area and technical potential solar

Type of area f	Optimal	Maximal
Territory, km^2	325839	325839
Black soil municipal owners, km^2	8100	8100
Black soil private owners, km^2	235113	na
Black soil state owners, km^2	40940	40940
Potential industrial area, km^2	49786	276799
Technical potential, GW	8,2	45,4

Table 4. Potential industrial area and technical potential wind

Type of area f	Optimal	Maximal
Territor, km^2	325023	325023
Black soil municipal owners, km^2	8100	8100
Black soil private owners, km^2	235113	na
Black soil state owners, km^2	40940	40940
Potential industrial area, km^2	48970	275983
Technical potential, GW	33,8	190,4

Having considered the proportion of installed renewable energy capacity over the past 12 years, it was found that, under equal economic conditions, more solar projects were installed than wind projects. This implementation was primarily influenced by the difference in the cost of the technologies themselves and the speed of designing solar parks, which made it possible to start construction under high "green tariff" conditions. We emphasize that in Ukraine, along with the existence of a simplified system of environmental impact assessment, there is a rejection by society and private farmers of the construction of large wind generation, which may be the subject of further research and discussion.

4 Summary and Conclusion

The paper provides a scientific justification for the need to develop an environmental policy and implement energy decentralization in Ukraine to reduce carbon emissions and promote sustainable economic growth. The study demonstrates the significant potential of renewable energy sources in Ukraine, particularly solar and wind energy, which can be harnessed through the development of grid and off-grid projects at the local level.

The focus on environmental policy development and energy decentralization in Ukraine, together with the even distribution of solar and wind potential, provides an opportunity for the decarbonization of industry and the economy. The significant potential of renewable energy and the scale of its distribution will lead to the development of on-grid and off-grid projects at the local level. The feasibility of using agricultural land for energy is decided at the local level, so the study on the economic and social replacement of agricultural land with energy land will continue.

The use of geospatial analysis allowed us to determine the Potential industrial area and calculate the technical potential for the introduction of solar and wind energy in Ukraine.

1) The assessment of previously recommended potentials was tied either to domestic market objectives and oil substitution or to continental assessment methods without taking into account local issues.
2) The most favourable areas that can be used for the implementation of renewable energy infrastructure have been identified on a quantitative scale.
3) The technical potential of the sun at the maximum extraction level of 6,84 GW and the optimal extraction level of 45,43 GW and, respectively, wind - 28,76 GW and 190, 99 GW - is determined.

The results of the study have several potential applications and extensions. Policy makers can use the technical potential estimates provided in the study to set goals and targets for renewable energy development in Ukraine. Investors can use the identification of favourable areas for the prioritisation of their investments in renewable energy infrastructure. Moreover, the study methodology can be replicated in other regions to identify favourable renewable areas, making it a valuable tool for sustainable energy planning. Furthermore, future research could provide insight into the feasibility and perception of such projects at the local level, and expand on the economic and social impacts of replacing agricultural land with energy.

References

1. International Renewable Energy Agency. Renewable Energy Prospects for Ukraine – IRENA. https://www.irena.org/publications/2015/Apr/Renewable-Energy-Prospects-for-Ukraine. 09 Jan 2023
2. International Renewable Energy Agency. REMAP - 2030. Prospects for renewable energy development in Ukraine until 2030. https://saee.gov.ua/sites/default/files/UKR%20IRENA%20REMAP%20_%202015.pdf. 04 Jan 2023
3. Cabinet of Ministers of Ukraine, Order on the National Renewable Energy Action Plan for the period up to 2020, p. 902, Kyiv, October 1, 2014. https://zakon.rada.gov.ua/laws/show/902-2014-%D1%80?find=1&text=%D0%BF%D0%BE%D1%82%D0%B5%D0%BD%D1%86%D1%96%D0%B0%D0%BB#w1_1. 04 Jan 2023
4. National commission for state regulation in the areas of energy and community services. Resolution On Approval of the Procedure for Establishing, Revising and Terminating the "Green" Tariff for Electricity for Business Entities, Electricity Consumers, including Energy Cooperatives, and Private Households Whose Generating Units Produce Electricity from Alternative Energy Sources No. 1817 of 30.08.2019. https://zakon.rada.gov.ua/laws/show/v1817874-19#Text. 05 Jan 2023
5. Verkhovna Rada of Ukraine. Law of Ukraine on Environmental Impact Assessment, May 23, 2017, No. 2059-VIII. https://zakon.rada.gov.ua/laws/show/2059-19#Text. 11 Dec 2022
6. Doronina, I., Kryshtof, N., Moskalenko, S.: Green-coal paradox in Ukraine and in the world as a challenge for state regulation in the energy industry. Eur. J. Econ. Manage. **6**(6), 39–51 (2020). https://doi.org/10.46340/eujem.2020.6.6.4

7. International Renewable Energy Agency. Cost-Competitive Renewable Power Generation: Potential Across South East Europe. International Renewable Energy Agency (IRENA), Abu Dhabi (2017)
8. Atlas of Renewable Energy Potential of Ukraine. Kyiv, Institute of Renewable Energy of the National Academy of Sciences of Ukraine (2020). https://www.ive.org.ua/wp-content/upl oads/atlas.pdf. 21 Dec 2022
9. The Cabinet of Ministers of Ukraine. The State Agency on Energy Efficiency and Energy Saving has developed a draft National Action Plan for the Development of Renewable Energy for the period up to 2030. https://www.kmu.gov.ua/news/derzhenergoefektivnosti-rozrob leno-proekt-nacionalnogo-planu-dij-z-rozvitku-vidnovlyuvanoyi-energetiki-na-period-do-2030-roku. 03 Jan 2023
10. Skrypnyk, A., Klymenko, N., Talavyria, M., Goray, A., Namiasenko, Y.: Bioenergetic potential assessment of the agricultural sector of the Ukrainian economy. Int. J. Energy Sect. Manage. **14**(2), 468–481 (2019). https://doi.org/10.1108/IJESM-04-2019-0015
11. Matviychuk, A., Novoseletskyy, O., Vashchaiev, S., Velykoivanenko, H., Zubenko, I.: Fractal analysis of the economic sustainability of industrial enterprise. CEUR Workshop Proceedings **2422**, 455–466 (2019)
12. Abebe, Y.M., Rao, P.M., Nak, M.G.: Load flow analysis of a power system network in the presence of uncertainty using complex affine arithmetic. Int. J. Eng. Manufact. **7**(5), 48–64 (2017). https://doi.org/10.5815/ijem.2017.05.05
13. Davydenko, N., Buriak, A., Titenko, Z.: Financial support for the development of innovation activities. Intellectual Economics **13**(2), 144–151 (2019). https://doi.org/10.13165/IE-19-13-2-06
14. Oliskevych, M., Beregova, G., Tokarchuk, V.: Fuel consumption in Ukraine: evidence from vector error correction model. Int. J. Energy Econ. Policy **8**(5), 58–63 (2018)
15. Skrypnyk, A., Nehrey, M.: The formation of the deposit portfolio in macroeconomic instability. CEUR Workshop Proceedings **1356**, 225–235 (2015)
16. Tkachenko, R., Kutucu, H., Izonin, I., Doroshenko, A., Tsymbal, Y.: Non-iterative neural-like predictor for solar energy in Libya. CEUR Workshop Proceedings **2105**, 35–45 (2018). http://ceur-ws.org/
17. Kobets, V., Yatsenko, V.: Influence of the fourth industrial revolution on divergence and convergence of economic inequality for various countries. Neuro-Fuzzy Modeling Techniques in Economics **8**, 124–146 (2019). https://doi.org/10.33111/nfmte.2019.124
18. Dimitrov, I., Davydenko, N., Lotko, A., Dimitrova, A.: Comparative study of main determinants of entrepreneurship intentions of business students. In: 2019 International Conference on Creative Business for Smart and Sustainable Growth (CREBUS) 2019 Mar 18, pp. 1–4. IEEE (2019). https://doi.org/10.1109/CREBUS.2019.8840050
19. Miroshnychenko, I., Kravchenko, T., Drobyna, Y.: Forecasting electricity generation from renewable sources in developing countries (on the example of Ukraine). Neuro-Fuzzy Modeling Techniques in Economics **10**, 164–198 (2021). https://doi.org/10.33111/nfmte.2021.164
20. Klymenko, N., Nehrey, M.: Electricity tariff structures modeling for reengineering Ukrainian energy sector. Lecture Notes on Data Engineering and Communications Technologies **135**, 493–502 (2022). https://doi.org/10.1007/978-3-031-04809-8_45
21. Prokopenko, O., Chechel, A., Sotnyk, I., Omelyanenko, V., Kurbatova, T., Nych, T.: Improving state support schemes for the sustainable development of renewable energy in Ukraine. Polityka Energetyczna **24**(1), 85–100 (2021)
22. Sabishchenko, O., Rębilas, R., Sczygiol, N., Urbański, M.: Ukraine energy sector management using hybrid renewable energy systems. Energies **13**(7), 1776 (2020). https://doi.org/10.3390/en13071776

23. Ostapenko, O., Olczak, P., Koval, V., Hren, L., Matuszewska, D., Postupna, O.: Application of geoinformation systems for assessment of effective integration of renewable energy technologies in the energy sector of Ukraine. Appl. Sci. **12**(2), 592 (2022). https://doi.org/10.3390/app12020592
24. Praiselin, W.J., Edward, J.B.: A review on impacts of power quality, control and optimization strategies of integration of renewable energy based microgrid operation. Int. J. Intell. Syst. Appl. (IJISA) **10**(3), 67–81 (2018). https://doi.org/10.5815/ijisa.2018.03.08
25. Aliyu, S.O., Okwori, M., Onwuka, E.N.: A prototype automatic solar panel controller (ASPC) with night-time hibernation. Int. J. Intell. Syst. Appl. **8**, 18–25 (2016). https://doi.org/10.5815/ijisa.2016.08.03
26. Shakhovska, N., Montenegro, S., Kryvenchuk, Y., Zakharchuk, M.: The neurocontroller for satellite rotation. Int. J. Intell. Syst. Appl. **11**(3), 1–10 (2019)
27. Mochurad, L., Kotsiumbas, O., Protsyk, I.: A model for weather forecasting based on parallel calculations. In: Advances in Artificial Systems for Medicine and Education VI 2023 Jan 21, pp. 35–46 (2023). Springer Nature Switzerland, Cham. https://doi.org/10.1007/978-3-031-24468-1_4
28. Hermann, S., Miketa, A., Fichaux, N.: Estimating the renewable energy potential in Africa. IRENA-KTH working paper. International Renewable Energy Agency, Abu Dhabi (2014)
29. Joerin, F., Thériault, M., Musy, A.: Using GIS and outranking multicriteria analysis fo land - use suitability assessment. Int. J. Geogr. Inf. Sci. 15, 153-174 (2001)
30. Global landcover 2000, Europe (2020). https://www.eea.europa.eu/data-and-maps/data/global-land-cover-2000-europe. 15 Dec 2022
31. Terrestrial and Inland Waters Protected Areas https://www.protectedplanet.net/country/UKR. 13 Dec 2022
32. World Bank. Soil Fertility to Increase Climate Resilience in Ukraine (2014). https://www.worldbank.org/en/news/feature/2014/12/05/ukraine-soil. 28 Dec 2022
33. Law of Ukraine. On the moratorium on changing the designated purpose of certain recreational land plots in cities and other settlements. No. 3159-VI of March 17, 2011. https://zakon.rada.gov.ua/laws/show/3159-17#Text
34. Law of Ukraine. On Amendments to Certain Legislative Acts of Ukraine on the Terms of Turnover of Agricultural Land, No. 552-IX, March 31, 2020. https://zakon.rada.gov.ua/laws/show/552-20#Text

Complex Approach for License Plate Recognition Effectiveness Enhancement Based on Machine Learning Models

Yakovlev Anton[(⊠)] and Lisovychenko Oleh

Igor Sikorsky Kyiv Polytechnic Institute, Peremohy Prospect 37, Kyiv 03056, Ukraine
liferunner@gmail.com

Abstract. OCR is an effective method when it comes to license plate recognition. However, the variability of the input data quality could reduce detection precision and may require extensive computation costs. Modern machine learning approaches for pattern recognition in the field of license plate recognition is an effective method to improve the traditional OCR only approaches. But effectiveness of machine learning models in terms of precision depends on train techniques from one side and quality, amount, variability of input train data from another. This paper addresses proposed methodology and complex approach for license plate recognition system development involving all components of the process like hardware, machine learning and a software implementation with field approbation. Approach is based on the Yolo v5 detection system which uses machine learning and includes research, synthesis and practical software and hardware implementation for all process constituents. Methodology being proposed allows higher detection precision comparingly to classical OCR-only approaches. Statistical methods are used to prove validity of license plate recognition results obtained. A set of 20 000 input image sets used in the experiment leads to 95,57% detection precision at average 1,168 s per image proving effectiveness of the method being proposed.

Keywords: Decision support system · machine learning · artificial intelligence · pattern recognition · Yolo v5

1 Introduction

The work is devoted to improvement of approaches to build detection systems (DS) and vehicle license plate recognition (LPR) using machine learning models. Complexity of the approach for mentioned task solving lies in full hardware-software cycle implementation. This includes all sequences from hardware setup in accordance with the standard [6] till DS learning methodology with a custom annotated images dataset usage. Solving similar tasks using machine learning systems and artificial intelligence methods is included into the scope of definitions and field of decision support systems (DSS) study. Yolo v5 is a machine learning models based image pattern recognition system. It provides a learning interface with user defined custom datasets and characterized with high

© The Author(s), under exclusive license to Springer Nature Switzerland AG 2023
Z. Hu et al. (Eds.): ICAILE 2023, LNDECT 180, pp. 416–425, 2023.
https://doi.org/10.1007/978-3-031-36115-9_38

performance [7]. Artificial intelligence methods in pattern recognition are widely used in various fields, like road safety [8], health care, production, signal processing, forecast systems, big data processing, etc.

Problems of nonlinear machine learning methods usage are widely researched in scientific studies nowadays [5, 14–26]. These studies are focusing on methods of increasing accuracy and false detections reduction, increasing the time characteristics of the detection process [5]. LPR with machine learning methods usage is implemented in some modern commercial systems [32]. However, points like accuracy indicators, input data variability, computational ability growth, capturing devices resolution growth and new highly efficient machine learning models development provokes scientific interest in the raised issue. The problem of approaches with OCR-only method usage still persist as these methods are limited and outdated [24], which could be eliminated or altered using modern artificial intelligence based methods [35].

The purpose of the work is to develop a complex approach and build vehicle license plate DS that will increase effectiveness of its functioning by precision, speed and commercial implementation accessibility indicators.

2 Problem Statement

To build the system it is mandatory to determine its components and justify effectiveness and applicability of its usage. Vehicle license plate DS consists of hardware (camera, computers, communication channels) and software (image processing software, training interfaces, detection interfaces, system for storage, reporting and visualization, etc.). Basic requirements for the hardware part of the system are defined in standard [6], which will be used during build [12]. The rest of the work is targeted to graphical pattern recognition detection system train methodics, train process application and results evaluation with confirmation or rejections of theses being proposed.

It is necessary to:

- To define and to engage hardware components in accordance with the requirements;
- Setup hardware environment to form real world variable input images dataset for recognition model train and evaluation;
- Create software for automated dataset annotation and determine optimality of parameters for Yolo v5 model train with given input dataset;
- To process and analyze system functioning results in scope of precision, speed, applicability for commercial implementation.

3 Problem Solving

Methodology of detection system development was created for stated problem solving. It covers all stages of the detection process from input image obtaining to string represented data view. Complex approach to LPR system development with artificial intelligence based DSS was created within the proposed methodology (Fig. 1).

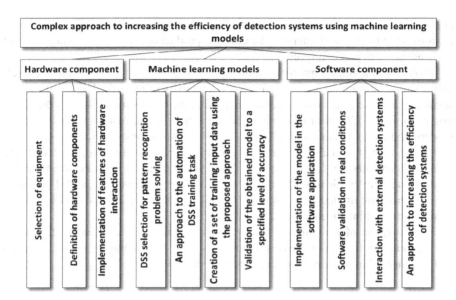

Fig. 1. Complex approach to increasing the efficiency of detection systems using machine learning models structural diagram

3.1 Hardware Component

Hardware component role in stated recognition problem solving comes down to two main functions. Firstly is to provide technical conditions to obtain images in accordance with the standard, which regulates requirements for road safety systems with a function of photo and video capturing [6, 12]. Secondly, ensuring the receipt of the original graphic image that will meet the quality characteristics for further processing (size, resolution, etc.). During this stage existing systems were analyzed as well as available on the market solutions which conformed minimal requirements for stated problem solving. Hardware components interaction features were also defined on all the "image capture - processing - storage" chain and implemented into the hardware system. For stated task accomplishment Axis P1344-E surveillance camera was chosen as a capturing and primary image processing device[9].

3.2 Analysis of DSS for Pattern Recognition Problem Solving

The system for responding to road events when working with road infrastructure must meet the requirements for real-time systems, since it can respond to events in the environment external to the system within the necessary (given) time limits [13]. That is why, when selecting the DSS for solving pattern recognition problems, one of the requirements is the ability to solve the task within the time limits Δt, which is defined by the Δt_s standard [6] or faster. The temporal representation of the requirements for the process looks like this $\Delta t \leq \Delta t_s$, Also, the input data processing time Δt_i should not change or increase significantly when the number of recognition objects within the image being processed increases $\Delta t_i \leq \Delta t_i n$. The Yolo v5 pattern recognition system based

Complex Approach for License Plate Recognition Effectiveness 419

on machine learning models was chosen based on the specified criteria [7].To solve the task of recognizing specific patterns, this DSS should be trained accordingly and at a sufficient (according to [6] requirements) level of accuracy to cover the variability of the external environment, which is the source of input data.

3.3 An Approach to the Automation of a DSS Training Task

The peculiarity of training the chosen pattern recognition DSS consists in the annotation of images with the definition of the boundaries of the detection region(s) in a normalized form. Available software tools for automating the process of annotation of detection regions were investigated, and it was found that some user actions remain unautomated and impose additional time costs [8]. When forming the input dataset for training, the process can involve thousands of images to achieve the required level of accuracy. The lack of automation in the processes that can be automated by modern software engineering tools entails human cost overruns and the impossibility of achieving the required level of accuracy in certain cases. The process of obtaining an annotated dataset D_a for training was defined as $D_a = \sum_{i=1}^{n} f(i) \sum_{r=1}^{m} f(r)$, where i is the input image, r is the annotation region(s) in the image, and n and m are the number of images and regions in the image, respectively. As a result of the defined process, the software application YoloAnno was developed, which implements the proposed automation approaches [10]. Practical use of the application in creating a set of input annotated training data confirmed its effectiveness.

3.4 Creation of a Set of Training Input Data Using the Proposed Approach

Using the built hardware system, 20,000 images were collected under different environmental conditions, from which 2,000 images were selected and annotated. For this, the proposed software application was used, which practically implemented all annotation tasks [10]. As a result, a set of graphic data and annotation data of detection regions was formed, ready for use in the Yolo v5 training system. Empirically, the optimal training parameters were established, which made it possible to perform training on the available equipment in the predicted time and obtain the resulting training data. The model of the pattern recognition system was obtained, which was **validated to a given level of accuracy** on random data from the built hardware graphic information retrieval system.

3.5 Software Validation in Real Conditions

To check the effectiveness of the expressed hypotheses, an experiment was conducted using the created software. For the experiment, a set of 20,000 images, which were not involved in any way in the learning process, was collected using the built hardware complex from real road conditions. This set was processed by the built system and distributed according to the specified detection accuracy criteria. The distribution of results is presented in Table 1.

420   Y. Anton and L. Oleh

Table 1. Result distribution for testing dataset input on license plate region detection

Type of dataset	Correct detection [%]	Semi-correct detection [%]	False detection [%]
All images	98,25	1,11	0,64
Plate present only	95,57	3,19	1,24

3.6 Interaction with External Detection Systems

Using software engineering tools, the standardized API interface was developed for interaction with existing systems currently in operation. With the specified implementation, the created software received graphic data as an input and returned the detection result. Due to the non-linear dependence of the detection time when using machine learning algorithms, it was experimentally confirmed that the detection efficiency increased 5–8% compared to the use of classical approaches where only OCR algorithms are used [30, 31].

A processing time cost factor Δt was considered in comparison to the "OCR-only" approach Δt_{OCR} for an all image area and the proposed "Plate region detection" approach Δt_{RD} combined with "Detected region OCR" approach Δt_{RO}. The results of 1500 images processing within all Δt factor approaches are shown in Table 2.

Table 2. Time cost for testing dataset input recognition with various approaches

Approach	t_{min}[sec.]	t_{max}[sec.]	t_{avg}[sec.]
Δt_{OCR}	0,592	19,969	3,989
Δt_{RO}	0,333	2,125	0,517
Δt_{RD}	0,555	0,867	0,651
$\Delta t_{RD} + \Delta t_{RO}$	0,888	2.992	1,168

An Approach to Increasing the Efficiency of Detection Systems. The result of the theoretical justification of the proposed complex approach to increasing the efficiency of license plate recognition in combination with the practical implementation of the hardware and software complex confirmed the proposed theses. The investigated parameters of the detection process, such as license plate region detection accuracy and error rate, detection time, computational cost of the process, implementation cost were improved in the proposed approach and confirmed experimentally using statistical methods to process the obtained results. The obtained results were compared with the available performance data of the existing systems and the comparable efficiency was confirmed, which is presented in Figs. 2 and 3.

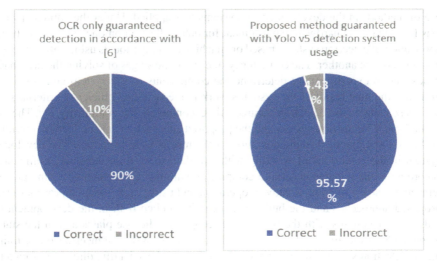

Fig. 2. Distribution comparison chart of the threshold for guaranteed correct detection of license plates region for existing systems (1) and using the proposed approach (2). Where higher value of correct detections is a positive trend. (See Table 1)

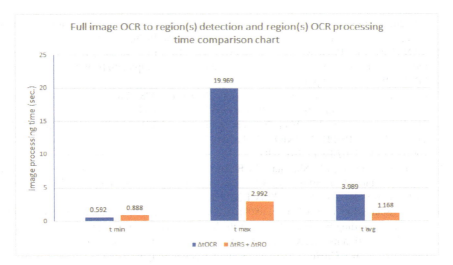

Fig. 3. OCR processing only to OCR combined with proposed method region detection processing results comparison. Where image processing time increasing is a negative trend (See Table 2)

4 Conclusion

Traffic safety systems in operation use OCR approaches to solve the task of license plate recognition, which are inefficient nowadays. A complex approach and methodology for the construction of the DSS was developed when solving the problem of increasing the efficiency of license plate recognition. In the proposed approach, the complex use of methods in the development of systems was applied. Under the approach work was focused on the hardware environment for data collection on one side and software application of detection systems based on machine learning models using software engineering tools on another. The consistency of each of the stages of solving the task made it possible to experimentally determine the components of the system and positively influence the final result of the system's work in the specified efficiency criteria, such as license plate region detection accuracy, time, computational complexity, etc. The use of the Yolo V5 detection system and the proposed approach to data collection, training, evaluation and practical implementation of the results were proposed. The developed system in the conditions of working with real data showed an increase in efficiency according to the selected criteria in comparison with the requirements for similar systems and systems that are already in operation and using OCR based approaches. The proposed approaches and the built system can be applied both in the development of traffic safety systems with the function of recognizing license plates and in the study of methods of creating such systems with modern software engineering tools, training of detection systems in solving similar tasks and the scientific direction of pattern recognition by methods of machine learning in general.

References

1. ISO. 2021. ISO 39001:2012 - Road traffic safety (RTS) management systems — Requirements with guidance for use (2023). https://www.iso.org/standard/44958.html
2. RANKING EU PROGRESS ON ROAD SAFETY: 16th Road Safety Performance Index Report, June 2022 (2022). https://etsc.eu/wp-content/uploads/16-PIN-annual-report_FINAL_WEB_1506_2.pdf
3. European regional status report on road safety 2019, ISBN 978 92 890 5498 0 (2019). https://apps.who.int/iris/bitstream/handle/10665/336584/9789289054980-eng.pdf)
4. The European Automobile Manufacturers' Association: NEW PASSENGER CAR REGISTRATIONS, EUROPEAN UNION. 18 January 2023 (2023). https://www.acea.auto/files/20210916_PRPC_2107-08-FINAL.pdf)
5. Shan, Du., Ibrahim, M., Shehata, M., Badawy, W.: Automatic License Plate Recognition (ALPR): a state-of-the-art review. IEEE Trans. Circuits Syst. Video Technol. 23(2), 311–325 (2013). https://doi.org/10.1109/TCSVT.2012.2203741
6. DSTU (State standard of Ukraine) 8809:2018 Metrology. Traffic control devices with photo and video recording functions. Remote vehicle speed meters, remote space-time parameters of vehicle location meters. Metrological and technical requirements; (Ukr.) (2018)
7. Ultralytics/yolov5. GitHub (2020). https://github.com/ultralytics/yolov5
8. Yakovlev, A., Lisovychenko, O.: Підхід до автоматизації анотування зображень для навчання моделей штучного інтелекту. Адаптивні системи автоматичного управління 1(36), 32–40 (2020). https://doi.org/10.20535/1560-8956.36.2020.209755
9. AXIS P1344-E Network Camera (2023). https://www.axis.com/en-gb/products/axis-p1344-e
10. Yakovlev, A., Yakovlev/Yoloanno, A.: GitHub (2020). https://github.com/AntonYakovlev/Yoloanno
11. Massoud, M.A., Sabee, M., Gergais, M., Bakhit, R.: Automated new license plate recognition in Egypt. Alex. Eng. J. (2013). https://doi.org/10.1016/j.aej.2013.02.005
12. Anton, Y., Oleh, L.: Automated license plate recognition process enhancement with convolutional neural network based detection system to improve the accuracy and reliability of vehicle recognition. In: Hu, Z., Dychka, I., Petoukhov, S., He, M. (eds.) Advances in Computer

Science for Engineering and Education. ICCSEEA 2022. Lecture Notes on Data Engineering and Communications Technologies, vol. 134. Springer, Cham (2022). https://doi.org/10.1007/978-3-031-04812-8_21

13. Decision support system - Wikipedia (2023). https://en.wikipedia.org/wiki/Decision_support_system

14. Lin, C.-H., Lin, Y.-S., Liu, W.-C.: An efficient license plate recognition system using convolution neural networks. In: 2018 IEEE International Conference on Applied System Invention (ICASI). IEEE (2018) https://doi.org/10.1109/icasi.2018.8394573

15. Rhead, M., Gurney, R., Ramalingam, S., Cohen, N.: Accuracy of automatic number plate recognition (ANPR) and real world UK number plate problems. In: 2012 IEEE International Carnahan Conference on Security Technology (ICCST). IEEE (2012). https://doi.org/10.1109/ccst.2012.6393574

16. Koval, V., Turchenko, V., Kochan, V., Sachenko, A., Markowsky, G.: Smart license plate recognition system based on image processing using neural network. In: Proceedings of the Second IEEE International Workshop on Intelligent Data Acquisition and Advanced Computing Systems: Technology and Applications, 2003 (2003) https://doi.org/10.1109/idaacs.2003.1249531

17. Fahmy, M.M.M.: Automatic number-plate recognition: neural network approach. In: Proceedings of VNIS'94 - 1994 Vehicle Navigation and Information Systems Conference (1994). https://doi.org/10.1109/vnis.1994.396858

18. Shashirangana, J., et al.: License plate recognition using neural architecture search for edge devices. Int. J. Intell. Syst. 37, 10211–10248 (2021). https://doi.org/10.1002/int.22471

19. Shashirangana, J., Padmasiri, H., Meedeniya, D., Perera, C.: Automated license plate recognition: a survey on methods and techniques. IEEE Access. (2021). https://doi.org/10.1109/access.2020.3047929

20. Shrivastava, S., Singh, S.K., Shrivastava, K., Sharma, V.: CNN based automated vehicle registration number plate recognition system. In: 2020 2nd International Conference on Advances in Computing, Communication Control and Networking (ICACCCN) (2021). https://doi.org/10.1109/icacccn51052.2020.9362737

21. Akther, M., Ahmed, M., Hasan, M.: Detection of vehicle's number plate at nighttime using iterative threshold segmentation (ITS) algorithm. Int. J. Image, Graph. Sig. Process. 5(12), 62–70 (2013). https://doi.org/10.5815/ijigsp.2013.12.09

22. Kaur, S.: An automatic number plate recognition system under image processing. Int. J. Intell. Syst. Appl. 8(3), 14–25 (2016). https://doi.org/10.5815/ijisa.2016.03.02

23. Ramshankar, Y., Deivanathan, R.: Development of machine vision system for automatic inspection of vehicle identification number. Int. J. Eng. Manuf. 8(2), 21–32 (2018). https://doi.org/10.5815/ijem.2018.02.03

24. Aung, M.M.: Study for license plate detection. Int. J. Image, Graph. Sig. Process. 11(12), 39–46 (2019). https://doi.org/10.5815/ijigsp.2019.12.05

25. Sakharkar, Y.A., Singh, M., Kumar, K.A., Aju, D.: A reinforcement learning based offload decision model (RL-OLD) for vehicle number plate detection. Int. J. Eng. Manuf. (IJEM) 11(6), 11–18 (2021). https://doi.org/10.5815/ijem.2021.06.02

26. Pirgazi, J., Sorkhi, A.G., Kallehbasti, M.M.P.: An efficient robust method for accurate and real-time vehicle plate recognition. J. Real-Time Image Proc. 18(5), 1759–1772 (2021). https://doi.org/10.1007/s11554-021-01118-7

27. Calitz, A., Hill, M.: Automated license plate recognition using existing university infrastructure and different camera angles. The African Journal of Information Systems 12(2), 4 (2020). https://digitalcommons.kennesaw.edu/ajis/vol12/iss2/4

28. Ma, L., Zhang, Y.: Research on vehicle license plate recognition technology based on deep convolutional neural networks. Microprocess. Microsyst. 82, 103932 (2021). https://doi.org/10.1016/j.micpro.2021.103932

29. Silvano, G., et al.: Synthetic image generation for training deep learning-based automated license plate recognition systems on the Brazilian Mercosur standard. Des. Autom. Embed. Syst. **25**(2), 113–133 (2020). https://doi.org/10.1007/s10617-020-09241-7

30. Schegolihin, Y., Mitrohin, M., Sazykina, V., Semenkin, M.: Gradual labeling of the training set to improve the efficiency of image detection by a neural network on the example of license plate recognition. In: 2021 29th Conference of Open Innovations Association (FRUCT) (2021). https://doi.org/10.23919/fruct52173.2021.9435602

31. "VIDEOKONTROL-Rubezh" complex. Ollie.com.ua (2020). http://www.ollie.com.ua/videocontrol/index.html. (Ukr.)

32. USIT, Produkciya - USI. Ukrsi.com.ua (2020). http://ukrsi.com.ua/products/. (Ukr.)

33. Yampolskyi, L.S., Lisovichenko, O.I., Oliynyk, V.V.: Neurotechnologies and neurocomputer systems: textbook, p. 576. Dorado-Druk, Kyiv (2016). (Ukr.)

34. Joshi, P., Escrivá, D., Godoy, V.: OpenCV by Example. Packt Publishing, Birmingham, UK (2016)

35. Omar, N., Sengur, A., Al-Ali, S.G.S.: Cascaded deep learning-based efficient approach for license plate detection and recognition. Expert Syst. Appl. **149**, 113280 (2020). https://doi.org/10.1016/j.eswa.2020.113280

36. Sahoo, A.K.: Automatic recognition of Indian vehicles license plates using machine learning approaches. Materials Today: Proceedings **49**, 2982–2988 (2022). https://doi.org/10.1016/j.matpr.2020.09.046

37. Jamtsho, Y., Riyamongkol, P., Waranusast, R.: Real-time license plate detection for non-helmeted motorcyclist using YOLO. ICT Express **7**(1), 104–109 (2021). https://doi.org/10.1016/j.icte.2020.07.008

38. Qian, Y., et al.: Spot evasion attacks: Adversarial examples for license plate recognition systems with convolutional neural networks. Comput. Secur. **95**, 101826 (2020). https://doi.org/10.1016/j.cose.2020.101826

39. Silva, S.M., Jung, C.R.: Real-time license plate detection and recognition using deep convolutional neural networks. J. Vis. Commun. Image Represent. **71**, 102773 (2020). https://doi.org/10.1016/j.jvcir.2020.102773

40. Slimani, I., Zaarane, A., Okaishi, W.A., Atouf, I., Hamdoun, A.: An automated license plate detection and recognition system based on wavelet decomposition and CNN. Array **8**, 100040 (2020). https://doi.org/10.1016/j.array.2020.100040

41. Xiang, H., Zhao, Y., Yuan, Y., Zhang, G., Hu, X.: Lightweight fully convolutional network for license plate detection. Optik **178**, 1185–1194 (2019). https://doi.org/10.1016/j.ijleo.2018.10.098

42. Björklund, T., Fiandrotti, A., Annarumma, M., Francini, G., Magli, E.: Robust license plate recognition using neural networks trained on synthetic images. Pattern Recogn. **93**, 134–146 (2019). https://doi.org/10.1016/j.patcog.2019.04.007

43. Kessentini, Y., Besbes, M.D., Ammar, S., Chabbouh, A.: A two-stage deep neural network for multi-norm license plate detection and recognition. Expert Syst. Appl. **136**, 159–170 (2019). https://doi.org/10.1016/j.eswa.2019.06.036

44. Asif, M.R., Qi, C., Wang, T., Fareed, M.S., Raza, S.A.: License plate detection for multi-national vehicles: an illumination invariant approach in multi-lane environment. Comput. Electr. Eng. **78**, 132–147 (2019). https://doi.org/10.1016/j.compeleceng.2019.07.012

45. Rao, W., Yao-Jan, Wu., Xia, J., Jishun, Ou., Kluger, R.: Origin-destination pattern estimation based on trajectory reconstruction using automatic license plate recognition data. Transp. Res. Part C: Emerg. Technol. **95**, 29–46 (2018). https://doi.org/10.1016/j.trc.2018.07.002

46. Li, H., Wang, P., You, M., Shen, C.: Reading car license plates using deep neural networks. Image Vis. Comput. **72**, 14–23 (2018). https://doi.org/10.1016/j.imavis.2018.02.002

47. Wang, R., Sang, N., Wang, R., Jiang, L.: Detection and tracking strategy for license plate detection in video. Optik **125**(10), 2283–2288 (2014). https://doi.org/10.1016/j.ijleo.2013.10.126
48. Öztürk, F., Özen, F.: A new license plate recognition system based on probabilistic neural networks. Procedia Technol. **1**, 124–128 (2012). https://doi.org/10.1016/j.protcy.2012.02.024
49. Yuren, Du., Shi, W., Liu, C.: Research on an efficient method of license plate location. Phys. Procedia **24**, 1990–1995 (2012). https://doi.org/10.1016/j.phpro.2012.02.292
50. Erdinc Kocer, H., Kursat Cevik, K.: Artificial neural networks based vehicle license plate recognition. Procedia Comput. Sci. **3**, 1033–1037 (2011). https://doi.org/10.1016/j.procs.2010.12.169

The Analysis and Visualization of CEE Stock Markets Reaction to Russia's Invasion of Ukraine by Event Study Approach

Andrii Kaminskyi[1] and Maryna Nehrey[2(✉)]

[1] Taras Shevchenko National University of Kyiv, Kyiv 01033, Ukraine
[2] National University of Life and Environment Science of Ukraine, Kyiv 03041, Ukraine
marina.nehrey@gmail.com

Abstract. The "Black swan" event which raised from Russia's invasion of Ukraine in February 2022, has caused an extensive external shock for world financial markets. Our paper examines the scale of its impact on the three largest stock markets in Central and Eastern Europe: the Czech, Polish, and Hungarian. The choice of markets was due to the close proximity to Ukraine. To do this, we created a set of indicators that allowed us to display the specificity of reaction to shock. Indicators were split into three groups. The first group included indicators that reflect the «hustle» of investors and the reformatting of investment portfolios in the first ten days after the invasion event. The second group's metrics consider shock in the context of the price-based dimension "deepness of fall – level of recovery". The third group allows displaying changes in "risk-return" correspondence in the context of ESG score. The K-ratio tool was applied for considering consistency. The result indicated a strong impact on researched markets and different consequences for stock index constituents, especially for companies with high ESG scores.

Keywords: Event study approach · stock market data; shock · ESG investing · Russia-Ukraine war · CEE countries · K-ratio

1 Introduction

Russia invaded Ukraine at 24 February 2022, and hostilities began. This event had a tremendous geopolitical impact which was expressed in military, real and financial economic, social and many other aspects. This event, which can be characterized as a strong shock, was negatively affected by three directions. The first direction includes the socially specific economic consequences. The second direction is due to the impact on multiple sectors through global supply chain disruption. This is especially related to sharp fluctuations in oil, gas and number of different commodities prices. The third direction is highlight sanctions, which have a direct impact on the aggressor but at the same time have spillover effects.

The reaction of stock markets was significant. Investors, under the influence of a shock, started actively buying and selling various assets, which leads to sharp price

© The Author(s), under exclusive license to Springer Nature Switzerland AG 2023
Z. Hu et al. (Eds.): ICAILE 2023, LNDECT 180, pp. 426–436, 2023.
https://doi.org/10.1007/978-3-031-36115-9_39

fluctuations and high volatility of returns. Our study aims was to apply an event study approach to this "Black Swan" shock for Polish, Czech and Hungarian stock markets. These markets are the most capitalized and liquid in the CEE (Central Eastern European) group. Moreover, these countries are geographically close to Ukraine (Poland and Hungary share a border with Ukraine). Stocks of companies which are constituents of baskets of stock indices PX, WIG20, and BUX were object of investigation. The scale of such reaction and their peculiarities are in focus in our research.

For our research goals, we have formed a set of indicators. They have allowed us to display the specificity of the shock reaction and to analyze the specificities of the different companies' stock reactions. Indicators were combined into three groups. The first group included indicators that reflect the «hustle» of investors and the reformatting of investment portfolios in the first ten days after the invasion event. The second group's indicators described dimension "deepness of fall – level of recovery". The third group allows to display impact in the context of ESG (Environment, Social, Governance) score. The actuality of this connected with the process of quite actively promoting the implementation of ESG criteria in companies.

The K-ratio tool was applied for investigation of stock returns consistency changes. The result indicated a strong impact shock to such consistency. Moreover, the consistency changing through shock pipeline demonstrated interconnection with ESG scores. Thus, most negative K-ratio changes were revealed to high ESG scores companies.

The paper is structured as follows. Section 2 presents the literature review. Section 3 presents the methodology, which applied in our investigation. The Sect. 4 shows the results of applying the methodology to the CEE stock market. The special visualization of indicators values displays shock influences. Finally, Sect. 5 presents the conclusions and controversial issues.

2 Literature Review

Complex and deep analysis of investigated shock and its development demonstrates significance it to modern economic and, particularly, to stock markets. During the year after the Russian invasion of Ukraine, a lot of literature on the economic consequences of this has been published. The complexity of this aggression-based impact is examined in [1]. In particular, the author highlights such after-effects: global supply chain disruption, growth of oil and gas prices, changes in the banking segment, global inflation acceleration and cost of living and some others. The basic statistical tool of the author's research is a two-stage least square regression analysis. About global stock markets, the author indicates his findings about the falling of several world stock indices within five days (from February 18 to February 25). The choice of such interval is interesting. Although, from our point of view, it assumes the presence of a certain anticipating, which is discussible. Our point of view is more to tend the "Black swan" case.

The paper [2] provides an in-depth statistical analysis of the reaction of the G20 stock markets to the Russian-Ukrainian conflict. The paper analysis shows that abnormal returns and cumulative abnormal returns show that the European and Asian regions have been affected the most by the launch of the Russian aggression against Ukraine. Moreover, the author's estimation indicates Poland, Hungary, and Turkey plummeted essentially more than others (disregarding the Russian stock market). Abnormal returns on event day were Poland (-10.374%), Hungary (-9.488%), and Turkey (-8.262%).

An extensive analysis of the impact of the Russian-Ukrainian war on financial markets is presented in [3]. The authors analyzed the effect on markets of 73 countries and found significant results. It turned out that the war has created considerable instability in the markets, which is reflected in the variability of yields (up to 20%). At the same time, the authors estimate the impact on average stock prices to be relatively low.

There are many empirical studies (including event study methods applying) is devoted to the analysis of Russian aggression-based shocks in various segments of the investment market. This applies to different types of investment assets, both traditional and alternative. Also, several studies already present the results of the behavior of financial instruments produced by companies from different sectors of the economy. In addition, a few studies describe the response of stock markets in individual countries.

There are many empirical studies (including event study methods applying) devoted to the analysis of Russian aggression-based shocks in various segments of the investment market [4–7]. This applies to different types of investment assets, both traditional and alternative [8–14]. Also, a few studies already present the results of the behavior of financial instruments produced by companies from different sectors of the economy [15–17]. In addition, several studies describe the response of stock markets in individual countries [18–22].

Thus, considering that Russia is the significant exporter of oil and gas. Authors of [23] verified hypothesis about Russia aggression affecting the to the energy and metal markets. Authors found that war had a significant influence on gold, platinum, palladium and nickel, markets. Moreover, renewable energy industry also demonstrates a significant increasing in the anomalous returns for the with the renewable energy industry. Methodologically they use an event research technique and highest deviation was on the t + 1 day.

We find a study [24] that examines the impact of Russian aggression on the yields and risks in the commodity market very fascinating. The authors estimate an rising in volatility from 35% to 85%. This, according to the authors, exceeds the level observed during the pandemic COVID-19. Authors grounded that crude oil turns into a net transmitter of side effects, while wheat and soybeans become net recipients of side effects. Such markets commodities as silver, gold, copper, platinum, aluminum and sugar become pure transmitters of volatility.

Considering the impact of Russian aggression on alternative assets, it is necessary to mention the article [25], which justifies the impact on the pricing of cryptocurrencies. This is explained by an increase in geopolitical risk (expressed by an index) caused by aggression. The authors substantiate the statistical impact of the role of geopolitical risk on cryptocurrency markets.

The impact on the aerospace and defense industries was quite strong, as substantiated in [26]. The authors examined new sentiments from war-related news from October 2021 to June 2022. The results showed a negative impact of war sentiment on market behavior. The authors' conclusion indicates a negative effect of the war on the air transportation market and a positive effect on the defense market.

The paper [27] also considers the event study method for application to 24 Feb 2022 shock also. Author examines 209 stocks from Borsa Istanbul (BIST) in terms of pre-and post-historical period. The abnormal returns were found in the second, third, fourth, and fifth days they were negative and statistically significant.

3 Methodology

3.1 Research Data

We used the event study method to analyze the data and to calculate the indicators we are considering. The event is on 24.02.2022. On this date, Russia started a military aggression against Ukraine called a "special military operation". To apply the event study method, we considered a time interval of one year: 24.08.2021 - 24.08.2022. The "before event" interval was 24.08.2021–23.02.2022. We considered the event as a short time interval of 24.02.2024 - 09.03.2024. It was defined as the ten working days that our study took as "subject for methods applying" The "after event" interval was 10.03.2022–24.08.2022.

Our sample includes shares of companies included in the stock indices of the three CEE countries. They are the index of the Czech stock market PX (9 companies), the index of the Polish stock market WIG20 (20 companies), and the index of the Hungary stock market BUX (16 companies). A total of 45 companies were used for the analysis.

The following data from investing.com were used about the companies in question:

1) Daily (closed) prices.
2) Daily high and low prices.
3) Daily trading volume.

The daily values of these three indices were also used.

The ESG scores were a separate data block used for the analysis. We used ESG scores from the company S&P Global [28]. Score values involved four numbers: ESG score, E score, S score, and G score. At the time of the survey, these values were calculated for 25 companies out of 46. Therefore, the ESG survey was implemented for limited samples.

3.2 Shock Estimation by "Price Spread – Volume Volatility" Indicators

We used the hypothesis that in shock environment (the "Black Swan") investors are suddenly faced with uncertainty and this affects their behavior. Specifically, they reformat their portfolios actively by selling and buying stocks. Based on this logic, we have introduced the following two indicators based on daily trading. The first indicator is "shock deepness volume" (SDV) which is defined as:

$$SDV = \frac{Average\ daily\ volume\ over\ shock\,(10\,days)}{Average\ daily\ volume\ before\ shock\,(time\ interval\ before\ event)} - 1 \quad (1)$$

The second indicator is "recovery rate volume" (RRV) which is defined as:

$$RRV = \frac{Average\ daily\ volume\ after\ shock\,(time\ interval\ "after\ event")}{Average\ daily\ volume\ before\ shock\,(time\ interval\ "before\ event")} \quad (2)$$

These are dimensionless indicators. SDV shows the percentage increase in daily trading volume at the time of a shock. Since a shock is characterized by the abruptness of change, the higher the SDV, the more pronounced the shock. RRV ratio shows to what extent the average daily trading volume after the shock exceeds (or is already equal to) that before the shock. In other words, if it is well above 1, the recovery has not yet taken place. This indicator essentially reflects the long memory of shock affecting. Of course, it depends on the selected interval.

Thus, our first shock indicator is a pair (SDV; RRV).

The following two indicators reflect investors' hustle (or in slang "rush asunder"). In case of a shock, there is a rapid increase in uncertainty regarding present momentum and future returns. Therefore, there is no classical equilibrium price. The change in price during the day can be very significant. This occurrence we formalize in our next indicator:

$$HLDD = \frac{Highprice - Lowprice}{0,5 * (Highprice + Lowprice)}(Day) \tag{3}$$

This indicator integrates two effects. The first, reflected in the numerator, shows the investor's hustle. Large fluctuations during the day increase the values. The second effect, reflected in the denominator, can show the decrease in the prices and influence (increase) its values. Thus, according to our approach, the higher the HLDD, the stronger the shock manifests itself.

The HLDD is calculated for each day. The logic of SD and RR construction can be applied to its values. As a result, we obtain two similar indicators SDHL and RRHL.

3.3 Shock Estimation by Price Changing

The following hypothesis in our study was the hypothesis of a sharp decline in share price followed by a recovery. In our research, the fall-recovery pair characterizes the "risk-return" correspondence. For this purpose, we used indicators methodologically consistent with SDV and RRV. Only instead of trading volume it uses price. The first indicator is "shock deepness price":

$$SDP = \frac{Average\ proce\ through\ 10\ days\ of\ shock}{Average\ price\ during\ time\ interval\ "before\ event")} - 1. \tag{4}$$

The second indicator is "recovery rate price" (RRP) which is defined as:

$$RRP = \frac{Average\ price\ after\ shock\ (during\ time\ interval\ "after\ event")}{Average\ price\ before\ shock\ (during\ time\ interval\ "before\ event")} \tag{5}$$

SDP has the nature of a classical rate of return with some specifications linked to average prices. It was supposed that such an approach nihilate price volatility before the shock to before the shocking price. The logic of using such a form of RRP is to desire an estimate comparison with the before shock period, not with the "bottom price".

3.4 ESG-based Analysis

One of the most dynamic areas in the financial investment market is ESG investing. Today, approximately 40% of assets under management meet ESG criteria. The level of ESG itself is determined by ESG scores, which have been implemented in the market by major analytical agencies - S&P Global, Refiniti, Sustainalytics and others. The main question is how to combine three criteria - risk minimization, return maximization and ESG maximization. The companies from CEE markets step-by-step move towards receiving ESG scores. Our research question in this regard was to compare how companies with different ESG scores passed the shock.

4 Results

4.1 Markets' Overall Visualization of Shock: Indexes Handling

Markets received a significant shock as a response to Russian invasion in Ukraine. This shock was visualized in the following stock indices of CEE countries: PX, WIG, and BUX (Fig. 1).

The shock visualization shows the integral impact of the Russian invasion on the CEE stock market. Figures on the left show the volatility of returns. Figures on the right show the 30-day rolling standard deviation. Data visualization analysis allows us to conclude the significant event's impact on the stock market.

4.2 Shock Estimation by "PRICE Spread – Volume Volatility" Indicators

We estimated shock by "price spread – volume volatility" indicators. For all except one company (Photon), the SDHL figure increased (Fig. 2). The average increase is 111% (i.e., more than double the pre-shock value). At the same time, the recovery over the period under review showed: 27% of stocks began to have an HLDD value less than the pre-shock value, and 73%. We can conclude that the shock and further military developments are reflected in high investor fussiness in the face of uncertainty.

A similar examination of the indicators related to daily trading volume shows a slightly different picture. Thus, five companies showed lower volume during the shock (Fig. 3). The average increase is 132%. Moreover, the number of companies that have "returned" to pre-shock level is also different - 44%.

4.3 Shock Estimation by Price Changing

A price trend analysis shows that the average value of the fall (in fact, the yield expressed by SDP) was -12.46% (Fig. 4). The decline is not evenly distributed across the markets. In particular, Poland -14.84% Czech -4.52% Hungary -13.95%. The average RRP was 88.08%. This suggests that the markets have not yet recovered from the shock of six months.

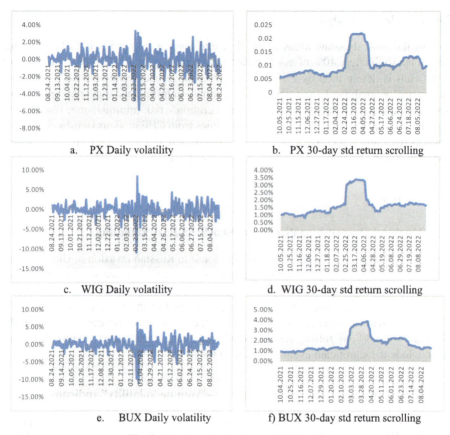

Fig. 1. Stock market indexes visualization

4.4 Investigation of Shock Transition by Stocks with Different ESG Scores

The K-ratio was used to project the consistency analysis. It was calculated for the entire period before and after the event. The event impact was estimated as the difference between the K-ratio before and after the event. In Fig. 5, the columns show that in most cases the K-ratio decreased.

We further compared changes in K-ratio and ESG scoring. It turned out that they had opposite directions. The K-ratio changed the most for companies with high ESG scores.

5 Summary and Conclusion

The Russian invasion of Ukraine had a great impact on the global economy. One of these strong impacts concerns stock markets. In our research, we investigated the response to this event of the three largest CEE stock markets. For research, we applied three approaches. Each approach was considered as some modification of event study logic. The first approach involved analyzing the behavior of investors in the moment of shock. We studied event at 10 days after it appearance. Two indicators related to trading volume

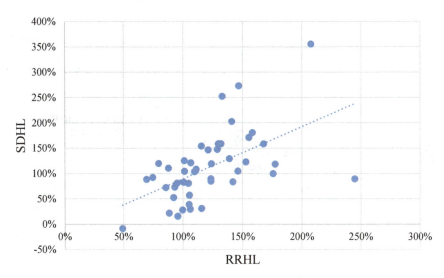

Fig. 2. Shock visualization in SDHL-RRHL scale

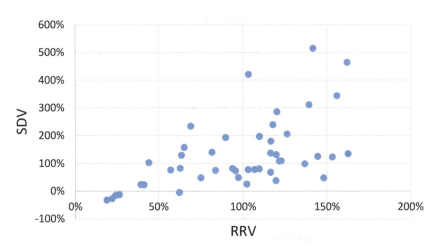

Fig. 3. Shock visualization in the SDV-RRV scale

and price spread showed huge deviations. This is due to the great uncertainty that caused the shock. Uncertainties have led to the impossibility of precise future pricing and the reformatting of investment portfolios. Within six months of the shock, processes were normalized, but the stock parameters of most companies did not reach before shock levels.

There were also applied indicators directly based on price movements showing the depth of the fall and the level of price recovery. It shows that the average value of the fall (in fact, the return expressed by SDP) was -12.46% (Fig. 4). The decline is not evenly

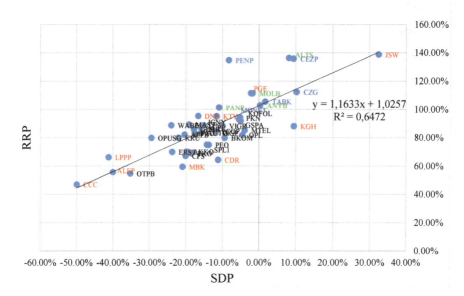

Fig. 4. Price trend analysis

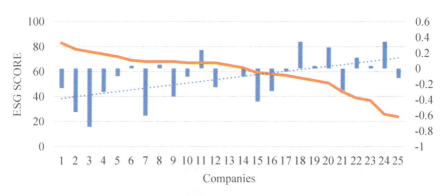

Fig. 5. Correspondence between ESG and changes in K-ratio value

distributed across the markets. In particular, Poland -14.84% Czech -4.52% Hungary -13.95%. These figures demonstrate comparable value with results presented in [2].

The average RRP was 88.08%. This suggests that the markets have not yet recovered from the shock of six months.

K-ratio was used to analyze the effect of shock affecting on risk-return correspondence in the context of ESG score values. The change of this indicator during the transition was overwhelmingly negative. An interesting finding was that the K-ratio change is inversely related to the ESG score. The largest declines in K-ratio were seen in stocks with a large ESG score.

References

1. Ozili, P.K.: Global Economic Consequence of Russian Invasion of Ukraine (2022). SSRN. https://doi.org/10.2139/ssrn.4064770
2. Yousaf, I., Patel, R., Yarovaya, L.: The reaction of G20 stock markets to the russia-ukraine conflict 'black-swan' event: evidence from event study approach. J. Behav. Experimental Fin. (2022). https://doi.org/10.2139/ssrn.4069555
3. Lo, G., Marcelin, I., Bassène, T., Sène, B.: The Russo-Ukrainian war and financial markets: the role of dependence on Russian commodities. Finance Research Lett. **50**(103194) (2022). https://doi.org/10.1016/j.frl.2022.103194
4. Diaconaşu, D., Mehdian, S., Stoica, O.: The reaction of financial markets to Russia's invasion of Ukraine: evidence from gold, oil, bitcoin, and major stock markets. Applied Econ. Lett. (2022). https://doi.org/10.1080/13504851.2022.2107608
5. Fang, Y., Shao, Z.: The Russia-Ukraine conflict and volatility risk of commodity markets. Finance Res. Lett. **50** (103264)
6. Gaio, L., Stefanelli, N., Pimenta, T., Bonacim, C., Gatsios, C.: The impact of the Russia-Ukraine conflict on market efficiency: evidence for the developed stock market. Finance Research Lett. **50** (103302) (2022)
7. Joshipura, M., Lamba, A.: Asymmetric Impact of Russia-Ukraine War on Global Stock Markets (2022). https://ssrn.com/abstract=4273419 or http://dx.doi.org/https://doi.org/10.2139/ssrn.4273419
8. Zulqarnain, M., Ghazali, R., Ghouse, M.G., Hassim, Y.M.M., Javid, I.: Predicting financial prices of stock market using recurrent convolutional neural networks. Int. J. Intell. Syst. Appl. (IJISA) **12**(6), 21–32 (2020)
9. Guryanova, L., Bolotova, O., Gvozdytskyi, V., Olena, S.: Long-term financial sustainability: an evaluation methodology with threats considerations. Rivista di Studi sulla Sostenibilita **1**, 47–69 (2020)
10. Izonin, I., Tkachenko, R., Vitynskyi, P., Zub, K., Tkachenko, P., Dronyuk, I.: Stacking-based GRNN-SGTM ensemble model for prediction tasks. In: 2020 International conference on decision aid sciences and application (DASA). IEEE, 2020, pp. 326-330. https://doi.org/10.1109/DASA51403.2020.9317124
11. Dey, P.P., Nahar, N., Hossain, B.M.: Forecasting stock market trend using machine learning algorithms with technical indicators. Int. J. Information Technol. Comput. Sci. **12**(3), 32–38 (2020)
12. Kaminskyi, A., Butylo, D., Nehrey, M.: Integrated approach for risk assessment of alternative investments. Int. J. Risk Assessment Manage. **24**(2–4), 156–177 (2021). https://doi.org/10.1504/IJRAM.2021.126413
13. Mahedy, A.S., Abdelsalam, A.A., Mohamed, R.H., El-Nahry, I.F.: Utilizing neural networks for stocks prices prediction in stocks markets. Int. J. Inf. Technol. Comput. Sci. **12**(3), 1–7 (2020)
14. Sova, Y., Lukianenko, I.: Theoretical and empirical analysis of the relationship between monetary policy and stock market indices. In: 2020 10th International Conference on Advanced Computer Information Technologies, ACIT 2020 - Proceedings, pp. 708–711, 9208926 (2020)
15. Jain, V.R., Gupta, M., Singh, R.M.: Analysis and prediction of individual stock prices of financial sector companies in NIFTY50. Int. J. Information Eng. Electronic Bus. **12**(2), 33 (2018)
16. Kaminskyi, A., Nehrey, M.: Changing risk-return correspondence during the COVID-19 turmoil: evidence from polish stock market. Research on Enterprise in Modern Economy theory and practice, **1**(32), 18–33 (2021). https://doi.org/10.19253/reme.2021.01.002

17. Davydenko, N., Buriak, A., Titenko, Z.: Financial support for the development of innovation activities. Intellectual Economics **13**(2), 144–151 (2019)
18. Lukianenko, D., Strelchenko, I.: Neuromodeling of features of crisis contagion on financial markets between countries with different levels of economic development. Neuro-Fuzzy Modeling Techniques in Economics, **10**, 136–163 (2021). https://doi.org/10.33111/nfmte.2021.136
19. Kaminskyi, A., Baiura, D., Nehrey, M.: ESG investing strategy through COVID-19 turmoil: ETF-based comparative analysis of risk-return correspondence. Intellectual Econ. **16**(2), 5–23 (2022)
20. Caldara, D., Iacoviello, M.: Measuring geopolitical risk. American Economic Rev. **112**(4), 1194–1225 (2022). https://doi.org/10.1257/aer.20191823
21. Matviychuk, A., Lukianenko, O., Miroshnychenko, I.: Neuro-fuzzy model of country's investment potential assessment. Fuzzy Economic Rev. **24**(2), 65–68 (2019). https://doi.org/10.25102/fer.2019.02.04
22. Kaminskyi, A., Nehrey, M., Fedchun, A.: ESG-score effect in risk assessment of direct and portfolio investment: evidence from CEE markets. The Journal of V. N. Karazin Kharkiv National University Series: "International Relations. Economics. Country Studies. Tourism" (IRECST), (15), 49–62 (2022)
23. Umar, M., Riaz, Y., Yousaf, I.: Impact of Russian-Ukraine war on clean energy, conventional energy, and metal markets: evidence from event study approach. Resour. Policy **79**, 102966 (2022). https://doi.org/10.1016/j.resourpol.2022.102966
24. Wang, Y., Bouri, E., Fareed, Z., Dai, Y.: Geopolitical risk and the systemic risk in the commodity markets under the war in Ukraine. Financ. Res. Lett. **49**, 103066 (2022)
25. Long, H., Demir, E., Będowska-Sójka, B., Zaremba, A., Jawad, S., Shahzad, H.: Is geopolitical risk priced in the cross-section of cryptocurrency returns?. Finance Research Letters **49** (103131). https://doi.org/10.1016/j.frl.2022.103131
26. Le, V.H., von Mettenheim, H.-J., Goutte, S., Liu, F.: News-based sentiment: can it explain market performance before and after the Russia-Ukraine conflict? J. Risk Finance **24**(1), 72–88 (2023). https://doi.org/10.1108/JRF-06-2022-0168
27. Doğan, M.: The impact of the Russia-Ukraine war on stock returns. Social Sci. Res. J. **11**(1), 1–9 (2022)
28. S&P Global Sustainable. 2022 (2023). https://www.spglobal.com/esg/

The Same Size Distribution of Data Based on Unsupervised Clustering Algorithms

Akbar Rashidov[✉], Akmal Akhatov, and Fayzullo Nazarov

Samarkand State University, Samarkand 140107, Uzbekistan
researcher.are@gmail.com

Abstract. It is known that dividing data into groups based on certain rules not only helps to separate meaning from the data but also increases the efficiency of the large-volume data processing process. The same size distribution of data is especially effective in approaches such as distributed computing or parallel processing. The main reason for this is that dividing the data into as equal clusters as possible allows achieving the highest efficiency results in these approaches. But the distribution of data in the same size based on the human factor is a complicated process due to the impossibility of pre-planning the data contained in the data flow and the size of the data. In such a situation, unsupervised clustering algorithms are one of the main solutions to the problem of uniform data distribution. In this research work, hierarchical, K-means, Bisecting K-means, and DBSCAN unsupervised clustering algorithms are analyzed in order to solve the given problem, and the results of research on equal clustering of different groups of data are presented using them. Moreover, a new unsupervised equal-size clustering algorithm for equal clustering of data is proposed. During the research, the efficiency indicator of this algorithm for dividing data into equal groups is compared with the indicators of existing algorithms, and the conclusions of the experimental results are presented.

Keywords: Unsupervised clustering algorithms · hierarchical clustering · K-means · Bisecting K-means · DBSCAN · new equal clustering algorithm

1 Introduction

Nowadays, data analysis, which is, identifying hidden contents from a data set, and separating them into useful, reliable, and interrelated data groups is one of the topics of modern research [1]. In the modern digital age, the increase in the volume of data flow increases the relevance of this research topic. Because as the flow of information increases, people's natural data analysis capabilities are limited, and the process of extracting meaningful information from big data becomes more complicated [2–4]. In such a situation, the most effective solution for finding hidden patterns in data is unsupervised learning based on data clustering approaches according to certain criteria. Unsupervised clustering is an approach to grouping data by grouping or separating those with similar patterns, and those with different patterns from the data without human intervention [5].

© The Author(s), under exclusive license to Springer Nature Switzerland AG 2023
Z. Hu et al. (Eds.): ICAILE 2023, LNDECT 180, pp. 437–447, 2023.
https://doi.org/10.1007/978-3-031-36115-9_40

438   A. Rashidov et al.

In these clustering approaches, the grouping of data mainly focuses on characteristics such as similarity and uniqueness of the data, and usually ignores or does not consider the number of elements in the groups to be very important. But in real life, there are such problems that not only the similarity and difference of data are important in solving these problems, but also the equality of the number of elements in clusters is an important factor [6, 7]. For example, when storing and processing data using parallel and distributed computing mechanisms. In these mechanisms, the more evenly the data is distributed to the parts of the system, the higher the efficiency [8–10]. Because in this case, all distributed functions within the system perform the same number of tasks, and all parts of the system complete their tasks at the same time. In other words, they do not wait for each other's tasks to be completed. This is the most important indicator of achieving the highest efficiency in the work process. Therefore, achieving equal size distribution based on unsupervised clustering algorithms was taken as the major research objective.

In order to solve the problem presented during the research, the capabilities of hierarchical, K-means, Bisecting K-means, and DBSCAN existing unsupervised clustering algorithms to divide data into clusters of equal size are analyzed. However, due to the limited capabilities of existing algorithms, a new unsupervised clustering algorithm is proposed to achieve uniform distribution of data. At the end of the study, the efficiency indicator of the proposed algorithm for dividing data into groups of equal size is compared with the indicators of existing algorithms, and opinions and comments are given on the results of the experiment.

2 Methodology

In this research work, first of all, the literature related to the solution of the problem was analyzed. It was found that most of the literature covered only the content of unsupervised clustering algorithms [13–19]. Only in some literatures, studies related to the size of clusters have been conducted [11, 12]. However, in these literatures, the problem of clustering of the same size has not been considered.

During the research, the selection of the characteristics of the data was also considered in order to divide the data into clusters. In this process, the ASCII codes of the characters in the data were used and the following features were extracted. The ASCII code of the first character of the data, the number of vowels and consonants, the sum of their ASCII codes, the number of total characters of the data, and the sum of its ASCII codes.

At the next stage of the research, the data were distributed using existing unsupervised clustering algorithms based on the features extracted above, and the sizes of the clusters were determined. The research process showed that existing unsupervised clustering algorithms are not always effective in solving the given problem. Therefore, a new vector module square-based clustering algorithm was proposed during the research. These steps are described in the following sections of this article.

3 Opportunities of Existing Unsupervised Clustering Algorithms to Distribute Data into Clusters of Equal Size

3.1 Hierarchical Clustering

Hierarchical clustering is one of the widely used unsupervised algorithms for clustering data into groups. Hierarchical clustering mainly uses two approaches. These are agglomerative hierarchical clustering and divisive hierarchical clustering [11, 13]. In both approaches, the clustering result can be illustrated using a dendrogram (Fig. 1).

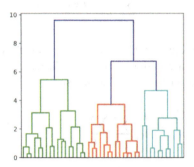

Fig. 1. Representation of the result of hierarchical clustering using a dendrogram

In hierarchical clustering algorithms, grouping is done after the complete cluster hierarchy is built. That is, after all the elements in the collection are formed as branches of a single tree, the branches of the tree corresponding to the required number of clusters are separated. This process can be visualized using the dendrograms shown in Fig. 2.

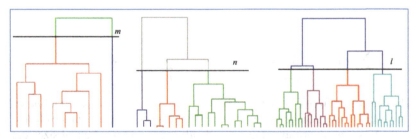

Fig. 2. Line m divides the first dendrogram into 2 clusters, line n divides the second dendrogram into 3 clusters, line l divides the third dendrogram into 4 clusters

In Fig. 2, line m in the first dendrogram divides objects into 2 clusters, line n in the second dendrogram divides objects into 3 clusters, and line l in the third dendrogram divides objects into 4 clusters. As can be seen, only for the objects in the third dendrogram, the size of the clusters is almost equal. It can be concluded that it is not always possible to divide the data into groups of (almost) the same size by means of hierarchical clustering, which usually depends on the state of the objects to be clustered.

3.2 K-means Clustering

The K-means clustering algorithm is a center-based grouping method, and the objects in the set are clustered according to which of the central points known as centroids are close [14–17]. The number of centroids is the same as the number of clusters requested by the user. Centroids are initially chosen arbitrarily. Therefore, clustering will not be perfect in the initial state. Minimization of the sum of squares of the errors of the objects in the clusters relative to the centroids is used to make the clusters perfect (1).

$$S = \sum_{i=1}^{k} \sum_{x_j \in C_i} dist(c_i, x_j)^2 \tag{1}$$

where k- is the number of clusters, C_i- is the cluster, c_i is the centroid coordinate of the i cluster, x_j is the j element coordinate in the C_i cluster. It is known that S reaches its minimum value when the derivative of S with respect to c_i is equal to 0 (2).

$$\frac{\partial S}{\partial c_i} = \frac{\partial S}{\partial c_i} \sum_{i=1}^{k} \sum_{x_j \in C_i} dist(c_i, x_j)^2 = 0 \tag{2}$$

Based on this formula, the optimal value of the centroid of the nth cluster can be found as follows.

$$\frac{\partial S}{\partial c_n} = \frac{\partial S}{\partial c_n} \sum_{i=1}^{k} \sum_{x_j \in C_n} (c_n - x_j)^2 = 0 \tag{3}$$

If the number of objects in the nth cluster is m, then the following solution comes from Eq. 3

$$2 \cdot \sum_{j=1}^{m} c_n = 2 \cdot \sum_{j=1}^{m} x_j \Rightarrow m \cdot c_n = \sum_{j=1}^{m} x_j \Rightarrow c_n = \frac{\sum_{j=1}^{m} x_j}{m} \tag{4}$$

The conclusion from formula 4 is that the centroids reach optimal coordinates when they are equal to the average value of the coordinates of the objects in the cluster.

In the K-means algorithm, objects are divided into clusters of equal size compared to hierarchical clustering. This is especially evident when objects are located at the same density in the coordinate system.

In Fig. 3a, the objects are divided into 3 clusters, and the size of each cluster is almost the same. But if these objects are divided into 2 or 4 clusters, the sizes of the clusters will be very different from each other. In Fig. 3b, it can be seen that two clusters are large in size, and the other two are small in size. Figure 3c shows a perfect set of objects, i.e., a set of objects that are relatively evenly distributed over space. When this set is divided into an arbitrary number of clusters, clusters of almost the same size are formed.

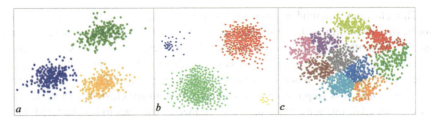

Fig. 3. The result of object clustering using K-means

3.3 Bisecting K-means Clustering

The Bisecting K-means algorithm is an improved interpretation of the K-means algorithm, which is more efficient than K-means in splitting into clusters of equal size. Because the main idea in Bisecting K-means is to create a new cluster by dividing a cluster with a large size or a large sum of squared errors [18]. This process is shown in Fig. 4 can be seen.

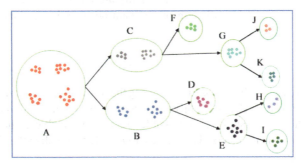

Fig. 4. Clustering of objects using the Bisecting K-means algorithm

In Fig. 4, the first set A is given. Collection A is divided into clusters B and C based on the Bisecting K-means algorithm. When the number of clusters is equal to 3, cluster B is divided into two since there are more objects in cluster B than in cluster C. Then three clusters C, D, and E are formed. When the number of clusters is equal to 4, cluster C is divided into two since cluster C has more objects than clusters D and E. After that and 4 clusters named D, E, F, G are formed. As the number of clusters increases, the same processes are continued.

3.4 DBSCAN Clustering

DBSCAN is a density-based clustering algorithm that combines objects that are adjacent to each other at a distance of *max_dist* (maximum distance) into one cluster [19, 20]. The second important parameter of DBSCAN is *min_points* (minimum number of objects). It means the minimum number of objects forming a cluster. That is, if the number of objects located at *max_dist* distance is less than *min_points*, these objects do not belong to any cluster and they are considered as noise.

442   A. Rashidov et al.

This clustering method is more powerful than other clustering methods. In other words, this algorithm is the most effective approach to identifying hidden meanings. But this algorithm is ineffective in solving the problem of equal clustering. Because in the DBSCAN clustering method, it is impossible to specify the number of clusters in advance by the user. When the number of clusters changes depending on *max_dist* and *min_points*, it is impossible to say exactly how many clusters will be formed in advance. In addition, some information contained in the data to be clustered may not belong to any clusters, that is, they may be considered as noise. Taking into account these shortcomings, it can be said that it is impossible to use the DBSCAN algorithm in solving the problem of equal-sized clustering of data streams.

4   Clustering Algorithm Based on Square of Vector Modulus

In this research, a new clustering algorithm is proposed in order to solve the problem of distributing the data flow into clusters of equal size. This clustering algorithm is based on the construction of vectors directed from the coordinate origin to the feature coordinates of objects. It is known that any n-dimensional vector $\vec{a}$ with coordinates $(x_1; x_2; x_3; \ldots; x_n)$ has a length of a certain positive number and it is found by formula 5.

$$|\vec{a}| = \sqrt{x_1^2 + x_2^2 + x_3^2 + \ldots + x_n^2} \tag{5}$$

This proposed algorithm is based on dividing the squares of vector lengths into groups based on certain rules. After the objects are converted to numerical values, clustering based on length values makes it even possible to separate two objects that are very close to each other into other clusters. This makes it possible to equalize the sizes of clusters during clustering.

The proposed clustering algorithm based on the square of the vector modulus is implemented by performing the following steps:

Step 1. Read the data (objects);
Step 2. Transfer the specific characteristics of each data to vector coordinates in digital form;
Step 3. Calculate the square of the modules of the resulting vectors;
Step 4. Sort the squares of the modules of the vectors in ascending order;
Step 5. Write the squares of the modules of the vectors sorted in ascending order into the array A;
Step 6. Input k which number of clusters;
Step 7. Divide array A into k equal parts;
Step 8. Determine the boundaries of the pieces;
Step 9. Find the boundaries of the cluster through the boundaries of the slices;
Step 10. Separate into clusters the data using defined thresholds.

As a result of the execution of this algorithm, the elements of k clusters separated have their own geometric meaning. The elements of the cluster 1 consist of the points of a circle with a radius R, the elements of clusters 2, 3,... k − 1 are the points of the rings

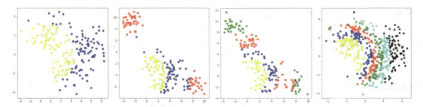

Fig. 5. Graphical representation of clustering based on the square of the vector modulus

with thicknesses d_2, d_3, ... d_{k-1}, and the elements of the cluster k are the points lying outside the circle with the radius R_1. Here $R_1 = R + d_1 + d_2 + ... + d_{k-1}$ equality will be appropriate. The results of the proposed algorithm can be seen in Fig. 5.

It can be seen from Fig. 5 that in the clustering based on the square of the vector module, the points that are very close to each other also belong to two other clusters. On the contrary, the points located far from each other belong to one cluster. This feature is another indicator that distinguishes the proposed algorithm from other clustering algorithms. But the main feature of this algorithm is that the sizes of the clusters formed as a result of clustering are almost equal.

5 Results

In order to check the ability of clustering algorithms to divide into clusters of equal size, 3 different data sets containing 1000 elements were used. Each data set was divided into 2, 3, 4 clusters. The elements within these clusters were enumerated and the results compared.

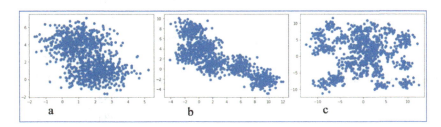

Fig. 6. 3 different data sets used in the experiments

3 types of data shown in Fig. 6 were used in the experiments. The data used in the experiment was generated by the random function and does not have any special meaning. But these data were created using the make_blobs data generation function of the sklearn.datasets library of the python programming language, and data were generated around 2, 6, and 30 central points, respectively.

A. Rashidov et al.

Table 1. The results of the distribution of a, b, and c data into 2 groups using clustering

The name of the algorithm	Clusters №	Clusters size		
		a	b	c
Hierarchical clustering	1	2	2	999
	2	998	998	1
K-means clustering	1	521	364	436
	2	479	636	564
Bisecting K-means clustering	1	508	364	596
	2	492	636	404
The proposed clustering	1	500	500	500
	2	500	500	500

Table 2. The results of the distribution of a, b, and c data into 3 groups using clustering

The name of the algorithm	Clusters №	Clusters size		
		a	b	c
Hierarchical clustering	1	2	2	997
	2	997	4	2
	3	1	994	1
K-means clustering	1	272	430	380
	2	257	329	339
	3	471	241	281
Bisecting K-means clustering	1	250	213	468
	2	259	423	211
	3	491	364	321
The proposed clustering	1	333	333	333
	2	333	333	333
	3	334	334	334

The Same Size Distribution of Data Based on Unsupervised Clustering   445

Table 3. The results of the distribution of a, b, and c data into 4 groups using clustering

The name of the algorithm	Clusters №	Clusters size		
		a	b	c
Hierarchical clustering	1	2	2	997
	2	6	4	1
	3	991	993	1
	4	1	1	1
K-means clustering	1	245	264	215
	2	246	232	146
	3	269	186	223
	4	240	318	416
Bisecting K-means clustering	1	216	364	365
	2	275	272	233
	3	250	151	202
	4	259	213	200
The proposed clustering	1	250	250	250
	2	250	250	250
	3	250	250	250
	4	250	250	250

As can be concluded from Tables 1, 2 and 3, the proposed vector length-based clustering algorithm recorded a good result in terms of the characteristic of equal-sized clustering of most data for 3 types of data. This situation can be seen in all 2, 3, and 4 clusters. The worst indicator belongs to the Hierarchical clustering algorithm, in all cases, the sizes of the clusters were very different from each other. In all cases, a large part of the data belonged to a certain cluster, and a small part belonged to other clusters. K-means and Bisecting K-means algorithms performed relatively well as expected, but in no case did they perform better than the proposed approach.

The result of the proposed algorithm for solving the given problem can be seen in Fig. 7.

Figure 7 shows the distribution of the data a, b, and c into 4 clusters based on the proposed algorithm, and the 4 clusters are shown in 4 different colors.

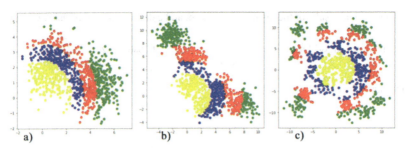

Fig. 7. The result of the distribution of the data set a, b, and c into 4 clusters based on the proposed algorithm

6 Conclusion

In conclusion, this research work was devoted to studying the possibility of clustering algorithms to divide data into clusters of equal size. For this purpose, this possibility of existing unsupervised clustering algorithms Hierarchical, K-means, Bisecting K-means and DBSCAN clustering algorithms was studied. The results of the research showed that existing unsupervised clustering algorithms have strong capabilities in identifying hidden content in data, but limited capabilities in dividing clusters of equal size. In order to overcome these shortcomings, a clustering algorithm based on the square of the length of the proposed vector was proposed and the steps of the algorithm were presented. The research results showed that this algorithm is very effective in solving the given problem.

References

1. Maheshwari, A.K.: Business Intelligence and Data Mining. Business Expert Press, LLC, 222 East 46th Street, New York, NY 10017, p. 162 (2015)
2. Akhatov, A., Nazarov, F., Rashidov, A.: Mechanisms of information reliability in big data and blockchain technologies. In: ICISCT 2021: Applications, Trends and Opportunities, 3-5.11.2021 (2021). https://doi.org/10.1109/ICISCT52966.2021.9670052
3. Akhatov, A., Nazarov, F., Rashidov, A.: Increasing data reliability by using bigdata parallelization mechanisms. In: ICISCT 2021: Applications, Trends and Opportunities, 3-5.11.2021 (2021). https://doi.org/10.1109/ICISCT52966.2021.9670387
4. Jumanov, I., Djumanov, O., Xolmonov, S.: Mechanisms of image recovery optimization in the system for recognition and classification of micro-objects. AIP Conf. Proc. **2686**, 020009 (2022). https://doi.org/10.1063/5.0113052
5. Tan, P.-N., Steinbach, M., Karpatne, A., Kumar, V.: Introduction to Data Mining, 2nd edn. Pearson Education, New York, NY (2019)
6. Akhatov, A., Sabharwal, M., Nazarov, F., Rashidov, A.: Application of cryptographic methods to blockchain technology to increase data reliability. In: 2nd ICACITE (2022). https://doi.org/10.1109/ICACITE53722.2022.9823674
7. Jumanov, I.I., Xolmonov, S.M.: Optimization of identification of non-stationary objects due to information properties and features of models. IOP Conf. Ser. Mater. Sci. Eng. (2021). https://doi.org/10.1088/1757-899X/1047/1/012064
8. Akhatov, A., Renavikar, A., Rashidov, A., Nazarov, F.: Development of the Big Data processing architecture based on distributed computing systems. Uzbek Journal of the Problems of Informatics and Energetics (1), 71–79 (2022)

9. Rashidov, A., Akhatov, A.: Big data and its application in various fields. Descendants of Muham-mad Al-Khwarizmi **4**(18), 135–144 (2021)
10. Rashidov, A., Akhatov, A., Renavikar, A.: Optimization of the database structure based on Machine Learning algorithms in case of increased data flow. In: ICABCS (2023)
11. Höppner, F., Klawonn, F.: Clustering with size constraints. In: Jain, L.C., Sato-Ilic, M., Virvou, M., Tsihrintzis, G.A., Balas, V.E., Abeynayake, C. (eds.) Computational Intelligence Paradigms, pp. 167–180. Springer Berlin Heidelberg, Berlin, Heidelberg (2008). https://doi.org/10.1007/978-3-540-79474-5_8
12. Ganganath, N., Cheng, C.T., Chi, K.T.: Data clustering with cluster size constraints using a modified k-means algorithm. Institute of Electrical and Electronics Engineers (IEEE) (2016). https://doi.org/10.1109/CyberC.2014.36
13. Kumari, P.L., Jeeva, M., Satyanarayana, C.: A novel hierarchical document clustering framework on large TREC biomedical documents. Int. J. Inf. Technol. Comput. Sci. (IJITCS) **14**(3), 16–22 (2022). https://doi.org/10.5815/ijitcs
14. Zhao, Y., Karypis, G.: Hierarchical clustering algorithms for document datasets. Data Min. Knowl. Disc. (2005). https://doi.org/10.1007/s10618-005-0361-3
15. Sinaga, K.P., Yang, M.-S.: Unsupervised K-means clustering algorithm. IEEE Access **8**, 80716–80727 (2020). https://doi.org/10.1109/ACCESS.2020.2988796
16. Jumanov, I.I., Safarov, R.A., Xurramov, L.Y.: Optimization of micro-object identification based on detection and correction of distorted image points. AIP Conf. Proc. **2402**, 070041 (2021). https://doi.org/10.1063/5.0074018
17. Fahim, A.: Finding the number of clusters in data and better initial centers for K-means algorithm. Int. J. Intell. Syst. Appl. (IJISA) **12**, 1–20 (2020). https://doi.org/10.5815/ijisa
18. Abuaiadah, D.: Using bisect K-means clustering technique in the analysis of Arabic documents. ACM Trans. Asian Low-Resource Lang. Inf. Process. **15**, 1–13 (2016). https://doi.org/10.1145/2812809
19. Maithri, C., Chandramouli, H.: Parallel DBSCAN clustering algorithm using hadoop mapreduce framework for spatial data. Int. J. Inf. Technol. Comput. Sci. (IJITCS) (2022). https://doi.org/10.5815/ijitcs
20. Laskhmaiah, K., Murali Krishna, S., Eswara Reddy, B.: An optimized K-means with density and distance-based clustering algorithm for multidimensional spatial databases. Int. J. Comput. Netw. Inf. Secur. **13**(6), 70–82 (2021). https://doi.org/10.5815/ijcnis.2021.06.06

Investigation of Microclimate Parameters in the Industrial Environments

Solomiya Liaskovska[1]($\boxtimes$), Olena Gumen[2], Yevgen Martyn[3], and Vasyl Zhelykh[4]

[1] Department of Artificial Intelligence, Lviv Polytechnic National University, Kniazia Romana Street, 5, Lviv 79905, Ukraine
solomiam@gmail.com
[2] Department of Descriptive Geometry, Engineering and Computer Graphics Peremohy pr, National Technical University of Ukraine "Igor Sikorsky Kyiv Polytechnic Institute", 37, Kyiv 03056, Ukraine
[3] Department of Project Management, Information Technologies and Telecommunications, Lviv State University of Life Safety, Kleparivska Street, 35, Lviv 79007, Ukraine
[4] Department of Heat and Gas Supply and Ventilation St. Bandery, Lviv Polytechnic National University, 12, Lviv-13 79013, Ukraine

Abstract. Researches on the organization and implementation of the project management of selection processes and the development of adequate models of interaction of many parameters of the microclimate of the technological process of raising the poultry meat breeds in industrial premises have been carried out. Actuality of the researches has been confirmed, the analysis and a choice of interaction of parameters research methods are proved. It is shown that the use of physical modeling with the involvement of the corresponding experimental equipment and the use of the experimental results obtained for geometric modeling on the basis of applied multidimensional geometry are effective. The significance of the study is in the established features of the nature of the heat exchange of air in the premises in view of the effectiveness of growing products and in various types of farming. The method of measuring the temperature in the room chosen in the process of geometric modeling allows to cover with lines of the same temperature all points of the living space of the production premises in which the bird is located.The proposed method of research can be practically adapted for the processing of the results of studies on the microclimate parameters of agro-industrial objects of various intended uses.

Keywords: Industrial premises · microclimate · parameters · physical and geometric modeling

1 Introduction

The effective implementation of technological processes of production in the industrial and agrarian sectors has big impact in development of plant. With regard to the food production, in particular, poultry, an important additional condition is the observance of the comfortable microclimate in the zone of its cultivation. For this, there are industrial

© The Author(s), under exclusive license to Springer Nature Switzerland AG 2023
Z. Hu et al. (Eds.): ICAILE 2023, LNDECT 180, pp. 448–457, 2023.
https://doi.org/10.1007/978-3-031-36115-9_41

buildings that contain poultry. The main elements of poultry keeping are such components of comfort as the air temperature and its purity. Heaters and ventilation systems are used to provide regulatory requirements. Their effective work is possible when rational modes of operation of the corresponding technological equipment are used. The development of appropriate approaches to the creation of industrial heating and ventilation systems for industrial buildings with the massive holding of poultry, the choice of means and the study of the parameters of the process constitute one of the promising directions of development of the agricultural sector for the holding of meat breed poultry.

Scientific research, however, is scattered on a particular aspect of the study of the comfortable stay of the poultry in the poultry house [1–14]. A free scientific niche is the question of a reasonable choice based on the principles of project management of selection and development of tools for processing graphical experimental data with the involvement of the apparatus of geometric modeling [15, 16].

The selection of previously unsettled parts of the general problem points to a limited number of scientific publications in its separate direction concerning the provision of the comfortable microclimate in industrial premises with the massive holding of poultry. In addition, the recent studies and publications have been analyzed. They testify to the absence of a unified methodological approach to the organization and analysis of experimental studies, taking into account the project approach.

1.1 Defining of the Peculiarities of Resource Support for the Technological Process of Poultry Breeding

In many countries, a project to grow agricultural products under hothouse conditions is successfully implemented and improved. This applies to both vegetable and livestock breeding. In particular, projects for the implementation of appropriate technological processes are being carried out for the breeding of meat breed poultry.

In the process of the project initiating, fundamental decisions are taken on the creation and arrangement of an industrial building (poultry house). Project planning defines the purpose of the project and the peculiarities of resource support for the technological process of poultry breeding. At this stage of the project management, it is important to develop appropriate measures and to take measures to ensure proper conditions for holding of poultry, which is one of the important moments in the implementation of the technological process of poultry breeding.

The aim of the study is to find out on the basis of the project approach the main components of the project on the research of the microclimate of industrial premises, to substantiate, develop and apply means of geometric analysis of the graphic dependencies of its parameters.

The process realization of research has been carried out on the example of determining the amount of heat utilized from the premises, the work of local tidal and exhaust ventilation. In accordance with the research objectives, a universal laboratory installation that can serve as a tool for conducting research is installed (Fig. 1).

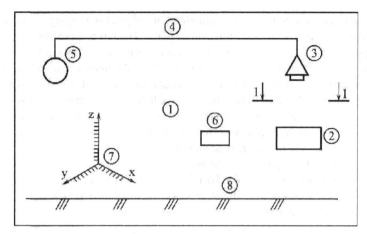

Fig. 1. The graphic visualization of the laboratory installation, 1—room; 2—infrared heater; 3—exhaust outlet; 4—air duct; 5—ventilator; 6—thermometer; 7—coordinate grid; 8—floor

The laboratory installation, located in the room 1, contains an infrared heater 2 over which there is an exhaust outlet 3 connected by an air duct 4 with a ventilator 5. To measure the temperature at an arbitrary point of space 1, a thermometer 6 with a coordinate grid 7 installed on the floor 8 is provided. In the scheme (Fig. 1) there are no commuting devices, conventionally not depicted, needed for the formation of the given equipment the scheme of intended use.

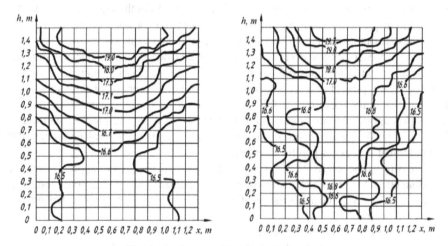

Fig. 2. Isotherms of the cutting plane 1–1

The work of local exhaust ventilation the temperature fields in the room space 1 1—room; 2—infrared heater; 3—exhaust outlet; 4—air duct; 5—ventilator; 6—thermometer; 7—coordinate grid; 8—floor. The temperature of the air we can measure with the thermometer 6 and use the coordinate meter 7. When studying the work of local exhaust ventilation the temperature fields in the room space 1 are compared with the ventilator 5 switched off and on. The results of measurements were given by the isotherms in the cutting plane with the measurements of the height of the room h and the coordinate x of the room 1 with a constant value of the coordinate y, that is the width of the room (Fig. 2 a, b).

2 The Comparative Analysis of the Insulation Isotherms Experimental Results

This density is characterized by Δh (the maximum value of exceeding the height of the isotherm) at a certain interval Δx (Fig. 2 a) and (Fig. 2 b). The interval $\Delta x = 0.3\ 0.8$ m is common to the isotherms.

Table 1 shows Δh in the section plane 1–1.

2.1 Solving Model Based on Graphic Dependencies

The values of Δh (some isotherms) plotted in the secant plane 1–1 are given in Table 1.

Table 1. The value of the slope of isotherms Δh

Isotherm's meaning	16.5	16.6	17	18	19
without a fan Δh, m	0.45	0.11	0.1	0.09	0.08
with a fan Δh, m	> 1.4	0.3	0.1	0.11	0.15

Graphic dependencies, constructed according to the data of Table 1, give an opportunity to determine the effectiveness of exhaust ventilation (Fig. 3).

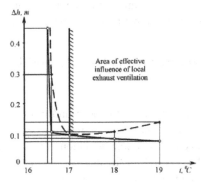

Fig. 3. Determination of the field of effective influence of the work of local exhaust ventilation: ○────── —the work of local exhaust ventilation; ●─ ─ ─ —without with the work of local exhaust ventilation

From analysis of graphic dependencies $\Delta h = f(t\ °C)$ we have that the area of the effective influence of the work of the local exhaust ventilation starts at a higher altitude then $h > 1.0$m (isotherm of 17°C).

Almost each of the isotherms has a minimum. In the case of local exhaust ventilation, there is a slight change in the height h of the location of the minimum height value of each isotherm (Table 2), in addition to the isotherm 17°C, as well as their condensation in the upper part of the room.

Height interval Δhb of the location of isotherms in the upper part of the room:

- without the work of local exhaust ventilation for isotherms of interval17∘C… 19°C value $\Delta hb = 0.42$m;
- with the work of local exhaust ventilation for the same isotherm interval of 17∘C… 19°C value $\Delta hb = 0.28$m.

Table 2. Minimum height values of the isotherm location

Isotherm	16.5	16.6	17	18	19
Δ h, m without a fan	0	0.51	0.83	1.18	1.25
Δ h, m with a fan	0	0	1.08	1.19	1.3

Height interval Δhb of the location of isotherms in the lower part of the room:

- without the work of local exhaust ventilation for isotherms of interval 0∘C 16.7°C value $\Delta hb = 0.68$m;
- with the work of local exhaust ventilation for the same isotherm interval of 0∘C 16.8°C value $\Delta hb = 0.23$m.

Thus, there is the decrease in the height interval in the upper part of the room in relation to the lower part of the room without the work of local exhaust ventilation 0.6 times and the increase in the height interval in the upper part of the room in relation to the lower part of the room with the work of local exhaust ventilation 1.2 times.

3 Processing Results by Geometric Method Using Complex Drawing

The research methodology consists in the deep and more thorough use of geometric modeling tools in the process of analyzing the results of experimental research through an effective project approach to the implementation of the described stages of research. In the process of scientific research of the thermal field using experimentally obtained isotherms (Fig. 2), it becomes necessary to use those isotherms that are not on the graph. It is possible to construct the necessary isotherms, if we consider such lines as the result of intersection of the surface of the temperature field with the horizontal cutting plane with a given required temperature value.

Let's show the sequence of construction of an isotherm at a temperature value, for example, 18.5°C in the case of local exhaust ventilation system work.

We isolate two neighboring isotherms, in this case the isotherms are with temperature values 18.0°C and 19.0°C (Fig. 4).

We construct a complex drawing of the section of the temperature field with values of isotherms 18.0°C and 19.0°C (Fig. 5).

Typically, the required lines of level of surface of the temperature field are constructed by passing the cutting plane 18.5 between the planes of level with traces 18 and 19 (Fig. 5).

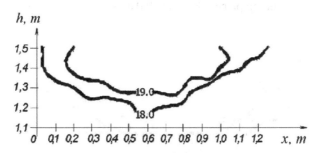

Fig. 4. Determination of the field of effective influence of the work of local exhaust ventilation

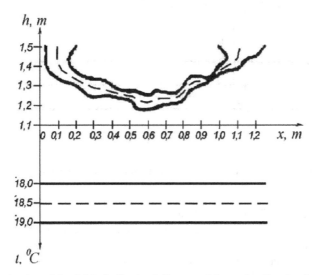

Fig. 5. Determination of the field of effective influence of the work of local exhaust ventilation

Taking into account that the surface of the temperature field is smooth, the line of intersection of the temperature field with the cutting plane of level 18.5 is equidistant to the lines, isotherms, with traces 18 and 19 (in Fig. 5 it is indicated by a dashed line).

The proposed complex drawing allows determining the temperature of the air at an arbitrary point on the surface of the thermal field.

Let's construct projections of a point of the temperature field with the value of temperature, for example, $t = 18.7°C$ at $x = 0.3$m (Fig. 6).

The construction is carried out in the following sequence:

1. First we conduct a trace of the cutting plane at 18.7 (in Fig. 6 it is indicated by a dotted-dashed line).
2. Then we conduct a projection link line at $x = 0.3$m perpendicular to the axis Ox. At the intersection of both lines we obtain point T1, which is a horizontal projection of the temperature value 18.7°C of surface of the temperature field.
3. Through the point T1 we pass a section a1 of an arbitrary straight line which intersects the traces of the cutting planes 18.5 and 19.0 at points 11 and 21 respectively.

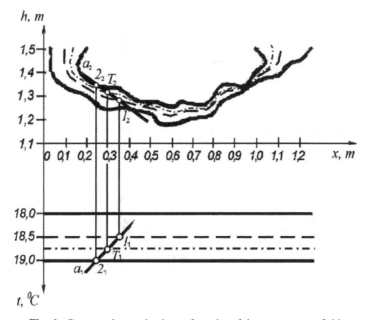

Fig. 6. Constructing projections of a point of the temperature field

4. Their projections 12 and 22 are found on isotherms 18.5 and 19.0, respectively. By connecting the points 12 and 22, we obtain the projection a2 of the section of an arbitrary line *a*.
5. Point T2 of the intersection of the projection lines and section a2 is the required point of the temperature field surface with the temperature value of 18.7°C.

The complex drawing in Fig. 5 gives two projections of the temperature field in Oxht0 space. According to such projections, a 3D model of the temperature field, for example, in the working range 18… 19°C of temperature changes in the room can be constructed by means of graphic information technologies (Fig. 7).

A visual computer model of temperature distribution in the room is also constructed when the coordinate is changed (Fig. 8).

The researches of the air temperature at change of coordinate value show constancy of the pattern of temperature distribution in cutting plane 1–1.

Effective project implementation of the technological process of raising poultry in the direction of conducting comprehensive scientific research also involves identifying the features of the influence of tidal ventilation on the microclimate in the production premises.

In the process of conducting research on the influence of tidal ventilation, the ventilator 5 of the laboratory installation provided air from the air channel 4 and the exhaust outlet 3 to the room 1. The overall picture of the location of the isotherms is the same: all lines have a minimum located within $3 < x < 4$ m. As with exhaust ventilation, there is an extension of all isotherms without exception. An increase in the speed of supply of inflow air causes a decrease in the height of the maximum value of the temperature in the room.

The trends in the influence of the exhaust ventilation on the nature of the distribution of heat in the room will be considered by the example of the isotherm displacement, in

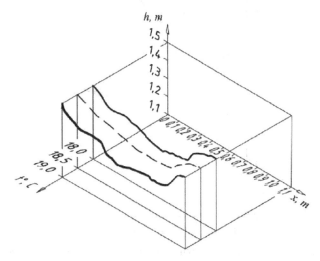

Fig. 7. Constructing projections of a point of the temperature field

When Δh exceeds the height of the 24 °C isotherm throughout the gap $\Delta x = 0...$ 7 m is equal *to 0.18*m for the speed of the supply air flow *v = 0.1 m/s; 0.45*m for *v = 0.2 m/s; 0.46 m* for *v = 0.35*m/s particular *24 °C* (Fig. 9).

With an increase in the velocity of the flow of tidal air from 0.1 m/s to 0.2 m/s, the isotherm of 24 °C dropped from a height h = 1.5 m to a height of h = 0.9 m, that is 60%. With an increase in the velocity v of the flow of tidal air from 0.2 m/s to 0.35 m/s it dropped to a height of 0.55 m. Exceeding Δh height of isotherm 24 °C across the entire gap $\Delta x = 0...$ 7 m is 0.18 m for the velocity of the flow of tidal air v = 0.1 m/s; 0.45 m for v = 0.2 m/s; 0.46 m for v = 0.35 m/s.

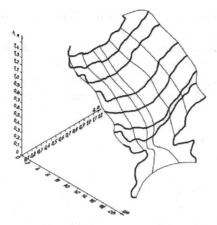

Fig. 8. 3D model of temperature distribution along the axis

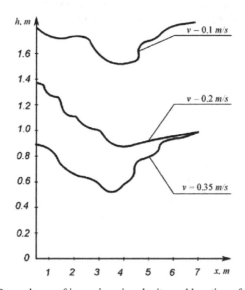

Fig. 9. Dependence of incoming air velocity and location of isotherms

4 Summary and Conclusion

In this paper, we analyzed that the involvement of that the involvement of geometric modeling allows to effectively study the parameters of the microclimate of industrial buildings. An example of the technological process of poultry breeding is considered.

A geometric method of processing experimental parameters has been implemented. Quantities that are difficult or impossible to determine under experimental conditions were also investigated.

The obtained scientific results demonstrate the use geometric modeling using multidimensional geometry. This method will allow research of the agro-industrial complex for various purposes.

References

1. Made, R.I., et al.: Experimental characterization and modeling of the mechanical properties of Cu–Cu thermocompression bonds for three-dimensional integrated circuits. Acta Mater. **60**(2), 578–587 (2012)
2. Zhou, H., Li, Q., Lin, T.: Power system disturbance and operation identification based on WAMS electric power. Automation Equip. **31**(02):,7-11 (2011)
3. Liu, X., Yang, M.: Simultaneous curve registration and clustering for functional data. Comput. Stat. Data Analysis **53**(4), 1361-1376 (2019)
4. Rama, A., Lekyasri, N., Rajani, K.: PID control design for second order systems. IJEM **9**(4), 45–56 (2019)
5. Qin, X., Bi, T., Yang, Q.: A new method for hybrid nonlinear state estimation with PMU. Automation of Power Syst. **31**(4), 28–32 (2017, in Chinese)
6. Pijush, D., Madhurima, M., Asok, K.: Parametric optimization of liquid flow process by ANOVA optimized DE, PSO & GA algorithms. IJEM **11**(5), 14–24 (2021)
7. Karimui, R.Y.: A new approach to measure the fractal dimension of a trajectory in the high-dimensional phase space. Solitons & Fractals **151**, 111–239 (2021)
8. Zarei, M., et al.: Employing phase trajectory length concept as performance index in linear power oscillation damping controllers. IEPE **98**, 442–454 (2018)
9. Syed, K., Ali, S., Waqas, M.: Smart home automation using IOT and its low cost implementation. IJEM **10**(5), 28–37 (2019)
10. Wang, L., Sharkh, S., Chipperfield, A.: Optimal decentralized coordination of electric vehicles and renewable generators. IJEPES **98**, 474–487 (2018)
11. Ali, S., Xie, Y.: The impact of industry 4.0 implementation on organizational behavior and corporate culture: the case of pakistan's retail industry. IJEM **10**(6), 20–31 (2020)
12. Zarei, M., et al.: Oscillation damping of nonlinear control systems based on the phase trajectory length concept: an experimental case study on a cable-driven parallel robot. Mech. Mach. Theory **126**, 377–396 (2018)
13. Yu, Q., Wang, X., You, J., et al.: Equality constraints two-step state estimation model based on phasor measurements. Power System Technol. **31**(10), 84-88 (2007, in Chinese)
14. Savelyev, A.V., Stepanyan, I.V.: Spiral flows at the cardiovascular system as the experimental base of new cardiac-gadgets design. IJEM **8**(6), 1–12 (2018)
15. Hovorushchenko, T., et al.: Development of an intelligent agent for analysis of nonfunctional characteristics in specifications of software requirements. Eastern-European J. Enterprise Technol. **1**(2), 6–17 (2019)
16. Liaskovska, S., Izonin, I., Martyn, Y.: Investigation of anomalous situations in the machine-building industry using phase trajectories method. ISEM **463**, 49–59 (2021)

Detection of Defects in PCB Images by Separation and Intensity Measurement of Chains on the Board

Roman Melnyk and Ruslan Tushnytskyy[(✉)]

Lviv Polytechnic National University, Lviv 79013, Ukraine
roman.a.melnyk@lpnu.ua, ruslan.tushnytskyy@gmail.com

Abstract. The subtraction requires constant comparison with the input images. The developed approach automatically demonstrates defects and chains containing them. Algorithms for K-means clustering, flood-filling the image traces with color, thinning of binary images are used to find the coordinates of the contacts and defects on the images of the printed circuit board. The clustering algorithm is used to reduce the number of colors, obtain uniform colors, and highlight the objects of the image. The thinning algorithm is used to construct skeletons and find special points indicating the contact pixels. Pixels of contacts are accepted as start positions to mark pixels of traces in the test image. The flood-fill algorithm is used to mark contacts, tracks, and background. The developed approach detects PCB defects of connectivity, and internal defects of traces.

Keywords: Printed circuit board · defect detection · chain · contact · pixel · switch · ending · flood-fill · thinning · K-means clustering

1 Instruction

There are many modern approaches for PCB defects detection which are in the published articles. They conditionally could be grouped into three classes. The first one unites works describing techniques for fault diagnostics, and artificial networks for image processing. One representant of this group is the works [1] in which the PCB surface inspection is realized by a powerful Deep Learning tool. Works of the second class are based on comparing a standard PCB image with a PCB image to be inspected. Approaches use different subtraction algorithms. For example, the work [2] consider a normalized histogram to select components of the PCB image and then separates defects by the subtraction algorithm. As a result defects are classified. The paper [3] presents an approach for defects detection based on the connected table of a reference image. In many cases this table is unknown. And coordinates of contacts are to be determined. In the publications [5–8] to compare two PCB images a subtraction algorithm to detect the defected regions was used. In the works [2, 9, 10] different methods of PCB defect detection are presented.

In this article detection of contacts coordinates in a PCB image is realized by segmentation with the K-means clustering algorithm, flood-filling of PCB chains, and thinning

© The Author(s), under exclusive license to Springer Nature Switzerland AG 2023
Z. Hu et al. (Eds.): ICAILE 2023, LNDECT 180, pp. 458–467, 2023.
https://doi.org/10.1007/978-3-031-36115-9_42

algorithm to find ending and switches in skeletons. The found pixels of contacts are input data for checking procedures and algorithms to find defects of connectivity, errors in contacts and traces and extra material on a PCB background.

2 PCB Image Preprocessing Algorithms

The main part of the defects of printed circuit boards is caused by incorrect shape and filling of contacts and traces.

Defects in the image of a printed circuit board can be conventionally grouped into three classes: the contact surface, the surface of traces, and metal residues in the background space.

Defects are associated with the wrong amount of conductive material and the wrong shape of contacts and traces. PCB images show defects as background colors or traces compared to a reference image. In one case, the defect is marked by background pixels. Otherwise, the defect is indicated by additional pixels of the trace (Fig. 1).

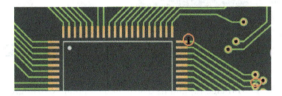

Fig. 1. PCB image with marked two defects: open and short

The number of pixel defects is very small and it is very difficult to find a tool to measure it. A very sensitive approach is used to illustrate this fact. The method discussed below constructs two images: $DCH(r)$ and $DCH(m)$. They are distributed cumulative histograms of the reference and manufactured PCB image. Then a special logical subtraction operation is applied to them: $DCH(r - m) = DCH(r) - DCH(m)$.

This operation is called special because it is applied to grayscale images and different pixels are marked with red or blue. The resulting image is shown in Fig. 2.

The $DCH(r - m)$ image contains two bands indicating places where $DCH(r)$ and $DCH(m)$ differ by 3–5 intensity units. The red bar indicates increased background pixels and the blue bar indicates additional trace pixels. Their OX coordinates correspond to the OX coordinates of the defects on the printed circuit board image. Thus, it is possible to measure the integral intensity of defects and determine the OX coordinates. Two more distributed cumulative histograms for the OY axis are required to detect the full defect positions.

This approach does not allow us to specify the type of defects, recording only irregularities in the distribution of pixels.

The previous image of a PCB is taken for discussion about used preprocessing algorithms. Their goal is to prepare the PCB image to be processed for reliable results. And then to apply algorithms for the detection of defects and measurement their intensity.

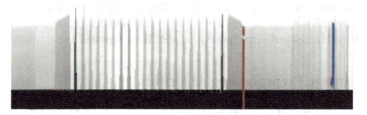

Fig. 2. Difference of two distributed cumulative histograms

The reference image is shown in Fig. 2a. It is the image from Fig. 1, but without defects. Also, for illustration in Fig. 3b the distributed cumulative histogram of the input image is shown. It takes part in the special operation of subtraction.

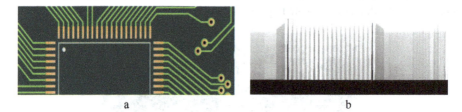

Fig. 3. PCB image (a) and its distributed cumulative histogram (b)

2.1 Distributed Cumulative Histogram

To build the image of a distributed cumulative histogram two sets of ordinary histograms of numbers N for columns and M for rows are calculated:

$$V_i(c) = \{V_{ij}(c)\}, j = 0, 255, i = 1, N \\ V_j(r) = \{V_{ji}(r)\}, i = 0, 255, j = 1, M \quad (1)$$

Then two distributed cumulative histograms as sets of frequencies sums:

$$V_j(cc) = \left\{\sum_{l=0}^{i} V_{li}(c), i = 0, 255\right\}, j = 1, N$$

$$V_j(cr) = \left\{\sum_{l=0}^{i} V_{l_i}(r), i = 0, 255\right\}, j = 1, M \quad (2)$$

where $V_i(cr)$, $V_j(cc)$ – are cumulative histograms in rows and columns, $V_{ij}(c)$, $V_{li}(r)$ – are an intensity frequencies in column and row, N, M – are numbers of columns and rows. The last equations are the mathematical models of distributed cumulative histograms of an image.

2.2 K-Means Clustering Slgorithm

The K-Means clustering algorithm approximates the PCB image by K segments of pixels with the same intensity. This algorithm assigns pixels into different clusters to minimize a sum of distances between the centroids and pixels within a cluster.

The criterion function is as follows:

$$S = \sum_{i=1}^{m} \sum_{k=0}^{K} w_{ik} |I - \bar{I}_k|, \; S = \sum_{i=1}^{m} \sum_{k=0}^{K} w_{ik} [I - \bar{I}_k]^2 \tag{3}$$

where $w_{ik} = 1$ if a pixel of intensity I_{ir} belongs to the cluster k; otherwise, $w_{ik} = 0$; $\bar{I}_k$ is the intensity of the i-th centroid.

The input image has three colors. So, the clustering is processed for $K = 3$. The clustered image is shown in Fig. 4a. Its $DCH(r)$ image is shown in Fig. 4b.

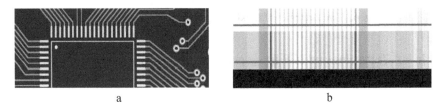

Fig. 4. Clustered PCB image (a) and its distributed cumulative histogram (b)

The $DCH(r)$ image distinctly shows three groups of pixels corresponding to contacts, traces, and background. Segmenting the PCB image by intensity levels marked with red in Fig. 4.b allows us to separate contacts and traces. After darkening for better visibility, they are shown in Fig. 5a and Fig. 5b.

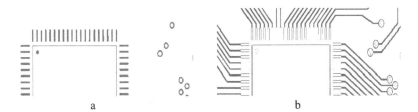

Fig. 5. Segmented contacts (a) and traces (b) in the clustered PCB image

Separation of contacts and traces is not the best solution because many connectivity defects arose in places of their touching. Often these defects are very thin and are of poor visibility.

2.3 Flood-Fill Algorithm

The developed flood-fill algorithm realizes four special functionalities: 1–2) a start pixel can be appointed by coordinates or by the intensity value, 3–4) colors can be distributed within a closed area or among all pixels with a target color. The last case is demonstrated in Fig. 6 by marking pixels of one level (a), or of a few levels.

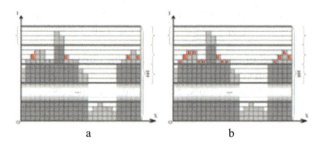

Fig. 6. Two strategies of flood-filling: one level (a) several levels (b)

In Fig. 7a the clustered PCB image presents some blue traces and red contacts. Then these elements are segmented and shown in Fig. 7b.

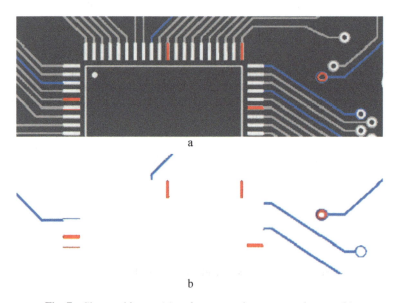

Fig. 7. Clustered image (a) and segmented contacts and traces (b)

These segmented elements are ready for further processing by algorithms of defects detection.

2.4 Thinning Algorithm

The Hilditch's Thinning Algorithm [4] works with binary (black and white) images. When applying the flood-fill algorithm to the clustered PCB image (white for background, black for contacts and traces) the binary PCB image in Fig. 8a is ready for further processing.

The thinning algorithm builds a skeleton shown in Fig. 8b. The algorithm realizes two additional functionalities: 1) it finds endings for lines, 2) it finds switches for nodes. The first and second elements are shown in Fig. 8b.

The last tool is very important because it finds coordinates of pixels which are accepted as contacts of chains and assigned in fact to identify chains. Also, they are used as starting points for the flood-fill algorithm to build and select chains in the reference and real PCB images.

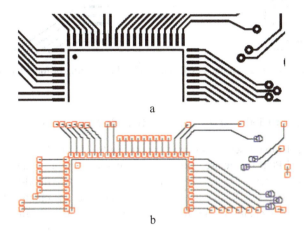

Fig. 8. Binary image (a) and its skeleton with endings and switches (b)

3 Selection and Separation of Chains

3.1 Connectivity Defects

After finding coordinates of endings and switches in the skeleton image the following step is to bind chains in the reference PCB image with their specific points: ending and switches. It is realized by a help of the flood filling algorithm. Arbitrary ending or switch pixels are taken as start points. The resulting PCB image is shown in Fig. 9. Note that not all chains are not flood-filled.

An order in which chains are flood-filled gives us their ID numbers after flood-filling all endings and switches are with corresponding colors and ID numbers of chains to which they belong. Thus, a set of pixels as specific points for the reference PCB image $E_e = \{E_1, E_2, \ldots, E_n\}$ is formed, where E_i – is a set of pixels belonging to the i-th chain.

All endings and switches are placed in middle geometrical positions of contacts and traces. So, their positions are not very sensitive from the manufacturing process. They are taken as starting points for the manufactured PCB to select chains in the image:

$$De = Ee, \quad \{D_1, D_2, \ldots, D_n\} = \{E_1, E_2, \ldots, E_n\}$$

where D_e is a set of specific points for the manufactured PCB image and D_i is a set for the i-th chain.

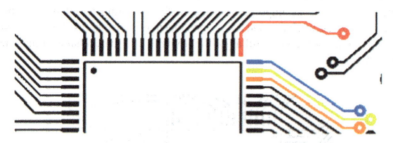

Fig. 9. The reference PCB image after flood filling of four chains

3.2 Separation of Chains and Connectivity Defect Detection

Taking elements from the set D_i as starting points for the flood-fill algorithm chains from the manufactured PCB image are selected and separated. Figure 10a shows two sets of starting points $D_1 = \{B, C, D, H\}$, $D_2 = \{A, E, F, G\}$, and two flood-filled chains in the reference image. Figure 10b shows one correct chain and one chain with open defect. Figure 10c shows two chains as one chain caused by short defect. It is in the manufactured PCB image.

To select the red chain with the open defect two iterations of flooding are used: for the point B and for the point H. Two iterations indicate available open defect. One color for two and more sets of starting points indicate available short defect. Two blue chains with the short defect are flooded and selected by one starting point and one iteration.

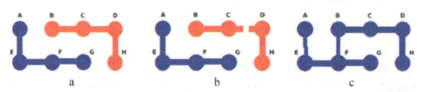

Fig. 10. Illustration of flood-filling and selection of of chains: chains in the reference (a), open (b), short (c)

The resulting PCB image after three iterations is shown in Fig. 11.

When defective chains are selected and separated on a basis of starting points the places of defects can be found by new application of the thinning algorithm. In Fig. 12a two chains from the reference image and its skeleton are shown. The image of selected defective chains and its skeleton are shown in Fig. 12b.

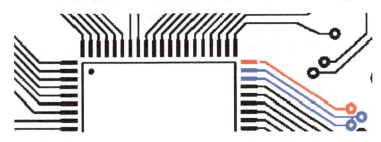

Fig. 11. The manufactured PCB image after flood filling of two chains

Fig. 12. Chains after flood-filling and thinning: of the reference PCB (a), of the manufactured PCB (b)

The coordinates of the found specific points directly indicate the places of defects: endings indicate a circuit break, and switches indicate a short circuit.

3.3 Measurement of Intensity Defects

To measure the chain inaccuracy a mean intensity function in the W columns and H rows of the image matrix is used:

$$\bar{I}(i) = 1/W \left(\sum_{j=1}^{W} I(i,j) \right), i = 1, 2, ..., H, \qquad (7)$$

$$\bar{I}(j) = 1/H \left(\sum_{i=1}^{H} I(i,j) \right), j = 1, 2, ..., W, \qquad (8)$$

where $I(i,j)$ is pixel intensity in the i-th row and the j-th column ($1 \leq i \leq H, 1 \leq j \leq W$).

The mean intensity functions of the reference and real components in Fig. 13 illustrate the different numbers of pixels in two defect places.

3.4 Evaluation by Distributed Cumulative Histogram

To visualize defects a distributed cumulative histogram is more productive. It is possible to notice changes of a difference between pixels on the etalon and real samples.

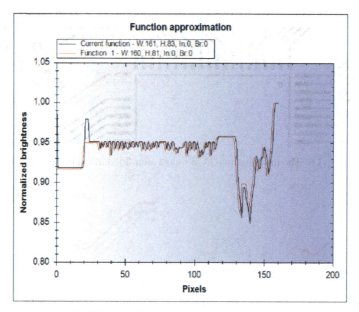

Fig. 13. Mean intensity functions of two chains

Figure 14 contains two DCH of the etalon and defective chains, and their difference overlayed with a chain.

Fig. 14. Difference of DCH images and an overlay

4 Summary and Conclusion

Developed approach allows us to inspect all open and short defects visible for a camera and innoticable for the user. The K-Means clustering algorithm, flood-filling, thinning, and distributed cumulative histograms are used to detect PCB defects and to visualize them.

The developed approach automatically demonstrates defects and chains containing them. The developed software on the basis of the created mathematical apparatus confirmed the expediency and effectiveness of the developed processing concept.

References

1. Jungsuk, K., Jungbeom, K., Hojong, C., Hyunchul, K.: Printed circuit board defect detection using deep learning via a skip-connected convolutional autoencoder. Sensors **21**(15), 4968 (2021)
2. Chaudhary, V., Dave, I., Upla, K.: Automatic visual inspection of printed circuit board for defect detection and classification. International Conference on Wireless Communications, Signal Processing and Networking (WiSPNET), pp. 732–737 (2017)
3. Tatibana, M.H., Lotufo, R.A.: Novel automatic PCB inspection technique based on connectivity. In: Proceedings X Brazilian Symposium on Computer Graphics and Image Processing, pp. 187–194 (1997)
4. Hilditch's Thinning Algorithm (2023). http://cgm.cs.mcgill.ca/~godfried/teaching/projects97/azar/skeleton.html
5. Chauhan, A.P., Bhardwaj, S.C.: Detection of bare PCB defects by image subtraction method using machine vision. In: Proceedings of the World Congress on Engineering, vol. II, WCE, pp. 68 (2011)
6. Bond. An Efficient and Versatile Flood Fill Algorithm for Raster Scan Displays (2011). www.crbond.com
7. Kamalpreet, K., Beant, K.: PCB defect detection and classification using image processing. Int. J. Emerging Res. Manage. Technol. **3**(8), 42–46 (2014)
8. Moganti, M., Ercal, F., Dagli, C.H., Tzumekawa, S.: Automatic PCB inspection algorithms: a review. Comput. Vis. Image Underst. **63**(2), 287–313 (1996)
9. Sarath, A.K., Kumar, N.S.: A review of PCB defect detection using image processing. Int. J. Eng. Innovative Technol. (IJEIT) **4**(11), 188–192 (2015)
10. Putera, S.H., Ibrahim, Z.: Printed circuit board defect detection using mathematical morphology and MATLAB image processing tools. In: 2nd International Conference on Education Technology and Computer, **5**, pp. 359–363 (2010)

Tropical Cyclone Genesis Forecasting Using LightGBM

Sabbir Rahman[⊠], Nusrat Sharmin, Md. Mahbubur Rahman,
and Md. Mokhlesur Rahman

Military Institute of Science and Technology, Mirpur, Dhaka 1216, Bangladesh
sabbir.2331@gmail.com

Abstract. Cyclones, or tropical storms, are a major threat to coastal communities and industries around the world. Accurate and timely forecasts of cyclone strength and trajectory are critical for minimizing their impact. In recent years, machine learning algorithms have shown promise in predicting cyclone genesis by analyzing large amounts of data and identifying patterns and correlations. In this paper, we propose a new machine learning approach using the LightGBM technique for cyclone genesis forecasting. Our experiments were conducted using the Best Track dataset, which includes wind speed, sea level pressure, and surface moisture flux data. We compare our method to the Random Forest algorithm and provide a detailed analysis of the performance of both approaches. Our findings demonstrate that the LightGBM method outperforms Random Forest and offers improved accuracy and efficiency in cyclone forecasting. Our study contributes to the field of cyclone prediction by offering a novel machine learning approach that can help mitigate the impact of these extreme weather events on society.

Keywords: Machine Learning · Weather Prediction · LightGBM · Random Forest

1 Introduction

Cyclones are dreadful weather phenomena that can kill people and seriously harm infrastructure. To predict cyclones' potential impact and reduce the potential damage they can do, accurate forecasting is necessary. [1] Presently, the Saffir-Simpson scale, which has categories 1 through 5, is used to categorize cyclones according to wind speed. [2] For better catastrophe management and prediction, more precise classification systems might be needed because this system might not always be able to capture the complexity of cyclone data.

Machine learning algorithms have shown promise in cyclone genesis forecasting, with various algorithms such as decision trees, logistic regression, random forest, AdaBoost, support vector machine, support vector regression, and multiple linear regression applied for tropical cyclone genesis forecasting. However, previous studies have not explored the use of the LightGBM algorithm in this context, and there is no concrete conclusion on the use of different parameters from various datasets.

© The Author(s), under exclusive license to Springer Nature Switzerland AG 2023
Z. Hu et al. (Eds.): ICAILE 2023, LNDECT 180, pp. 468–477, 2023.
https://doi.org/10.1007/978-3-031-36115-9_43

In this study, we propose a new cyclone categorization system that classifies cyclones into six categories, ranging from 0 to 1. This new system aims to provide a more nuanced understanding of cyclones, with improved accuracy and relevance to disaster management. Categories 2 to 6 are further categorized according to the Saffir-Simpson scale.

In order to categorize cyclones, we employed the LightGBM algorithm, a well-liked machine learning technique that has shown effectiveness in a number of classification applications. We found that LightGBM performed better than the Random Forest algorithm in terms of precision and accuracy, demonstrating that it is a better algorithm for this particular application.

In the following sections, we provide an overview of related work on cyclone categorization and machine learning algorithms. We describe our dataset and the preprocessing steps taken to prepare it for the machine learning models. We present our experimental results, including a detailed analysis of the performance of LightGBM and Random Forest. Finally, we conclude our study and discuss future directions for research on cyclone categorization. Our study aims to provide a more accurate and nuanced approach to cyclone categorization, which can help improve disaster management efforts.

2 Literature Review

Several studies have explored the use of machine learning algorithms for weather prediction, including LightGBM and Random Forest models. In this section, we describe relevant works related to cyclone prediction and evaluate their relevance to our research objectives.

A model based on the LightGBM algorithm was created by Xinwei Liu et al. [3] to identify and predict severe convective events, such as hail, brief periods of very heavy rain, and convective gusts. The authors classified various types of severe weather using ground data and C-band radar echo outputs. This study shows that LightGBM is excellent in forecasting extreme weather even if it focuses on diverse kinds of weather occurrences.

To improve forecasting performance and rectify numerical prediction findings, Rongnian Tang et al. [4] suggested a spatial LightGBM model. By utilizing a single-station, single-time approach and capturing the local spatial information of stations, the authors were able to give high-performance correction of medium-range forecasts. Although cyclone prediction is not the primary focus of this study, the proposed model's capacity to capture local spatial information may be helpful in predicting cyclone movement.

To estimate dominating wave periods in oceanic waters, Pujan Pokhrel et al. [5] suggested a forecasting model based on the LightGBM algorithm. To build the model, the authors used a dataset of oceanic wave data collected from five separate locations. Despite the fact that this work primarily focuses on oceanic waves, cyclone behavior could be predicted by using LightGBM to forecast wave patterns.

Pingping Wang et al. [6] proposed a novel intensity applicable conformal prediction framework using the Random Forest model as the underlying algorithm. The authors used satellite infrared images of tropical cyclones to extract 71 intensity-related features that may be divided into four categories: eye features, circle features, texture features, and time-series features. This study is directly relevant to our research objectives, as it demonstrates the effectiveness of Random Forest in predicting the intensity of tropical cyclones.

Gregory R. Herman et al. [7] used historical forecasts from NOAA's Second Generation Ensemble Forecast System Reforecast ensemble to provide probabilistic forecasts of severe weather in the contiguous United States. The authors found that SPC outlooks performed somewhat better on Day 1 than RF outlooks, but much better on Days 2 and 3. Although this study focuses on severe weather in general, the use of Random Forest in probabilistic forecasting could be useful in predicting the path of cyclones.

In Tamil Nadu, India, Rajasekaran Meenal et al. [8] investigated the application of machine learning to estimate global sun radiation and wind speed. The authors tested the Random Forest machine learning model with statistical regression and SVM models using data on recorded wind and sun radiation from IMD, Pune. Although cyclone prediction is not the main emphasis of this study, using Random Forest to anticipate wind speed may help in predicting cyclone behavior.

These studies demonstrate the accuracy of the LightGBM and Random Forest models at forecasting wind speed, dominant wave durations, and severe weather events. The proposed models can capture local geographical data, extract features linked to intensity, and make probabilistic projections. Although some studies don't specifically address cyclone prediction, they can be changed to do so by changing the machine learning algorithms and forecasting techniques they employ.

3 Methodology

The proposed methodology aims to address the research objective of accurately predicting the category of a new, unseen cyclone in the North Indian Ocean region. To achieve this objective, the study uses machine learning techniques, specifically feature selection and a mapping function to predict the cyclone category based on input features.

3.1 Problem Formulation

We can represent the input features of a cyclone as a matrix X, where each When analyzing data related to cyclones, we can organize the various factors that make up a cyclone into a matrix called X. Each column in this matrix indicates a distinct characteristic of the cyclones in each row, such as wind speed or temperature. As an illustration, Cyclone A's wind speed would be listed in one column, while Cyclone B's temperature would be listed in another column.

We may make use of a vector called Y to keep track of the cyclone types. Each component of this vector represents a particular cyclone's categorization. For instance, if Cyclone C is designated as a Tropical Storm, the vector would include an element to reflect that information.

The issue of cyclone categorization can be described as follows:

Given a dataset $D = (X, Y)$ containing n cyclones, where X is an nxm matrix of input features and Y is an $nx1$ vector of output categories, we seek to learn a mapping $f : X \rightarrow Y$ that can accurately predict the category of a new, unseen cyclone. We can represent this mapping as a function that takes as input a vector of input features x_j for a given cyclone j, and outputs a predicted category y'_j. This function can be defined as:

$$y'_j = f(x_j)$$

3.2 Dataset

The North Indian Ocean 5-day predictor dataset contains historical data on cyclones in the North Indian Ocean region from 1990 to 2017. The dataset is used to anticipate the track and intensity of upcoming cyclones based on a variety of meteorological factors. The variables in the dataset are likely to contain details about the location, wind direction and speed, pressure, and surface moisture flux. The data also include information on the cyclone's timing, length, and impact on coastal economy and communities. The dataset's structure is likely in the form of a time-series, with each row representing a single observation of a cyclone at a specific moment. Most likely, the observations are arranged in chronological order, with the earliest observations coming first and the most recent observations coming last in the dataset. Very likely a large and varied collection of data regarding cyclones in the North Indian Ocean region makes up the dataset's content. The information could be used to look for patterns and connections between different meteorological traits and cyclone intensity and course. Also, it can be used to evaluate the socioeconomic effects of cyclones on coastal economies and communities.

3.3 Feature Selection

Choosing features is an important step in any machine-learning activity, especially when working with large and complex datasets. In the context of cyclone prediction, feature selection is crucial in defining the model's accuracy and efficiency. We will explore the feature selection process for cyclone prediction using different methods. This study's dataset contains four main features: VMAX, MSLP, CFLX, and TWAC. VMAX is the maximum sustained wind speed, MSLP is the minimum central pressure, CFLX is the net heat exchange of the cyclonic storm with the ocean, and TWAC is the total water vapor transport into the storm. These four features were selected because they are the main determinants of tropical storm behavior and intensity. We seek to uncover the most relevant and critical features that have a significant influence on cyclone prediction throughout the feature selection phase. To do this, we employed both the Pearson correlation coefficient and SelectKBest method from sklearn, a python based machine learning framework to assess the relevance of each characteristic. The relevance of a feature is measured by its ability to predict the target variable using each feature. The greater the feature priority score, the more vital the feature.

The results of the feature selection procedure revealed that VMAX was selected as the most essential feature by both the Pearson correlation coefficient and SelectKBest module. This suggests that the maximum sustained wind speed is the most important feature in determining tropical storm strength and behavior followed by MSLP, CFLX, and TWAC. This finding is consistent with prior research that demonstrated a substantial link between wind speed and tropical storm intensity.

3.4 LightGBM Algorithm

LightGBM is a framework for gradient boosting that use tree-based learning techniques. [9] It is optimized for large-scale, high-dimensional data and is meant to be efficient and scalable. LightGBM's main concept is to develop a model by iteratively adding decision trees to capture complicated, non-linear connections in data. The approach fits a decision tree to the negative gradient of a loss function, which evaluates the difference between predicted and actual values for the target variable, at each iteration. Each tree's output is then integrated to generate a final forecast for the target variable. The tree-building approach, which employs a histogram-based representation of the data to effectively split the feature space and limit the number of candidates for split points, lies at the heart of LightGBM's efficiency. This histogram-based approach also enables LightGBM to efficiently and robustly handle categorical features and missing data.

LightGBM uses the gradient boosting framework to minimize a loss function, $L(y,f(x))$, where y is the goal variable and $f(x)$ represents the prediction provided by the model. The purpose of gradient boosting is to obtain the best prediction function, $f(x)$, by including decision trees into the model, $T_1, T_{2\_},...,T_m$, so that:

$$f(x) = \sum_{i=1}^{m} f(x)T_i(x)$$

where each decision tree, T_i, is fit to the negative gradient of the loss function.

$$T_i(x) = -\alpha_i h_i(x)$$

where α_i and $h_i(x)$ are the learning rate and decision tree fit to the negative gradient of the loss function, respectively.

3.5 Train with LightGBM

We can train a LightGBM or Random Forest model to learn this mapping by minimizing a loss function that measures the discrepancy between the predicted categories y'_j and the true categories y_j for the training set. We can define a loss function L as:

$$L(Y, Y') = \frac{1}{n} \sum_{k=1}^{n} l(y_i, y'_i)$$

where L is a loss function that measures the discrepancy between the true category y_i and the predicted category y'_j for a single cyclone i.

The goal is to find a set of model parameters that minimize this loss function, which can be done using an optimization algorithm such as gradient descent. Once the model is trained, we can use it to make predictions on new, unseen cyclones by applying the learned function f to their input features.

Figure 1 provides us with the overview of our working Model.

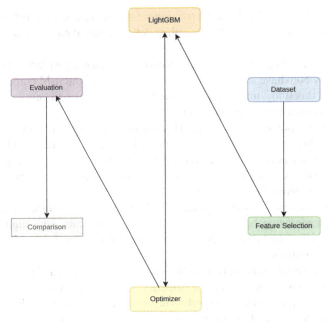

Fig. 1. Working Diagram of Proposed Model

4 Result and Discussion

In this study, LightGBM outperformed Random Forest in terms of classification accuracy, precision, recall, and f1-score. LightGBM achieved an accuracy of 1.00, precision of 1.00, recall of 1.00, and f1-score of 1.00, while Random Forest had an accuracy of 0.94, precision of 0.92, recall of 0.93, and f1-score of 0.92. These results suggest that LightGBM may be a better choice for this classification task. Using LightGBM we were able to predict all the six cyclone categories correctly. From the confusion matrix in Fig. 5 and 6 we observe that LightGBM predicted category 0 as 156 out of 156. On the other hand, Random forest was able to predict category 0 as a number of 107 out of 156 in the provided dataset and the same is applicable for all other categories as well.

In Fig. 2, the plot displays the variation of multi_logloss over the course of the training process, with the number of iterations on the x-axis and the loss on the y-axis. Both the train loss and the validation loss exhibit a steady decrease from an initial value of 1.2 to a final value of 0.1, within a span of 100 iterations. This trend suggests that the LightGBM algorithm is effective at reducing the multi_logloss metric for the training and validation datasets, implying that the model has been trained to generalize well and perform accurately on unseen data.

In this study, we analyzed the performance of the model in terms of the train score and cross-validation score as the training set size increased. In Fig. 3, the train score, which represents the model's accuracy on the training set, began at 0.860 and gradually increased as the training set size increased from 600 to 1000. This indicates that the model can learn more from the data as the size of the training set grows. Additionally, the cross-validation score, which represents the model's accuracy on a validation set, increased from 0.860 to 0.920 as the training set size increased from 600 to 1000. This suggests that the model is able to generalize better to new data as the training set size increases.

LightGBM and Random Forest are both commonly used algorithms for classification tasks, and they both have advantages and disadvantages. LightGBM outperformed Random Forest in this study, implying that it may be a better choice for similar classification tasks, at least under the conditions of this study.

The accuracy, precision, recall, and f1-score of the models were used to assess their performance. The percentage of correct predictions made by the model is known as accuracy, while the percentage of correct positive predictions is known as precision, and the percentage of correctly identified positive examples is known as recall. The F1-score is the harmonic mean of precision and recall.

According to the study's findings, LightGBM was more effective than Random Forest at correctly recognizing both positive and negative situations, with higher values for accuracy, precision, and recall. Furthermore, LightGBM outperformed Random Forest in probability prediction..

Overall, the results indicate that LightGBM is a more precise and accurate algorithm for this specific classification task, with superior ability to differentiate between positive and negative examples and forecast probability. However, keep in mind that the algorithm of choice can vary depending on the specifics of the data and the task at hand. This is why it's always a good idea to compare multiple models and choose the one that best serves our needs.

Table 1 depicts the comparison results between LightGBM and Random Forest in terms of accuracy, precision, recall, and f1-score. As the problem is based on classification tasks, we used cross-entropy as the loss function.

Figure 2 and 3 shows us the pictorial view of training and validation loss with learning curve of LightGBM algorithm respectively.

Figure 4 resembles the root of the tree that represents the entire dataset, and the tree branches out from there. A decision point based on a particular feature value is represented by each split. The final prediction or choice is made at each leaf node, or the end of the tree branch. The majority class of the training instances that reached that leaf node provides the basis for the prediction at each leaf node.

Table 1. Performance metrics of two machine learning model

Ml Models	Accuracy	Precision	Recall	f1-Score
LightGBM	1.00	1.00	1.00	1.00
Random Forest	0.94	0.92	0.93	0.92

The confusion matrices from the two models, LightGBM and Random Forest, are shown in Figs. 5 and 6, respectively. Through comparing the instance of the models, it is observed that the LIghtGBM model performs better on the specified categorization job. Additionally, accuracy, precision, recall, and F1 score are metrics that can be obtained

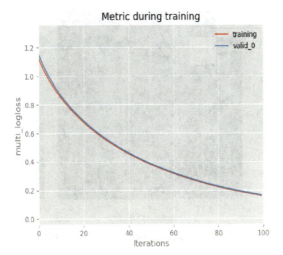

Fig. 2. Train and validation loss

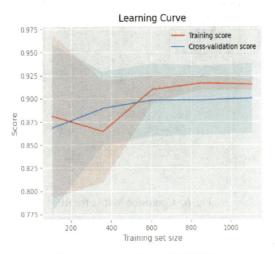

Fig. 3. Learning Curve LightGBM

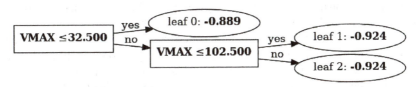

Fig. 4. Tree Generated by LIghtGBM

from the confusion matrix and used to assess how well the model performs in terms of true positives, true negatives, false positives, and false negatives.

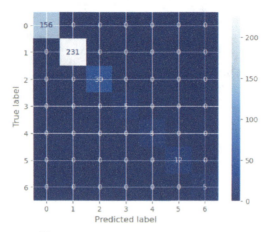

Fig. 5. Confusion Matrix for LightGBM

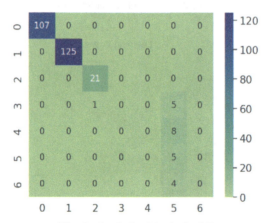

Fig. 6. Confusion Matrix for RF

5 Conclusion

The research focused on cyclone categorization, which is critical for forecasting their potential impact and minimizing damage. By comparing the performance of two machine learning algorithms, LightGBM and Random Forest, the study discovered that LightGBM outperformed Random Forest in terms of accuracy and precision in this specific task. This finding is significant because it suggests that LightGBM may be a better choice for similar classification tasks, at least under the conditions of this study.

From a scientific standpoint, this study contributes to the field of cyclone research by providing a new approach to categorizing cyclones that may be more accurate and effective than existing methods. This is important because accurately categorizing cyclones can help emergency responders and disaster management teams prepare for and respond to cyclone-related events more effectively.

In terms of future research, there are several avenues that could be explored. One possibility is to investigate the use of other machine learning algorithms or ensembles to further improve cyclone categorization. Additionally, with the potential impacts of climate change on cyclones, it would be interesting to explore how the categorization of cyclones may change in the future and how machine learning algorithms can be adapted to these changes. Finally, incorporating real-time data could improve the accuracy and timeliness of cyclone categorization, which would have important implications for emergency response and disaster management.

Acknowledgment. This research is supported by Military Institute of Science and Technology, Dhaka, Bangladesh.

References

1. https://public.wmo.int/en/our-mandate/focus-areas/natural-hazards-and-disaster-risk-reduction/tropical-cyclones
2. https://www.nhc.noaa.gov/aboutsshws.php
3. Classified Early Warning and Forecast of Severe Convective Weather Based on LightGBM Algorithm. written by Xinwei Liu, Haixia Duan, Wubin Huang, Runxia Guo, Bolong Duan, published by Atmospheric and Climate Sciences 11(2) (2021)
4. Tang, R., Ning, Y., Li, C., Feng, W., Chen, Y., Xie, X.: Numerical forecast correction of temperature and wind using a single station single time spatial LightGBMMethod. Sensors **22**, 193 (2022).https://doi.org/10.3390/s22010193
5. Pokhrel, P., Ioup, E., Hoque, M., Abdelguerfi, M., Simeonov, J.: A LightGBM based Forecasting of Dominant Wave Periods in Oceanic Waters (2021)
6. Wang, P., Wang, P., Wang, D., Xue, B.: A conformal regressor with random forests for tropical cyclone intensity estimation. IEEE Trans. Geoscience Remote Sensing, p. 1 (2021). https://doi.org/10.1109/TGRS.2021.3139930
7. Hill, A., Herman, G., Schumacher, R.: Forecasting severe weather with random forests. Mon. Weather Rev. **148**, 2135–2161 (2020). https://doi.org/10.1175/MWR-D-19-0344.1
8. Meenal, R., Michael, P., Pamela, D., Rajasekaran, E.: Weather prediction using random forest machine learning model. Indonesian J. Electrical Eng. Comput. Sci. **22**, 1208 (2021). https://doi.org/10.11591/ijeecs.v22.i2.pp1208-1215
9. Ke, G., et al.: Lightgbm: A highly efficient gradient boosting decision tree. Advances in Neural Inf. Processing Syst. **30** (2017)
10. Nti, I., Nyarko-Boateng, O., Aning, J.: Performance of machine learning algorithms with different K values in K-fold cross-validation. Int. J. Inf. Technol. Comput. Sci. **6**, 61–71 (2021). https://doi.org/10.5815/ijitcs.2021.06.05
11. Saber, M., et al.: Examining lightGBM and catboost models for wadi flash flood susceptibility prediction Geocarto Int. **37** (2021). https://doi.org/10.1080/10106049.2021.1974959
12. Meng, F., et al.: IEEE international geoscience and remote sensing symposium IGARSS. Brussels, Belgium **2021**, 8476–8479 (2021). https://doi.org/10.1109/IGARSS47720.2021.9555156
13. Eisenstein, L., Schulz, B., Qadir, G.A., Pinto, J.G., Knippertz, P.: Identification of high-wind features within extratropical cyclones using a probabilistic random forest – Part 1: method and case studies. Weather Clim. Dynam. **3**, 1157–1182 (2022)
14. Acula, D.: Classification of disaster risks in the philippines using adaptive boosting algorithm with decision trees and support vector machine as based estimators. J. Modeling Simulation Materials **4**, 7–18 (2021). https://doi.org/10.21467/jmsm.4.1.7-18

Optimization of Identification and Recognition of Micro-objects Based on the Use of Specific Image Characteristics

Isroil I. Jumanov and Rustam A. Safarov[✉]

Samarkand State University, Samarkand 140104, Uzbekistan
`rustammix.rs@gmail.com`

Abstract. Models and algorithms for optimizing the identification and recognition of micro-objects have been developed based on the combination of dynamic models with neural networks and mechanisms for extracting statistical, dynamic, specific characteristics of images, extracting and segmenting the contour, selecting reference points and reducing redundant points, as well as setting model variables. Identification optimization mechanisms are implemented that use the correlation and dynamic characteristics of a sequence of points on the image contour. A dynamic model has been implemented that performs the functions of forming a matrix with the coordinates of distorted points, approximating image segments, taking into account its deformation. The effectiveness of image pre-processing tools, recognition and classification algorithms was studied using examples of a large amount of images presented about wheat pollen grains. When solving the problems of identification and recognition of pollen grains, the mechanism of non-linearity of the influence of factors, as well as the conditions of a priori insufficiency and uncertainty, were used. A software package for visualization, recognition, classification of micro-objects based on the use of an interpolation spline-function 7, a three-layer neural network with learning algorithms for forward and back propagation of errors, training with and without a teacher, mechanisms for forming a training sample with vector quantization, segmentation, and the formation of a "sliding window"» and tracking control points along the contour of the image segment.

Keywords: Micro-object · Identification · Recognition · Classifications · Detection of a Distorted Point · Characteristics and Properties · Image

1 Instruction

Research and development of methods for identification, recognition and classification of micro-objects are relevant and in demand in the fields of palynology, medicine, environmental protection and ecology, etc. [1, 3]. Solutions to the problem of recognition and classification of micro-objects are being carried out by specialists from a number of foreign countries such as England, France, Germany, Italy, Spain, Austria, New Zealand, etc. [2, 6].

© The Author(s), under exclusive license to Springer Nature Switzerland AG 2023
Z. Hu et al. (Eds,): ICAILE 2023, LNDECT 180, pp. 478–487, 2023.
https://doi.org/10.1007/978-3-031-36115-9_44

Recognition and classification systems have been created to solve the problems of identifying and counting pollen grains in air samples [4, 8]. Nineteen European countries are connected to the EAN/EPI European Airborne Allergy Network, which collects information on pollen. Pollen scientists currently operate only with the available conventional standards [3, 5, 6].

However, the standards used are not sufficient to take them as a basis for research on the creation of an automated image visualization system for recognition, classification and accounting of micro-objects [7–9]. Recognition and classification of micro-objects are preceded by the tasks of removing foreign particles, etc. [10–12]. The problem of detecting and correcting distorted image points due to interference, noise, and blur is becoming increasingly important [13–15].

In classical algorithms for identifying micro-objects, little attention is paid to the use of mechanisms for detecting distorted points in image identification [16–18].

It is necessary to improve and develop methods for identifying micro-objects based on the use of mechanisms for detecting and correcting distorted points against the background of noise in conditions of a priori insufficiency, uncertainty and low data reliability [19–21]. This study is devoted to the creation and application of microobject image identifiers, which use a multilayer neural network, radial-basic networks, as well as mechanisms for extracting redundant information structures - histological, morphological, fractal, geometric and other specific characteristics of images [22–24].

2 Methodology for Identification and Recognition of Microobjects Based on the Use of Redundant Information Structures

Let the input of the mechanism of identification, recognition and classification of microobjects be given the implementation of a signal with noise, the distortion of the points of the image in which is given in the form [25, 26] $\xi(t) = Qs(t) + n(t)$, where $s(t)$ - a signal of a known form; $n(t)$ - Gaussian noise with a spectral density of N_0; $E = \frac{2}{N_0} \int_0^T \xi(t)s(t)dt = \int_0^T s^2(t)dt$ - spectrum power, $\xi(t) = s(t)$; Q - the value that takes the value "1" if the distorted point is present, "0" if the distorted point is absent. The control of distorted points on the image contour is evaluated by the criterion of maximum likelihood [24, 27] $P_{err}(z) = \frac{\alpha(z)+\beta(z)}{2} = 1 - F\left(\frac{z}{2}\right)$, where $z = \sqrt{2E/N_0}$ - signal-to-noise ratio. If the sequence of image points is characterized by stationary behavior, then the problem of detecting distorted points is not difficult.

If the sequence of image points is characterized by a dynamic and non-stationary process, then adaptive control mechanisms are required that use the main characteristics of the signals.

The simulation of the process was carried out using multi-stage correlators combined with the NN, which are aimed at performing mathematical operations similar to the operations of algorithms for the statistical detection of distorted image points.

In a modified version of this mechanism, the synaptic weight coefficient of neurons w_i is used. The weighted sum of neuron weights of the correlator built on the basis of a multilayer NN is given as $U = \sum_{i=0}^{N} w_i \xi_i$, where ξ_i - input realization counts; N - number of inputs. At the second stage of operation of the mechanisms, the sum of the products of the input implementation ξ_i and its weight coefficients w_i is compared with a given threshold. The term "0" of the coefficient and the imaginary "1" input are distinguished.

A weighting factor with a zero index acts as a decision threshold, the value of which is compared with the weighted sum. Weight coefficients are used for fixing and identifying "reference" (informative) points of micro-object images.

An algorithm for learning a supervised NN is studied, at the output of which the network reactions are determined in the form $w_i = w_i + \eta(d - y)\xi_i$, where η - coefficient selected in [0, 1]; d - expected network reactions [25, 28].

Implementations of the correlator based on the multilayer NN have been tested in the MATLAB environment.

10 NNs are investigated, each having 11 inputs, the weight coefficients are set randomly. The coefficients are normalized with respect to w_0.

The weight coefficients are calculated by $w_{i\ cp} = \frac{1}{10} \sum_{j=1}^{10} w_{ij}$, where w_{ij} is the weight coefficient of the j-th network (Fig. 1).

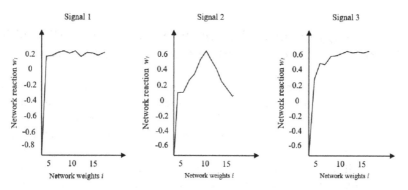

Fig. 1. Weight coefficients of the NN correlator of the image signal

It is determined that the required stability of the neural network learning algorithm is achieved when noise with a power 5–7 times less than the power of the signal itself is superimposed on the training sample.

The dependencies of the probability of detecting distorted points on the parameter - signal-to-noise for all given three signals are investigated. Estimates of the error function of the neural network correlator of image signals are obtained $e(U) = d - y(U) = d - H(U)$, где $U = \sum_{i=0}^{N} w_i \xi_i$; $H(U)$ - Heaviside function.

Figure 2 shows the dependence of the probability of detecting distorted points for signals 1 (Fig. 2. a)) and signal 2 (Fig. 2. b)): 1 - solid line, corresponds to the traditional algorithm; 2 – dash-dotted line, corresponds to the algorithm of the modified mechanism.

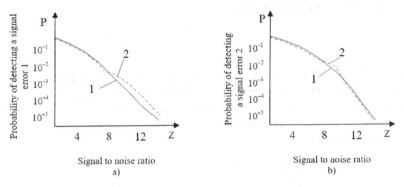

Fig. 2. Distorted Point Detection Efficiency

It is determined that the value of the probability of detecting distorted image points varies depending on the training steps of the NN. The error function $H(U)$ has a discontinuity of the first kind at the threshold point of the neuronal activation function. It is determined that it is impossible for the weighting coefficients to achieve an exact repetition of the waveform.

Graphs of the probability of detecting distorted points depending on the parameter - signal/noise almost converge with the curve of the modified mechanism. The efficiency of the mechanism was analyzed by the maximum likelihood criterion. The main features of the mechanism are superimposed on training samples, size variations that depend on the initial values of the NN weights.

3 Optimal Image Identification with Detection and Correction of Distorted Points Based on Wavelet Filtering

To improve the processes of detecting and correcting distorted image points, it is proposed to use a generalized filtering mechanism based on wavelet-transform algorithms and a multi-parameter threshold function under the influence of additive noise. The useful signal summation mechanism $S(t)$ with noise $n(t)$ includes sampling, binarization, coding, and wavelet filtering procedures. At its output, a filtered signal $S'(t)$ is issued.

The wavelet function in a certain finite time and frequency domain is represented by an infinitely oscillating function, which distinguishes this approach from the classical Fourier transform and encoded information is used.

Another feature of the mechanism is that it works with orthogonal functions and has the ability to use a smaller number of expansion coefficients.

Analysis of the results of the implementation of such a mechanism allows us to establish the following positive aspects of the study:

– a greater gain in the speed of information processing is achieved with less time spent;

482   I. I. Jumanov and R. A. Safarov

- Discrete Wavelet Transformer (DWT) is obtained from a continuous time scaled with a factor of a and a factor of b, like $a = 2^m$, $b = k \cdot 2^m$, where m and k are some integers;
- the continuous plane a and b is divided into a certain grid and the DWT characterizes the shift of the next function in time

$$\psi_{mk}(t) = \frac{1}{\sqrt{a}}\psi\left(\frac{t-b}{a}\right) = \frac{1}{\sqrt{2^m}}\psi\left(2^{-m}t - k\right) \tag{1}$$

Direct DWT has the form $c_{mk} = (S(t) \cdot \psi_{mk}(t)) = \int_{-\infty}^{+\infty} S(t) \cdot \psi_{mk}(t)dt$, where c_{mk}- conversion coefficients. The inverse DWT is defined as $S(t) = \sum_{m,k} c_{mk} \cdot \psi_{mk}(t)$.

Implementations of the generalized mechanism have been tested. Signal $S(t)$ with an amplitude of A_m, with a duration of T_s is considered. Analytical expressions for them are given in the form:

$$S(t) = \begin{cases} A_m(1 - \exp(-\alpha t)), & at \ 0 \leq t \leq T_s/2; \\ A_m(1 - \exp(\alpha(t - T_s))), & at \ T_s/2 \leq t \leq T_s. \end{cases} \tag{2}$$

For DWT, it is necessary to represent a continuous signal $S(t)$ in the form of sample $S_d(t) = \{S_i\}$, where $i = 0, 1, ..., N - 1$ with a frequency of f_m and with time steps of $\Delta t = \frac{1}{2f_m}$. The discrete signal is given as $S_d(t) = \sum_{i=1}^{N-1} s(i\Delta t)\delta(t - i\Delta t)$.

The efficiency of the $E(.)$ - fractal-correlation functional of the correlator of the identification mechanism is studied based on the following calculation $E(x, y, n) = (1 - (D_x + D_y)/d(n))^{1/3} \approx r(x, y)$, where $E(.)$- fractal-correlation functional of the correlator; $d(n)$ - scaling factor depending on the size of the test sample $-n$; D_x, D_y - sums of modules of the difference of adjacent data x and y.

Table 1. Results of the fractal-correlation functional

n	d(n)	σ(Fr)	σ(r)
6	5.05	0.513	0.365
8	5.61	0.492	0.360
10	6.12	0.501	0.349
12	6.75	0.452	0.345
14	7.16	0.471	0.317
16	7.52	0.487	0.314
18	8.43	0.462	0.295

Table 1 shows the values of normalizing coefficients and standard deviations for fractal-correlation functionals, with different test sample sizes and zero correlation of the tested data pairs. Standard deviations of the distribution of errors, obtained by calculations using the standard formula.

It has been established that DWT allows one to obtain effective filtering results in an area with distorted image points.

Optimization of Identification and Recognition 483

The efficiency of the generalized mechanism for selecting objects (points, features, fragments) belonging to rectangular areas with the procedures of wave segmentation and image identification has been studied. Each point p of the original image I is taken as the origin of the local coordinate system and it is checked whether its value belongs to a rectangular area.

For this, a rectangular area d_p centered at the selected point is defined. Area d_p is characterized by a set width of one raster unit, a given length of l, and a rotation angle of α between d_p and the selected x-axis. The rotation angle α is changed in the range of values $[0; \pi)$ in increments of ϕ. The length l of area d_p is selected taking into account the scale of the original image I. Thus, d_p is a rectangular segment in the original image I.

The mean-square error (MSE) of identification, represented by a function of $f(.)$ points belonging to d_p of all possible d_p relative to point p, was studied. If at a certain angle of rotation, d_p takes a positive value, then with a certain probability it can be argued that point p belongs to the rectangular area of the image. The mean-square error (MSE) of identification, represented by a function of $f(.)$ points belonging to d_p of all possible d_p relative to point p, was studied. The uniformity of the distribution of the brightness function in the area of the extension and the sharpness of the contours of the object is checked.

4 Analysis of the Results of the Study

Symmetric spectral density functions are implemented, in which d_p is measured along the abscissa. Its calculation is made for the first half d_p, which includes points with positive offsets. And the second half is d_p with negative offsets of points along the ordinate axis. For each value of the rotation angle α of area d_p, the first and second half d_p, respectively, σ_1 and σ_2 are calculated.

Of all the values σ_1, the minimum $\sigma_{1\,min}$ and the corresponding value of the rotation angle α_1 are determined. The average values of α_2, and $\overline{a}_2$ and mean square deviations (MSD) of $\overline{\sigma}_1$, and $\sigma_{2\,min}$ are calculated. The belonging of points to a rectangular area is checked as $(|\overline{\sigma_1} - \sigma_{1\,min}| > T_p) \wedge (|\overline{\sigma_2} - \sigma_{2\,min}| > T_p)$.

Threshold T_p is set by the operator, taking into account the brightness of the dots and the contrast of the original image.

The uniformity of the brightness distribution function on a rectangular area of the image is checked as $(\sigma_{1\,min} \to 0) \wedge (\sigma_{2\,min} \to 0)$. An additional check is due to the division of each half of area d_p in the form $(|\sigma_{1\,min} - \sigma_{2\,min}| \to 0) \wedge (\alpha_1 \approx \alpha_2)$.

The MSD value must correspond to one position d_p or an adjacent position, which is performed at a low value of ϕ of the angle of rotation. The minimum values of the MSD brightness of the points correspond to the position of area d_p, coinciding with the stretch of the object. The presence of clear indices of image points reflects the transition of the brightness level of the points on the border of the fragments of the object.

If at a certain value of the rotation angle α, area d_p is a spread, then the brightness value of point d_p will become high. If at point p it takes a positive value, then for p the values σ_{min} and the corresponding rotation angle α_σ are calculated. Approximately the same value of angles α_1 and α_2 are replaced by a threshold value of $|\alpha_1 - \alpha_2| < T_\alpha$, where T_α is a threshold value that depends on the rotation angle step.

When area d_p is a straight line segment, then it becomes necessary to calculate the position coordinates and brightness values of points in area d_p at the nodes of the discrete grid. To calculate the MSD, it is necessary to determine the brightness $f(.)$ at points belonging to area d_p centered at p.

The decompression of a compressed grayscale image is also performed in two parts: the first, a two-level contour of the image, and the second, the wavelet-transform coefficients, the inverse discrete wavelet-transform of the masked image. The masked halftone image was reconstructed.

A generalized mechanism for selecting image points, checking whether they belong to rectangular areas allows you to detect distorted points. The following designations have been applied:

- $X = \{X_i\}$ is a set of image point vectors, each of which is given as $X_i = < x_{ij} > = < x_{i1}, ..., x_{iN} >$, where $i = 1, ..., M$;
- M - power set X, $j = 1, ..., N = 5$, and $j = 1$ which corresponds to the value of the concentration (optical density) of the background component of the image: $j = 2$, $j = 3, j = 4$;
- $A = \{A_p\}$ - set of standards A_p images $p = 1, ...N_p$.

To check the quality of image identification, the following are set: $A = \{A_{pg}\}$ is a set of standards of gradations A_{pg} in each of the fragments A_p of the image, where $A_{pg} = \frac{1}{N_g} \sum_{i=1}^{N_g} x_{ig}^p$ is the standard of the g-th gradation of the fragment A_p; N_g- number of historical data, $g = 1, ..., 7$. The cross-correlation coefficient ρ_{xy} is defined as the ratio of the covariance $\text{cov}(X_i, Y_i)$ of two vectors X_i and Y_i. The means μ_x and μ_y, as well as the RMS σ_x and σ_y, are given as: $\rho_{xy} = \frac{\text{cov}(X_i, Y_i)}{\sigma_x \cdot \sigma_y}$; $\text{cov}(X_i, Y_i) = \frac{1}{n} \sum_{i=1}^{n} (x_i - \mu_x)(y_i - \mu_y)$, where x and y is respectively x_{kj} and y_{lj}.

The values of membership functions of image points are calculated by

$$\rho(A, B) = \sum_{i=1}^{n} |\mu_A(x_i) - \mu_B(x_i)| \tag{3}$$

The distance is interpreted as $h_{kl} = \sum |x_{kj} - x_{lj}|$, where x_{kj} and x_{lj} are compared analyzes of parameters k and l, $i = 1, ..., M$, $n = M$.

The software package (PC) for visualization, identification, recognition and classification of images of micro-objects includes the following functional modules: detection, selection of contours in the image, segmentation of the contour, determination of the boundaries of segments; morphological processing of the contour of the image, dilatation of the contour two-level image; compression of one-dimensional, two-level contours of the image and layout of the compressed halftone image file; reconstruction of the masked halftone image based on the scanned masked image using a contour image; contour bilinear interpolation with a bidirectional weighting of control points; masking of image contour points, line-by-line representation, and interpolation based on wavelet transformations and biorthogonal polynomials.

Compressed halftone image files are generated in two parts. The first, is a two-level contour of the image, and the second, is wavelet coefficients - transformations. Functional modules are tested in a Matlab environment.

5 Conclusion

A methodology has been developed for optimizing identification, recognizing the classification of micro-objects based on combining dynamic models with neural networks, extracting statistical, dynamic, and specific characteristics.

Methods and algorithms use adaptive correlators, synthesized with a Fourier transform, a Gaussian filter, mechanisms for detecting and correcting distorted points on the contour of an image of a micro-object and allow you to design transparent software tools that complement existing standards for pollen recognition in palynology, environmental protection and others.

A software package based on the use of an interpolation spline - function 7, a three-layer neural network, algorithms for direct and inverse learning, learning with and without a teacher has been implemented. The identification mechanisms are modified based on the synthesis of algorithms for forming a training sample with vector quantization, segmentation, the formation of a "sliding window" and tracking of control points.

It is proved that when solving the problems of selection and seed production of wheat grain, medical diagnosis of patients with tuberculosis, the effectiveness of the implemented models for identifying micro-objects increases based on the mechanisms of using histological, morphological, fractal, geometric characteristics.

The required stability of the neural network learning algorithm is achieved when noise with a power of 5–7 times less than the power of the signal itself is applied to the training sample, as well as with mechanisms for detecting and correcting distorted points, filtering based on wavelet transforms and a multi-parameter threshold function.

References

1. Gonzalez, R., Woods, R.: Digital Image Processing, M.: Technosfera. (2005)
2. Kuleshov, S.V., Aksenov, Y.A., Zaitseva, A.A.: An approach to identifying the source of images from digital cameras. Innovative Sci. **5**, 82–86 (2015)
3. Jumanov, I.I., Djumanov, O.I., Safarov, R.A.: Methodology of optimization of identification of the contour and brightness-color picture of images of micro-objects. International Russian Automation Conference, pp. 190–195 (2021). https://doi.org/10.1109/RusAutoCon52004.2021.9537567
4. Khaikin, S.: Neural Networks: Full Course, p. 1104. Williams, Moscow (2006)
5. Philist, S.A., Tomakova, R.A., Zar, D.Y.: Universal network models for problems of classification of biomedical data, SWGU. Part 2, **4**(43), 44–50 (2012)
6. Jumanov, I., Safarov, R.: Optimization of the processing of images of pollen grains based on the use of their specific characteristics and geometric features. AIP Conference Proceedings, 2637, 040016 (2022). https://doi.org/10.1063/5.0118857
7. Kohonen, T.: Self-Organizing Maps. Kohonen, T. BINOM. Knowledge Lab, p. 655 (2013)

8. Jumanov, I.I., Djumanov, O.I., Safarov, R.A.: Optimization of identification of images of micro-objects taking into account systematic error based on neural networks. International Russian Automation Conference, pp. 626–631 (2020). https://doi.org/10.1109/RusAutoCon49822.2020.9208164

9. Osovsky, S.: Neural Networks for Information Processing. Osovsky, S. M.: Finance and statistics, p. 344 (2002)

10. Jumanov, I.I., Djumanov, O.I., Safarov, R.A.: Recognition of micro-objects with adaptive models of image processing in a parallel computing environment. Journal of Physics: Conference Series, (1791). Omsk, Russia (2020). https://doi.org/10.1088/1742-6596/1791/1/012099

11. Bouguet, J.Y.: Pyramidal implementation of the Lucas Kanade feature tracker. Intel Corporation, Microprocessor Research Labs **9** (2000)

12. Jumanov, I.I., Djumanov, O.I., Safarov, R.A.: Mechanisms for optimizing the error control of micro-object images based on hybrid neural network models. AIP Conf. Proc. **2402**, 030018 (2021). https://doi.org/10.1063/5.0074019

13. Bovik, A.C., Clark, M., Geisler, W.S.: Multichannel texture analysis using localized spatial filters. IEEE Trans. PAMI. **12**(1), 55–73 (1990)

14. Dash, R., Majhi, B.: Motion blur parameters estimation for image restoration. Optik **125**(5), 1634–1640 (2014)

15. Ibragimovich, J.I., Isroilovich, D.O., Abdullayevich, S.R.: Optimization of identification of micro-objects based on the use of characteristics of images and properties of models. International Conference on Information Science and Communications Technologies, ICISCT, 9351483 (2020). https://doi.org/10.1109/ICISCT50599.2020.9351483

16. Dunn, D., Higgins, W.E.: Optimal Gabor filters for texture segmentation. IEEE Trans. Image Processing **4**, 947–964 (1995)

17. Haralick, R.M., Shapiro, L.G.: Computer and Robot Vision. **2**. Addison-Wesley, Reading, MA (1993)

18. Jumanov, I., Djumanov, O., Safarov, R.: Improving the quality of identification and filtering of micro-object images based on neural networks. E3S Web of Conferences, **304**, 01007 (2021). https://doi.org/10.1051/e3sconf/202130401007

19. Hoang, M.A., Gcuscbrock, J.M., Smoulders, A.W.M.: Color texture measurement and segmentation. Signal Process **85**(2), 265–275 (2005)

20. Maltoni, D., Maio, D., Jain, A.K., Prabhakar, S.: Handbook of Fingerprint Recognition. Springer-Verlag, N. Y. **496** (2009).

21. Ibragimovich, J.I., Isroilovich, D.O., Abdullayevich, S.R.: Recognition and classification of pollen grains based on the use of statistical, dynamic image characteristics, and unique properties of neural networks. Adv. Intell. Syst. Comput. **1323**, 170–179 (2021). https://doi.org/10.1007/978-3-030-68004-6_22

22. Sakthi Bharathi, D., Manimegalai, A.: 3D digital reconstruction of brain tumor from MRI scans using Delaunay triangulation and patches. ARPN J. Eng. Applied Sci. **10**(20), 9227–9232 (2015)

23. Singh, M., Kumar, S., Singh, S., Shrivastava, M.: Various image compression techniques: lossy and lossless. Int. J. Comput. Appl. **142**(6), 23–26 (2016)

24. Jumanov, I.I., Safarov, R.A., Djumanov, O.I.: Mechanisms for using image properties and neural networks in identification of micro-objects. In: The 16th IEEE International Conference, Application of Information and Communication Technologies, 22541907.l Washington DC (2022). https://doi.org/10.1109/AICT55583.2022.10013633

25. Jumanov, I.I., Safarov, R.A.: Optimization of recognition of micro-objects based on reducing excessive information structures of images. J. Phys.: Conf. Ser. 2373, 072030 (2022). https://doi.org/10.1088/1742-6596/2373/7/072030

26. Bramesh, S.M., Anil Kumar, K.M.: An efficient and scalable technique for clustering comorbidity patterns of diabetic patients from clinical datasets. I.J. Modern Education Comput. Sci. **14**(6), 35–52 (2022)
27. Panda, M.: Developing an efficient text pre-processing method with sparse generative naive bayes for text mining. Modern Educ. Comput. Sci. **10**(9), 11-19 (2018)
28. Jadhav, A.N., Dharwadkar, N.V.: A speaker recognition system using gaussian mixture model, EM algorithm and K-means clustering. I.J. Modern Educ. Comput. Sci. **10**(11), 19–28 (2018)

Development and Comparative Analysis of Path Loss Models Using Hybrid Wavelet-Genetic Algorithm Approach

Ikechi Risi[1], Clement Ogbonda[2], and Isabona Joseph[3](✉)

[1] Department of Physics, River State University, Port Harcourt, Nigeria
[2] Department of Physics, Ignatius Ajuru University of Education, Port Harcourt, Nigeria
[3] Department of Physics, Federal University, Lokoja PMB 1154, Nigeria
Joseph.isabona@fulokoja.edu.ng

Abstracts. The received signal strength of any cellular network system at the user equipment terminal is dependent on the propagation environment and the path loss model used during the network planning stage. One of the most commonly used empirical model for predictive path loss analysis and estimation is the COST231 model. Despite the popularity of this model, its precision performance is usually poor when engaged for propagation prediction modelling outside the intended environment wherein it was originally developed. To cater for such limitation, an adaptive path loss model for optimal path loss prediction is developed in this paper using a hybrid Wavelet-Genetic genetic algorithm (Wavelet-GA). For the purpose of comparative analysis between wavelet-GA, GA, and COST231 models, three evaluation indicators based on root mean square error (RMSE), mean absolute error (MAE), correlation coefficient (R), were engaged. The results showed that the estimated RMSEs attained with the proposed wavelet-GA are 2.896, 4.715, 1.945, and 3.498, whereas those of GA and COST231 models are 3.47, 5.49, 3.69, 4.55 and 78.13, 74.74, 84.30, 76.54 respectively. Also, the estimated MAE of wavelet-GA, GA, and COST231 models are 2.38, 3.92, 1.45, 2.95; 2.83, 4.66, 2.84, 3.84 and 78.04, 74.37, 84.36, 76.39 respectively. The analyzed results also showed that wavelet-GA, GA, and COST231 models correlate with the measured data by 93.3%, 90.8%, 90.8%; 83.9%, 79.6%, 79.6%; 92.1%, 76.9%, 76.9%; and 86.3%, 78.4%, 78.4% for site 1,2,3 and 4 respectively. Again, the analysed validation results also proved that the developed wavelet-GA model is 99.2% and suitable to be applied in Awka, Nigeria. The developed wavelet-GA based path loss model has the capacity to care of future planning and development cellular wireless systems in similar environments.

Keywords: Wavelet · Genetic Algorithm · Hybrid Wavelet-GA · Path loss · COST231 model

1 Instruction

Predicting the path loss has been a crucial task in the design and planning of mobile communication networks. This is as a result of various physical mechanisms such as reflection, diffraction and scattering of the signal during propagation, as well as the

© The Author(s), under exclusive license to Springer Nature Switzerland AG 2023
Z. Hu et al. (Eds.): ICAILE 2023, LNDECT 180, pp. 488–500, 2023.
https://doi.org/10.1007/978-3-031-36115-9_45

phenomenon of multipath and the constant movement of mobile communication users [1]. Therefore, the prediction of the electromagnetic signal power loss with minimal error during signal propagation from transmitter to receiver is an important topic in planning and optimization of telecommunications networks. When deploying a new mobile technology such as a Long Term Evolution (LTE) network with the goal of increasing network capacity and speed with broader coverage and strong Quality of Service (QoS), it is important to identify these factors that affect the network quality and capacity of the mobile network [2].

Signal path loss estimation is an influential factor in network planning because it allows network engineers to perform various configuration tests before the changes are physically implemented. However, path loss prediction for radio coverage is a complex task, hence the need for an accurate and computationally efficient prediction tool [3]. Accurate and reliable models are crucially important for the prediction of radio channel parameters in the cellular network systems.

Path loss models are generally the empirical mathematical formulation of the signal propagation behavior of an environment. Empirical models essentially explain the relationship between environment and path loss. It does not require large amounts of computation and it's easy to implement, but are less sensitive to the physical and geometric configuration of the environment.

The need for proper cellular network planning, determination of base stations (BS) locations and proper operating frequency during upgrade or deployment of a new communication network such as LTE to ensure improved QoS is stressed in many previous works [4–13]. However, understanding and building these networks relies on the knowledge of signal path loss over distance in a pragmatic environment.

Path loss is one of the most important characteristics of communication networks. It is the received signal power relative to transmitted [14]. The drop in electromagnetic signal power density as it travels through space can be as a result of diffraction, reflection, scattering, etc., and depends on different environments [15]. Factors such as topography, urban planning, population density, rainfall, vegetation, etc. contribute to path loss. Variations in height of the transmitter and receiver antennas also create losses. An accurate estimate of the path loss provides a good basis for correct BS positioning and appropriate frequency plan determination [13]. There is also a need for network engineers to have a suitable method for mapping the extent of coverage of both existing and planned networks.

2 Literature Review

Eichie et al. (2017) [16], developed ANN model that estimated the path loss within suburban and rural paths in Minna. Different parameters such as transmitter and receiver antennas distance, transmitter and receiver powers, and the potential height of the measurement locations. Comparative analysis were made between path loss estimated through the developed ANN model achieved by geometric algorithm and path loss calculated from existing predicted models among the existing predicted models considered, Hata model proved the best achievement along the rural paths where the RMSE estimated varied from 5.05 to 9.30 dB, whereas, in the suburban paths, the RMSE calculated were from 9.50 to 82.14 dB. The best performed existing predicted model in the suburban paths was Egli's model with an estimated RMSE that ranged from 3.81 to 8.18 dB. However, Ericsson and COST 2.31 models showed that highest path loss calculated in the rural and suburban paths. The ANN paths loss model developed, showed better performance than the existing predicted models in the environments were the RMSE achieved ranged from 3.96 to 7.07 dB and 1.22 to 4.82 dB in rural and suburban paths respectively. The formulated ANN path loss model proved to be suitable for the estimation at GSM signals in both rural and suburban areas within Minna and its surroundings.

The authors in [17, 18], analyzed and presented commonly used propagation loss models. The analyzed and simulated results were performed to identify the propagation loss by varying base transmitter station antenna height, mobile station antenna height and distance between transmitter and receiver. The relationship between propagation loss and other wireless propagation parameters, such as transmitter–receiver antenna spacing, antenna heights, and operating frequency were presented to enhance cellular network performance optimization. The data provided were analyzed in the MATLAB software to predict signal propagation losses; radio coverage estimation; avoid interference; and determine the received power level. The propagation loss decreases due to the increase in BTS antenna height for all models. These interpretations obviously showed the impact of propagation loss on cellular network. The models selected simulated were; freespace, Okumura, Okumura-Hata, and COST-231 propagation loss models in suburban areas. The demonstrations apparently confirmed that the Okumura model performed better than Okumura-Hata and COST-231 in terms of propagation loss reduction. The Okumura model could be used to study 5G radio network planning in cellular network.

2.1 COST231 Model

COST231 model is an enhancement of Okumura-Hata model and it is mostly used for path loss determination and can be applicable in range of frequency between 1.8 GHz to 2 GHz.

$$P_L(dB) = 46.3 + 33.9\log(f) - 13.82\log(h_t) - \alpha(h_r) \\ + \left[44.9 - 6.55\log(h_t)\right]\log(d) + C_m \tag{1}$$

where
F is the transmitting frequency.
h_t is the transmitter height.
hr is the receiver height.
d is the distance

$$\alpha(h_r) = 3.2\left[\log(11.75h_m)\right]^2 - 4.97 \, for \, large \, cities, \, f \geq 300\,MHz$$

$$\alpha(h_r) = 8.29\left[\log(1.54h_m)\right]^2 - 1.1 \, for \, large \, cities, \, f < 300\,MHz$$

$$\alpha(h_r) = (1.1\log f - 0.7)h_m - (1.56\log f - 0.8) \, for \, medium \, to \, small \, cities$$

$$C_m = \begin{cases} 0\,dB; & medium \, sized \, cities \, and \, suburban \, areas \\ 3\,dB; & metropolitan \, areas \end{cases}$$

3 Methodology

In this research, the locations in which field works are conducted were first visited to survey the area (that is, have good knowledge of the terrain) and know which the drive test routes to take when collecting signal strength data. It also provided means of knowing the number of eNodeBs antennas the LTE cellular systems network provider have in the study locations (Fig. 1).

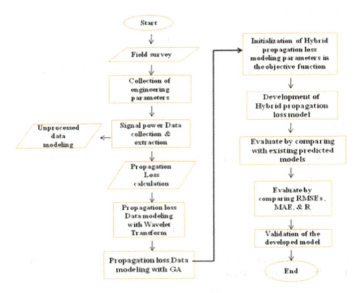

Fig. 1. The flowchart of path loss models development

3.1 Method of Data Collection

With the aid of professional and reliable TEMS equipped labtop and TEMS pocket Sony Ericson phone with map info software, field measurements were conducted over a popular commercial LTE cellular networks air interface, propagating on the 2600 MHz band in Port Harcourt, in River State and Awka in Anambra State, Nigeria. The building clusters in the area are a mixture of residential/commercial bungalows, two-story or three-story buildings encompassed with medium density user and vehicular traffics as earlier described. Precisely, the measurement routes were selected along the main streets and sideway of the roads of the area, where the LTE eNodeB transceivers are deployed.

One of the main LTE radio networks data collected during measurement is RSRP (i.e., Reference Signal received Power) data. Technically, the RSRP is an indicator of signal power level at the UE terminal in LTE networks. Generally, the stronger RSRP level received at UE, better signal coverage quality can be achieved in the radio network. There exist sundry factors that can impact the RSRP levels at the UE terminals, among which are transmitter–receiver (Tx-Rx) communication distance, RF channel conditions, signal propagation loss, UE location, total radiated eNodeB power, etc.

3.2 Proposed Hybrid Wavelet-GA Method

Contained in Fig. 1 is the summarized flowchart employed to develop and implement the proposed hybrid Wavelet-GA method. It started by conducting field signal data ass described above. This followed by routing the measured signals through wavelet processing platform. The resultant processed signal is then engaged to calculated the propagation loss, which in turn supplied to GA optimization scheme to tune the the existing log-distance model in correspondence with measured loss data. The log-distance loss model is defined as [6].

$$P_L(dB) = a_1 + a_2 \log f + a_3 \log d \tag{2}$$

where a_1, a_1 and a_1 are the parameters to be optimized using the proposed wavelet-GA method.

4 Results

Table 1 is the measured RSRP from four commercial cell sites deployed in Awka and Table 2 contain the developed mean wavelet-GA and GA path loss models of the four cell sites.

Table 1. Measured Signal Strength from Four Sites in Awka

eNodeB Site 1		eNodeB Site 2		eNodeB Site 3		eNodeB Site 4	
Dist (m)	RSRP(dBm)	Dist (m)	RSRP(dBm)	Dist (m)	RSRP(dBm)	Dist (m)	RSRP(dBm)
100	−76.01	100	−80.19	200	−82.44	100	−72.3
120	−74.31	130	−76.81	220	−74.38	120	−71.21
140	−75.24	160	−71.81	240	−71.81	130	−73.33
150	−69.5	190	−75.38	260	−71.94	150	−81.13
160	−64.56	200	−77.88	270	−73.38	170	−76.56
170	−71.38	230	−73.63	300	−72.63	190	−76.38
180	−76.63	250	−70.81	310	−74.13	200	−76.75
190	−72.81	270	−77	320	−75	210	−77.94
200	−81.69	300	−78.94	340	−76	220	−85.25
210	−77.81	320	−80.69	360	−77.31	230	−85.06
220	−77.94	340	−82.81	380	−74.94	260	−82.44
230	−74.19	360	−80.13	400	−75.38	280	−81.13
250	−78.38	390	−81.56	410	−79.69	300	−79.25
260	−77.94	400	−84.56	420	−84.25	310	−77.69
280	−77.31	410	−86	440	−85.25	340	−73.56
300	−77.44	420	−91.63	460	−90	360	−77.5

(continued)

Development and Comparative Analysis of Path Loss Models

Table 1. (*continued*)

eNodeB Site 1 Dist (m)	RSRP(dBm)	eNodeB Site 2 Dist (m)	RSRP(dBm)	eNodeB Site 3 Dist (m)	RSRP(dBm)	eNodeB Site 4 Dist (m)	RSRP(dBm)
310	−76.56	440	−90.38	480	−78.06	400	−86.44
320	−77.56	460	−95.88	500	−80.69	420	−93.25
330	−74.63	480	−87.38	520	−79.56	450	−92.19
350	−83.88	490	−91.69	540	−86.69	480	−94.06
360	−78.88	500	−87.38	560	−80.94	500	−86.13
380	−84.25	510	−97.06	570	−76.94	520	−92.38
400	−89.88	520	−101.06	590	−81.94	550	−85.75
420	−91.44	540	-96.25	600	−86.81	570	−87.19
440	−87.75	550	−97.81	610	−90.31	600	−97.06
460	−86.5	560	−93.88	630	−83.69	620	−90.56
480	−87.75	580	−90.13	660	−84.31	640	−91.88
500	−91.25	600	−94	680	−79.19	670	−96.52
530	−88.44	609	−78.13	700	−81.44	700	−95.69
550	−84.44	647	−77.25	720	−83.94	720	−94.32
570	−85.06	685	−78.31	740	−84.38	760	−86.56
600	−86.94	724	−77.88	760	−84.19	780	−90.5
610	−87.5	762	−18.31	780	−82.44	800	−87.25
620	−90.19	792	−78.31	800	−89.81	830	−86.31
630	−91.38	832	−79.13	820	−85.88	860	−87.19
660	−90.13	870	−79.19	840	−88.25	900	−87.75
670	−86.63	900	−80.06	860	−93.94	904	−70.88
700	−88.88	929	−79.00	880	−83	908	−68.69
710	−92.44	959	−75.44	900	−85.63	913	−70.88

The propagation loss data used in this paper were obtained from the measured RSRP values by subtracting the total transmit power of the various eNodeB antennas from the measured RSRP values [17, 18] (Figs. 2, 3, 4, 5, 6, 7, 8, 9, 10, 11, 12, 13 and 14).

Table 2. The Developed Propagation loss Model using the proposed hybrid Wavelet-GA approach and the existing standard GA approach

Cell sites	Proposed Hybrid Wavelet-GA Model	Standard GA Model
Awka 1	$-8.14 + 24.46\log(d) + 24.56\log(f)$	$9.92 + 26.38\log(d) + 117.89\log(f)$
Awka 2	$-2.59 + 30.0\log(d) + 19.64\log(f)$	$-0.59 + 23.63\log(d) + 23.83\log(f)$
Awka 3	$4.98 + 26.45\log(d) + 17.74\log(f)$	$-3.45 + 23.79\log(d) + 22.29\log(f)$
Awka 4	$0.48 + 21.53\log(d) + 24.68\log(f)$	$9.81 + 21.81\log(d) + 21.74\log(f)$
Mean	**$-1.32 + 25.6\log(d) + 21.66og(f)$**	**$3.92 + 23.27\log(d) + 21.36\log(f)$**

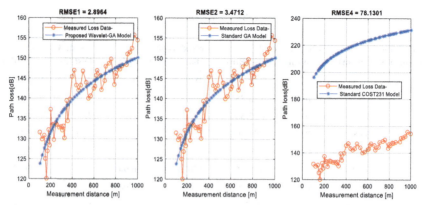

Fig. 2. Predicted RMSE Difference using hybrid Wavelet-GA model, GA Model, and COST231-Hata, in eNode Site 1 of Awka

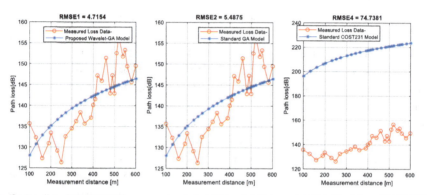

Fig. 3. Predicted RMSE Difference using hybrid Wavelet-GA model, GA Model, and COST231-Hata, in eNode Site 2 of Awka

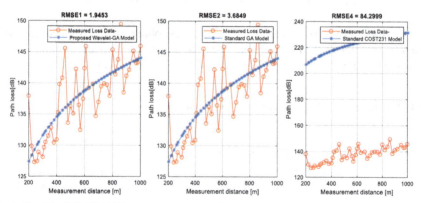

Fig. 4. Predicted RMSE Difference using hybrid Wavelet-GA model, GA Model, and COST231-Hata, in eNode Site 3 of Location 2

Development and Comparative Analysis of Path Loss Models   495

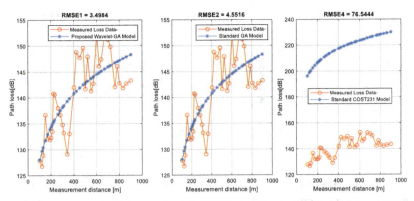

Fig. 5. Predicted RMSE Difference using hybrid Wavelet-GA model, GA Model, and COST231-Hata, in eNode Site 4 of Awka

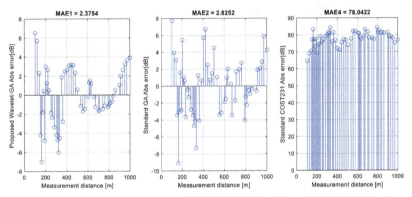

Fig. 6. Predicted MAE difference using the hybrid Wavelet-GA method, GA Method, and the COST231-Hata model in Site 1 of Location 2

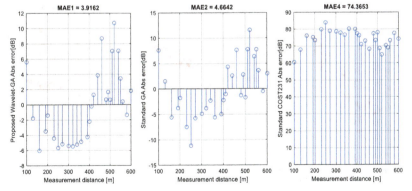

Fig. 7. Predicted MAE difference using the hybrid Wavelet-GA method, GA Method, and the COST231-Hata model in Site 2 of Awka

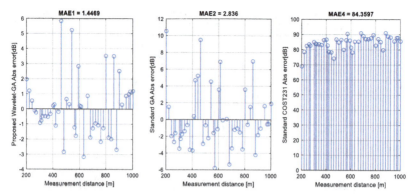

Fig. 8. Predicted MAE difference using the hybrid Wavelet-GA method, GA Method, and the COST231-Hata model in Site 3 of Location2

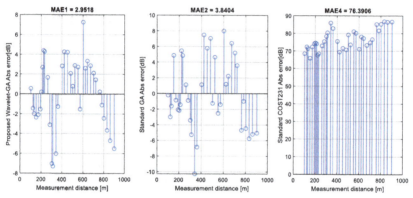

Fig. 9. Predicted MAE difference using the hybrid Wavelet-GA method, GA Method, and the COST231-Hata model in Site 4 of Awka

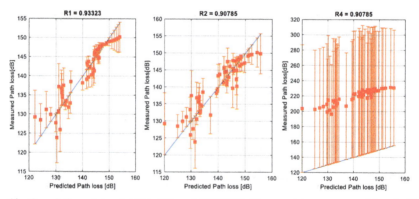

Fig. 10. Correlation of wavelet-GA, GA and COST231-Hata models to measured data in Site 1 of Awka

Development and Comparative Analysis of Path Loss Models 497

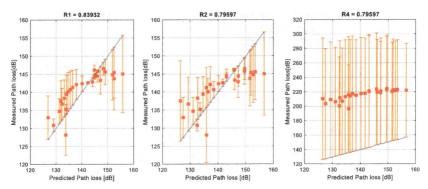

Fig. 11. Correlation of wavelet-GA, GA and COST231-Hata models to measured data in Site 2 of Awka

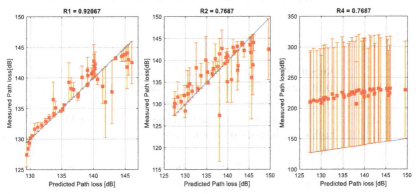

Fig. 12. Correlation of wavelet-GA, GA and COST231-Hata models to measured data in Site 3 of Awka

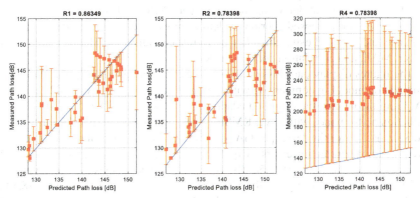

Fig. 13. Correlation of wavelet-GA, GA and COST231-Hata models to measured data in Site 4 of Awka

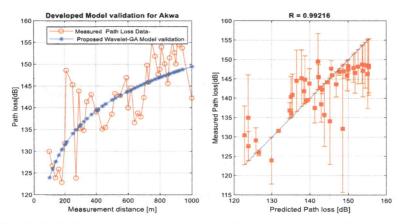

Fig. 14. Validation of proposed hybrid wavelet-GA model using another Site in Awka

5 Discussion

COST231 model attained path loss values of 220-250 dB, 210-248 dB, 210-238 dB, 220-241 dB as compared to measured path loss whose values are 138-184 dB, 120-156 dB, 126-155 dB, and 130-142 dB, respectively.

From the results, it is noticeably clear that path loss values attained by the COST231 model is quite too high than the ones obtained through field measurements. The higher path loss values produced by the existing model may be ascribed to the physical terrain and the topographical differences between locations where the measurement loss data were conducted or the terrain characterization where the COST231 model was developed. Thus, the needs to fine tune the existing model to reliably fit the measured path loss data.

The results attained by applying measured path loss with hybrid Wavelet-GA, GA model achieved without denoising the signal, and COST 231 model are provided. The performance of the proposed hybrid wavelet-GA model over the standard approaches are evaluated and revealed using three key performance indicators. The indicators include RMSE, MAE, and correlation coefficient (R).

The figures display path loss prediction accuracies attained using the Wavelet-GA optimization model in comparison with the ones attained using COST 231 model and GA. From the results, the proposed adaptive hybrid wavelet-GA model attained better prediction performance with lower RMSE values, which are 2.89 dB, 4.71 dB, 1.94 dB, and 3.49 dB. But for the developed GA optimization model, without denoising, the measured signal, the prediction errors are quite higher in terms of RMSE values. The RMSE values are 3.93 dB, 4.62 dB, 4.56 dB and 7.31 dB. The analysed result showed that COST231 model has the highest RMSEs. The values of COST231 model are 78.13 dB, 74.74 dB, 84.30 dB, 76.54 dB respectively. The enhanced path loss prediction performance attained using the proposed wavelet-GA model over standard models are also presented in the figures and tables respectively, using MAE, and Coefficient of correlation (R) performance. It showed that the developed hybrid wavelet-GA method proved to estimate the lowest MAE, whereas the COST 231 model achieved the highest value.

The MAE achieved are 2.38 dB, 3.92 dB, 1.45 dB, and 2.958 dB. The COST231 model achieved the highest values of MAE as 78.04 dB, 74.37 dB, 84.36 dB, 76.39 dB respectively. Lastly, the measured path loss where also compared with the developed hybrid wavelet-GA model, GA model, and COST231 model in terms of the correlation coefficient (R). It showed that the developed hybrid wavelet-GA model achieved the highest values of R, which means that there is correlation between the measured path loss and the developed hybrid wavelet-GA optimization model. The values of R on wavelet-GA are 93.32%, 83.93%, 92.07% for site 1, 3, 4 of respectively.

To validate a developed path loss model means to test the model prediction performance using another cell site data other than the one used for wavelet-GA model developed. Thus, validation provides means of assessing the performance of any newly developed path loss model in another eNodeB cell in order to establish its prediction dynamism and efficacy. The respective mean resultant developed model using the proposed Wavelet-GA.

It showed that the validation prediction performances of proposed Wavelet-GA model 0.99 (99%) correlate with the measured data and as such valid. This again validates the efficacy of the hybrid signal path loss predictive modeling approach, which is proposed in this paper.

6 Conclusion

The proposed adaptive hybrid path loss prediction modeling performance is compared with the standard genetic algorithm and other theoretical methods in literature such as COST 231 model. The results show that the proposed adaptive hybrid path loss prediction modeling method gives better prediction accuracy in terms of RMSE, MAE, and correlation coefficient (R). For instance, the proposed Wavelet-GA optimization model attained better prediction performance with lower RMSE values of 2.89 dB, 4.71 dB, 1.94 dB, and 3.49 dB. But for standard GA optimization model, the prediction errors are quite higher in terms of RMSE values, which are 3.93 dB, 4.62 dB, 4.56 dB and 7.31 dB. From the results, the proposed Wavelet-GA model for better path loss prediction over other existing techniques is clearly justified.

Particularly, the ability of wavelet transform to extract both local spectral and temporal real signal information component from the noisy one assisted the GA to adaptive model and predicts the signal path loss values more effectively, than using only the GA optimization model. The power of genetic algorithms comes from their ability to combine both exploration and exploitation in an optimal way.

References

1. Japertas, S., Grimaila, V:. Mobile signal path losses in microcells behind buildings. Radio Eng. **26**(1), 191–197 (2017). https://doi.org/10.13164/re.2017.019.
2. Coinchon, M., Salovaara, A. P., Wagen, J. F.: The impact of radio propagation predictions on urban UMTS planning. In: 2002 International Zurich Seminar on Broadband Communications Access - Transmission - Networking: Meeting the Challenge of High-Speed Communications, IZS 2002 - Proceedings, vol. 1, no. 1, pp. 321–326 (2001). https://doi.org/10.1109/IZSBC. 2002.991775

3. Nationale, E., Sa-bedja, K., Nationale, E.: Propagation models calibration in mobile cellular networks : a case study in togo. Future Technollogies Conf. **2**(11), 923–927 (2017)
4. Joseph, I., Divine, O.O.: Application of Levenberg-Marguardt algorithm for prime radio propagation wave attenuation modelling in typical urban, suburban and rural terrains. Int.J. Intell. Syst. Appl. **7**, 35–42 (202)
5. Ojuh, D.O., Isabona, J.: Field electromagnetic strength variability measurement and adaptive prognostic approximation with weighed least regression approach in the Ultra-high radio frequency band. J. Intell. Syst. Appl. **2021**(4), 14–23 (2021)
6. Isabona, J., Ojuh, D.O.: Adaptation of propagation model parameters toward efficient cellular network planning using robust LAD algorithm. Int.J. Wireless Microwave Technol. **5**, 13–24 (2020)
7. Isabona J., Konyeha C.C.: Urban area path loss propagation prediction and optimisation using Hata Model at 800MHz. IOSR J. Appl. Phys. (IOSR-JAP) **3** (4), 8–18, 2013.
8. Isabona, J., Kehinde, R.: Multi-resolution based discrete wavelet transform for enhanced signal coverage processing and prediction analysis. FUDMA J. Sci. **3**(1), 6–15 (2019)
9. Ebhota, V.C., Isabona, J., Srivastava, V.M.: Environment-Adaptation Based Hybrid Neural Network Predictor for Signal Propagation Loss Prediction in Cluttered and Open Urban Microcells. Wireless Pers. Commun. **104**(3), 935–948 (2019)
10. Isabona, J., Imoize, A.L.: Terrain-based adaption of propagation model loss parameters using Non-linear square Regression. J. Eng. Appl. Sci. **68**(1), 1–19 (2021). https://doi.org/10.1186/s44147-021-00035-7
11. Almalki, F. A., Sciences, P.: Optimisation of a Propagation Model for Last Mile Connectivity with low Altitude Platforms Using Machine Learning Brunel University London, 1–129 (2017)
12. Oguejiofor, O.S.: Pathloss Prediction for a typical mobile communication system in Nigeria using empirical models. Int. J. Comput. Netw. Wireless Commun. **3**(2), 207–211 (2015)
13. Khare, A., Saxena, M., Tiwari, S.: Multimedia networks based dynamic WCDMA system proposal for QoS. Int. J. Eng. Adv. Technol. **1**(1), 52–55 (2011)
14. Ebhota, C., Isabona, J., Srivastava, V.M.: Improved adaptive signal power loss prediction using combined vector statistics based smoothing and neural network approach. Progress Electromagnet. Res. C **82**(1), 155–169 (2018). https://doi.org/10.2528/pierc18011203
15. Jakborvornphan, S.: Analysis of path loss propagation models In mobile communication. J. Theor. Appl. Inf. Technol. **98**(4), 725–730 (2020)
16. Echie, J.O., Oyedum, O.D., Ajewole, M.O., Aibinu, A.M.: Comparative analysis of basic models and artificial neural network based model for path loss prediction. Progress Electromagnet. Res. **61**(9), 133–146 (2017)
17. Isabona, J., Zhimwang, J.T., Risi, I.: Cascade forward neural networks-based adaptive model for real-time adaptive learning of stochastic signal power datasets. Int. J. Comput. Netw. Inf. Secur. **2**(3), 63–74 (2022). https://doi.org/10.5815/ijcnis.2022.03.05
18. Joseph, I., Ituabhor, O., Timothy, J., zhimwang, Risi Ikechi,: Achievable throughput over mMobile broadband network protocol layers: practical measurements and performance analysis. Int. J. Adv. Netw. Appl. **13**(04), 5037–5044 (2022). https://doi.org/10.35444/IJANA.2022.13404